ACS SYMPOSIUM SERIES **565**

Coordination Chemistry

A Century of Progress

George B. Kauffman, EDITOR

California State University, Fresno

Developed from a symposium sponsored
by the Divisions of the History of Chemistry,
Chemical Education, Inc., and Inorganic Chemistry, Inc.,
at the 205th National Meeting
of the American Chemical Society,
Denver, Colorado,
March 28–April 2, 1993

American Chemical Society, Washington, DC 1994

Library of Congress Cataloging-in-Publication Data

Coordination chemistry: a century of progress : developed from a symposium / sponsored by the Divisions of the History of Chemistry, Chemical Education, Inc., and Inorganic Chemistry, Inc., at the 205th National Meeting of the American Chemical Society, Denver, Colorado, March 28–April 2, 1993; George B. Kauffman, editor.

p.   cm.—(ACS symposium series, ISSN 0097–6156; 565)

Includes bibliographical references and indexes.

ISBN 0–8412–2950–3 (cloth).—ISBN 0–8412–2958–9 (pbk.)

1. Coordination compounds—Congresses.

I. Kauffman, George B., 1930–   . II. American Chemical Society. Division of the History of Chemistry. III. American Chemical Society. Division of Chemical Education. IV. American Chemical Society. Division of Inorganic Chemistry. V. American Chemical Society. Meeting (205th: 1993: Denver, Colo.) VI. Series.

QD474.C663   1994
541.2′242—dc20   94–34445
                                                                CIP

The paper used in this publication meets the minimum requirements of American National Standard for Information Sciences—Permanence of Paper for Printed Library Materials, ANSI Z39.48–1984.   ∞

PRINTED IN THE UNITED STATES OF AMERICA

Foreword

THE ACS SYMPOSIUM SERIES was first published in 1974 to provide a mechanism for publishing symposia quickly in book form. The purpose of this series is to publish comprehensive books developed from symposia, which are usually "snapshots in time" of the current research being done on a topic, plus some review material on the topic. For this reason, it is necessary that the papers be published as quickly as possible.

Before a symposium-based book is put under contract, the proposed table of contents is reviewed for appropriateness to the topic and for comprehensiveness of the collection. Some papers are excluded at this point, and others are added to round out the scope of the volume. In addition, a draft of each paper is peer-reviewed prior to final acceptance or rejection. This anonymous review process is supervised by the organizer(s) of the symposium, who become the editor(s) of the book. The authors then revise their papers according to the recommendations of both the reviewers and the editors, prepare camera-ready copy, and submit the final papers to the editors, who check that all necessary revisions have been made.

As a rule, only original research papers and original review papers are included in the volumes. Verbatim reproductions of previously published papers are not accepted.

M. Joan Comstock
Series Editor

Contents

SPECIALIZED ASPECTS

ISOMERISM

COMPOUNDS OF VARIOUS ELEMENTS

APPLICATIONS

Preface

COORDINATION CHEMISTRY IS ONE of the most active research fields in chemistry today. Known since ancient times, coordination compounds attracted as much attention during the nineteenth century for their tremendous importance to the general problem of chemical bonding as for their own unique and fascinating properties. Thus Alfred Werner (1866–1919), the founder of coordination chemistry, was trained as an organic chemist and entered inorganic chemistry through the back door, so to speak. Yet, even before he began his quarter-century-long *tour de force* of research on what were known as "molecular compounds," he became vitally concerned with one of the most basic problems of chemistry—the nature of affinity and valence. Coordination compounds provided him with an exciting and challenging means to this end. In 1913 Werner was awarded the Nobel Prize in chemistry, the first Swiss chemist to attain this honor. In the words of the Royal Swedish Academy of Sciences, he received the prize for "his work on the linkage of atoms in molecules, by which he has thrown fresh light on old problems and opened new fields of research, particularly in inorganic chemistry."

However, as the chapters in this volume illustrate, the theoretical and practical aspects of coordination chemistry in general and Werner's work in particular have significance in fields far removed from classical inorganic chemistry. Because of the importance of Werner's research on coordination compounds, they are sometimes referred to as Werner complexes, and the coordination theory is colloquially called Werner's theory. Werner's work was so complete and all-encompassing that for many years coordination chemistry was neglected because most chemists thought that all the important research had been done. It remained for chemists such as John C. Bailar, Jr., in the United States and Kai Arne Jensen and Jannik Bjerrum in Denmark to revive interest in the field as part of what became known as "the Renaissance in inorganic chemistry." Evaluations of the lives and works of these recently deceased chemists appear as Chapters 6–9 in this volume.

Werner was born in Mulhouse, France, on December 12, 1866. To commemorate the 100th anniversary of his birth, I organized and chaired a symposium at the 152nd National Meeting of the American Chemical Society held in New York City in 1966. Forty-two of the papers presented at that symposium, the longest in ACS history up to that time, were published as a book, *Werner Centennial* (Advances in Chemistry Series No. 62, American Chemical Society: Washington, DC, 1967).

To commemorate the 100th anniversary of the publication of Werner's "Beitrag zur Konstitution anorganischer Verbindungen"(Z. anorg. Chem. **1893**, 3, 267–330), in which he first proposed his monumental coordination theory, a Coordination Chemistry Centennial Symposium (C_3S), was held for four full days at the 205th National Meeting of the American Chemical Society in Denver, Colorado. Fifty-one papers on the historical, educational, review, and research aspects of the field were presented by speakers from 17 countries (Australia, Denmark, England, Finland, France, Germany, Hungary, India, Israel, Italy, Japan, New Zealand, Russia, South Africa, Switzerland, the United States, and Wales).

The well-attended symposium was opened by ACS President Helen M. Free and the famous trumpet fanfare from Ludwig van Beethoven's opera *Fidelio* played by symposium participant John G. Verkade. The distinguished participants included three 1993 ACS awardees (Robert H. Crabtree, Robert W. Parry, and me) and 10 past ACS awardees (Daryle H. Busch, Malcolm H. Chisholm, Gregory R. Choppin, Harry B. Gray, James A. Ibers, R. Bruce King, Steven J. Lippard, Arthur E. Martell, Alan M. Sargeson, and Jean'ne M. Shreeve).

One of the papers was coauthored by the 1951 Nobel chemistry laureate Glenn T. Seaborg. The 92-year-old Linus Pauling, a two-time Nobel laureate (for chemistry in 1954 and for peace in 1963), made a rare public appearance at the opening session. His entrance was heralded by another fanfare by Professor Verkade—this one from Benjamin Britten's operetta *Noah's Ark,* in which God's words to Noah are always preceded by this fanfare—appropriately, in view of Professor Pauling's megavitamin therapy research, in the key of C. Pauling's presentation (which is now Chapter 5 in this volume) featured his reminiscences and interpretations.

This volume contains 37 chapters: 28 were presented at the symposium on which this book is based; 9 are additional chapters. It includes biographical chapters and those dealing with the history of coordination chemistry as a whole and the history of various aspects of the field as well as reviews, research of more than ordinary interest, and chapters about the applications of coordination compounds. Therefore this book should be useful to historians of chemistry and of science, practicing chemists, students, and anyone concerned with the past, present, and future of one of the most intriguing and productive areas of chemistry. Eleven of the experimental research papers from the symposium appeared in a special "Coordination Chemistry Centennial Symposium" issue of *Polyhedron: The International Journal for Inorganic and Organometallic Chemistry* (Volume 13, Number 13, July 1994, for which I was the guest editor).

Acknowledgments

For financial support of foreign speakers attending the symposium, I am pleased to acknowledge the donors of the Petroleum Research Fund, administered by the American Chemical Society; the Chemical Heritage Foundation; and the ACS Divisions of the History of Chemistry and of Chemical Education, Inc. I am also indebted to my wife Laurie for her constant encouragement and helpful advice from the initial organization of the symposium to the completion of this volume.

GEORGE B. KAUFFMAN
California State University, Fresno
Fresno, CA 93740

February 2, 1994

*This book is dedicated to the memory
of John C. Bailar, Jr., Jannik Bjerrum,
Joseph Chatt, Kai Arne Jensen, and Linus Pauling,
pioneers in coordination chemistry.*

Introduction

I GREATLY APPRECIATE THE HONOR of being invited to provide an introduction to the proceedings of this very comprehensive and well-attended symposium on coordination chemistry. My main regret is that I was not able to deliver it in person. I am reminded that I made a similar request to Professor Nevil Vincent Sidgwick of Oxford University when he attended the first International Conference on Coordination Chemistry held at Welwyn, England, in 1950 (1). He gave a brief history of the subject, as he said, to around 1925. He discussed the well-known Jørgensen–Werner controversy and Werner's extensive paper of 1893 (2), in which he expounded his famous theory, the centennial of which we now commemorate. Sidgwick, who in 1893 was an undergraduate at Oxford, commented that this theory was little regarded by chemists in general. Chemists were convinced that Werner's views were to be taken seriously only after 1911, when he resolved 6-coordinate compounds such as $[CoCl(NH_3)(en)_2]Cl_2$ into enantiomers in accordance with his prediction that such compounds should be optically active if the groups are located at the vertices of an octahedron (3).

Two types of chemical compounds were apparent—those in which the number of an atom's covalent bonds is equal to the atom's valency, as in most organic compounds, and those in which the number of bonds is equal to the atom's coordination number. G. N. Lewis, in 1916, finally reconciled these two patterns (4). He explained for the first time the nature of non-ionized bonds (covalencies) as resulting from the sharing of two electrons in some way between the two bonded atoms. The electrons could be provided either one from each or both from one of the bonded atoms. It was soon realized that the latter bonding explained Werner's coordination compounds (5).

Why did Sidgwick in 1950 limit his theory to pre-1925? Did the subject die in 1925? Not completely, but almost so. There were useful correlations from Russian chemists who developed Ilya Il'ich Chernyaev's ideas of directing effects to cis or trans positions in the substitution of one ligand by another (6), rather like ortho–para and meta substitution in benzene. Kai Arne Jensen in Copenhagen, by measurements of dipole moments, showed that many planar complexes of the type $[PtCl_2L_2]$ (where L is a *tert*-phosphine or other ligand) had been assigned incorrect configurations on the basis of their colors alone (7). Frederick George Mann in Cambridge had shown that Sugden's parachor was useless for determining the nature of bonding in chemical complexes (8). Of course, a few chemists prepared new and interesting complexes, especially those complexes soluble in organic solvents. Nevertheless, the subject was essentially ignored by the great body of chemists who associated it with the metal-ammines whose chemistry they regarded as dull and useless. That view took a long time to die. It was still very much alive when I was a student in Cambridge during the period 1935–1940 (9).

Inorganic chemistry was taught in a monotonous and descriptive manner. I remember one of our mature chemistry tutors whom we asked about Werner's theory stuttering through it and finishing, "That's what they say. Take it or leave it." It was fortunate that Mann, one of our organic lecturers, gave us a short and excellent course on complex compounds. He was my inspiration. In it he mentioned Zeise's salt $K[PtCl_3(C_2H_4)] \cdot H_2O$, stating that the ethylene pi electrons were believed to serve the function of the lone pair on the usual donor molecules. The term *ligand* was not used then (10), but when I first became aware of it after the war it seemed self-evident. *To lig,* in my local Cumbrian dialect, meant *to lie down,* and obviously, groups lying with the central atom were ligands.

William Christoffer Zeise discovered his salt in 1827 (11), but no one had yet firmly established the nature of its bonding. My aim was to teach in a university, and I decided to make olefin complexes my area of research, but World War II intervened. Six years elapsed before I could start on my project in a newly established fundamental research laboratory of Imperial Chemical Industries, Ltd. (ICI). Even then, most organic chemists were unable to appreciate the importance of Werner's coordination numbers and tried to formulate complex compounds so that the number of bonds to the metal atom was equal to what they called its valency (actually its oxidation number). Thus such nonsensical formulations as $Me_2S^+-Pt-S^+Me_2Cl_2^-$ and $Cl_2Pt(-CH_2CH_2-)_2PtCl_2$ were still appearing in reputable journals such as the *Journal of the Chemical Society* and the *Journal of the American Chemical Society* as late as the mid-1930s.

One problem delaying the development of complex chemistry was that coordination compounds were usually salts, either insoluble in water, or, if soluble, excessively labile. Their lability led to the determination of "stability constants" relating to the study of the equilibria between the simple salts and various nitrogen ligands, notably by Jannik Bjerrum (12) and other Scandinavian chemists, Gerold Schwarzenbach in Zürich, and Harry M. N. H. Irving and Robert J. P. Williams in Oxford. Complex chemistry was slow to grow in the United States but was greatly nourished there by John C. Bailar, Jr. (13). On my first visits to the United States, I noted that there were only two types of coordination chemists there, those who had been John Bailar's students and those who had not, and the former appeared greatly to outnumber the latter.

The spark that led to the outburst of activity in coordination chemistry after World War II was the budding, but still small, prewar interest in the chemistry of nonionic complexes, soluble in organic solvents, whose study was less impeded by complex lability. Such activity was favored by the very rapid development of nondestructive physical methods of structure determination by X-rays, magnetic and dipole moment measurements, infrared spectroscopy, and the many other spectroscopic techniques appearing over the past half-century. Another factor encouraging such research was the growing interest in transition-metal-based catalysts used in the conversion of mineral oil into raw materials for the plastics and similar new industries. This potential for applications ensured adequate finances for fundamental research. Hence I was allowed an open remit in the laboratories of ICI Ltd., Welwyn.

Which direction will future research on coordination compounds take? Many complex enzymic reactions depend on the unique qualities of transition metal atoms at their reaction centers. Complex chemistry holds their key, and they will become its next growing edge. Their solution will demand close cooperation among all types of chemists.

References

1. Chatt, J. *Pure Appl. Chem.* **1977**, *49*, 815–826.
2. Werner, A. *Z. Anorg. Chem.* **1893**, *3*, 267–330. For a discussion and annotated English translation, *see* Kauffman, G. B. *Classics in Coordination Chemistry, Part 1: The Selected Papers of Alfred Werner*; Dover: New York, 1968; pp 5–88. For a detailed discussion of the Jørgensen–Werner controversy, *see* Chapter 1 in this book.
3. Werner, A. *Ber. Dtsch. Chem. Ges.* **1911**, *44*, 1887–1898. For a discussion and annotated English translation, *see* Kauffman, reference 2, pp 155–173.
4. Lewis, G. N. *J. Am. Chem. Soc.* **1916**, *38*, 762–785.
5. Sidgwick, N. V. *J. Chem. Soc.* **1923**, *123*, 725–730. For a discussion and annotated reprint, *see* Kauffman, G. B. *Classics in Coordination Chemistry, Part 3: Twentieth-Century Papers (1904–1935)*; Dover: New York, 1978; pp 127–142.
6. Chernyaev, I. I. *Izv. Inst. Izuch. Platiny Drugikh Blagorodn. Met. Akad. Nauk SSSR* **1927**, *5*, 118–156. For a discussion and annotated English translation, *see* Kauffman, reference 5, pp 143–195.
7. Jensen, K. A. *Z. Anorg. Chem.* **1935**, *225*, 97–114; 115–141; **1936**, *229*, 252–264. For details on Jensen's life and work, *see* Chapter 7 in this book.
8. Mann, F. G.; Purdie, D. *J. Chem. Soc.* **1935**, 1549–1563.
9. Leigh, G. J. *Coord. Chem. Rev.* **1991**, *108*, 1–25.
10. The term *ligand* was first proposed by Alfred Stock, of borane and silane fame, in a lecture on November 27, 1916, at the Kaiser-Wilhelm Institut für Chemie in Berlin-Dahlem, but it did not come

into extensive use among English-speaking chemists until the 1940s and 1950s, largely through the popularity of Jannik Bjerrum's doctoral dissertation. For detailed accounts of the origin, dissemination, and eventual adoption of the term *ligand, see* Brock, W. H.; Jensen, K. A.; Jørgensen, C. K.; Kauffman, G. B. *Ambix* **1981**, *28*, 171–183; *J. Coord. Chem.* **1982**, *11*, 261–263; *J. Chem. Inf. Comput. Sci.* **1982**, *22*, 125–129; *Polyhedron* **1983**, *2*, 1–7; *J. Chem. Educ.* **1983**, *60*, 509–510.

11. Zeise, W. C. *Danske Vidensk. Selskabs. Overs.* **1825–1826**, 13; *Pogg. Ann. Phys.* **1827**, *9*, 632. Zeise's more detailed report on the preparation and properties of his salt are found in his paper "Von der Wirkung zwischen Platinchlorid und Alcohol, und von den dabei entstehenden neuen Substanzen." *Pogg. Ann. Phys.* **1831**, *21*, 497–541. For a discussion and annotated English translation, *see* Kauffman, G. B. *Classics in Coordination Chemistry, Part 2: Selected Papers (1798–1899)*; Dover: New York, 1976; pp 17–37.

12. For discussions of Jannik Bjerrum and his work, *see* Chapters 8 and 9 in this book.

13. For discussions of Bailar's life and work, *see* Chapter 6 in this book.

JOSEPH CHATT†
16 Tongdean Road
Hove, East Sussex BN3 6QE
England

December 6, 1993

†Joseph Chatt died on May 19, 1994.

HISTORICAL ASPECTS

*Alfred Werner (1866–1919). (Reproduced with permission
from reference 7, frontispiece. Copyright 1966 Springer-Verlag.)*

Chapter 1

Theories of Coordination Compounds
Alfred Werner's Triumph

George B. Kauffman

Department of Chemistry, California State University,
Fresno, CA 93740

The supersession of the most successful pre-Werner theory of the structure of coordination compounds, the so-called Blomstrand-Jørgensen chain theory, by Alfred Werner's coordination theory constitutes a valuable case study in scientific method and the history of chemistry. The highlights of the Werner-Jørgensen controversy and its implications for modern theories of chemical structure are sketched in this article.

Coordination compounds are of great practical importance. Coordinating agents are used in metal-ion sequestration or removal, solvent extraction, dyeing, leather tanning, electroplating, catalysis, water softening, and other industrial processes too numerous to mention here. In fact, new practical applications for them are found almost daily. They are of tremendous importance in biochemistry. For example, vitamin B_{12} is a coordination compound of cobalt, the hemoglobin of our blood is a coordination compound of iron, the hemocyanin of invertebrate animal blood is a coordination compound of copper, and the chlorophyll of green plants is a coordination compound of magnesium. Yet, as we shall see, their primary importance for chemistry lies elsewhere.

Theories of Coordination Compounds

In most fields of science theory generally lags behind practice. In other words, sufficient experimental data must be accumulated before attempts are made to explain these experimental facts and predict new phenomena. During the first half of the 19th century discoveries of coordination compounds were few, sporadic, and often accidental, and it was not until after Gibbs and Genth's classic memoir of 1856 that chemists began to devote themselves in earnest to a systematic study of this field. We might therefore think that few theories of coordination compounds were advanced until late in the second half

This chapter reprinted with permission from *Spectrum (Pretoria)*
© 1987 Foundation for Education, Science and Technology

of the 19th century, but this was not the case. In coordination chemistry the lag of theory behind practice was not a great one because of the tremendous importance of coordination compounds to the general question of chemical bonding. In the words of Alfred Werner himself, "Of all inorganic compounds, [metal-ammines] are best suited to the solution of constitutional problems. . . . It was through the investigation of metal-ammines that the decisive basic principles involved in the constitutional conception of inorganic compounds could first of all be clearly recognized."

Because of the importance of coordination compounds to chemical bonding, theories of their structure were advanced by some of chemistry's brightest luminaries. In 1841 the Swedish chemist Jöns Jacob Berzelius proposed his conjugate theory, using terms and ideas that he borrowed from the French chemist Charles Gerhardt. According to this theory, he viewed metal-ammines as conjugated or copulated compounds consisting of ammonia and a conjugate or copula. The latter cannot be removed by reaction with an acid and neither increases nor decreases the saturation capacity of a base. In other words, a metal in conjugation with ammonia is still capable of combining with other substances.

In 1837 the Scottish chemist Thomas Graham proposed the so-called ammonium theory, in which metal-ammines are considered as substituted ammonium compounds. Although Graham's ammonium theory could be applied only when the number of ammonia molecules in the coordination compound was equal to the electrovalence of the metal, it met with a fair degree of success and was generally accepted until Werner's time, largely because of the modifications of it that were proposed by other chemists such as Gerhardt, Wurtz, Reiset, von Hofmann, Boedecker, and Weltzien.

In 1854 the German-Russian chemist Carl Ernst Claus (Karl Karlovich Klaus) rejected the ammonium theory and suggested a return to the earlier view of complexes as conjugated compounds. He designated coordinated ammonia as "passive, in contrast to the active, alkaline state in the ammonium salts, where it can easily be detected and replaced by other bases."

Claus's close parallel between metal hydrates and metal-ammines was attacked on the grounds that many hydrates were known for which corresponding ammines were unknown. As we shall see shortly, all of Claus's three postulates reappeared in modified form almost four decades later in Alfred Werner's coordination theory.

Constant Valency and Kekulé's "Molecular Compounds." The next theory of coordination compounds that we shall examine in more detail was also applicable to a wide variety of substances. It was proposed by none other than the patriarch of structural organic chemistry, August Kekulé, professor of chemistry at the universities of Ghent and Bonn. It may come as a surprise to some of you that Kekulé's valence theory, which was so flexible and fruitful in the realm of organic chemistry, proved to be a virtual straitjacket when applied to inorganic compounds. Yet, by his own admission, Kekulé's concept of constant valence proved, in his own words, "embarrassing to the chemist." However, instead of abandoning this obviously untenable belief, he compounded his error by invoking a still more unsatisfactory concept in order to maintain it, namely, the concept of "molecular compounds."

Most of the pioneers in the theory of valence, such as the Englishman Edward Frankland and the Scot Archibald Scott Couper, readily admitted the possibility of variable valence. In other words, they felt that a given element could exhibit one valence in one compound and a different valence in another compound. On the other hand, Kekulé, from his first statements on the self-linking of carbon in 1858 until his death in 1896, adopted and rigidly adhered to the principle of constant valence. In spite of the mass of data that soon accumulated to contradict such a simple and admittedly attractive assumption, Kekulé dogmatically insisted that atomicity, which was the term that he used for valence, was, in his own words, "a fundamental property of the atom, which is just as constant and unchangeable as the atomic weight itself." The simplicity of this principle, however, was more than outweighed by the complicated and unrealistic formulas required to maintain it, and eventually the stubborn Kekulé stood virtually alone in its defense. Once again, the liberal of one generation had become the conservative of the next.

Listen to Kekulé's dichotomy of compounds into "atomic compounds" and "molecular compounds," an attempt to buttress his theory of constant valence. According to Kekulé, "Compounds in which all the elements are held together by the affinities of the atoms which mutually saturate one another could be called *atomic compounds*. They are the only ones which can exist in the vapor state. . . . We must distinguish a second category of compounds that I shall designate *molecular compounds*."

A few examples should suffice to illustrate Kekulé's concept of "molecular compounds," and they are shown in Figure 1. Since Kekulé regarded the valences of nitrogen, phosphorus, and cobalt as invariably three, and of copper as invariably two, he was forced to consider phosphorus(V) chloride, ammonium chloride, copper(II) sulfate pentahydrate, and hexaamminecobalt(III) chloride as "molecular compounds" with the formulas shown in Figure 1. Today, Kekulé's mysterious noncommittal dot has all but disappeared in writing the formulas of coordination compounds. When we occasionally still use it to write the formulas of metal hydrates or of hydrochlorides of organic bases, we unwittingly invoke the ghost of Kekulé and his now defunct doctrine of constant valency.

Figure 2 shows a page from a holograph book of Alfred Werner's elementary chemistry notes, in which we see PCl_5 formulated as a "molecular compound" in accordance with Kekulé's doctrine of constant valency. This 127-page book in Werner's handwriting dates from 1883-84 when he was between seventeen and eighteen years old. A decade later, in his coordination theory of 1893, Werner was destined to offer an alternative and much more satisfactory explanation for the constitution and configuration of what were then called "molecular compounds."

In a sense, Kekulé's concept of "molecular compounds" was a revival of Berzelius's dualistic theory whereby "secondary compounds" (in Kekulé's terminology, "atomic compounds") containing a small excess of electrical charge could still combine with other "secondary compounds" containing a small excess of opposite charge to form "tertiary compounds" (in Kekulé's terminology, "molecular compounds"). At most, Kekulé's artificial division of compounds into "atomic compounds," which obeyed the rules of classical valence theory, and into "molecular compounds," which did not obey these rules, had some limited value as a formal classification. However, in no way did it explain the nature or operation of the forces involved in the formation of "molecular

PCl_5 (Phosphorus(V)Chloride) $\quad PCl_3 \cdot Cl_2 \xrightarrow{\Delta} PCl_3 + Cl_2$

NH_4Cl (Ammonium Chloride) $\quad NH_3 \cdot HCl \xrightarrow{\Delta} NH_3 + HCl$

$CuSO_4 \cdot 5H_2O$ (Copper(II) Sulfate Pentahydrate) $\quad CuSO_4 \cdot 5H_2O \xrightarrow{\Delta} CuSO_4 + 5H_2O$

$[Co(NH_3)_6]Cl_3$ (Hexaamminecobalt(III)Chloride) $\quad CoCl_3 \cdot 6NH_3 \xrightarrow{\Delta}$ No Reaction

$CoCl_3 \cdot 6NH_3 \xrightarrow{NaOH} [Co(NH_3)_6](OH)_3$ (No $Co(OH)_3$ formed)

$CoCl_3 \cdot 6NH_3 \xrightarrow{H_2SO_4} [Co(NH_3)_6]_2(SO_4)_3$ (No NH_4^+ salt formed)

Figure 1. Kekulé's "molecular compounds" and constant valency. (Reproduced from reference 25, p 15. Copyright 1977 American Chemical Society.)

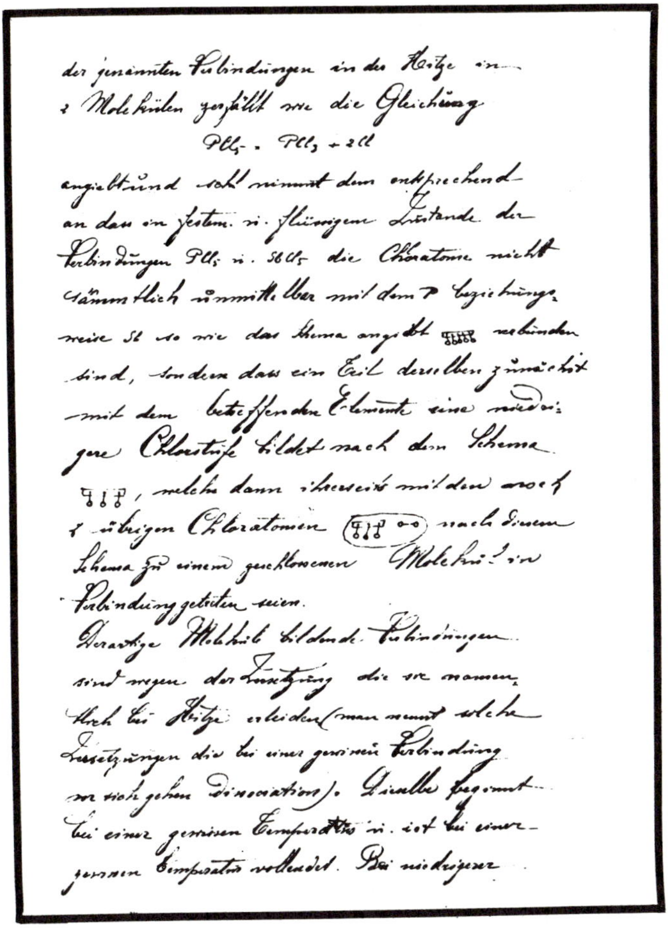

der genannten Verbindungen in der Hitze in
2 Molekülen zerfällt wie die Gleichung
$PCl_5 = PCl_3 + 2Cl$
angiebt und soll nimmt dann entsprechend
an dass im festen u. flüssigen Zustande der
Verbindungen PCl_5 u. $SbCl_5$ die Chloratome nicht
sämmtlich unmittelbar mit dem P beziehungs-
weise Sb so wie das Schema angiebt verbunden
sind, sondern dass ein Teil derselben zunächst
mit dem betreffenden Elemente eine niedri-
gere Chlorstufe bildet nach dem Schema
, welche dann ihrerseits mit den noch
2 übrigen Chloratomen nach diesem
Schema zu einem geschlossenen Molekül in
Verbindung getreten seien.
Derartige Molekül bildende Verbindungen
sind wegen der Zersetzung die sie namen-
tlich bei Hitze erleiden (man nennt solche
Zersetzungen die bei einer gewissen Verbindung
vor sich gehen Dissociation). Dieselbe beginnt
bei einer gewissen Temperatur u. ist bei einer
gewissen Temperatur vollendet. Bei niedrigerer

Figure 2. Page 111 of the 17-year-old Alfred Werner's *Einleitung in die Chemie, Mulhouse, 1883-84* showing the formulation of PCl_5 as a "molecular compound" ($PCl_3 \cdot Cl_2$) (Reproduced with permission from reference 9, p 46. Copyright 1967 American Chemical Society.)

compounds," except to assume that the forces were acting between molecules rather than between atoms.

Inasmuch as the forces acting between molecules were supposedly weaker than the forces acting between atoms, according to Kekulé, the resulting "molecular compounds" should be less stable than "atomic compounds." Indeed, as we can see in Figure 1, some of the substances of limited thermal stability cited by Kekulé as prototypes of "molecular compounds," such as phosphorus(V) chloride, ammonium chloride, and copper(II) sulfate pentahydrate, did decompose in the vapor state. However, this was a relative rather than an absolute phenomenon. For example, under certain conditions, such as the use of lower temperatures or the addition of decomposition products, many of these substances could be vaporized without decomposition. Therefore many chemists began to regard Kekulé's classification as meaningless. The great Russian chemist, Dmitriĭ Ivanovich Mendeleev, the discoverer of the periodic law, wrote, "Kekulé's division of chemical compounds into 'atomic' and 'molecular' types is artificial, arbitrary and unsound. . . . No practical test exists by which the two categories may be sharply separated."

But Kekulé's stability criterion, or to be more accurate, instability criterion failed completely in the case of many coordination compounds, especially the metal-ammines, which were classified as "molecular compounds" by sheer dint of necessity even though they were extremely resistant to heat and chemical reagents. For example, look at Figure 1. Although hexaamminecobalt(III) chloride contains ammonia, it neither evolves this ammonia on mild heating nor does it react with acids to form ammonium salts. Also, despite its cobalt content, addition of a base to its aqueous solution fails to precipitate hydrated cobalt(III) hydroxide. It remained for Alfred Werner to explain successfully the constitution of such compounds, but the time was not yet ripe. Before considering Werner's coordination theory, we must examine one more theory of coordination compounds, perhaps the most successful of the pre-Werner theories, namely, the Blomstrand-Jørgensen chain theory.

The Blomstrand-Jørgensen Chain Theory. Christian Wilhelm Blomstrand was professor of chemistry and mineralogy at the University of Lund. Living in Sweden during a transition period between the older and newer chemistry and being a scientific as well as a political conservative, he tried to reconcile Berzelius's old dualistic theory with the newer unitary and type theories. In fact, his best known work, the book *Die Chemie der Jetztzeit* (Contemporary Chemistry), in which he first proposed his chain theory, bears the subtitle "from the viewpoint of the electrochemical interpretation developed from Berzelius's theory." Despite his conservatism, Blomstrand was opposed to Kekulé's dogma of constant valency, and he tried to establish a sound and complete theory of variable valency. For him, Kekulé's dichotomy of compounds into "atomic compounds" and "molecular compounds" was completely unacceptable. In his book of 1869 Blomstrand asserted: "It has become the principal task of the newer chemistry to explain atomistically, *i.e.*, from the saturation capacity of the elements, compounds which previously have been conceived of more or less definitely as molecular." This statement was later chosen by no less an authority on coordination compounds than Alfred Werner (1866-1919) as the motto for the frontispiece of his monumental, but modestly titled textbook *Neuere Anschauungen auf dem Gebiete der anorganischen*

Chemie (Newer Views in the Field of Inorganic Chemistry). Levi Tansjö's article, "While Waiting for Werner: Chemistry in Chains," which appears in this symposium volume, briefly discusses *Die Chemie der Jetztzeit* and Blomstrand's chain theory.

Sophus Mads Jørgensen (Figure 3) was professor of chemistry at the University of Copenhagen. He was also Alfred Werner's primary scientific adversary. Except for some early isolated research, Jørgensen devoted himself exclusively to investigating the coordination compounds of cobalt, chromium, rhodium, and platinum. This work, upon which his fame securely rests, forms an interconnected and continuous chain from 1878 to 1906. Although Werner's ideas eventually triumphed, this in no way invalidated Jørgensen's experimental observations. On the contrary, his experiments, performed with meticulous care, have proven completely reliable. They provided the experimental foundation not only for the Blomstrand-Jørgensen chain theory but for Werner's coordination theory as well.

As a research worker, Jørgensen was methodical, deliberate, painstaking, and solitary. Although he could have delegated much routine work to assistants, he insisted on personally performing all his analyses. In fact, he reserved one day a week for this task. Werner, on the other hand, allowed many of the details of his syntheses to be worked out by his assistants or students. Consequently, almost all of Jørgensen's experimental work is reproducible, whereas some of Werner's work in this area leaves much to be desired. In view of Jørgensen's passion for perfection, his research output was tremendous, and we are indebted to him for many of the basic experimental facts of coordination chemistry.

The latter half of the 18th century was a period of tremendous progress in organic chemistry, and organic chemistry exerted a predominant influence on other branches of chemistry. Thus Blomstrand suggested that ammonia molecules could link as $-NH_3-$ chains, analogous to $-CH_2-$ chains in hydrocarbons. The number of ammonia molecules associated with the metal, that is, the length of the chain, depended upon the metal and its valence. This point was later accounted for more adequately by Werner's concept of the coordination number. Jørgensen made provision for different reactivities of various atoms and groups. For example, halogen atoms that could not be precipitated immediately by silver nitrate were called "nearer" and were considered to be bonded directly to the metal atom. Halogen atoms that could be precipitated immediately by silver nitrate were called "farther" and were considered to be bonded through the ammonia chains. These two different kinds of bonding were later explained more satisfactorily by Werner's terms "non-ionogenic" and "ionogenic," respectively, and by his concepts of inner and outer spheres of coordination. Despite the chain theory's admitted limitations, it permitted the correlation of a considerable amount of empirical data.

In 1883 Jørgensen demonstrated that tertiary amines, which, of course, contain no hydrogen, are capable of forming compounds that are completely analogous to the metal-ammines both in their composition and their properties. Recall that the ammonium theory conceived of metal-ammines as salts in which some of the hydrogen atoms of the ammonium group were replaced by metal atoms. Since tertiary amines contain no hydrogen, Jørgensen's discovery effectively eliminated the ammonium theory from serious consideration as an explanation for the constitution of the metal-ammines. Therefore chemists at that time were forced to assume the existence of ammonia chains

Figure 3. Sophus Mads Jørgensen (1837-1914) (Reproduced with permission from reference 25, p 12. Copyright 1977, American Chemical Society.)

copied from hydrocarbons or to conceive of metal-ammines as "molecular compounds." Since Kekulé's theory really explained nothing and only "substituted a beautiful word for a confused concept," to use Werner's words, the Blomstrand-Jørgensen chain theory became the most popular and satisfactory way of accounting for metal-ammines.

Wilhelm Ostwald has divided scientific geniuses into two types—the classic and the romantic. Jørgensen seems the embodiment of the classic type—the conservative, slow, and deep-digging completer who produces only after long deliberation and who methodically develops a traditional theory to new consequences. Jørgensen's strong and conservative sense of history caused him to regard Werner's new theory as an unwarranted break in the development of theories of chemical structure. He regarded it as an *ad hoc* explanation insufficiently supported by experimental evidence.

Although Jørgensen created no new structural theory of his own, he logically and consistently extended and modified Blomstrand's chain theory to interpret the many new series of complexes that he, Jørgensen, had succeeded in preparing for the first time. Just as the medieval astronomers tried to force an explanation for the motion of the planets in terms of the old geocentric Ptolemaic theory by postulating more and more complicated epicycles, so did Jørgensen strain to the breaking point the theory of his mentor Blomstrand in his attempt to account for his newly-prepared coordination compounds from a unified, theoretical point of view. Finally, in 1893, the Copernican figure of Alfred Werner appeared on the scene to challenge the old system with a revolutionary new theory based, according to Werner's own admission, upon the sturdy foundation of Jørgensen's painstaking experimental investigations. Ironic as it may seem, Jørgensen's work bore the seeds of the Blomstrand-Jørgensen theory's destruction, for, as we shall soon see, many of the compounds first prepared by Jørgensen later proved instrumental in demonstrating the validity of Werner's views. It is tempting to compare this situation with Priestley's discovery of oxygen, which led to Lavoisier's classic experiments on the nature of combustion and to the subsequent collapse of the Phlogiston theory. However, unlike Priestley, who staunchly defended the Phlogiston theory until his death, Jørgensen finally became convinced of the correctness of Werner's theory and acknowledged its worth. Since we shall compare in some detail the predictions of the Blomstrand-Jørgensen chain theory with those of the Werner coordination theory, we shall postpone a consideration of the chain theory until we have examined the basic postulates of the coordination theory.

Alfred Werner's Coordination Theory. In 1893 a comparatively unknown twenty-six-year-old *Privat-Dozent* or unsalaried lecturer at the Eidgenössisches Polytechnikum in Zürich came forth to challenge and discard the confining rigidities of both the Kekulé constant valence theory and the Blomstrand-Jørgensen chain theory. Like a modern Alexander the Great, Alfred Werner cut the Gordian knot that for decades had caused confusion in structural inorganic chemistry.

The circumstances surrounding the creation of Werner's coordination theory provide us with a classic example of the "flash of genius" that ranks with August Kekulé's famous dreams of the self-linking of carbon atoms and of the benzene ring. At the time, Werner's primary interest lay in the field of organic chemistry, and his knowledge of inorganic chemistry was extremely limited. Perhaps there is some truth after all in Albert Einstein's statement that "imagination is more important than knowledge," for one night in late 1892 Werner awoke at 2 AM with the solution of the constitution of "molecular compounds," which had come to him like a flash of lightning. He arose from his bed and wrote furiously and without interruption. By 5 PM of the following day he had finished his most famous paper titled "Contribution to the Constitution of Inorganic Compounds." (*Z. anorg. Chem.* **1893**, *3*, 267). Its publication marked the beginning of modern coordination chemistry, the centenary of which we are celebrating at this symposium.

This event might lead us to consider Werner to be the prototype of Ostwald's second type of genius—the romantic—the liberal, even radical, impulsive, and brilliant initiator who produces prolifically and easily during his youth, in short, the exact opposite of his adversary Jørgensen. Yet Werner's personality was too complex and self-contradictory to be accommodated by Ostwald's oversimplified dichotomy. At the time of its inception, Werner's theory was largely without experimental verification. After giving birth to the coordination theory, the typical romantic genius of the stereotype would have diverted his attention elsewhere, possibly on devising new, additional theories, and left to others the long and arduous task of painstakingly accumulating the experimental data necessary for its rigorous proof. But Werner combined the impulsive, intuitive, and theoretical brilliance of the romantic with the thorough, practical, and experimental persistence of the classicist. Firmly convinced of the correctness of his views, he devoted the remainder of his career to an almost unprecedented series of experimental researches which explored nearly every conceivable aspect of coordination chemistry and simultaneously verified his original theory in virtually every particular.

In Figure 4 we see a small part of the fruit of this quarter-century's whirlwind of research activity on the part of Werner and his more than two hundred doctoral students. I would estimate that the total number of preparations contained in carefully labeled tubes stored in the Chemical Institute of the University of Zürich where Werner worked is in excess of eight thousand! Much of this research was performed in the substandard basement laboratories nicknamed "The Catacombs" by Werner's students. Looking at one of these laboratories, shown in Figure 5, we may be tempted to agree with Helmholtz that "the best works come out of the worst laboratories." Confronted with the physical evidence of an almost superhuman capacity for work, carried out under primitive conditions, we can only regard Werner's achievement as that of a man obsessed with a dream, the dimensions of which we can never fully realize.

In fact, Werner played such a central and almost monopolistic role in coordination chemistry that his name is virtually synonymous with the field. Even today, almost 75 years after his death in 1919, coordination compounds, particularly metal-ammines, are still colloquially called Werner complexes. The coordination theory not only provided a logical explanation for known "molecular compounds," but also predicted series of unknown compounds, whose eventual discovery lent further weight to Werner's controversial ideas. He showed how ammonia could be replaced by water or other groups, and he demonstrated the existence of transition series between ammines, double salts, and hydrates. Werner recognized and named many types of inorganic isomerism such as coordination isomerism, polymerization isomerism, ionization isomerism, hydrate isomerism, salt isomerism, coordination position isomerism, and valence isomerism. He also postulated explanations for polynuclear complexes, hydrated metal ions, hydrolysis, and acids and bases. His view of the two types of chemical

Figure 4. Partial view of Werner's collection of coordination compounds at the Universität Zürich. (Reproduced from reference 9, p 82. Copyright 1967 American Chemical Society.)

Figure 5. "The Catacombs," Universität Zürich. (Reproduced with permission from reference 7, p 65. Copyright 1966 Springer-Verlag.)

linkage, "ionogenic" and "non-ionogenic," did much to clarify ideas of chemical bonding a generation before the views of Kossel and Lewis in 1916 led to our present concepts of ionic and covalent bonding. Thus, beginning with a study of metal-ammines, hydrates, and double salts, Werner's ideas soon encompassed almost the whole of systematic inorganic chemistry and even found application in the organic realm. Today, when the practical and theoretical significance of modern structural inorganic chemistry is unquestioned, it is clear that the foundations of this field were erected largely by Werner, who, for this reason, is sometimes honored by the designation, "the inorganic Kekulé."

We are now ready to examine the basic postulates of the coordination theory. Instead of regarding Werner's concepts as proven facts, let us try to go back in time and compare as objectively as possible the Blomstrand-Jørgensen chain theory with Werner's coordination theory. If we can regard ourselves as contemporary spectators, we may be able to see the advantages and disadvantages of each theory instead of regarding the view currently accepted, namely, Werner's view, as correct and any other views as hopelessly naïve. If history teaches us anything, it teaches us that the latest view is not always the best and that change is not always progress.

In his revolutionary theory, which marked an abrupt break with the classical theories of valence and structure, Werner postulated two types of valence—*Hauptvalenz*, primary or ionizable valence, and *Nebenvalenz*, secondary or nonionizable valence. According to Werner, every metal in a particular oxidation state, that is, with a particular primary valence, also has a definite *coordination number*, that is, a fixed number of secondary valences that must be satisfied. Now, whereas primary or ionizable valences can be satisfied only by anions, secondary or nonionizable valences can be satisfied not only by anions but also by neutral molecules containing donor atoms such as nitrogen, oxygen, sulfur, and phosphorus. Typical ligands which can be coordinated to the central atom in this manner include ammonia, organic amines, water, organic sulfides, and phosphines. The secondary valences are directed in space around the central metal atom, and the combined aggregate forms a "complex," which usually exists as a discrete unit in solution. Typical configurations are octahedral for coordination number six and square planar or tetrahedral for coordination number four.

In the following comparison between the chain theory and the coordination theory, we shall confine ourselves to the most common types of complexes, namely, the octahedral hexacovalent ammines of cobalt(III). Although we are concentrating on coordination number six, Werner used similar arguments to prove the constitution and configuration for compounds of coordination number four. This survey will be organized on the basis of compound type, that is, in a logical, rather than strict chronological sequence. First we will consider type MA_6 in which the coordination number of the central metal atom is satisfied by six ammonia molecules. We shall then proceed to replace these ammonia molecules one at a time with other groups.

The acknowledged test of a scientific theory is its ability to explain known facts and to predict new ones. In examining the comparative successes of the chain theory and the coordination theory in meeting these criteria, we shall examine the metal-ammines under two aspects—(i) *constitution*, that is, the manner of bonding of the constituent atoms and groups, and (ii) *configuration*, that is, the spatial arrangement of these atoms and groups.

Constitution. In Table I is shown a comparison of the formulas and predicted numbers of ions for a series of cobalt-ammines according to the two theories. We have already mentioned hindsight and foresight and the necessity for historical perspective. Please forget that modern orbital theory teaches that nitrogen can form at most four bonds and overlook the fact that the Blomstrand-Jørgensen formulas involve quinquevalent nitrogen. Notice that according to the Blomstrand-Jørgensen chain theory, the top compound, hexaamminecobalt(III) nitrite, to use modern IUPAC nomenclature, should dissociate to form three nitrite ions in solution, because all three nitrite groups are bonded through the ammonia chains. Three nitrite ions plus the remainder of the compound, which forms a tripositive cation, results in the total formation of four ions. Jørgensen regarded four as the maximum number of ammonia molecules that could enter into a chain. He regarded such a chain as particularly stable, and his formulas contained a chain of four ammonia molecules whenever possible. Loss of an ammonia molecule from the hexaammine results in formation of the second compound, penta-amminenitrocobalt(III) nitrite, which, according to the Blomstrand-Jørgensen formulation, should furnish a total of only three ions—two nitrite ions, which are bonded through the ammonia chains, plus the remainder of the compound which functions as a dipositive cation. The uppermost nitrite group does not ionize, because it is bonded directly to the cobalt atom. Loss of an ammonia molecule from the pentaammine results in formation of the third compound, tetraamminedinitrocobalt(III) nitrite, which should furnish a total of only two ions—one nitrite ion, which is bonded through the ammonia chain, plus the remainder of the compound, which functions as a monopositive cation. The two uppermost nitrite groups do not ionize because they are bonded to the cobalt atom. Loss of an ammonia molecule from the tetraammine merely shortens the one remaining ammonia chain by one member, and therefore the resulting compound, triamminetrinitro-cobalt(III), should form two ions in solution. Loss of further ammonia molecules would result in the formation of nonexistent compounds.

Now let us look at the formulas and predicted numbers of ions according to Werner's theory. According to Werner, loss of ammonia molecules from ammines is not actually a simple loss, but rather a substitution in which a change in function of the anion occurs simultaneously, that is, as each molecule of ammonia leaves the coordination sphere, shown by square brackets, its place is taken by an anion which is no longer bonded by a primary or ionizable valence, but instead by the secondary or nonionizable valence vacated by the departing ammonia molecule. The charge of a complex should be equal to the algebraic sum of the charges of the central metal ion and of the coordinated groups. Consequently, as neutral molecules of ammonia (A) in a metal-ammine (MA_6) are successively replaced by anions (B), the number of ions in the resulting compounds should progressively decrease until a nonelectrolyte is formed and then increase as the complex becomes anionic. How do the predictions of the two theories jibe with the experimental facts?

Werner's first published experimental work in support of his coordination theory was a study of conductivities carried out during the years 1893 to 1896 in collaboration with his friend and former fellow-student Arturo Miolati, who was later to become professor of chemistry at the universities of Turin and Padua (Figure 6).

Kohlrausch's principle of the additivity of equivalent conductivities of salts provided Werner and Miolati with a convenient method for determining the number of

Table I. Constitution of cobalt(III) coordination compounds

Class of compound	BLOMSTRAND–JØRGENSEN		WERNER	
	Formula	No. of ions	Formula	No. of ions
Hexaammines MA_6	$Co\begin{cases} NH_3-NO_2 \\ NH_3-NH_3-NO_2 \\ NH_3-NH_3-NH_3-NH_3-NO_2 \end{cases}$ $\xrightarrow{-NH_3}$	4	$[Co(NH_3)_6](NO_2)_3$ $\xrightarrow{-NH_3}$	4
Pentaammines MA_5B	$Co\begin{cases} NO_2 \\ NH_3-NO_2 \\ NH_3-NH_3-NH_3-NH_3-NO_2 \end{cases}$ $\xrightarrow{-NH_3}$	3	$[CoNO_2(NH_3)_5](NO_2)_2$ $\xrightarrow{-NH_3}$	3
Tetraammines MA_4B_2	$Co\begin{cases} NO_2 \\ NO_2 \\ NH_3-NH_3-NH_3-NH_3-NO_2 \end{cases}$ $\xrightarrow{-NH_3}$	2	$[Co(NO_2)_2(NH_3)_4]NO_2$ $\xrightarrow{-NH_3}$	2
Triammines MA_3B_3	$Co\begin{cases} NO_2 \\ NO_2 \\ NH_3-NH_3-NH_3-NO_2 \end{cases}$	2	$[Co(NO_2)_3(NH_3)_3]$ $\xrightarrow{-NH_3}$	0
Diammines MA_2B_4	Unaccountable	—	$K[Co(NO_2)_4(NH_3)_2]$ $\xrightarrow{-NH_3}$	2
Monoammines MAB_5	Unaccountable	—	Unknown for cobalt $\xrightarrow{-NH_3}$	(3)
Double salts, MB_6	Unaccountable	—	$K_3[Co(NO_2)_6]$	4

Reprinted with permission from ref. 8. Copyright 1967 The Royal Society of Chemistry, London.

Figure 6. Arturo Miolati (left) and Alfred Werner (right) in front of the Old Chemical Laboratory on the Rämistrasse, Universität Zürich, January or February 1893. (Reproduced with permission from reference 7, p 68. Copyright 1966 Springer-Verlag.)

ions in a variety of complexes. After having established the ranges of conductivities to be expected for salts of various types, they were able to demonstrate the complete agreement in magnitude, variation, and pattern between their experimentally measured conductivities, shown in Figure 7, and the numbers of ions predicted according to the coordination theory, shown in Table I. Their conductivity results were also concordant with the number of precipitable halogen atoms.

For compounds of the first three classes, hexaammines, pentaammines, and tetraammines, the electrolytic character as predicted by the two theories is in complete agreement, and conductivity data do not permit a choice between the two. For triammines, however, the ionic character differs radically according to the two theories, and the conductivities of these compounds became an important and bitterly contested issue. For some nonelectrolytes, unfortunately, Werner and Miolati's conductivity values were not always zero because of aquation reactions such as:

$$[CoCl_3(NH_3)_3] + H_2O \rightleftharpoons [CoCl_2(NH_3)_3(H_2O)]^+ + Cl^-$$

| Nonelectrolyte | | Cation | Anion |

| Nonconductor | | Conductor | |

Jørgensen immediately seized upon such so-called "discrepancies" in an attempt to discredit their results. But in its explanation of anionic complexes and its demonstration of the existence of a continuous transition series (*Übergangsreihe*) between metal-ammines (MA_6) and double salts (MB_6), the Werner theory succeeded in an area in which the Blomstrand-Jørgensen theory could not pretend to compete.

Configuration. Now let us examine the means used by Werner to establish the configuration of cobalt-ammines. The technique of "isomer counting" that he used as a means of proving configuration admittedly did not originate with Werner. The idea of an octahedral configuration and its geometric consequences with respect to the number of isomers expected had been considered as early as 1875 by Jacobus Henricus van't Hoff, and the general method is probably most familiar through Wilhelm Körner's work of 1874 on disubstituted and trisubstituted benzene derivatives. Yet the technique of comparing the number and type of isomers actually prepared with the number and type theoretically predicted for various configurations probably reached the height of its development with Werner's work. By this method, Werner was able not only to discredit completely the rival Blomstrand-Jørgensen chain theory but also to demonstrate unequivocally that tripositive cobalt possesses an octahedral configuration rather than another possible symmetrical arrangement such as hexagonal planar or trigonal prismatic.

The method is indirect but basically simple. Look at Figure 8. For coordination number six, if all six positions are equivalent, four configurations are possible—hexagonal pyramidal, hexagonal planar, trigonal prismatic, and octahedral. Now look at Table II, which shows the predicted number of isomers theoretically possible for selected compound types according to each of the different configurations (Hexagonal planar is a special case of hexagonal pyramidal). For compound type MA_6 or hexaammines, each of the three configurations should result in the same number of

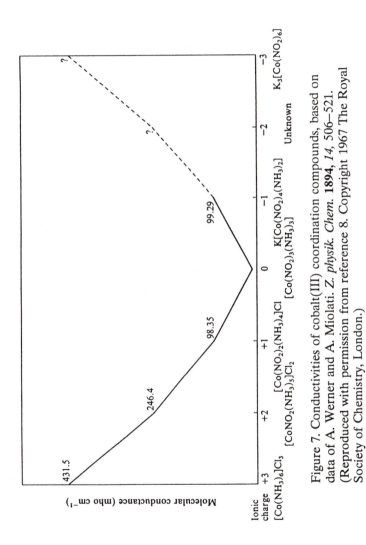

Figure 7. Conductivities of cobalt(III) coordination compounds, based on data of A. Werner and A. Miolati. *Z. physik. Chem.* **1894**, *14*, 506–521. (Reproduced with permission from reference 8. Copyright 1967 The Royal Society of Chemistry, London.)

Table II. Proof of Configuration for Coordination Number Six by "Isomer Counting"

Compound type	Theoretically predicted isomers				Experimentally found isomers	Result VI
	I Hexagonal pyramidal	II Hexagonal planar (a special case of I)	III Trigonal prismatic[b]	IV Octahedral	V	
MA_6	One form only	One form only	One form only	One form only	One form only	None
MA_5B	One form only	One form only	One form only	One form only	One form only	None
$M(\overline{AA})B_4$[a]	One form only	One form only	Two geometric	One form only	One form only	Provisionally eliminates trigonal prismatic (III/I)
MA_4B_2	Three geometric (1,2; 1,3; 1,4)	Three geometric (1,2; 1,3; 1,4)	Three geometric (1,2; 1,3: 1,4)	Two geometric (1,2 *cis*; 1,6 *trans*)	Two or less geometric	Provisionally proves octahedral (IV); discovered 1907
MA_3B_3	Three geometric (1,2,3; 1,2,4; 1,3,5)	Three geometric (1,2,3; 1,2,4; 1,3,5)	Three geometric (1,2,3; 1,2,5; (1,2,6)	Two geometric (1,2,3 *facial*; 1,2,6 *meridional*)	Two or less geometric	Provisionally proves octahedral (IV)
$M(\overline{AA})_2B_2$[a] or $M(\overline{AA})_2BC$[a]	Two *geometric*	Two *geometric*	Four *geometric*, one of which is asymmetric	Two *geometric* (1,2 *cis*; 1,6 *trans*), the first of which is asymmetric	Two *geometric* (1,2 *cis*; 1,6 *trans*), the first of which was *resolved*	Unequivocally proves octahedral (IV); discovered 1911
$M(\overline{AA})_3$[a]	One form only	One form only	Two *geometric*	One asymmetric pair	One pair *optical* resolved	Unequivocally proves octahedral (IV); discovered 1912

[a] $\overline{AA}$ represents a symmetrical bidentate (chelate) ligand. Such ligands coordinate at two adjacent positions. They can span *cis* positions but not *trans* positions.

[b] Coordination compounds with this configuration have now been synthesized (R. Eisenberg and J. A. Ibers, *J. Am. Chem. Soc.* **87**, 3776–3778 (1965); *Inorg. Chem.* **5**, 411–416 (1966); E. I. Stiefel and H. B. Gray, *J. Am. Chem. Soc.* **87**, 4012–4013 (1965); H. B. Gray, R. Eisenberg and E. I. Stiefel, in *Werner Centennial*, G. B. Kauffman, Symposium Chairman, American Chemical Society, Washington, DC, 1967, pp. 641–650).

geometric isomers, namely one, as shown in Figure 9. For compound type MA_5B or pentaammines, each of the three configurations should also result in the same number of geometric isomers, namely one, as shown in Figure 10. Therefore the number of isomers actually found in these two cases does not permit a choice between the three configurations. On the other hand, as shown in Figure 11, in the case of compound type MA_4B_2 or tetraammines, the hexagonal planar and trigonal prismatic configurations would each result in three possible geometric isomers, whereas the octahedral configuration would result in only two possible geometric isomers. As shown in Figure 12, for compound type MA_3B_3 or triammines, again the hexagonal planar and trigonal prismatic configurations would each result in three possible geometric isomers, whereas the octahedral configuration would result in only two possible geometric isomers. Finally, as shown in Figures 13 and 14, for compound type $M(\overline{AA})_3$, that is, trisdidentate complexes or those containing three didentate chelate groups, each of the three possible configurations would result in different isomeric possibilities.

In most cases, the number and type of isomers prepared corresponded to the expectations for the octahedral arrangement, but there were a few exceptions, and Werner required more than 20 years to accumulate a definitive proof for his structural ideas. For example, the best known case of geometric or *cis-trans* isomerism was observed by Jørgensen not among simple tetraammines MA_4B_2, but among salts $M(\overline{AA})_2B_2$, in which the four ammonia molecules had been replaced by two molecules of the didentate chelate organic base, ethylenediamine (en), that is, among the so-called praseo (green) and violeo (violet) series of formula $CoCl_3 \cdot 2en$. As shown in Figure 15, Jørgensen regarded the difference in color as due to *structural* isomerism connected with the linking of the two ethylenediamine molecules, whereas Werner regarded the compounds as *stereoisomers*, that is, compounds composed of the same atoms and bonds, but differing only in the orientation of these atoms and bonds in space.

If this type of isomerism were merely a geometric consequence of the octahedral structure as Werner maintained, it should also be observed among simple tetraammines MA_4B_2 which do not contain ethylenediamine. Yet, for compounds $[CoCl_2(NH_3)_4]X$, only one series (praseo, green) was known. Jørgensen, a confirmed empiricist, quite correctly criticized Werner's theory on the ground that it implied the existence of unknown compounds. It was not until 1907 that Werner finally succeeded in synthesizing the unstable, highly crucial violeo tetraammines, *cis*-$[CoCl_2(NH_3)_4]X$, which were a necessary consequence of his theory but not of Jørgensen's (Figure 16). The structure of these crucial compounds is shown in Figure 17. His Danish opponent immediately conceded defeat.

However, even though the discovery of the long-sought violeo salts convinced Jørgensen that his own views could not be correct, Werner's success in preparing two—and only two—isomers for compounds of types MA_4B_2 and MA_3B_3 was not sufficient to prove conclusively his octahedral configuration. Despite such "negative" evidence, it could still be argued logically that failure to isolate a third isomer did not necessarily prove its nonexistence. A more "positive" type of proof was necessary.

As early as 1899, Werner recognized that the resolution into optical isomers of certain types of coordination compounds containing chelate groups, which can span adjacent or *cis* positions only, could provide the "positive" proof that he needed. After many unsuccessful attempts, in 1911 he succeeded with the help of his American

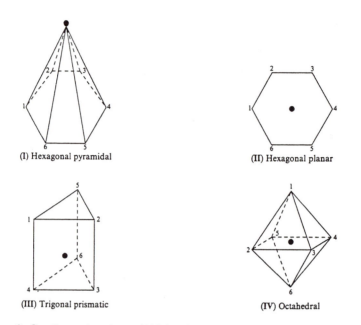

(I) Hexagonal pyramidal

(II) Hexagonal planar

(III) Trigonal prismatic

(IV) Octahedral

Figure 8. Configurational possibilities for coordination number six. (Reproduced with permission from reference 8. Copyright 1967 The Royal Society of Chemistry, London.)

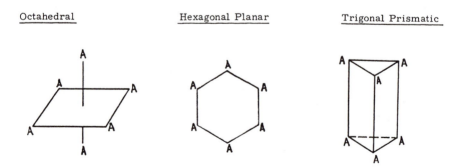

Octahedral

Hexagonal Planar

Trigonal Prismatic

Figure 9. Predicted isomers for compound MA_6. (Reproduced from reference 25, p 23. Copyright 1977 American Chemical Society.)

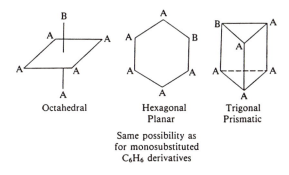

Octahedral   Hexagonal   Trigonal
                      Planar   Prismatic

Same possibility as
for monosubstituted
C_6H_6 derivatives

Figure 10. Predicted isomers for compound MA_5B. (Reproduced from reference 25, p 23. Copyright 1977 American Chemical Society.)

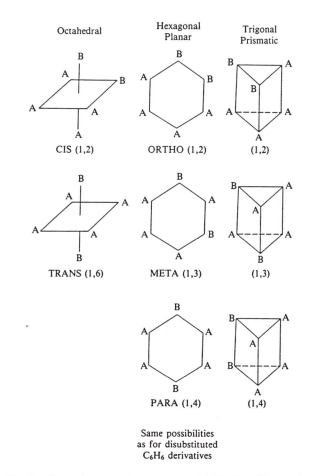

Same possibilities
as for disubstituted
C_6H_6 derivatives

Figure 11. Predicted isomers for compound MA_4B_2. (Reproduced from reference 25, p 24. Copyright 1977 American Chemical Society.)

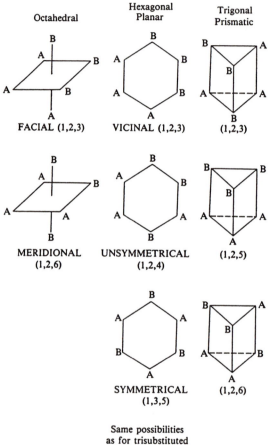

Figure 12. Predicted isomers for compound MA_3B_3. (Reproduced from reference 25, p 25. Copyright 1977 American Chemical Society.)

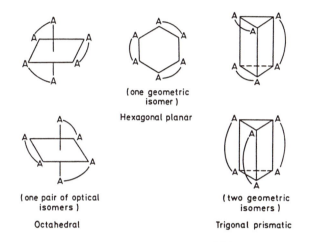

(one geometric
isomer)

Hexagonal planar

(one pair of optical
isomers)

Octahedral

(two geometric
isomers)

Trigonal prismatic

Figure 13. Predicted isomers for compound $M(\overline{AA})_3$. (Reproduced from reference 25, p 26. Copyright 1977 American Chemical Society.)

Figure 14. Alfred Werner's models of the enantiomorphs of an octahedral trisdidentate complex. (Reproduced with permission from reference 7, p 69. Copyright 1966 Springer-Verlag.)

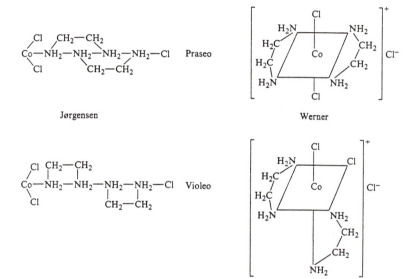

Figure 15. Jørgensen's and Werner's formulas for praseo and violeo ethylenediamine isomers. (Reproduced with permission from reference 8. Copyright 1967 The Royal Society of Chemistry, London.)

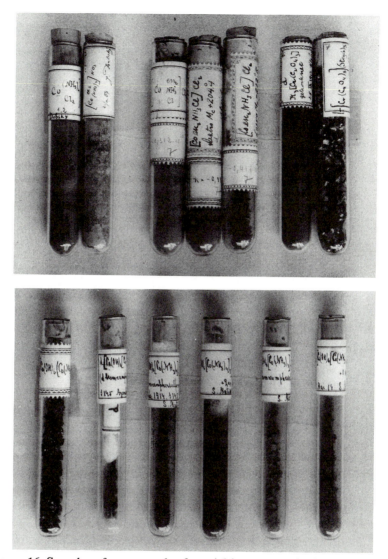

Figure 16. Samples of compounds of crucial importance in the history of coordination theory. Top photograph (left to right): *cis*-[CoCl₂(NH₃)₄]Cl, *cis*-[CoCl₂(NH₃)₄]NO₃ (violeo salts, 1907); racemic, *d*-, and *l*-[CoCl(NH₃)(en)₂]Cl₂ (the first coordination compound to be resolved, 1911); K₃*d*-[Cr(C₂O₄)₃], (strychninium)₃ *l*-[Cr(C₂O₄)₃] (the first coordination compound with an optically active anion to be resolved, 1912). Bottom photograph, salts of the [Co (OH)₂Co(NH₃)₄ ₃]⁶⁺ cation (the first purely inorganic complex ion to be resolved, 1914) (left to right) racemic bromide; *d,d*-bromocamphorsulfonate; *d,l*-bromocamphorsulfonate; *d*-bromide; *l,d*-bromocamphorsulfonate; *l*-bromide. (Reproduced with permission from reference 10, facing p 146. Copyright 1968 Dover Publications.)

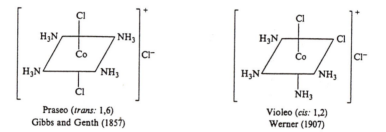

Praseo (*trans:* 1,6)
Gibbs and Genth (1857)

Violeo (*cis:* 1,2)
Werner (1907)

Figure 17. Structures of praseo and violeo ammonia isomers. (Reproduced with permission from reference 8. Copyright 1967 The Royal Society of Chemistry, London.)

doctoral student Victor L. King (Figure 18). Their resolution of *cis*-amminechlorobis-(ethylenediamine)cobalt(III) salts by means of the resolving agent silver *d-α*-bromo-camphor-π-sulfonate conclusively proved the octahedral configuration for cobalt(III) (Figure 19) and was probably the main reason that Werner was awarded the Nobel Prize in chemistry two years later in 1913. The structural formulas for the antipodes of this compound are shown in Figure 20.

Because of the prevalent view that optical activity was almost always connected with carbon atoms, a number of Werner's contemporaries argued that the optical activity of this and the many other mononuclear and polynuclear coordination compounds subsequently resolved by him was somehow due to the organic chelate groups present, even though these symmetrical ligands were all optically inactive. Any vestige of doubt was finally dispelled by Werner's resolution in 1914 of a compound first prepared by Jørgensen—a completely carbon-free coordination compound—the tris[tetraammine-μ-dihydroxocobalt(III)]cobalt(III) salt shown in Figure 19. These salts are compounds of the $M(\overline{AA})_3$ trisdidentate type (Figure 14), in which $\overline{AA}$ is the inorganic didentate ligand:

$$\left[Co \left\{ \begin{array}{c} HO \\ \diagdown \\ HO \end{array} Co(NH_3)_4 \right\}_3 \right] Br_6$$

Tris[tetraammine-μ-dihydroxocobalt(III)]cobalt(III) Bromide

By any standards, Werner's achievements are remarkable. At the very beginning of his career, he had destroyed the carbon atom's monopoly on geometric isomerism. In his doctoral dissertation he had explained the isomerism of oximes as due to the tetrahedral configuration of the nitrogen atom. Then, at the peak of his career, he had likewise forced the tetrahedron to relinquish its claim to a monopoly on optical isomerism. He had finally attained one of the major goals of his life's work—the demonstration that stereochemistry is a general phenomenon not limited to carbon compounds and that no fundamental difference exists between organic and inorganic compounds.

In conclusion, we must note that the validity of Werner's structural views was later amply confirmed by X-ray diffraction studies. Yet, despite the advent of more *direct* modern techniques, Werner's classical configurational determinations by simple *indirect* methods still remain today a monument to his intuitive vision, experimental skill, and inflexible tenacity. In the words of former American Chemical Society President Henry Eyring, "The ingenuity and effective logic that enabled chemists to determine complex molecular structures from the number of isomers, the reactivity of the molecule and of its fragments, the freezing point, the empirical formula, the molecular weight, *etc.* is one of the outstanding triumphs of the human mind."

Figure 18. Victor L. King (1886–1958), codiscoverer in 1911 with Alfred Werner of optically active coordination compounds, Universität Zürich, 1910. (Courtesy of Victor R. King.) (Reproduced from reference 19, p 132. Copyright 1975 American Chemical Society.)

Figure 19. The polarimeter used by Werner to measure the rotation of optically active coordination compounds (Franz Schmidt & Haensch model no. 8142). (Reproduced with permission from reference 10, facing p 178. Copyright 1968 Dover Publications.)

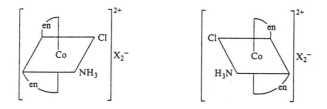

Figure 20. Structures of the enantiomorphs of *cis*-$[CoCl(NH_3)(en)_2]X_2$. (Reproduced with permission from reference 8. Copyright 1967 The Royal Society of Chemistry, London.)

Acknowledgments

The author thanks the Educational Activities Department of the American Chemical Society for permission to reproduce material from his audiotape and book *Coordination Chemistry: Its History Through the Time of Werner*; ACS Audio Courses, History of Chemistry Series, Schubert, L.; Kauffman, G. B., Eds., Washington, DC, 1977. He also thanks Professor Michael Laing of the University of Natal for his kind invitation to present this paper as a plenary lecture at the 75th Anniversary of the South African Chemical Institute, University of Natal, Durban, South Africa, July 7, 1987. This article is reprinted with permission from *Spectrum* **October 1987**, *25*(4), 5-14.

Suggested Readings

Detailed references to the facts presented in this paper are found in many of the books and articles listed below.

1.   Bailar, Jr., J. C. "The Early Development of the Coordination Theory," In *The Chemistry of the Coordination Compounds*; Bailar, Jr., J. C., Ed., Reinhold Publishing Corp.: New York, 1956; Chap. 2.

2.   Basolo, F.; Johnson, R. C. *Coordination Chemistry*; 2nd edn.; Science Reviews: Wilmington, DE, 1986.

3.   Jones, M. M. *Elementary Coordination Chemistry*; Prentice-Hall: Englewood Cliffs, NJ, 1964.

4.   Kauffman, G. B. "Sophus Mads Jørgensen (1837-1914): A Chapter in Coordination Chemistry History," *J. Chem. Educ.* **1959**, *36*, 521; reprinted in *Selected Readings in the History of Chemistry*; Ihde, A. J., Kieffer, W. F., Eds.; Journal of Chemical Education Publishing Co.: Easton, PA, 1965; pp 185-191.

5.   Kauffman, G. B. "Sophus Mads Jørgensen and the Werner-Jørgensen Controversy," *Chymia* **1960**, *6*, 180.

6.   Kauffman, G. B. "Alfred Werner—Architect of Coordination Chemistry," *Chemistry* **1966**, *39*(12), 14.

7.   Kauffman, G. B. *Alfred Werner: Founder of Coordination Chemistry*; Springer-Verlag: Berlin, Heidelberg, New York, 1966.

8.   Kauffman, G. B. "Alfred Werner's Co-ordination Theory," *Educ. Chem.* **1967**, *4*(1), 11.

9.   Kauffman, G. B., Ed. *Werner Centennial;* Advances in Chemistry Series No. 62; American Chemical Society: Washington, DC, 1967.

10.   Kauffman, G. B. *Classics in Coordination Chemistry, Part 1: The Selected Papers of Alfred Werner*; Dover: New York, 1968.

11.   Kauffman, G. B. "Arturo Miolati (1869-1956)," *Isis* **1970**, *61*, 241.

12.   Kauffman, G. B. "Werner, Kekulé, and the Demise of the Doctrine of Constant Valency," *J. Chem. Educ.* **1972**, *49*, 813.

13.   Kauffman, G. B. "Alfred Werner's Research on Polynuclear Coordination Compounds," *Coord. Chem. Rev.* **1973**, *9*, 339.

14.   Kauffman, G. B. "Alfred Werner's Research on Structural Isomerism," *Coord. Chem. Rev.* **1973**, *11*, 161.

15.   Kauffman, G. B. "Alfred Werner's Research on Optically Active Coordination Compounds," *Coord. Chem. Rev.* **1974**, *12*, 105.

16.   Kauffman, G. B. "Early Theories of Metal-Ammines: A Brief Historical Review from Graham (1837) to Claus (1856)," *J. Chem. Educ.* **1974**, *51*, 522.

17.   Kauffman, G. B. "Alfred Werner's Research on Geometrically Isomeric Coordination Compounds," *Coord. Chem. Rev.* **1975**, *15*, 1.

18.   Kauffman, G. B. "Christian Wilhelm Blomstrand (1826-1897), Swedish Chemist and Mineralogist," *Ann. Sci.* **1975**, *32*, 12.

19.   Kauffman, G. B. "The First Resolution of a Coordination Compound," In *Van't Hoff—Le Bel Centennial*; Ramsay, O. B., Ed., American Chemical Society: Washington, DC, 1975; pp 126-142.

20.   Kauffman, G. B. "The Discovery of Optically Active Coordination Compounds—A Milestone in Stereochemistry," *Isis* **1975**, *65*, 38.

21.   Kauffman, G. B. "Alfred Werner," In *Dictionary of Scientific Biography*; Gillispie, C. C., Ed.; Charles Scribner's Sons: New York, 1976; Vol. 14, pp 264-272.

22.   Kauffman, G. B. *Classics in Coordination Chemistry, Part 2: Selected Papers (1798-1899)*; Dover: New York, 1976.

23.   Kauffman, G. B. "Alfred Werner's Early Views of Valence," *J. Chem. Educ.* **1976**, *56*, 197.

24.   Kauffman, G. B. "Christian Wilhelm Blomstrand (1826-1897) and Sophus Mads Jørgensen (1837-1914): Their Correspondence from 1870 to 1897," *Centaurus* **1977**, *21*, 44.

25.   Kauffman, G. B. *Coordination Chemistry: Its History through the Time of Werner;* ACS Audio Course; American Chemical Society: Washington, DC, 1977.

26.   Kauffman, G. B. *Inorganic Coordination Compounds*; Heyden: London, Philadelphia, Rheine, 1981.

27.   Kauffman, G. B. "General Historical Survey to 1930," In *Comprehensive Coordination Chemistry: The Synthesis, Reactions, Properties & Applications of Coordination Compounds*; Wilkinson, G.; Gillard, R. D.; McCleverty, J. A., Eds.; Pergamon: Oxford, 1987; Vol. 1; pp 1-20.

28.   Martin, D. F.; Martin, B. B. *Coordination Compounds*; McGraw-Hill: New York, 1964.

29.   Reitzenstein, F., "Über die verschiedenen Theorien zur Klärung der Konstitution der Metallammoniaksalze," *Z. anorg. Chem.* **1898**, *18*, 152.

RECEIVED August 16, 1994

*Christian Wilhelm Blomstrand (1826–1897). (Reproduced
with permission from reference 10. Copyright 1954 Kemisk Forening.)*

Chapter 2

While Waiting for Werner
Chemistry in Chains

Levi Tansjö

Chemical Center, Lund University, P.O. Box 124, S–221 00 Lund, Sweden

The principal thoughts in Christian Wilhelm Blomstrand's chain theory of so-called complex inorganic compounds are briefly outlined. This theory held sway for roughly three decades until it was superseded by Alfred Werner's coordination theory.

In his opus magnum *Neuere Anschauungen auf dem Gebiete der anorganischen Chemie (1)* Alfred Werner used these words, borrowed from C.W. Blomstrand (*2*, p. 127), as a motto: "Es ist die Hauptaufgabe der neueren Chemie geworden, die früher mehr oder minder entschieden molecular aufgefassten Verbindungen atomistisch, d.h. aus der Sättigungscapacität der Grundstoffe zu erklären". (It has become the principal task of the newer chemistry to explain atomistically, *i.e.*, from the saturation capacity [valence] of the elements, compounds which earlier have been conceived more or less definitely as molecular).

Christian Wilhelm Blomstrand (1826-1897), Swedish chemist and mineralogist, was professor at Lund University 1862-1895 (*3*). He gained an international reputation for his inorganic experimental research and for his accurate analyses of minerals. He was also renowned for his views on the constitution of diazo compounds and metal complexes (coordination compounds), published in German periodicals during the 1870s but already developed in his book *Die Chemie der Jetztzeit* (Contemporary Chemistry), the 417-page treatise on the theoretical aspects of chemistry, from which Werner took his motto.

In 1869, when *Die Chemie der Jetztzeit* appeared, it was the consensus among chemists that chemistry had entered a new era – that the concepts and ideas introduced around 1860 were so far-reaching that one could talk about a *new* chemistry – and many chemists looked upon August Kekulé (1829-1896) as its true creator. Blomstrand, however, saw nothing *new* in this but had stated in the introduction to his book in bold type that the *new* chemistry was "only a consequential development of Berzelius' atomic theory necessarily evoked by the force of many newly discovered facts" (*2*, p. 2). These words he threw into Kekulé's camp where everyone considered Berzelius' dualistic theory obsolete when in the 1830s it became evident that electronegative chlorine atoms could be substituted for

0097–6156/94/0565–0035$08.00/0

electropositive hydrogen atoms in organic compounds without drastic changes in the properties of the compounds. Blomstrand did not accept this view of the ideas of his compatriot and idol Berzelius, and he began his book with a panegyric to Berzelius' teaching. He then gave a very learned, keen, and penetrating account of the development of theoretical chemistry from Berzelius via the unitary and type theories to the valence theory, especially emphasizing the importance of Edward Frankland's and Hermann Kolbe's contributions. He recognized only one concept as *new* in the *new* chemistry, namely the principle of atomicity (*Atomigkeit, Atomizität, Sättigungscapazität*) (valence), but he opposed Kekulé's doctrine that valence is "a fundamental property of the atom which is just as constant and unchangeable as the atomic weight itself" (*4*).

To preserve the constant monovalence of chlorine and divalence of sulfur Kekulé had to postulate chains of oxygen atoms in, *e.g.,* perchloric acid, sulfur trioxide, and sulfuric acid:

H-O-O-O-O-Cl   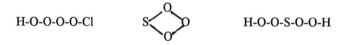   H-O-O-S-O-O-H

To maintain his doctrine Kekulé also had to devise an artificial division of compounds into *atomic* compounds with the atoms held together by normal affinities and *molecular* compounds composed of atomic ones combined by odd forces acting between molecules rather than between atoms. For example, Kekulé classified sal-ammoniac (NH_4Cl) and phosphorus pentachloride (PCl_5) as molecular compounds and assigned them the formulas NH_3,HCl and PCl_3,Cl_2 to preserve the trivalence of nitrogen and phosphorus.

Blomstrand rejected Kekulé's dogma of constant valence, asserting that, in accordance with Dalton's law of multiple proportions, the valence of most atoms is variable. As was his usual practice, he took Berzelius' teachings as his point of departure. Berzelius had divided the elements into two groups – the oxygen group (O, S, Se,C, Si, Sn, *et al.*) and the hydrogen group (H, F, Cl, N, P, As, *et al.*). According to Berzelius, atoms of the elements in the O-group could combine with other atoms individually, but those belonging to the H-group only doubled. Consequently, he wrote, for example, the formula of ammonia as N̶H̶³ and that of hydrogen chloride as H̶C̶l̶ with the barred symbols N̶, H̶, and C̶l̶ representing double atoms. Blomstrand considered this idea to be Berzelius' primary, tragic error, which he had made in 1812 as a result of his refusal to accept fully Avogadro's hypothesis. Blomstrand eliminated this error by allowing atoms in the H-group to combine individually, and he used formulas such as NH^3 and HCl – with superscripts as Berzelius had done –but he preserved Berzelius' dichotomy of elements because the law of multiple proportions suggested the valences 1, 3, 5, and 7, *i.e., odd* ones in the H-group and 2, 4, and 6, *i.e., even* ones in the O-group. Hydrogen, however, always seemed to be monovalent, and oxygen, which he designated "king of the elements" (*2*, p. 210), always divalent. Armed with these regularly variable valences, he harshly criticized Kekulé's formulas, occasionally being polemical and ironical, before he resolutely proclaimed in large type: "Die wechselnde Sättigungscapazität ist eine Grundeigenschaft der Atome" (Variable valence is a basic property of atoms) (*2*, p.185).

Blomstrand's confidence in variable valence was far from original. For example, as early as 1852 Frankland had ascribed to nitrogen, phosphorus, arsenic, and antimony "the tendency ..... to form compounds containing 3 or 5 equivalents of other elements" (*5*) and in 1866 he explicitly had stated that "atomicity is *apparently at least* not a fixed and invariable quality" [emphasis added] (*6*). During the 1860s many chemists were sceptical of Kekulé's doctrine of constant valence. Many vacillated between the two viewpoints, but Blomstrand did not. For him, the question concerning constant or variable valence was the major, deciding factor, which had to be settled definitively and had to include not only gases but liquid and solid compounds as well. Blomstrand's answer, the doctrine of variable valence, later became defensible, but it was far from defensible in 1869, the year of the publication of *Die Chemie der Jetztzeit*. Gas densities of compounds like PF_5 and SF_6 were still unknown, and it was only 18 years later (1887) Jacobus Henricus van't Hoff and Svante Arrhenius found sugar in aqueous solution to behave like a gas and sodium chloride like a mixture of two gases. Then it finally became possible to determine molecular weights of nonvolatile but soluble compounds. In 1869, however, the doctrine of variable valence could be nothing more than a reasonable interpretation of the law of multiple proportions.

Frankland cautiously stated that atomicity "apparently at least" is not fixed, but Blomstrand threw caution to the winds and consequently committed a great error. In the 4th section of his book (*2*, pp. 186-389) he postulated a peculiar theory of variable valence that was so naive that the entire book was simply passed over in silence. The theory, based on a mysterious "Kraft des Gegensatzes" (force of the opposition), lived only as long as Blomstrand and had Blomstrand as its only adherent. It did not contain one single well-defined concept and was so diffuse that it was adaptable to all facts.

In the blurredness of his strange theory Blomstrand discussed the properties and constitution of all types of compounds in the second half of his book. In discussing nitrogen compounds he had to explain why nitrous acid decomposes to form nitrogen oxide (NO) with a divalent rather than an *odd*valent nitrogen atom:

$$3 \, HONO = HONO^2 + NO + H^2O$$

Blomstrand's explanation was that the driving force of the reaction – which he thought was the strong attraction between water and nitric acid – prevents two of the three nitrogen atoms from appearing with any of their ordinary *odd* valences. He then demonstrated the unwillingness of nitrogen atoms to form NO by citing the reaction of nitrous acid in the presence of an amine such as aniline:

$$C^6H^5NH^2 + HONO = N^2 + C^6H^5OH + H^2O$$

In this case the nitrogen atoms have the opportunity, Blomstrand stated, of forming nitrogen gas, N^2, instead of NO. At low temperatures, however, they prefer, as Peter Griess had found in 1858, to form "diazoamidobenzol" with a chain of 3-valent nitrogen atoms:

$$2 \, C^6H^5NH^2 + HONO = C^6H^5\text{-}N{=}N\text{-}NHC^6H^5 + 2 \, H^2O$$

Blomstrand then mentioned that Griess had found that in acidic solution, instead of diazoaminobenzene, a salt of "diazobenzol" is formed, a discovery which had led to azo compounds, which even in 1869 were already technically important. Of course Kekulé had theoretically explored all these new organic nitrogen compounds and, "with his ordinary perspicacity completely disentangled their constitution" (*2*, p. 271), as Blomstrand ironically expressed it. He then cornered Kekulé by comparing Kekulé's way of describing the the diazotization of aniline with his own:

$$C_6H_5NH_2,HCl + HONO = C_6H_5N=N-Cl + 2 H_2O \qquad \text{Kekulé}$$

$$C^6H^5NH^3Cl + HONO = C^6H^5\underset{\substack{||| \\ N}}{N}-Cl + 2 H^2O \qquad \text{Blomstrand}$$

Blomstrand delightedly pointed out that in this case the dogma of constant valence had forced Kekulé for the first time to ascribe to an organic nitrogen *salt* an atomic constitution formula and to allow the *molecular* compound aniline hydrochloride to be transformed into an *atomic* one. In Blomstrand's description the 3-valent nitrogen atom in nitrous acid was substituted quite simply for the three hydrogen atoms at the 5-valent nitrogen atom in anilinium chloride to give "the most interesting ammonium compound you can imagine" (*2*, p. 272).

After having thus "proved" the existence of not only bonds between 3-valent nitrogen atoms but between 3- and 5-valent ones as well, Blomstrand sought bonds between 5-valent nitrogen atoms and quickly found them in the salts of Reiset's 1st base ([Pt(NH_3)_4](OH)_2). Berzelius had described them as "copulated" (*gepaart*) compounds, and he assigned the formula Pt~~NH²~~ ~~NH⁴~~Cl to the chloride ([Pt(NH_3)_4]Cl_2). Blomstrand resolved the double atoms and obtained the formula

$$\begin{array}{l} NH^2.NH^4.Cl \\ Pt \\ NH^2.NH^4.Cl \end{array}$$

However, because 4- and 6-valent nitrogen atoms were forbidden, two hydrogen atoms had to be moved from the outer to the inner nitrogen atoms to give the constitutional formula

$$\begin{array}{l} NH^3.NH^3.Cl \\ Pt \\ NH^3.NH^3.Cl \end{array}$$

with bonds between two 5-valent nitrogen atoms. Blomstrand assumed that the compound formed on addition of chlorine to this salt ([Pt(NH_3)_4]Cl_4) contained 4-valent platinum and he assigned it the constitutional formula

$$\begin{array}{l} Cl \\ NH^3.NH^3.Cl \\ Pt \\ NH^3.NH^3.Cl \\ Cl \end{array}$$

Blomstrand proposed that other metals accepted *chains* of three 5-valent nitrogen atoms; in the case of cobalt, for example, he assigned $[Co(NH_3)_6]Cl_2$ the formula:

$$NH^3.NH^3.NH^3.Cl$$
$$Co$$
$$NH^3.NH^3.NH^3.Cl$$

The dogma of variable – but either *odd* or *even* – valence then easily resulted in constitutional formulas for compounds earlier described as double salts, *e.g.*:

$$
4\ KCN\ +\ Fe(CN)^2:\quad
\begin{matrix}
& C & C & C \\
K^2{=}N{-}N{=}N & & & \\
& & & Fe \\
K^2{=}N{-}N{=}N & & & \\
& C & C & C
\end{matrix}
\quad (\text{modern, } K_4[Fe(CN)_6])
$$

and

$$
4\ KCl\ +\ CuCl^2:\quad
\begin{matrix}
K^2{=}Cl{-}Cl{=}Cl \\
\qquad\qquad Cu \\
K^2{=}Cl{-}Cl{=}Cl
\end{matrix}
\quad (\text{modern, } K_4[CuCl_6])
$$

In theoretical questions Blomstrand was more an artist than a scientist and he did not find it inconsistent to reject Kekulé's chains of oxygen atoms but to postulate chains of 5-valent nitrogen atoms and chains of 3-valent chlorine atoms. He merely regarded Kekulé's chains as unnecessary but his own chains as indispensible "to explain atomistically ..... compounds which earlier have been conceived more or less definitely as molecular", *i.e.*, to live up to what he saw as "the principal task of the newer chemistry".

Blomstrand's chain theory was adopted, modified and developed by his close friend in Copenhagen, Sophus-Mads Jørgensen (1837-1914) (*7*), who brilliantly defended it until Werner in 1907 – ten years after Blomstrand's death – synthesized *cis*-tetraamminedichlorocobalt(III) chloride (*cis*-$[CoCl_2(NH_3)_4Cl]$) (*8*). Then, with the words: "Nu er striden afgjort" (Now the battle is over), Jørgensen graciously conceded defeat (*9*).

Literature Cited

1.   Werner, A. *Neuere Anschauungen auf dem Gebiete der anorganischen Chemie*; Friedrich Vieweg und Sohn: Braunschweig, 1905, 1909, 1913, 1920, 1923; transl. from 2nd German edition by E.P. Hedley; Longmans, Green & Co: London, 1911.

2.   Blomstrand, C. W. *Die Chemie der Jetztzeit vom Standpunkte der electro-chemischen Auffassung aus Berzelius Lehre entwickelt*; Carl Winter's Universitätsbuchhandlung: Heidelberg, 1869.

3. Kauffman, G. B. *Ann. Sci.* **1975**, *32*, 13-37; *Centaurus* **1977**, *21*, 44-64.
4. Kekulé, A. *Compt. Rend.* **1864**, *58*, 510-514; *Z. Chem.* **1864**, *7*, 689-694.
5. Frankland, E. *Phil. Trans. Roy. Soc. London* **1852**, *142*, 417-444.
6. Frankland, E. *J. Chem. Soc.* **1866**, *19*, 372-395.
7. Kauffman, G. B. *J. Chem. Educ.* **1959**, *36*, 521-527; *Chymia* **1960**, *6*, 180-204; *Platinum Metals Rev.* **1992**, *36*, 217-223.
8. Werner, A. *Ber.* **1907**, *40*, 4817-4825. For a discussion and annotated English translation see Kauffman, G. B. *Classics in Coordination Chemistry, Part 1: The Selected Papers of Alfred Werner*; Dover: New York, NY, 1968; pp 141-154.
9. Sørensen, S. P. L. *Fys. Tidskrift* **1914**, *12*, 217-241.
10. Bjerrum, N. *Proceedings of the Symposium on Co-ordination Chemistry, Copenhagen, August 9–13, 1953;* Kemisk Forening: Copenhagen, 1954; introductory lecture.

RECEIVED May 9, 1994

Beitrag zur Konstitution anorganischer Verbindungen.

Von

ALFRED WERNER.

Mit 17 Figuren im Text.

Unter Metallammoniaksalzen versteht man Verbindungen, welche aus Metallsalzen dadurch entstehen, dafs sich in ihr Molekül Ammoniakmoleküle einschieben, ·oder besser: Metallammoniaksalze sind Verbindungen, welche nach derselben Reaktion aus Ammoniak und Metallsalzen entstehen, nach der sich Chlorammonium aus Salz-säure (welche letztere ja das Haloidsalz des Wasserstoffes ist) und Ammoniak bildet.

Die Metallammoniaksalze nach ihrer Beständigkeit in verschiedene Verbindungsklassen einteilen zu wollen, von denen die beständigen atomistische Konstitutionsformeln, die unbeständigen sogenannte Molekülformeln erhalten würden, erscheint beim heutigen Stande der Wissenschaft als unzulässig; wir müssen nach einem anderen Einteilungsprinzip suchen. Ein solches ergiebt sich denn auch mit Leichtigkeit, wenn wir die empirische Zusammensetzung der Ver-bindungen und gewisse Eigenschaften der zu betrachtenden Körper als leitende Momente der Einteilung benutzen.

Als erste Klasse erhalten wir dann Verbindungen, welche auf ein Metallatom sechs Ammoniakmoleküle [1] enthalten oder sich von diesen ammoniakreichsten nach bestimmten, später zu besprechenden Regeln ableiten lassen.

Die zweite Klasse wird gebildet durch Verbindungen, welche auf ein Metallatom vier Ammoniakmoleküle enthalten, und solchen, die sich auch wieder von diesen Körpern in bestimmter Weise ableiten lassen.

---

[1] Verbindungen, welche mehr als sechs Moleküle Ammoniak auf ein Metall-atom enthalten, sind bis jetzt nur in äufserst wenigen Fällen nachgewiesen, und bedürfen die betreffenden Körper noch näherer Untersuchung.

Chapter 3

Werner's *Beitrag*, 1893

A Linguistic and Epistemological Analysis

Luigi Cerruti

Dipartimento di Chimica Generale ed Organica Applicata, Università di Torino, Corso Massimo D'Azeglio 48, 10125 Turin, Italy

The text of Werner's *Beitrag* is a good test for the analysis of scientific discourse. Some features of its context appear from a study of the *Zeitschrift für anorganische Chemie*, the journal in which the *Beitrag* appeared. An analysis of the text is performed at three linguistic levels: rhetorical, grammatical, and semantic. Werner's use of his epistemological lexicon is discussed. The analysis throws light on a few important points: the extreme complexity of theoretical chemical discourse, the theory of chemical structure as an epistemological obstacle (*à la* Bachelard), and the style of Werner's approach to new knowledge.

This symposium is a celebration. A century ago Werner's *Beitrag* was published in the third volume of the *Zeitschrift für anorganische Chemie* (*1*). All of us think that this paper is a great classic of scientific literature, both for its intrinsic value and thanks to the lasting and loving work of George B. Kauffman (*2-7*). It may be useful and rewarding to apply to this scientific classic some simple techniques of linguistic analysis. In recent years the analysis of scientific discourse has increasingly attracted linguists (*8,9*), sociologists (*10,11*), philosophers of science (*12,13*), and, to a lesser extent, historians of science (*14,15*). The causes of this interest are in themselves obvious: language and its uses are a producer of social relations, a place for social struggle, a mirror of nature, and a guideline for thought. The possible depth and extent of linguistic analysis was pointed out many years ago by the founders of modern hermeneutics.

Modern hermeneutics was opened up as a fertile field of research in Germany, between the end of the eighteenth and the beginning of the nineteenth century. Without further explaining the many consequences of hermeneutics as a foundation for historical work, it may be useful to read a few words of Wilhelm von Humboldt (1776-1835), originally published in 1836: "Language is not just a medium of exchange for mutual comprehension, but a true *world* which the *mind* must insert, by its own inner labour, between itself and *objects*". Language is a *Zwischenwelt*, an intermediate world, between the mind of the observer (the speaker) and the object observed (spoken of); moreover, "every independent author ... cannot avoid imposing his *individuality* upon his language", so that "clearly and distinctly does *individuality* make its presence felt. It ... permits us to see into the speaker's mind" (italics in the text) (*16*). To be able to see into an author's mind may seem somewhat mystical, but we may reassure ourselves with the lay thought of an important contemporary linguist, Michael Halliday: language is "simply part of ourselves – the label 'natural' is entirely apt"; the quotation marks used by Halliday remind us that "in the course of [our] semiotic activity, without becoming aware of it we have been constructing the two macrocosmic orders of which we ourselves are a part: the social order, and the natural order"(*17*). Every description of the natural order also pertains to the social order, and the study of Werner's paper may suitably begin with an analysis of the role played in the international chemical community by the journal that the young (26-year-old) Alsatian-Swiss author chose for the publication of his long essay.

The *Zeitschrift für anorganische Chemie*

Alfred Werner (1866 -1919) sent his now-famous paper to the *Zeitschrift für anorganische Chemie* in December 1892, when this journal was only one year old. In the first number of the *Zeitschrift* the publisher, L. Voss, and the editor, G. Krüss, made two points (*18*): that the new and important results of inorganic chemistry could find no fitting place in all-purpose journals such as Liebig's *Annalen* or the *Berichte der Deutschen Chemischen Gesellschaft*, which were dominated by organic chemistry, and that the new relevance of inorganic chemistry could thus be expressed only by a new scientific journal. And secondly, that as a result of the great theory of the periodic system of the elements, inorganic chemistry had recently renewed itself, thus making the founding of the new journal urgent. Explaining the scope of the journal, Voss and Krüss called for papers in three fields: inorganic chemistry, obviously, but also theoretical and analytical chemistry as they relate to the inorganic aspects of the science. We may compare this program with its realization in the first three volumes of the *Zeischrift* (Table I).

The international character of the journal is clear: papers were contributed from authors in eleven different countries. An important and perhaps political exception was France; the country was represented on the editorial board by two chemists (M. Berthelot and A. Ditte), but no papers were submitted by French researchers. It is also clear that Gerhard Krüss was able to facilitate foreign scientists' access to a German journal by providing contributors with a translation service. Authors from four out of the eleven countries made use of this service, the translations being made by Krüss himself and a few other German researchers, also contributors to the *Zeitschrift*.

Table I. The First Three Volumes of the *Zeitschrift für anorganische Chemie*

Nation	Authors	Papers	Transl.	Inorg.	Theor.	Anal.
United Kingdom	3	3	3	3		
Italy	3	4	4	4		
Sweden	4	3		3		
Germany	34	50		28	4	18
United States	12	25	21	21	1	3
Belgium	2	2	1	2		
Austria	1	2		2		
Switzerland	5	8		4	3	1
Russia	1	1		1		
Denmark	1	1		1		
Holland	1	1		1		
Totals	67	100	29	70	8	22

The Modernity of the *Zeitschrift*. It may seem surprising that the number of contributions from the United States concerning inorganic chemistry were comparable to those from German researchers (21 *vs.* 28). This 'balance' between the hegemonic scientific power of the time, Germany's Second Reich, and the rising power of the United States, is a conspicuous index of the journal's 'modernity'. This 'modernity' can also be explained by the fact that Krüss was one of the first – if not the first – German chemists to have undertaken postgraduate research in the United States; in other words, he had made many good contacts with American chemists.

Analysis of the different disciplinary fields shows that theoretical papers were less than 10% of the total, and I may add they were usually short or very short and almost invariably polemic. With regard to its length, Werner's paper, obviously listed among the Swiss contributions, was a remarkable exception, being more than sixty pages long. Werner's 'polemical' style will be treated later, in connection with the interesting answer from his principal (expected) adversary, the Danish chemist Sophus Mads Jørgensen (1837-1914).

Rhetorical Analysis

In their professional papers scientists wish to persuade their readers, almost exactly as a lawyer in a court of justice wishes to persuade his listeners. They therefore carefully prepare the rhetorical structure of their discourse, and even its details , in order to make the greatest impact on the audience. Werner's paper is no exception to this rule, and a simple rhetorical analysis provides some interesting results.

The rhetorical structure of Werner's paper is outlined in Table II. After a brief introduction of four pages. there are three sections, which the author letters A, B, and C. With the exception of the introduction, every section or subsection has a subtitle, and Table II gives some key words for each subsection, drawn from Werner's subtitles. Section C contains the author's conclusions , while sections A and B deal with two classes of compounds which contain, respectively, six ammonia molecules to one metal atom (first class) and four ammonia molecules to one metal atom (second class). It is clear from Table II that section A (six subsections) is more complex than section B, which has only three subsections, those dealing with the formation of compounds, the replacement of ammonia by water, and Werner's new theoretical conceptions about the constitution of complex compounds.

Table II. Rhetorical Structure of Werner's *Beitrag*

	Section A*		Section B**	
	Sub-section	No. pages	Sub-section	No. pages
Formation of Metal-Ammonia Salts	A I	11	B I	4.5
Double Salts	A II	3	–	
Replacement of Water	A III	6.5	B II	1.5
Hydrates of Metal Salts	A IV	2	–	
Salts in Aqueous Solutions	A V	3.5	–	
A Concept of the Constitution	A VI	5	B III	11

*Preceded by the Introduction of 4 pages.

** Followed by Section C (Conclusions) of 9 pages.

Thus   section B is simplified because double salts, hydrates, and solutions are only discussed for   the first class of compounds. It is easy to conclude that the experimental basis for the first class   of compounds alone supports the important relationships between complexes, double salts, hydrates, and the physical chemistry of solutions. Moreover, those topics common to both sections are allotted very different amounts of space: in section A, 16.5 pages are devoted to the formation of compounds (A I), the replacement of ammonia molecules by water molecules (A III) is allotted 17.5 pages, while the theoretical discussion (A VI) occupies five pages; in section B the experimental topics (B I and B II) take up six pages, whereas the theoretical discussion extends to eleven pages. It is fair to conclude that the first subsection is the actual foundation of the whole argument.

Dispositio. From the above analysis of the disposition and textual length of the topics, it is clear that Werner preferred a rhetorical order going from the strongest points to the weaker ones. This is actually one of the three possible orders (or *Dispositiones*) of classical rhetoric, but it was not considered to be the best (*19*). However, Werner's choice is confirmed in the critical A I subsection. The discussion of the complexes of trivalent metals (7 pages) is followed by those of tetravalent (2 pages) and divalent (2 pages) metals, with a completely different rhetorical order from the logical or 'cardinal' order that Werner spontaneously follows in his discourse. In the text there are three enumerations of the three types of metal cations, but two are in order of increasing valency (WB:281,282), and one is in order of decreasing valency (WB:271). (WB is Werner's *Beitrag*, followed by the page cited) . From the point of view of classical rhetoric, Werner, the young scientist, was not an able pleader.

Confirmation by a Witness. The results of my very simplified formal analysis are confirmed when we look at Werner's paper through the eyes of an exceptional witness. Jørgensen 's answer to the *Beitrag* was quick and interesting: he sent a long paper (51 pages) to the *Zeitschrift für anorganische Chemie* only two months after the publication of Werner's paper (*20*). Jørgensen quotes Werner's text *verbatim* nine times; all his quotations are from section A, but the critical subsection A I is quoted no less than six times, so clearly Jørgensen's attention is directed mainly toward the first subsection, on the *Bildung* (formation) of compounds, *i.e.* Werner's interpretation of the experimental foundation of the theory that he was proposing.

Another meaningful feature emerges if Jørgensen's quotations are linked to Werner's use of the important personal pronouns *ich* and *mir* (I and me). In the 31 pages of section A, Werner uses personal pronouns in the first person singular only four times, four words out of 7000.   The first occurrence (*mir*, WB:277) is quoted by Jørgensen, and two others are in close proximity to quotations. It is evident that Jørgensen wishes to talk to Werner, and that – in a sense – he is meeting the younger scientist. (The two scientists were never actually able to meet (*3*)). The grammatical trail of the personal pronouns is worthy of interest, and I will follow it in the next section.

Grammatical Analysis

From the viewpoint of the conscious control of speech, the grammatical level of the discourse is intermediate between the rhetorical and the semantic levels. Thus it is of interest to try to delve more deeply into Werner's scientific discourse. In this section I will treat some grammatical features of Werner's Beitrag, and in the next section I will go 'down' to the semantic level.

At the very beginning of the paper we find a clear example of way that certain grammatical features can reflect the author's epistemic preferences. In the introduction to his treatise, Werner presents the classes of compounds that will be treated in the following sections. The first class – six ammonia molecules to one metal atom – is introduced with these words: *Als erste Klasse erhalten wir dann Verbindungen* (WB:267), *i.e.*: "As the first class we thus obtain compounds" (KT:9, from Kauffman's excellent translation (*2*)). The second class – four ammonia molecules to one metal atom – enters the discourse thus: *Die zweite Klasse wird gebildet durch Verbindungen* (WB:267), *i.e.*: "The second class is formed by compounds" (KT:9-10), while a third class, neglected in Werner's paper, is mentioned with these words: *Eine dritte Klasse endlich bilden diejenigen ammoniakalischen Metallsalze,...* (WB:268), *i.e.*: "Finally, a third class is formed by those ammoniacal metal salts, ..." (KT:10). Here we have the grammatical proof of a hierarchy of scientific interest: for the first class – the most important – the active subject is *wir*, *i.e.*, we, the chemists; for the second class – less useful – we find a passive form, and the *Klass* becomes the grammatical subject; lastly, the third class – neglected – is the object of the grammatical action of certain *Metallsalze*. There is a grammatical trend from the protagonists of the profession, to one of their theoretical items, to chemical substances.

Who Is Speaking? As subjects of his statements Werner uses various pronouns, both personal and impersonal, but for almost the entire argumentation the preferred subject is *wir* (we). In the critical subsection A I 1, on the hexacoordinate complexes of trivalent atoms, the author uses the first person plural (*wir*) 16 times before presenting himself as a critic of the Blomstrand-Jørgensen theory: *dieselbe erscheint mir deshalb unhaltbar* (WB:277), *i.e.*: "therefore this [theory] seems to me to be untenable" (KT:21). This is the first time that Werner separates himself from the protective *wir*, which many times means "I, the writer, and you, the reader", sometimes indicates the pertinent audience (we, the chemists), and yet in other occurrences is simply referring to other parts of the text (WB:275,276).

As a German speaker, Werner uses the indefinite pronoun *man* (one). Many times in the scientific literature *man* is used as the subject of an active verb, where in English the passive construction is preferred. However, the German pronoun is very versatile: *man* embraces singular and plural concepts and can include a range of meanings extending from the personal "I" to the "whole of mankind" (*21*). Werner's uses of this 'umbrella' pronoun cover a large range of situations. *Man* may be Werner himself, when he is suggesting that the reader follow his train of reasoning (WB:289), or it may be all chemists (*e.g.*,WB:267,279,281,285, *etc.*), some chemists (WB:323), and sometimes a single chemist, the subject either of a possible action (WB:288) or of an impossible action (WB:303,323).

Werner Speaking. We have seen that in Werner's *Beitrag* the occurrences of personal pronouns in the first person singular are very rare. The text of the *Beitrag* consists of more than 40,000 words, but there are only 13 occurences of these pronouns (*cf.* the 179 occurrences of *wir*). It is obvious that his direct participation in the dialogue with the reader has a special meaning for the author. The appearance of *ich/mir* is very sporadic before the final considerations, where we find 8 occurrences in 9 pages, whereas in the preceding 55 pages we find only 5 occurrences. Of these, only one is trivial (WB:308); two others are at points where Werner is in some difficulty (WB:281,293), and the last two underline important steps in Werner's argument (WB:277,294). Without doubt, from this grammatical point of view, the conclusions are exceptional because the author speaks directly and responsibly to his colleagues. Thus the analysis of the occurrences of these pronouns affords a quick reading of Werner's theoretical and professional intentions.

The first occurrence is so relevant that I will discuss it in two other sections of this paper. Thus the whole passage deserves to be quoted: *glaube ich die Überzeugung aussprechen zu dürfen, dass diese Ansicht richtig ist, trotzdem sie sich in Gegensatz stellt zu unseren heutigen Ansichten über die Konstitution der Verbindungen, wie sich dieselben aus dem Studium der Kohlenstoffverbindungen herausgebildet haben* (WB:323), *i.e.*,"I believe that [I] may state the conviction that this view is correct, even though it stands in opposition to our present views on the constitution of the compounds, as these have evolved from a study of carbon compounds"(KT:80). The first *ich* in the conclusions puts the author in opposition to the chemical community; it is remarkable that Werner stresses that his complex theory (the first *view* of the passage) contradicts his own opinions (*our* views!) based on the theory of carbon compounds. In the following two occurrences the author is promising future development of the theory, both against the dominant theories (WB:324) and toward a unifying theoretical principle (WB:326, note). The fourth and fifth occurrences are when Werner defines two relevant concepts of his theory: the coordination number (WB:326) and the coordination positions at the corners of the complexes (WB:328). The sixth occurrence is again in a deictic context (an engagement for the future), and the seventh is close to a citation of Arthur Hantzsch (1857-1935), a form of acknowledgment to Werner's teacher. The last occurrence in the paper, like the first in the conclusions, deserves a long quotation: *Ich übergebe [diese Entwickelungen] dem Urteil der Fachgenossen als eine Auffassung der chemischen Verbindungen, welche ... auf breiter Basis aufgebaut ist, als die nur aus den Kohlenstoffverbindungen abgeleitete Theorie von Valenz als Einzelkraft* (WB:330), *i.e.*, "I deliver [these developments] to the judgment of my colleagues as a concept of chemical compounds which ... is constructed on a broader basis than the theory of valence as a single force, which is derived only from carbon compounds" (KT:88). This last *ich* closes both the paper and, at the same time, the gap which he had opened with the first *ich* of the conclusions. With the first *ich*, Werner had underlined the separation between his *Überzeugung* (conviction, persuasion) and that of other chemists; here, with the last *ich*, he delivers his results to the *Urteil* (judgment, opinion) of his colleagues. The gap is closed, and the hard, even harsh word *Gegensatz* (contrast, opposition, conflict) used with the first occurrence is forgotten.

Semantic Analysis

The professional dialect (*22*) of chemists is extremely complex, as all chemistry teachers know from their students' doubts and difficulties. Werner's language is not an exception, as a few examples will show. The following one is a typical interpretation of a chemical 'fact': "in LUTEOCOBALT CHLORIDE $Co(NH_3)_6Cl_3$ all three CHLORINE *atoms* behave as *ions* and are immediately precipitated as SILVER CHLORIDE by SILVER NITRATE at room temperature" (KT:15). In this passage we find the names of four macroscopic substances (in small capitals) and of two microscopic objects (in italics); these two different ontological levels are openly mixed in the laboratory operation of precipitation, a 'fact' that occurs at a macroscopic level but is described/interpreted at a microscopic level: *all three atoms are precipitated as silver chloride*. Moreover, in the passage a formula (in bold type) describes the chemical constitution of a molecule; the formula is a theoretical item and points to a third different ontological level, that of the chemical discourse itself (a particular 'universe of discourse'). The three levels − the macroscopic level of the substances, the microscopic level of the particles and the linguistic level of the theoretical descriptions − are always intermingled, often with ambiguous results and sometimes with humorous effect.

The Seventh Water. As an example of ambiguity we may read this statement, referring to $Pt(NH_3)_2X_2$ compounds: "simultaneously with the replacement of the *acid residue* by **OH** an intercalation of WATER takes place" (KT:58). Clearly, WATER is shorthand for WATER *molecule*, but it is true that in the quoted text all three ontological levels seem to interact very smoothly in the *simultaneous replacement/intercalation* of substances, particles, and descriptions. The ambiguity is resolved when we ask ourselves *where* the 'facts' happen. The answer is that the entire statement is decoded (by a chemist) at the linguistic level of the formulae. Many more examples could be given, but I think that the point is clear: in 1893 and in Werner's writing this ambiguity is already intrinsic.

As chemists, we are so accustomed to this ontological ambiguity that we read passages such as the following with no surprise: "Müller-Erzbach has shown that according to the dissociation tensions of the sulfates of magnesium, nickel, and cobalt the seventh water is bound more weakly than the six others" (KT:35). As usual, a macroscopic measure is interpreted at the microscopic level, but the *seventh water* (*siebente Wasser*, WB:288), with its six stronger sisters, is still somewhat absurd.

Laboratory Bench & Conceptual Space. Translation is one of the most diffuse and one of the most fruitful hermeneutic activities (*23*), and simply to open a dictionary may serve to provide a preliminary understanding of the different semantic fields in which a word works. Werner frequently uses the verb *ableiten:* of its several meanings, in the transitive form it means "to derive, to deduce, to infer", and in mathematics it may mean to differentiate; in chemistry, in the reflexive form, it means to be derived (from another substance). In certain contexts Werner is fond of this verb, with interesting consequences from the semantic point of view. We may take into account two occurrences separated by only 11 lines (WB:305); both the occurrences concern the chemical behavior of certain complexes. In the first passage we read: "the behavior of the platosammine salts does not correspond with that deduced (*sich ableitet*) from this formula" (KT:57); here the actions expressed by the verb are a set of deductions and analogies based on theoretical chemical knowledge. In the second passage we find: "our view [is supported] by the behavior of the compounds derived (*sich ableitenden*) from PLATOSAMMINE CHLORIDE by substitution of SO_3H for CHLORINE. These do not behave like SULFITES OF AMMONIUM SALTS" (KT:57). The second statement makes it clear that the behavior in question is that of chemical substances on a laboratory bench and that Werner is speaking about a chemical derivation. However the 'substitution' "SO_3H for CHLORINE" may only be 'performed' in an equation. This reference to different but coexisting 'spaces' is really ambiguous in other occurrences of *ableiten.*

No less than three occurrences with different meanings may be found on the same page (WB:303); the first time the reference is to the theoretical space of formulae: "compounds ... which can be derived (*sich ableiten*) from this formula" (KT:54). The second time, the space is that of chemistry laboratories: "those [salts] which are derived (*sich ableiten*) from divalent platinum ... have been very well investigated" (KT:54-55). But the third time, *three* spaces are completely mixed: "we were able to derive (*ableiten*) from compounds of the formula $M(NH_3)_6X_3$ a complete transition series to the DOUBLE SALTS by successive loss of AMMONIA and a change in function of the *negative radicals*"(KT:55). The *loss of ammonia* may happen only in the macroscopic laboratory, the *change in function* concerns microscopic entities, and – finally – the act of *ableiten* is here purely textual (or theoretical) because the subject of the (many) actions is the ubiquitous *wir, i.e.* , (here) Werner and the reader.

Semantics and Conceptual Reference. In point of fact Werner's *Beitrag* was (and is) clear for every native reader, that is, for chemists living in 1893 (also for all chemists now living) able to read German texts. The various ontological levels, or the various spaces referred to, are disentangled only when the reader's attention , or the writer's intention, is focused on a specific 'level' or 'space', *e.g.*, laboratory instructions, statements on the constitution of microscopic particles, or discussions about the cognitive value of formulae. The *natural* chemical discourse is understood by native speakers with no ambiguity because the reference is overwhelming by a "conceptual reference." These terms are used by Gilbert and Mulkay, who state that in scientific discourse "The pictures do not refer directly to empirical phenomena but to conceptual entities or idealized versions of observable phenomena" (*10*). As is stated in the title, Werner's paper is a *Beitrag zur Konstitution anorganischer Verbindungen*, a contribution on the constitution of inorganic compounds: the most important 'objects' referred to in the treatise are the now-famous octahedra (WB:298,300, *etc.*), depicting the constitution of hexacoordinate complexes. These octahedra were new, extraordinary landmarks in the unfamiliar conceptual space created by Werner.

An Epistemological Lexicon

Beginning in 1819, in his lecture course at the University of Berlin, Friedrich Schleiermacher (1768-1834) , one of the founders of modern hermeneutics, taught that "the precise interpreter must gradually derive all of his conclusions from the sources themselves." A hermeneutical reader must be very suspicious, in particular – said Schleiermacher – he has to consider that "all of the information about a language which dictionaries and other resource works supply represents the product of particular and often questionable interpretation"(*24*). In the last section we opened a German-English dictionary for a first hint about semantic fields, but a "precise interpreter" has to draw up at least fragments of the lexicon used by the particular author whom he is studying. I will now consider a few fragments of Werner's lexicon.

At a rough estimate, the number of different words in the *Beitrag* is more than a thousand. These words constitute a general lexicon in which we find many words used in normal, 'natural' speech, but it is easy to single out several more specialized sections of the lexicon. For example, in chemical discourse many words have a specific chemical meaning (*e.g.*, base, reaction, equation). In the context of my research I am particularly concerned with the epistemological section of that lexicon. As we know, epistemic discourse is – by definition – discourse about knowledge, its forms, order, structure, value, nature, and so on; its scope is vast, and its study presents some problems.

Difficult Words. Words may be 'problematic' from different points of view. In the preceding section I considered some different meanings of *ableiten*, referring to different levels of the reality. A word such as *Betrachtung* (consideration, examination, reflection) seems to be less doubtful. In many cases (WB:289, 290, 293, *etc.*) it functions as a rhetorical device: *bei der Betrachtung* (WB:285), *i.e.* , "in considering" (KT:31), but in other cases the meaning is strictly epistemic, as in certain Kauffman translations: observations (KT:46, twice), view (KT:58), considerations (KT:79). Thus this word functions at the border between rhetoric and epistemology.

Another type of problem is found when we try to understand Werner's use of the word *Vorstellung*. It is a key word because it appears in the titles of both the subsections in which Werner proposes his new theoretical views of complexes (WB:297,310). The word also reappears in the conclusions with 3 occurrences (WB:325). In all five occurrences that I have counted, Kauffman translates *Vorstellung* as "concept ," with a conscious (and conscientious) rigour (KT:47,63,83). As an Italian reader, when I read the titles of the subsections where all 17 figures of the text are concentrated, my silent translation was *rappresentazione* (representation, *i.e.*, *Darstellung*). My interpretation was supported by the presence in the same contexts of the verb *vorstellen*, which Kauffman aptly translated as "represent" (KT:47) and "imagine" (KT:64). Without going too far in triangular considerations of theoretical terms in German, English, and Italian, I should like to remind the reader that the title of Schopenhauer's famous work *Die Welt als Wille und Vorstellung* (1819) is translated into English as *The World as Will and Idea*, while in Italian the traditional translation is *Il mondo come volontà e rappresentazione* (French philosophers translate *représentation*). Perhaps every reader finds his or her own difficulties, bound to the personal *Zwischenwelt* in which he or she lives.

The Lexicon. Werner's epistemological lexicon is particularly rich (more than 250 words), but more than its simple extent, it is interesting to examine more closely how the author uses this thesaurus. We have already seen the distribution in the text of rare or very rare words (*ich, Vorstellung*). Now we may consider a few families of words, centered on important epistemic roots. From a single *stem** a set of words may be derived by substituting   the asterisk with suffixes, according to the inflexional properties of the language (*25*). The first family has the common stem *entwickel**; these German words mean to develop, development, developed, and so on. We see in Table III that this word-family appears frequently in the text, totaling 33 occurrences, with a strong concentration in the final part of the paper. (The last column of Table III corresponds to the final considerations of the *Beitrag*). Another important stem is *Beziehung** (relation, relationship), with 27 occurrences scattered through the text. Relationship is a key word for a paper with as wide a scope as the *Beitrag*: the high 'concentration' observed in the text between pages 282 and 294 (partially 'diluted' in Table III) is due to the discussion on the constitution of double salts, the ability of water to replace ammonia in metal-ammonia salts, and the relationships between hydrates and double salts. The third family considered derives from the stem *thatsach** (fact, actual, really); its distribution   is relatively uniform, with a meaningful concentration in the conclusions. Because the same thing happens to the other two families, we obtain two results: Section C of the *Beitrag* is epistemologically crowded, and the word 'fact' is laden with theoretical meanings. A third result comes from the distribution throughout the text of the three families because text linguistics assigns an important role to words recurring often in a text: from a semantic point of view, this type of word is a device for textual cohesion (*26*). These three families function as epistemic threads in the fabric of Werner's text.

Table III. Werner's Epistemological Lexicon

	Occurrences in Werner's *Beitrag*						
Pages	267-270	271-280	281-290	291-300	301-310	311-320	321-330
entwickel*	3	–	4	4	5	2	15
Beziehung*	1	2	9	5	1	1	8
thatsach*	1	4	2	3	4	4	9
hápaks	7	1	10	7	6	7	65

*German stems.

Hápaks legómenon. One of the most puzzling results of my linguistic research on the *Beitrag* is reported in the last line of Table III. While I was filing the text, to my disappointment the number of new cards for each page did not decrease; on the contrary, toward the end of the paper, new words were found very frequently. The fourth line of Table III shows the occurrences of epistemic words that appear in Werner's text once and only once. In classical philology such words are known as *hápaks legómenon* (Greek, said once) or, for short, *hápaks*. The epistemic *hápaks* that I have filed are 103, and 61 of these appear in the final nine pages of the paper (section C). Very important words, such as *glaube* (I believe), *Überzeugung* (convinction), *Urteil* (judgment), *Gegensatz* (opposition), are found only in the concluding section. The concentration is really enormous: Werner is using a special, new lexicon at the end of his writing.

When I convinced myself of the richness and of the prodigious availability of Werner's epistemological lexicon, I remembered a trait of his scientific personality well described by a biographer: "On his laboratory table stood several microburners and microfilters, together with several hundreds of small glass dishes, whose contents were of all colors. Although none of these dishes was labeled, [Werner] was quite sure that confusion was almost an impossibility"(*27*).

Sophus Mads Jørgensen and Other Readers

Normally speaking, when a scientist is reading a paper of a colleague's, he pays more attention to the content than to the form, but if the content clashes with his ideas or threatens him with a loss of prestige, the rhetorical form becomes important. This was the case with Sophus Mads Jørgensen, whose works provided the experimental foundation for Werner's theoretical construction (2,3).

Werner was keenly aware of his indebtness to Jørgensen, and in the *Beitrag* the many citations of the Danish chemist's name are surrounded with terms of high praise: "a large number of beautiful experimental investigations" (KT:12); "we know such compounds through Jørgensen's latest investigation" (KT:32); "as was undoubtedly proven by Jørgensen's fine works" (KT:37); "as we know from Jørgensen's beautiful works" (KT:49); "the recognition of the configurations ... as made possible by Jørgensen's beautiful works" (KT:76). In certain cases Werner's use of the senior scientist's results was somewhat impudent: "another observation of Jørgensen's also supports our view" (KT:72), but on the whole Werner's polemical rhetoric is extremely effective.

In a nutshell, if Werner's attack against Jørgensen's theories became too harsh, it was transformed, more or less consciously, into an assault against Jørgensen's-and-another's theories. Without going into detail, I quote three passages from subsection A I, where the Blomstrand-Jørgensen theory is discussed at length: "Jørgensen's formula for the luteo salts explains why ..." (KT:17); "So far the facts agree with Jørgensen's formula" (KT:18); and finally: "this behavior is no longer explained by Blomstrand-Jørgensen formulas" (KT:19). (In reality, Werner held C. W. Blomstrand (1826-1897) in high esteem (36).)

Returning now to Jørgensen's paper cited above (20), it is enlightening merely to read the first page. The Danish scientist speaks of "a new theory" (twice), of a "new point of view", "new light", "new interest", and of the young author's (Werner's) "unmistakeable talent". But two other points of this opening page are more 'philosophical' than Jørgensen's surprised praise. The first statement is strikingly modern: "Actually the new theory does not collide with that developed by Blomstrand and myself. They rather lie on different planes." Such a judgment on the incommensurability of scientific theories might have been quoted from a paper of the contemporary "epistemic libertines"(12) such as Paul Feyerabend (37). The second statement is more dated, being linked to the historicism prevailing in the German-speaking culture of the nineteenth century: "Even if valence is hitherto something unclear, however, it is introduced into science from historical necessity (*mit historischer Notwendigkeit*)." This "historical necessity" of the direct valence theory was one of the biggest cultural obstacles in the way of the renewal of inorganic chemistry. (See below on Mendeleev's and Werner's denial of this 'necessity').

Many years ago Kauffman aptly stated that the controversy between Werner and Jørgensen was "a fine example of an ideal scientific discussion" (3), and my linguistic trails simply confirm this judgment. But Kauffman has collected the opinions of other readers, which deserve to be reported here too. Victor Meyer (1848-1897): "I am truly delighted with the theoretical-inorganic paper of A. Werner! These are really new thoughts" (2); Emil Fischer (1852-1919): "this work bears witness to an extraordinary talent for treating a whole series of apparently divergent facts from new unified viewpoints", and, lastly, Werner's own mentor, Arthur Hantzsch: "the most significant [work] that has appeared in years in the field of pure chemistry; truly revolutionary"(4). It is a pity that the historian has to work so hard in order to gain the same enthusiasm felt by contemporary readers.

Werner's Epistemology

Every scientist, as well as every philosopher, has a twofold epistemology: one is implicit, used in his or her struggle to find new truths, and the other is explicit, openly declared or discussed in his or her writings. This second aspect of the scientist's epistemology is intrinsically bound up with his or her rhetoric, as we may see in the following example. At a crucial point of the *Beitrag*, Werner states that in the complexes of formula (MA_6) "*all ammonia molecules, all water molecules, and all acid residues are bound directly to the metal atom*"(KT:43; WB:294, emphasis in the text). *Before* this statement Werner says: "As a conclusion (*Schluss*) to all these considerations (*Betrachtungen*), I propose the following proposition;" soon *after* that striking theoretical statement the author refers to "the numerous facts (*Thatsachen*) which have led us to the above conclusion (*Schluss*)." As a matter of fact, Werner's rhetoric suggests an extraordinary change. Werner's *Betrachtungen* were personal ("I propose"); after ten lines they become *Thatsachen*, and it is said that these "have led" the author and the reader ("us") to a conclusion which breaks the dominant valence paradigm.

As a historian, I am more interested in the 'covert' epistemology of the scientist, which is much more variable than the 'overt' one. Werner's covert epistemology is really classic, as is easily seen in the epistemic structure of the *Beitrag*. At the opening of the paper, what seems "inadmissible" to the author is simply the classification of metal-ammonia salts according to their stability. He states that a new "principle of division ... easily emerges if we use the empirical composition of the compounds and certain characteristics of the substances in question as the leading features of the classification" (KT:9). The empirical composition is surely the most fundamental chemical property of substances, just as the problems of classification are the most classic ones in chemistry. Werner manages to bring forth his reasoning without the help of electrochemistry until almost exactly the midpoint of the paper. Conductivity comes on the stage on page 295, whereas the *Beitrag* begins on page 267 and ends on page 330. From a linguistic point of view it is interesting to note that Werner uses (KT:44,45,46) three differents words for "conductivity": *Leitvermögen* (WB:295), *Leitfähigkeit* (WB:296) and *Leitungfähigkeit* (*ibid.*). Epistemologically, this confirms that when he sent the paper to Krüss's *Zeitschrift*, Werner was more an enthusiast for electrochemistry than a professional electrochemist. (Here I am echoing the evaluation *eine geniale Frechheit* (an ingenious impudence) given by a colleague of Werner's to the *Beitrag* (*4*).) And it is exactly in this context that Werner begins to lay his siege to the dominant position of carbon chemistry in the science: "the more the element deviates from carbon in its electrochemical properties, the more easily will water insert itself into the bonding", "in such a way that direct bonding between the metal atom and the acid residue can no longer take place", and the electrolytic dissociation of the salt takes place (KT:45-46). At the end of this subsection the crucial *Vorstellung* of the octahedral constitution of the hexacoordinate complexes begins.

The first strong attack upon the concept of valence follows two pages later: "the difference in valences is a somewhat obscure concept because valence itself is not a clear concept" (KT:49). (This is one of the passages quoted *verbatim* by Jørgensen (*20*); see also (*7*).) The attack is prepared at a metaphorical level with many nouns, verbs, and adjectives belonging to the semantic field of the opposition *obscure/clear*, and is followed by a triumphant exclamation: "Thus a stereochemistry of cobalt compounds and platinum compounds appears alongside the stereochemistry of carbon compounds and nitrogen compounds!"(WB:300; KT:50).

An Epistemological Obstacle

The most formidable obstacle to the modern theory of complexes was the enormous success of the structural theory of organic compounds. In 1938 the French philosopher Gaston Bachelard wrote at length of the epistemological obstacles encountered by scientists on their ways toward new and wider horizons of knowledge. A decisive obstacle is always the content of scientific knowledge itself (*28*). With an obvious metaphor: every beam of light makes the objects that it illuminates throw shadows; the stronger the light, the deeper the shadows.

Werner was perfectly aware of the relative situation of his (and of his colleagues') knowledge. War against the supremacy of organic chemistry had been declared by Dmitrii Mendeleev (1834-1907) (29,30) many years before the *Beitrag*, but the Russian chemist's fight was not yet over (31,32), Thus we may compare the moves of the two scientists in their difficult epistemic game against the central role of carbon.

Checkmate in Three Moves. Both the senior scientist and the junior researcher placed carbon, as an element, against the background of the entire system of chemical knowledge. For them it had become clear that knowledge of this element had been in a sense misleading. In his Faraday Lecture of June 4th, 1889 Mendeleev said: "it is only to carbon, which is quadrivalent with regard both to oxygen and to hydrogen, that we can apply the theory of constant valency and of bond"(31). Immediately after having defined the coordination number, the most important theoretical concept of his paper, Werner writes: "the coordination number and the valence number ... are the same for carbon, and therefore it appears probable that this accidental coinciding of the two numerical values for carbon has prevented the differentiation of the two concepts" (KT:85).

Both scientists felt the heavy weight of the historical success of carbon chemistry. In a speech to the Royal Institution, delivered on May 31, 1889, Mendeleev admitted that "the modern teaching relating to atomicity, or the valency of the elements [is] to be ranked as a great achievement of chemical science", but he added, "this teaching as applied to the structure of carbon compounds, is not, on the face of [its successes], directly applicable to the investigation of other elements" (32). We have seen that Werner overcame the "historical necessity" of the structural theory, with a resounding *ich* and with the cutting word *Gegensatz*. However, in addition to their rebuttal of carbon as a model element and their rejection of the rush of applications of the structural theory, both scientists play a third move, aiming at the conquest of a higher vantage point for the chemical 'facts'.

Mendeleev, who was arguing for a chemical mechanics founded on Newton's laws, plainly said that "it [was] time to abandon the structural theory" because "it [was] possible to preserve to chemistry all the advantages arising from structural teaching, without being obliged ... to ascribe to atoms definite limited valencies, directions of cohesion, or affinities"(32). Werner played his third move at the end of the *Beitrag*, where he 'spent' the last *ich* "on a broader basis [for chemistry] than the theory of valence as a single force" (KT:88).

Masters of the Critical Tradition. Werner's epistemology was brave but flexible, and in this respect it was closer to Meyer's (33) than to Mendeleev's (34). However, the three masters were fully aware of the historical nature of scientific truth; in the five editions of his *Modernen Theorien der Chemie* Lothar Meyer (1830-1895) continually criticized and changed his views on the problem of constant or variable valence (33). In his Faraday Lecture Mendeleev affirmed that: "sound generalizations – together with the relics of those which have proved to be untenable – promote scientific productivity, and ensure the luxurious growth of science"(31). Werner, in the introduction to the *Beitrag*, wrote that "for a better understanding" a brief historical summary must be given, and that it was "not surprising that the views of the constitution of these compounds, which follow[ed] the development of theoretical chemistry, [had] undergone many changes with time"(KT:12).

Kauffman has remarked that Werner worked on several organic topics, and that 45 of his 174 publications were in organic chemistry (2, 35). The finest feature of the critical tradition in the chemistry of the nineteenth century was the awareness that the different branches of chemical science require at all times to be unified within a common, theoretical point of view. This endeavor never completely succeeded, but masters such as Mendeleev, Meyer, and Werner were scientifically omnivorous. Werner confessed: "chemical work was always a pleasure for me, and I have experienced the purest pleasure in the laboratory" (*4*). And organic chemistry was (and is) a splendid pleasure ground (for a master).

Conclusion

The results of the linguistic and epistemological analysis of Werner's *Beitrag* agree with what I have found by analyzing other texts, by Stanislao Cannizzaro (*38*), Dmitrii Mendeleev (*30*), and Lothar Meyer (*33*). The epistemic nature of scientific discourse requires that all the language levels – rhetorical, grammatical, semantic – become functional to the twofold aim of demonstrating in words new aspects of knowledge and of gaining acceptance from the peer community. The analysis of Werner's *Beitrag* also confirms that "each author's discourse is organized to display its own factuality" (*11*). The description of the 'facts' of inorganic chemistry in Werner's discourse is brand-new and deserves all the devices of language to be mobilized. From a less technical point of view the analysis has shown moments of difficulty, challenge, and achievement in Werner's writing.

    The grammatical trail of the pronoun *ich* has shown itself to be particularly fruitful, and in this connection I regret that in current written scientific language the pronoun "I" has almost disappeared: "prevailing custom, alas, is ... against the use of the first person singular" (*39*). In regard to Alfred Werner's 'style', I am not able to give a literary judgment so I quote Hantzsch's words about Werner's *Habilitationsschrift, Beitrag zur Theorie der Affinität und Valenz* (*6*). These words were written in December 1891: "punctuation and style are also frequently careless or at least seem so; for example, the subjects in main and secondary clauses change almost regularly. ... Such faults are ... excusable in the case of Herr Dr. Werner, an Alsatian" (*4*).

    We are speaking here in honour of a strong and energetic man. Werner deserves to be remembered by a Zen verse (*40*):

Nothing whatever is hidden;
From of old, all is clear as daylight.

This is true in many cases for the great theorists. This is certain in the case of Werner's *Beitrag*, because – as we know – it contains no new experimental facts, and the atomic model used by Werner is easily traced back to Jöns Jacob Berzelius (1779-1848) and to his electrochemical theory (*41*). As Thomas Kuhn has stated (*42*), many important developments in theoretical thought occur because the scientist experiences a sudden shift of perception of the pertinent field of knowledge; with a flash of insight, the form, the *Gestalt* of the field changes to the eyes of the observer. Werner experienced this *Einsicht* and the *Beitrag* was the result.

Acknowledgment

The author wishes to dedicate this contribution to Professors Giuseppe Cetini and Gaetano Di Modica on the occasion of their seventieth birthdays.

Literature Cited

1. Werner, A. *Z . anorg. Chem.* **1893**, *3*, 267-330.
2. *Classics in Coordination Chemistry. Part I: The Selected Papers of Alfred Werner;* Kauffman, G. B., Ed.; Classics of Science Series No. 4, Dover: New York, NY, 1968.
3. Kauffman, G. B. *J. Chem. Educ.* **1959**, *36*, 521-527.
4. Kauffman, G. B. *Alfred Werner: Founder of Coordination Chemistry;* Springer: Berlin, 1966.

5. Kauffman, G. B. In *Werner Centennial;* Kauffman, G. B., Ed.; Advances in Chemistry Series No. 62, American Chemical Society: Washington, DC, 1967; pp 41-69.
6. Kauffman, G. B. *Chymia* 1967, *12*, 183-216 (annotated transl. of Werner, A. *Vierteljahrsschrift der Naturforschenden Gesellschaft in Zürich* 1891, *36*, 129-169).
7. Kauffman, G. B. *J. Chem. Educ.* 1979, *56*, 496-499.
8. Prelli, L .J. *A Rhetoric of Science: Inventing Scientific Discourse*; University of South Carolina Press: Columbia, SC , 1989.
9. Gross, A. G. *The Rhetoric of Science*; Harvard UP: Cambridge, MA, 1990.
10. Gilbert, G. N.; Mulkay, M. *Opening Pandora's Box. A Sociological Analysis of Scientists' Discourse*; Cambridge University Press: Cambridge, 1984.
11. Mulkay, M. *The Word and the World,* Allen & Unwin: London, 1985.
12. Hacking, I. *Representing and Intervening;* Cambridge University Press: Cambridge, 1983.
13. Fuller, S. *Philosophy of Science and Its Discontents;* Westview Press: Boulder, CO, 1989.
14. Golinski, J. V. In *Companion to the History of Modern Science*; Olby, R.C. *et al.,* Eds.; Routledge: London, 1990; pp 110-123.
15. Melia, T. *Isis* 1992, *83*, 100-106.
16. von Humboldt, W. *On Language*; Cambridge University Press: Cambridge, 1988.
17. Halliday, M A. K... In *The Linguistics of Writing.,* Fabb, N. *et. al.,* Eds.; Manchester University Press: Manchester, 1987; pp 135-154.
18. Voss, L.; Krüss, G. *Z. anorg. Chem.* 1892, *1*, 1-3.
19. Perelman, C.; Olbrechts-Tyteca, L. *Traité de l'argumentation. La nouvelle rhétorique*; Editions de l'Université de Bruxelles: Brussels, 1970.
20. Jørgensen, S. M. *Z. anorg. Chemie* 1894, *5*, 147-196.
21. *Duden Grammatik der deutschen Gegenwartssprache;* Grebe, P., Ed.; Bibliographisches Institut: Mannheim, 1959.
22. *Language, Thought, and Reality. Selected Writings of Benjamin Lee Whorf;* Carroll, J. B., Ed.;The M.I.T. Press: Cambridge, MA, 1962.
23. Steiner, G. *After Babel;* Oxford University Press: Oxford, 1992.
24. Schleiermacher, F. In *The Hermeneutics Reader;* Mueller-Vollmer, K., Ed.; Blackwell: Oxford, 1986.
25. Lyons, J. *Semantics;* Cambridge University Press: Cambridge, 1979; Vol. 2.
26. Dresler, W. *Einfürung in die Textlinguistik;* Niemeyer: Tübingen, 1972.
27. Pfeiffer, P. In *Great Chemists;*Farber, E., Ed.; Interscience: New York, NY, 1961.
28. Bachelard, G. *La formation de l'esprit scientifique;* Vrin: Paris, 1980.
29. Mendelejeff, D. *Ann. Supplementband* 1871, *8*, 133-229.
30. Cerruti, L. *Chim. Ind. (Milan)* 1985, *66*, 500-507.
31. Mendeléeff, [D.] *J. Chem. Soc.*1889, *45*, 634-656.
32. Mendeleeff, D. *Chem. News* 1889, *60*, 1-5, 15-17, 31-32.
33. Cerruti, L. *Rend. Acc. Naz. Scien.* 1990, ser. 5, *14*, 281-301.
34. Cerruti, L. In *Secondo Seminario di Chimica Inorganica e Metallorganica;* Cesarotti, E., Ed.; CLUED: Milano 1986; pp 197-206.
35. Kauffman, G. B. *Naturwiss.*1976, *63*, 324-327.
36. *Classics in Coordination Chemistry. Part II: Selected Papers (1798-1899);* Kauffman, G.B., Ed. ;Classics of Science Series No. 7; Dover: New York, NY, 1976.
37. Feyerabend, P. *Farewell to Reason;* Verso: London, 1987.
38. Cerruti, L. In Cannizzaro, S. *Sunto di un corso di filosofia chimica;* Sellerio: Palermo; 1991, pp 73-282.
39. Schoenfeld, R. *The Chemist's English;* VCH: Weinheim, 1985.
40. *The Gospel According to Zen. Beyond the Death of God;* Sohl R.; Carr, A., Eds.; New American Library: New York, NY, 1970.
41. Berzelius, J.J.*Traité de chimie;* Didot: Paris, 1836; Vol. 4.
42. Kuhn, T.S. *The Essential Tension. Selected Studies in Scientific Tradition and Change;* University of Chicago Press: Chicago, IL, 1977.

RECEIVED February 22, 1994

*J. V. Dubský (1882–1946). (Reproduced
with permission from reference 5. Copyright Czech Chemical Society.)*

Chapter 4

J. V. Dubský and His Participation in Werner's Coordination Theory

František Jursík[1] and George B. Kauffman[2]

[1]Department of Inorganic Chemistry, University of Chemical Technology, Prague, 166 28 Prague 6, Czech Republic
[2]Department of Chemistry, California State University, Fresno, CA, 93740

The so-called violeo salts (salts of the cis-$[CoCl_2(NH_3)_4]^+$ ion), which played a key role in Alfred Werner's coordination theory, were first prepared in 1907 by the Czech chemist J. V. Dubský (1882-1946), who began his career in 1904 as one of Werner's *Doktoranden* in Zürich. The first attempt to resolve an octahedral complex into enantiomers is also connected with Dubský's name. Dubský's life and work, especially his contributions to coordination chemistry, are briefly discussed in this paper.

Two series of compounds played a crucial role in the development of coordination chemistry: those of *racemic*-$[CoCl(NH_3)(en)_2]^{2+}$ (en = ethylenediamine) and those of cis-$[CoCl_2(NH_3)_4]^+$. The first series of complexes were resolved by Victor L. King (1886-1958) into enantiomers, and while this can be considered an elegant proof of Werner's conception of molecular structure, the preparation of salts of the labile cation cis-$[CoCl_2(NH_3)_4]^+$ represents the laying of the foundation stone of coordination chemistry (*1, 2*). Although biographical data concerning Alfred Werner (1866-1919) and some of his students or co-workers have appeared in the literature (*3, 4*), the work of J. V. Dubský from the viewpoint of his part in the development of the coordination theory has not been fully appreciated. This Coordination Chemistry Symposium provides an excellent opportunity for discussing the life and work of Dubský (*5-7*), one of the most outstanding Czech chemists and Alfred Werner's student, private assistant, and *Dozent*, who participated in the synthesis of cis-$[CoCl_2(NH_3)_4]^+$ and the resolution of *racemic*-$[CoCO_3(en)_2]^+$. Dubský was a versatile chemist, however, so an appreciation of his work would not be complete without a brief mention

0097–6156/94/0565–0059$08.00/0

of his other scientific activities in the fields of organic and analytical chemistry.

Dubský's Life

Jan Václav Dubský was born on June 18, 1882 in Řehnice (Rehnitz) in eastern Bohemia (then part of the Austro-Hungarian Empire). After attending the Technical High School there he worked for some time as a chemist in a sugar factory. However, because he yearned for a higher education, this work did not satisfy him. In 1904 he went to Switzerland to study at the University of Zürich. Because living was expensive there, his father helped with his expenses. When this money was gone, Dubský's brother, who had inherited a farm from his father, supported him. After receiving his PhD in 1908 for a study of basic chromium(III) salts (he was the first to recognize that these salts are actually coordination compounds), he served as Werner's private assistant. Here he found the true *milieu* for his further scientific career (At that time the faculty of the University of Zürich included numerous prominent scientists, *e.g.*, Richard Willstätter, Max von Laue, Georg Lunge, Albert Einstein, Hermann Staudinger, Paul Pfeiffer, *etc.*). In 1909 Dubský left Zürich to spend three years as an assistant of Prof. Antoine Paul Nicolas Franchimont (1844-1919) at the University of Leyden in the Netherlands, where he acquired experience in organic chemistry. In 1912 he returned to the University of Zürich, where he was appointed Head of Werner's laboratories for inorganic and organic synthesis. In the same year he visited Prof. Fritz Pregl's laboratory in Innsbruck, Austria, to become familiar with the new analytical micro methods.

Influenced by Werner's conception of the geometry of coordination compounds, in Zürich Dubský searched for isomers whose existence was predicted by Werner, who explained this phenomenon primarily by the saturation of primary and secondary valences. This research, together with a study of the action of acetic anhydride on hydroxo complexes, formed the principal ideas of Dubský's *Habilitationsschrift* (*Die Affinitätsabsättigung der Haupt- und Nebenvalenzen in den Verbindungen höherer Ordnung*). In 1904 Dubský became a *Dozent*, and he continued his work at the University of Zürich throughout World War I. According to Dubský, the progress of Werner's disease increasingly interfered with work so in 1919 Dubský left Zürich for the University of Groningen in the Netherlands, where he worked with his friend Prof. Hilmar Johannes Backer. After the presentation of his *Habilitationsschrift* on microanalysis, Dubský became *Dozent* at Groningen, where he remained until 1922, when he was appointed Professor of Analytical Chemistry at the recently founded (1919) Masaryk University of Brno (then Czechoslovakia). Even though he was a native Czech, at the time of his appointment, he had problems with the Czech language. When all Czech universities were closed by the German occupation administration on November 17, 1939 for the duration of World War II, Dubský protested, even at the risk of being arrested. With the exception of this wartime period, he worked at Masaryk University until his sudden death on March 25, 1946. A few years before his death he was elected a member of the Czech Academy of Sciences and Arts.

Dubský was a thoughtful, prolific worker and a modest, deeply pious man for whom both scientific work and faith in God were everything. His sojourns at different universities, resulting in his training in coordination, analytical, and organic chemistry, made him a broadly learned chemist whose scientific ideas exceeded the working possibilities of his laboratory. Dubský published more than 200 papers and created around himself in Brno a school of analytical chemists. Although most of his papers were devoted to analytical chemistry, he remained a coordination chemist for his entire life.

Dubský's Work

Coordination Chemistry. Dubský's first paper described the preparation and characterization of diamminediaquadihydroxochromium(III) salts (8). He prepared series of salts of formula $[Cr(OH)_2(NH_3)_2(H_2O)_2]X$ (X = Br, Cl, I, SCN, or SO_4), formerly formulated as $[Cr(NH_3)_2(H_2O)_4]X_n \cdot H_2O$. He also noted that the hydroxo group can be transformed into an aqua group by the action of acids. This "neutralization" of a coordinated OH⁻ group was used in the preparation of *cis*-(violeo) $[CoCl_2(NH_3)_4]X$ (violeo salt), an isomer whose existence had been predicted by Werner (9).

Sophus Mads Jørgensen (1837-1914), influenced both by his own success in the preparation (10, 11) of *cis*- and *trans*-$[CoCl_2(en)_2]^+$ and his lack of success in synthesizing the labile *cis*-$[CoCl_2(NH_3)_4]^+$ cation, stated that for the existence of *cis* and *trans* isomers the presence of ethylenediamine is necessary. Werner rejected this argument (12), claiming that the existence of both isomers is a general phenomenon caused by different spatial orientations of the coordinated atoms or groups around the central atom. He therefore put Dubský in charge of synthesizing the crucial *cis*-$[CoCl_2(NH_3)_4]^+$ isomer.

Although Werner published the paper on violeo salts alone (2), Dubský's active role is shown by the methodological link to his previous work on chromium(III) hydroxo complexes (8) and to his first *Habilitationsschrift* as well as by the outstanding acknowledgment to him at the end of Werner's paper: "Meinem Asistenten, Hrn. J. Dubsky, spreche ich für seine eifrige Unterstützung bei vorliegender Untersuchung meinem besten Dank aus." In a similar manner Werner (13) also appreciated Dubský's work on the preparation of $[CoI(NH_3)_5]X_2$, the experimental details of which excited the admiration of Lev Aleksandrovich Chugaev (1873-1922), who invited Dubský to come to Moscow to collaborate with him (5).

Dubský obtained the *cis*-[CoCl$_2$(NH$_3$)$_4$]X isomer by the hydrolysis of a binuclear complex at low temperature, isolating the violeo salt in the form of the dithionate:

The *cis*-[CoCl$_2$(NH$_3$)$_4$]$^+$ ion is subjected to acid hydrolysis in aqueous solution:

$$\text{\textit{cis}-[CoCl}_2\text{A}_4]^+ + \text{H}_2\text{O} \underset{2}{\overset{1}{\rightleftharpoons}} \text{[CoClA}_4\text{(H}_2\text{O)]}^{2+} + \text{Cl}^-$$

$$\text{[CoClA}_4\text{(H}_2\text{O)]}^{2+} + \text{H}_2\text{O} \underset{4}{\overset{3}{\rightleftharpoons}} \text{[CoA}_4\text{(H}_2\text{O)}_2]^{3+} + \text{Cl}^-$$

which prevented its isolation by Jørgensen's procedure (*10, 11*) for the preparation of *cis*-[CoCl$_2$(en)$_2$]$^+$ salts, which form during the evaporation of an aqueous solution of the *trans* isomer to dryness. In both cases (A = NH$_3$ or 1/2 en) in the initial stages acid hydrolysis predominates (equations 1 and 3), while during evaporation, when the concentration of Cl$^-$ ions increases, reactions 2 and 4 occur (*14a*). Although the course of acid hydrolysis and replacement of coordinated water by Cl$^-$ ions are common for both the *cis*-[CoCl$_2$(NH$_3$)$_4$]$^+$ and *cis*-[CoCl$_2$(en)$_2$]$^+$ isomers, the result of these reactions, *i.e.*, the separation of the particular isomer from solution is substantially different. For example, *cis*-[CoCl$_2$(en)$_2$]Cl is less soluble in water than the *trans* isomer, which is not the case with *cis*- and *trans*-[CoCl$_2$(NH$_3$)$_4$]Cl because of the solvation differences. Furthermore, as shown in Table I *(14b)* the rate of acid hydrolysis of *trans*-[CoCl$_2$(NH$_3$)$_4$]$^+$ is faster than that of *trans*-[CoCl$_2$(en)$_2$]$^+$. In addition to this, both the starting isomers and the products of their hydrolyses undergo rearrangement (*14c*) via a common trigonal bipyramidal intermediate.

Hence the preparation of the *pure cis*-[CoCl$_2$(NH$_3$)$_4$]$^+$ isomer is difficult. To obtain it in pure form, knowledge of the principles of retention of configuration (the hydrolysis of OH bridges proceeds without the breaking of Co-O bonds), recognized later, together with the rules governing the suppression of both competitive and isomerization reactions (low temperature and isolation of the isomer in the form of the insoluble dithionate) was applied as early as 1907. Thus, from the present point of view, the preparation of the *cis* isomer can be considered as stereotactic;

Table I. Rate of Aquation of $[CoCl_2A_4]^+$ Complexes

Ion	k (x 10⁴, min⁻¹)	Ion	k (x 10⁴, min⁻¹)
cis-$[CoCl_2(NH_3)_4]^+$	very fast	*cis*-$[CoCl_2(en)_2]^+$	150
trans-$[CoCl_2(NH_3)_4]^+$	1100	*trans*-$[CoCl_2(en)_2]^+$	19

and according to Prof. A. Okáč, Dubský's student and successor in the Department of Analytical Chemistry in Brno: "... by the synthesis of the long-sought *cis*-$[CoCl_2(NH_3)_4]^+$ isomer the young Dubský convinced the Master of this branch of chemistry, Jørgensen, that he was mistaken in his assumptions and showed the second Master, Werner, the way to the preparation of intuitively anticipated violeo salts" (*5*).

The successful preparation of the crucial *cis* isomer was criticized by Werner's primary scientific adversary Jørgensen as an indirect proof of octahedral geometry of the cobalt(III) atom. Werner considered the resolution of an octahedral cobalt(III) complex into enantiomers as an elegant and definitive proof of his postulated theory of molecular structure. His correspondence shows that as early as 1897 he was seeking suitable compounds for the resolution (*15*). The first attempt is documented in the collection of Werner's compounds stored in the Department of Chemistry at the University of Zürich, where one can find a tube of sample labeled: "Attempt to resolve $[CoCO_3(en)_2]Br$ using silver *d*-tartarate, 20.1.1908, Dubský" (*16*). Dubský's attempt as well as King's later experiments (*15, 17*) was unsuccessful, and Werner ascribed it to the "insufficient asymmetry" of the *cis*-$[CoCO_3(en)_2]^+$ cation (*18*). Werner therefore decided to resolve the "less symmetrical" *cis*-$[CoCl(NH_3)(en)_2]Cl$ isomer (CoA_4BC type rather than the earlier attempted CoA_4B_2 type), this time using (+)-3-bromocamphor-9-sulfonic acid (*1*), which, compared to *d*-tartaric acid, is a stronger acid and yields salts which are more stable and better crystallized. These facts, which, together with the salts' differences in solubility that make the use of the camphoric acid for resolution experiments more plausible, explain the reason for the unsuccessful resolution of the *cis*-$[CoCO_3(en)_2]^+$ isomer better than do considerations of the degrees of asymmetry of metal complexes.

Influenced by Werner's ideas, Dubský also studied the saturation of primary and secondary valences (*19-21*) and stereoisomerism (*22, 23*) of coordination compounds. As a result of his work on double salts, he stated that in symmetrically constructed molecular compounds there is no difference between primary and secondary valences (*19*). Furthermore, he recognized (*24*) that the copper(II) atom in iminodiacetatocopper(II) is not coordinatively saturated, and he prepared its ammonia adduct, which, according to the concept at that time, forms the link between addition compounds and inner complex salts:

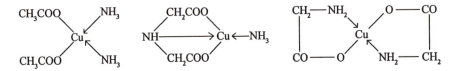

Dubský's paper on metal complexes of nitro-, nitroso-, and phenyliminodiacetic acids should be also mentioned in this connection (*25*).

Looking for different types of isomerism, Dubský also prepared numerous addition compounds (*26, 27*). He designated the isomerism arising from the existence of the following forms (L = dicyandiamide)

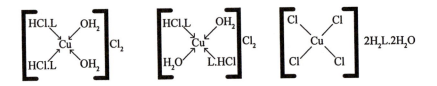

as second order isomerism of neutral salts (*28, 29*).

Dubský's research activity as a coordination chemist was extensive. According to Prof. Okáč: "The influence of Werner's school was too strong for him to surmount. This school, which remained a perennial feature of Dubský's further scientific work, was an inspiration for him all his life" (*6*). In addition to original research papers, Dubský contributed to several monographs (*30-33*), and he published papers on methodology (*34, 35*). Of his papers that are not directly related to coordination theory those concerning amino acid complexes are valuable contributions (*36, 37*).

Organic and Analytical Chemistry. As mentioned previously, Dubský was trained in organic chemistry during his stay in Leyden and Groningen. At Leyden he studied acetylation and nitration, especially of imines, which resulted in several papers (*38*). At the same time he extensively studied diketopiperazines, publishing a series of fourteen papers on these compounds (*39*). At Groningen Dubský devoted his effort to the sulfonation of carboxylic acids (*40*), resulting in his interest in metal salts of acids sulfonated in the α-position (*41*).

From the beginning Dubský's interest in analytical chemistry was quite practical, dictated only by the need for a rapid determination of both metal and nitrogen. In 1912 he visited future (1923) Nobel chemistry laureate Fritz Pregl (1869-1930) in Innsbruck, and in 1916 he published (*42*), at Pregl's request as he later remembered (*43*), a simplified micro method adapted for a need in Werner's laboratories. Over the years Dubský gradually changed his interest to analytical chemistry. In his own words, "as a student of Werner's I was primarily interested in studies of metal complexes. Since these play such an important role in analytical chemistry, I expended much time and effort on this branch of chemistry" (*5*). His experiences both in coordination chemistry and organic chemistry destined him to become a pioneer in the use of organic reagents in analytical chemistry. His main contribution to this branch of chemistry consists of his systematic work on the relation between the character of organic groups and the selectivity of analytical reagents (*44*).

After the German authorities confiscated Fritz Feigl's (1891-1970) now classic monograph on the use of spot tests in chemical analysis (Feigl was Jewish), some German chemists asked Dubský to write an analytical chemistry textbook emphasizing organic analytical reagents to be published by Springer Verlag. Although Dubský began work on this book, it was never completed because the wartime closure of all Czech universities, including their libraries, prevented him from collecting, consulting, and studying the pertinent literature.

Literature Cited

1. Werner, A. *Ber.* **1911**, *44*, 1887. For a discussion and annotated English translation see Kauffman, G. B. *Classics in Coordination Chemistry, Part 1: The Selected Papers of Alfred Werner*; Dover: New York, 1968; pp 155-173.
2. Werner, A. *Ber.* **1907**, *40*, 4017. For a discussion and annotated English translation see Kauffman, ref. 1, pp 141-154.
3. Kauffman, G. B. *Alfred Werner: Founder of Coordination Chemistry*; Springer-Verlag: Berlin, 1966; Kauffman, G. B. *Inorganic Coordination Compounds*; Heyden: London, Philadelphia, Rheine, 1981.
4. Kauffman, G. B. *Classics in Coordination Chemistry, Part 3: Twentieth-Century Papers (1904-1935)*; Dover: New York, 1978.
5. Okáč, A. *Chem. Listy* **1942**, *36*, 145.
6. Okáč, A. *Chem. Listy* **1946**, *40*, 198.
7. Jursík, F. *Chem. Listy* **1983**, *77*, 625.
8. Werner, A.; Dubský, J. V. *Ber.* **1907**, *40*, 4085.
9. Kauffman, G. B. *Coord. Chem. Rev.* **1975**, *15*, 1.
10. Jørgensen, S. M. *J. prakt. Chem.* **1889**, [2] *39*, 15.
11. Jørgensen, S. M. *J. prakt. Chem.* **1890**, [2] *41*, 448.
12. Werner, A. *Z. anorg. Chem.* **1893**, *3*, 267.

13. Werner, A. *Ber.* **1908**, *41*, 3007.
14. Basolo, F.; Pearson, R. G. *Mechanisms of Inorganic Reactions*; Wiley: New York, 1965; (a) p 242; (b) p 116; (c) p 243.
15. Kauffman, G. B.; Bernal, I. *J. Chem Educ.* **1989**, *66*, 293.
16. Kauffman, G. B. In *van't Hoff-Le Bel Centennial*; Ramsay, O. B., Ed.; American Chemical Society: Washington, DC, 1975; p 126.
17. King, V. L. *Über Spaltungsmethoden und ihre Anwendung auf Komplexe Metall-Ammoniakverbindungen*; PhD Thesis, J. J. Meier: Zürich, 1912.
18. Werner, A. *Ber.* **1911**, *44*, 2445.
19. Dubský, J. V. *J. prakt. Chem.* **1914**, [2] *90*, 61.
20. Dubský, J. V.; Beer, T.; Frank, H. *J. prakt. Chem.* **1916**, [2] *93*, 142.
21. Dubský, J. V. *J. prakt. Chem.* **1921**, [2] *103*, 109.
22. Dubský, J. V.; Vretos, S. *Schw. Chem. Ztg.* **1914**, *1*, 5.
23. Dubský, J. V. *Chem. Weekbl.* **1919**, *16*, 984.
24. Dubský, J. V.; Spritzmann, M. J. *J. prakt. Chem.* **1917**, *96*, 112.
25. Dubský, J. V.; Spritzmann, M. J. *J. prakt. Chem.* **1917**, *96*, 105.
26. Dubský, J. V.; Rabas, A. *Coll. Czech. Chem. Commun.* **1929**, *1*, 528.
27. Dubský, J. V.; Trtílek, J. *J. prakt. Chem.* **1934**, *140*, 185.
28. Dubský, J. V.; Strnad, M. *Chem. Obzor* **1942**, *17*, 69.
29. Dubský, J. V. *Chem. Obzor* **1943**, *18*, 169.
30. Dubský, J. V. *Vereinfache Quantitative Mikroelementaranalyse organischer Substanzen*; Veit: Leipzig, 1917.
31. Dubský, J. V. "Halbmikroelementaranalyse." In *Handbuch der biologischen Arbeitsmethoden*; Abderhalden, A. J., Ed.,; Urban and Schwarzenberg: Berlin, 1922.
32. Dubský, J. V. "Organische Mikroelementaranalyse." In Houben-Weyl. *Die Methoden der organischen Chemie*, 3rd ed.; G. Thieme: Leipzig, 1923.
33. Dubský, J. V. "*Selbsttätige Filtrationsapparate*; G. Thieme: Leipzig, 1931.
34. Dubský, J. V.; Rabas, A. *Chem. Weekbl.* **1928**, *25*, 293.
35. Dubský, J. V.; Dostál, V. *Chem. Obzor* **1930**, *6*, 268.
36. Dubský, J. V. *Z. med. Chem.* **1927**, *5*, 37.
37. Dubský, J. V.; Rabas, A. *Spisy přírodověd. fak. v Brně. No.123*, 1936.
38. Franchimont, A. P. N.; Dubský. J. V. *Rec. trav. chim. Pays-Bas* **1916**, *36*, 80.
39. Dubský, J. V.; Trtílek, J. *Ber.* **1933**, *66*, 1497 (Paper No. XIV in the series "Contribution to Diketopiperazines").
40. Backer, H. J.; Dubský, J. V. *Rec. trav. chim. Pays-Bas* **1922**, *41*, 125.
41. Dubský, J. V.; Trtílek, J. *J. prakt. Chem.* **1934**, *140*, 47.
42. Dubský, J. V. *Chem. Ztg.* **1916**, *40*, 201.

43. Dubský, J. V. *Chem. Ztg.* **1916**, *40*, 201.
44. Dubský, J. V.; Vrbová, J. *Chem. Obzor* **1941**, *16*, 115 (Paper No. XVII in the series "Systematic Building Up of Atomic Groups in Analytical Chemistry").

RECEIVED January 21, 1994

Linus Pauling (1901–1994) with George B. Kauffman (1930–   ).
(Reproduced from Chemical & Engineering News.
Copyright 1993 American Chemical Society.)

Chapter 5

Early Structural Coordination Chemistry

Linus Pauling†

Linus Pauling Institute of Science and Medicine, 440 Page Mill Road, Palo Alto, CA 94306

The history of coordination chemistry may in a sense be said to have begun with the work of Werner. The early crystal-structure determinations by W. L. and W. H. Bragg showed that in crystals such as sphalerite, ZnS, there is tetrahedral coordination around both zinc and sulfur, and in crystals such as sodium chloride there is octahedral coordination about both the anion and the cation. The modern period may be said to have begun in 1921, with the determination of a crystal containing an octahedral complex by Wyckoff and of crystals containing tetrahedral and square planar complexes (1922) by Dickinson. Later developments include application of quantum mechanics, discussion of hybrid orbitals especially suited to bonding, and detailed interpretation of interatomic distances found by careful X-ray diffraction studies.

Some of the substances that we now call coordination compounds have been known for hundreds of years. For example, potassium ferrocyanide, K_4 [FeCN)$_6$], was made by strongly heating a mixture of a nitrogenous material, iron filings, and wood ashes or potassium carbonate, eluting with water, and crystallizing the yellow substance (1,2). It was used as a source of hydrocyanic acid, Prussian blue, and other compounds containing cyanide.

Until about 100 years ago coordination compounds were usually called double salts. The formula of potassium ferrocyanide was written as 4KCN•Fe(CN)$_2$.

By the 1870s efforts were being made to assign structural formulas to coordination compounds. In his Nobel address, delivered in 1913 (3), Alfred Werner mentions some of the efforts that were being made, in particular those by the Swedish chemist C. W. Blomstrand and the Danish chemist S. M. Jørgensen. For example, the compound cobalt hexammine trichloride was assigned by Blomstrand a structure in which there were three chains attached to the cobalt atom, each chain consisting of two ammonia molecules and a chlorine atom. Jørgensen, on the basis of differences in chemical reactivity of the different ammine groups, assigned to this substance a structure with two short chains and one long chain, two chains with a single ammonia and a chlorine atom and a long chain with four ammonia molecules and a chlorine atom. These chains of ammonia molecules were suggested by hydrocarbon chains and were assumed to involve quinquevalent nitrogen atoms.

†Linus Pauling died on August 19, 1994.

0097–6156/94/0565–0069$08.00/0

The great step forward in coordination chemistry was taken over a period of years by Alfred Werner himself, beginning in 1893 in his classic work, "Beitrag zur Konstitution anorganischer Verbindungen" on the subject of the spatial arrangement of atoms in the ammonia complexes (4). He rejected the idea of directed primary valence bonds; instead, he suggested that the central metal atom exerts a spherically symmetrical force of attraction on surrounding atoms, a strong force for atoms in the first coordination shell and weaker forces for those in the outer coordination shells. He introduced the concept of coordination and coordination number (the number of atoms in the first coordination shell).

He made extensive studies of complexes with coordination number 6. At first he distinguished between primary and secondary valences but later suggested that there was little or no difference between them.

The observation that an octahedral complex with two ligands of one kind and four of another kind occurs in two forms characterized by different colors (essentially the same for corresponding complexes of tripositive chromium and cobalt) permitted a decision to be made between the octahedral and the trigonal prismatic coordination polyhedron for coordination number (ligancy) 6. Two isomers are expected for the octahedron, and three for the trigonal prism. Werner and his students culminated this effort by resolving pairs of chiral complexes, with the central atom chromium, cobalt, or rhodium and with two ethylenediamine groups and two chlorine atoms in the first coordination shell.

In addition to octahedral complexes, Werner also discussed complexes in which the central atom has ligancy 4. Some of these complexes were found not to form isomers with composition MA_2B_2, while others formed pairs of isomers, cis and trans. The tetraligated complexes of bipositive palladium and platinum are in the second group, to which Werner assigned the square planar configuration, with the other ligancy 4 complexes being tetrahedral. The Werner-Jørgensen controversy and the supersession of the Blomstand-Jørgensen chain theory by Werner's coordination theory is discussed in more detail by George B. Kauffman in his paper, "Theories of Coordination Compounds: Alfred Werner's Triumph," in this symposium volume.

In other complexes the central atom has ligancy 2 (as in the dicyanoargentate(I) anion), 8 (as in the octacyanomolybdate (IV) or (V) anion, or 9 (as in the enneahydridorhenate(VI) anion) (5).

The last of these complexes was identified through the X-ray diffraction determination of the structure of the salt K_2ReH_9; the nine hydrogen atoms are at the single-bond distance from the rhenium (6).

An important contribution to the understanding of the structure of these complexes was made through X-ray diffraction studies. I arrived in Pasadena, California, in September 1922 to begin my graduate work in the California Institute of Technology. The X-ray crystallographer Roscoe G. Dickinson immediately began teaching me the experimental techniques of X-ray crystallography and the ways of analyzing the diffraction data to locate the atoms in the crystal under study. Ralph W. G. Wyckoff had worked in the laboratory during the preceding year but had returned to the East coast before I had arrived. Both Wyckoff and Dickinson were pioneers in the X-ray crystallography field. They were the only researchers in this field who made great use of Laue photographs, with the result that neither of them published structure determinations that were later shown to be wrong, whereas many other workers in this field, mainly in Europe, published many incorrect structure determinations. The reason for this difference in reliability is that the Laue photographs showed a great number of diffraction maxima, including many very weak ones that were needed to determine the correct structure, and these very weak diffraction maxima were often not found by the other experimental techniques. I had become interested in Werner coordination complexes in 1919, and I was pleased to learn, when I arrived in Pasadena, that during the preceding year Wyckoff and

Dickinson had obtained X-ray evidence verifying the structures that Werner had described. The first such study was that of $(NH_4)_2$ [PtCl$_6$], reported by Wyckoff and Posnjak in 1921 (7). This was followed in 1922 by Dickinson's report on K$_2$ [SnCl$_6$] and $(NH_4)_2$ [SnCl$_6$] (8). In each case the Werner octahedron was found, and it was verified that the six ligands lie at the same distance from the central atom. In the same year, 1922, Dickinson reported his verification of the tetrahedral structure of the zinc tetracyanide complex in K$_2$ [Zn(CN)$_4$] (9) and the square planar structure of the palladium tetrachloride and platinum tetrachloride complexes in K$_2$[PdCl$_4$], $(NH_4)_2$ [PdCl$_4$], and K$_2$[PtCl$_4$] (10). Since this pioneer work was done, scores of octahedral, tetrahedral, and square planar complexes have been studied by X-ray crystallography, as well as complexes involving other values of the coordination number.

Another step forward in the understanding of coordination complexes was taken by me in 1931. For several years I had been applying the principles of quantum mechanics to various aspects of chemistry, and in 1928 I had published a short paper containing the sentence "It has been found that as the result of the resonance phenomenon a tetrahedral arrangement of the four bonds of the quadrivalent carbon atom is the stable one" (11). I also stated that a detailed account would be published later, but this detailed account did not appear until 1931 (12). The delay of nearly three years resulted from my dissatisfaction with the complexity of the calculations on which the 1928 statement had been based. Then one day in December 1930 or early January 1931 I saw that a tremendous simplification could be achieved by ignoring the overlap integrals of orbitals with different radial wave functions (s, p, d orbitals) and concentrating on the surface harmonics of the angular functions. With this simplification it took only a few minutes to show that the best sp hybrid bond orbitals are the tetrahedrally directed sp^3 orbitals (as for the carbon atom and the tetracyanozinc(II) complex), that if one d orbital is also available for bond formation four square planar bond orbitals are the most stable (as in the tetrahalopalladium(II) and tetrahaloplatinum(II) complex, and that if two d orbitals are available six bonds directed toward the corners of an octahedron are stable. Some predictions (later verified) were also made, such as that paramagnetic compounds of bipositive nickel should show tetrahedral or octahedral coordination and diamagnetic compounds should show square planar coordination (12).

Coordinated structures similar to Werner's coordinated complexes were also found in crystals in the very early days of X-ray crystallography. W. H. and W. L. Bragg reported in 1913 that in the rock salt crystal each Na is surrounded octahedrally by 6 Cl (also each Cl by 6 Na) (13) and that in the mineral sphalerite each Zn is surrounded tetrahedrally by 4 S (also each S by 4 Zn) (14). Cubic coordination of 8 Cl around Cs and 8 Cs around Cl was soon discovered. Also, X-ray diffraction studies have revealed other types of coordination complexes, such as 12 Al icosahedrally coordinated about Mo in the compound MoAl$_{12}$ (15).

It is interesting that an Englishman, William Barlow, working independently in Manchester, had described several of these crystal structures two decades before the X-ray work was done and had ascribed them correctly, on the basis of face development and other properties, to various substances: the sodium chloride structure to sodium chloride, the cesium chloride structure to cesium chloride, the sphalerite and wurtzite structures, respectively, to the cubic and hexagonal forms of zinc sulfide, cubic closest packing to copper and some other metals, and hexagonal closest packing to magnesium (16). His work seems not to have had much of an impact on crystallography or chemistry.

When I read his papers 70 years ago, I was astonished that he had not ascribed the diamond structure (closely related to the sphalerite structure) to diamond. Barlow surely knew about the tetrahedral carbon atom, which had been discovered twenty years earlier and was of course known to chemists and other scientists. Some decades later I searched the literature, as did also Paul Ewald at my request, to see if anyone had pointed out in the period before 1914 that diamond might have the diamond structure. Our search was unsuccessful. The structure was determined by W. H. and W. L. Bragg in 1914 (*17*).

It is clear now as it was 80 years ago that it was thoroughly justified to award the 1913 Nobel prize in chemistry to Alfred Werner of the University of Zürich "in recognition of his work on the linkage of atoms in molecules by which he has thrown new light on earlier investigations and opened up new fields of research especially in inorganic chemistry" (*3*).

Literature Cited

1. Woodward, J. *Phil.Trans. Roy. Soc. London* **1724**, *33*, 15 (in Latin). For an English translation see Powell, H. M. *Proc. Chem. Soc.* **1959**, 13.
2. Browne, J. *Phil. Trans. Roy. Soc. London* **1724**, *33*, 17.
3. Nobel Foundation, *Nobel Lectures, Chemistry, 1901-1921*; Elsevier: Amsterdam-London-New York, 1966; pp 251-272.
4. Werner, A. *Z. anorg. Chem.* **1893**, *3*, 267. For a discussion and annotated English translation see Kauffman, G. B. *Classics in Coordination Chemistry, Part 1: The Selected Papers of Alfred Werner*; Dover: New York, 1968; pp 5-88.
5. Abrahams, S. C.; Ginsberg, A. P.; Knox, K. *Inorg. Chem.* **1964**, *3*, 558.
6. Teller, R. G.; Bau, R. *Struct. Bonding* **1981**, *44*, 1.
7. Wyckoff, R. W. G.; Posnjak, E. *J. Am. Chem. Soc.* **1921**, *43*, 2292. For a discussion and annotated reprint see Kauffman, G. B. *Classics in Coordination Chemistry, Part 3: Twentieth-Century Papers (1904-1935)*; Dover: New York, 1978; pp 85-111.
8. Dickinson, R. G. *J. Am. Chem. Soc.* **1922**, *44*, 276.
9. Dickinson, R. G. *J. Am. Chem. Soc.* **1922**, *44*, 774.
10. Dickinson, R. G. *J. Am. Chem. Soc.* **1922**, *44*, 2404. For a discussion and annotated reprint see Kauffman, ref. 7, pp 112-126.
11. Pauling, L. *Proc. Natl. Acad, Sci. USA* **1928**, *14*, 359.
12. Pauling, L. *Proc. Natl. Acad. Sci. USA* **1931**, *53*, 1367.
13. Bragg, W. H.; Bragg, W. L. *Proc. Roy. Soc. London* **1913**, *88*, 428.
14. Bragg, W. H.; Bragg, W. L. *Proc. Roy. Soc. London* **1913**, *A89*, 248.
15. Adam, J.; Rich, J. B. *Acta Crystallog.* **1954**, *7*, 813.
16. Barlow, W. *Z. Kristallog.* **1898**, *29*, 433.
17. Bragg, W. H.; Bragg, W. L. *Proc. Roy. Soc. London* **1913**, *A88*, 277.

RECEIVED December 6, 1993

John C. Bailar, Jr. (1904–1991).

Chapter 6

John C. Bailar, Jr. (1904–1991)

Father of U.S. Coordination Chemistry

George B. Kauffman[1], Gregory S. Girolami[2], and Daryle H. Busch[3]

[1]Department of Chemistry, California State University,
Fresno, CA 93740
[2]School of Chemical Sciences, University of Illinois
at Urbana–Champaign, Urbana, IL 61801
[3]Department of Chemistry, University of Kansas,
Lawrence, KS 66045

John Christian Bailar, Jr., Professor Emeritus of Inorganic Chemistry at the University of Illinois, 1959 ACS President, 1964 Priestley Medalist, and dean of coordination chemists in the U.S., died on Oct. 17, 1991 in Urbana, Illinois at the age of 87. Born in Golden, Colorado on May 27, 1904, John received his B.A. (*magna cum laude*) (1924) and M.A. (1925) degrees from the University of Colorado and his Ph.D. (1928) from the University of Michigan under Moses Gomberg. Although trained in organic chemistry, he became fascinated with inorganic isomerism, especially among coordination compounds, and he trained several generations of coordination chemists (90 doctorates, 38 postdoctoral fellows, and numerous master's and bachelor's degree candidates). He was co-founder and first Chairman (1957) of the ACS Division of Inorganic Chemistry and played crucial roles in founding the journals *Inorganic Syntheses* and *Inorganic Chemistry*. More than anyone, he was responsible for the renaissance in inorganic chemistry in the U.S. Had he lived, John undoubtedly would have played an active role in this Coordination Chemistry Centennial Symposium.

On October 17, 1991 John Christian Bailar, Jr., Professor Emeritus of Inorganic Chemistry at the University of Illinois and dean of coordination chemists in the United States, died of a heart attack in Urbana, Illinois at the age of eighty-seven. John was responsible, more than any other single individual, for the continuing emphasis on research in inorganic chemistry, especially in the United States, from a near hiatus beginning sometime in the thirties and extending through the Second World War, and he was a central figure in the ultimate renaissance of that field during the late forties and fifties. Much of today's research in coordination chemistry has roots in his early studies. Had John lived, he would undoubtedly have played an active role in this symposium especially because he was born only a few miles away from Denver.

John was a tall (6 feet, 3 inches) imposing man, a devout Presbyterian (he served as a church elder) and politically conservative, although, contrary to

This chapter reprinted with permission from *The Journal of Coordination Chemistry*
© 1993 Gordon & Breach

the usual trend, he became distinctly more liberal in his later years. John symbolized all that is best in a university professor. He was a kind man with a charming sense of humor and an engaging laugh; he knew a million stories and told them all very well. Despite his strong views, he was never judgmental, and he believed in letting people live their own lives and make their own decisions. He was open with others and generous with his time and energies. John was truly respected and admired and was a valued friend, colleague, scientist, and teacher.

John was born on May 27, 1904 at Golden, Colorado. His parents, John C. Bailar, Instructor in Chemistry at Golden's Colorado School of Mines, and Rachel Ella Work Bailar, were the first married couple to enroll in and graduate from the University of Colorado, from which John received his B.A. degree (*magna cum laude*) in 1924 and his M.A. degree in 1925. His master's thesis, "Nitrogen Tetrasulfide and Nitrogen Selenide: Preparation, Molecular Weight, and Some Properties," carried out under Horace B. Van Valkenburgh's direction, resulted in his first publication (*1*). Like a number of the most prominent late 19th- and early 20th-century coordination chemists such as Sophus Mads Jørgensen, James Lewis Howe, Lev Aleksandrovich Chugaev, and Alfred Werner himself, John was trained as an organic chemist, entering inorganic chemistry "through the back door," so to speak (*2*). He worked under the supervision of Moses Gomberg (*3*), best known for his work on triphenylmethyl radicals, at the University of Michigan at Ann Arbor, where he received his Ph.D. degree in 1928 with a dissertation on substituted pinacols (*4*).

In that same year John accepted a position as Instructor in Organic Chemistry at the University of Illinois at Urbana, where he remained for 63 years -- almost half the period that the university had been in existence. He was assigned teaching duties in general chemistry, which was then primarily descriptive inorganic chemistry. He rose through the ranks, becoming Full Professor in 1943; he also served as Secretary of the Chemistry Department (1937-1951) and Head of the Division of Inorganic Chemistry (1941-1967). On August 8, 1931 he married his former graduate teaching assistant Florence Leota Catherwood. The couple, whose marriage lasted almost 44 years (until Florence's death on March 13, 1975), were the parents of two sons -- John Christian Bailar, III, currently Professor in the School of Medicine at McGill University, Montreal, and Benjamin Franklin Bailar, currently Dean of the Graduate School of Administration at Rice University, Houston and former Postmaster General of the United States. On June 12, 1976 John married Katharine Reade Ross, whom he had known since childhood, for Kay had been his babysitter.

During his graduate studies John had become interested in organic isomerism, but it was only during his second or third year at Illinois, while teaching a general chemistry class that he realized that isomerism was a phenomenon that could also exist in the inorganic realm (*5,6*). A student logically but incorrectly referred to SbOCl, the product of the hydrolysis of antimony(III) chloride, as antimony hypochlorite rather than as antimony oxychloride or antimonyl chloride. In his literature search for examples of inorganic isomers John soon encountered coordination chemistry, and in his own words, "My entire feeling toward the chemical profession changed. I had found my niche."

John considered his very first work on coordination chemistry (*7*), carried out with senior undergraduate student Robert W. Auten, to be his most significant achievement. He said, "It opened up a field. One doesn't often have the opportunity to do that" (*8*). The monumental, wide-ranging work of Alfred Werner, the founder of coordination chemistry (*9*), was so comprehensive that many chemists assumed that all the important work in this field had already been done, and therefore coordination chemistry was relatively neglected until John's entry into the field. John educated several generations of coordination chemists (90 doctorates, 38 postdoctoral fellows, and many candidates for bachelor's and master's degrees), earning him the universally acknowledged title, "father of American coordination chemistry." A list of his former students, many of whom, inspired by his love of teaching, entered academia, reads like a "Who's Who" of inorganic and coordination chemistry. Not only was John elected President of the American Chemical Society (1959), but also three of his former students -- Robert W. Parry (1982), Fred Basolo (1983), and Clayton F. Callis (1989) -- attained this honor.

John's work with Auten established the inorganic counterpart of the well-known organic Walden inversion *(10)*. Werner observed that the following reaction proceeds with a change in the sign of rotation *(11,12)*, although Werner realized that inversion of configuration need not take place (in this case, it does not): (-)-cis-[CoCl$_2$en$_2$]Cl + K$_2$CO$_3$ → (+)-[CoCO$_3$en$_2$]Cl + 2KCl. Because Walden had converted (+)-chlorosuccinic acid to the hydroxy acid (malic acid) with KOH or Ag$_2$O and had achieved an inversion in one of the cases, John hoped to repeat Werner's experiment with Ag$_2$CO$_3$ instead of K$_2$CO$_3$ to see if an inversion would result. At the time the Walden inversion mechanism had not been completely established, and John hoped that such an inversion with an octahedral configuration rather than a tetrahedral configuration might throw light on the mechanism of the Walden inversion. Bailar and Auten obtained an inversion of configuration in what was the first analogue of a Walden inversion among inorganic compounds (although the mechanism of this reaction was later shown to be quite different). Their 1934 article *(7)*, the first in a continuing 37-part series titled "The Stereochemistry of Complex Inorganic Compounds" extending to 1985, has attained the status of a classic. In John's words, "I owe a great deal to Robert Auten, for it was this piece of my work that originally caught the attention of the chemical public." Throughout his long career John continued his interest in inversion reactions *(13-16)*.

At the time that John began his career, not only coordination chemistry but also inorganic chemistry in general was languishing. In the United States inorganic chemists were exceedingly few, and most, like John, were overloaded with general chemistry teaching. There were few inorganic courses beyond the freshman course, little inorganic research was being carried out, and avenues for publication were limited.

At the Fall 1933 meeting of the American Chemical Society in Chicago, five inorganic chemists -- Harold S. Booth, Ludwig F. Audrieth, W. Conard Fernelius, Warren C. Johnson, and Raymond E. Kirk -- decided that there was a vital need for a series of volumes giving detailed, independently tested methods for the synthesis of inorganic compounds along the lines of *Organic Syntheses*, the series established by John's colleague at Illinois, Roger Adams. The five, who were soon joined by John, became the Editorial Board of the new journal, *Inorganic Syntheses*, the first volume of which appeared in 1939. Since its inception John was an active participant and motivating force in its affairs, contributing 16 syntheses and checking 5 others, especially in the early years when the fledgling publication needed considerable support. John served as Editor-in-Chief of Volume IV (1953), and seven of his former students and three of his academic grandchildren (students of his former students) later served in the same capacity for other volumes.

Despite the growing success of *Inorganic Syntheses*, in the American Chemical Society inorganic chemistry did not achieve divisional status until 1957 when the Division of Inorganic Chemistry was finally established largely through the efforts of John Bailar, who became the First Divisional Chairman and who has written a divisional history *(17)*. Similarly, *Inorganic Chemistry*, the first journal in the English language devoted exclusively to the field, began publication in 1962, again, largely through John's efforts. Thus the resurgence of the field after World War II, which the late Sir Ronald S. Nyholm *(18)* dubbed the renaissance in inorganic chemistry, owed much to John's pioneering labors.

Although the stereochemistry of coordination compounds was John's primary interest, he and his students worked on a wide variety of subjects -- synthesis;

inversion reactions; resolution into optical isomers of complexes of various coordination numbers and configurations; macrocyclic ligands; electrochemistry; electrodeposition and electroplating; polarography; homogeneous and heterogeneous catalysis; kinetics; stabilization of unusual oxidation states; dyes; spectra; unusual coordinating agents; coordination polymerization; solid state reactions, etc. With later (1990) Nobel chemistry laureate Elias J. Corey, he published a classic paper on octahedral trisdiamine complexes that led to applications of conformational analysis to coordination compounds *(19)*. Besides his 338 publications (which include two patents and 58 reviews) he wrote, cowrote, or edited nine monographs, texts, or laboratory manuals. His 834-page book, *The Chemistry of the Coordination Compounds (20)*, written with 24 of his former students, summarized almost every aspect of the field, which, largely due to his teaching and research, was attracting more and more scholars. A complete bibliography of John's publications along with lists of his Ph.D. students and postdoctoral students are found in a recent biographical article *(21)*. In addition to his work with research students, John was the Director of the university's General Chemistry and Student Placement Programs, and he spent considerable time finding suitable positions for his undergraduate students.

John devoted considerable time and thought to teaching. He once said, "I would throw away my notes after every lecture so that I wouldn't do the same thing the next year. I tried to talk about things that would interest students, to help them see the connection between chemistry and everyday life" *(8)*.

John was a consultant to several industrial companies and government laboratories, and during World War II he served as Official Investigator for the U.S. National Defense Research Committee (N.D.R.C.), devoting his efforts to studying screening smokes and nerve gases. In 1944 he was overseeing one of his students, Robert W. Parry, who was developing the fuel for a device that would generate colored smoke to aid in finding downed aircraft in the Pacific. Work had progressed quite rapidly, and a device meeting specifications was produced. A major test to demonstrate the device to top military leaders and N.D.R.C. officials was held in Urbana on the South Farm of the University of Illinois campus. The big moment came, and the smoke generator was ignited. For half an hour its performance was perfect; great clouds of bright red smoke rose up to high heaven. Everyone was delighted. But delight turned to apprehension when the generator emitted an unexpectedly large "puff" and a vigorous burst of red smoke. Things settled down for a few minutes, and people breathed a sigh of relief. Then a second large "puff" suggested that all was not rosy. Finally, on the third "puff" the whole device suddenly became airborne and headed directly toward the assembled top brass, who, by ducking and scattering, escaped with their lives but with considerable loss of dignity. The "rocket" hit a tree, broke open, and burned quietly on the ground. Professor Fraser Johnstone, Director of the University of Illinois Project, could not contain his disappointment. He went over to Parry, who had responsibility for the fuel, and made his displeasure clearly visible. When Johnstone stopped for breath, John, who never got excited, defused the tense situation by calmly telling Parry in the hearing of the disgruntled officials, "But Bob, you don't understand, you're supposed to kill the *other* side, not our own men."

On another occasion, in Parry's laboratory next door to John's office, a bottle of a smoke-generating fuel containing $FeCl_3$ picked up some of the atmospheric moisture for which Urbana is famous and not only ignited but set off a number of neighboring bottles, filling the room with smoke. John calmed down his distraught student by saying, "Now Bob, let's see if we can get the smoke out." He also remarked about a book salesman waiting in his office, who promptly disappeared, "I don't think that he'll be back." John's books bore the yellow-brown stains as long as he had them.

Throughout his career John was active in the American Chemical Society, being Chairman of the Divisions of Chemical Education (1947), Physical and Inorganic Chemistry (1950), and Inorganic Chemistry (1957); Chairman of the Divisional Officers Group (1949); Chairman or member of numerous national committees; and Director (1958-1960). He was a member of Alpha Chi Sigma (from 1922), the National Research Council (member of various committees), the Electrochemical Society (1948-1962), the International Union of Pure and Applied Chemistry (IUPAC) (Treasurer, 1963-1972; Conference Delegate, 1959, 1961, and 1963), Phi Beta Kappa, Sigma Xi, Phi Lambda Upsilon, and other scientific and fraternal organizations. He was elected an Honorary Fellow of the Indian Chemical Society (1974) and an Honorary Member of the Illinois State Academy of Sciences (1976) and the Chemical Society of Japan (1985). He was a member of the editorial or publication boards of twelve journals and the holder of honorary doctorates from four universities. He delivered innumerable lectures in North and South America, Europe, Asia, and Australia; held more than a dozen lectureships; and was a visiting lecturer at five U.S. and three foreign universities as well as a plenary lecturer at three International Conferences on Coordination Chemistry (Rome, 1957; Krakow, 1970; and Moscow, 1973).

John's numerous honors include the Scientific Apparatus Makers' Association Award in Chemical Education (ACS, 1961), John R. Kuebler Award (named after the father of one of his former students) (Alpha Chi Sigma, 1962), Priestley Medal (the ACS's highest award, 1964), Frank P. Dwyer Medal (Royal Society of New South Wales, 1965), Manufacturing Chemists Association Award for Excellence in College Chemistry Teaching (1968), Award for Distinguished Service in the Advancement of Inorganic Chemistry (ACS, 1972), J. Heyrovský Medal (Czechoslovakian Academy of Sciences, 1978), Monie Ferst Award for Education through Research (Sigma Xi, 1983), and Chernyaev Jubilee Medal, N. S. Kurnakov Institute, USSR Academy of Sciences (Moscow, 1989).

John was the first recipient of the University of Illinois' John C. Bailar, Jr. Medal, named in his honor, and his alma mater, the University of Michigan, named him Distinguished Alumnus for 1967. In recognition of his inestimable contributions to coordination chemistry, the Schweizerische Chemische Gesellschaft in Zürich, on September 3, 1966, at the centennial celebration of Alfred Werner's birth (IX ICCC), presented John with the only Werner Gold Medal ever to be awarded, a fitting tribute to the elder statesman and prime mover of American coordination chemistry *(16)*.

John officially retired in 1972 but continued to spend seven hours per day in his office instead of his previous twelve hours. John's published works survive him, but we shall all miss his wise advice, patient counsel, and cheerful personality.

We think it appropriate to conclude this tribute to John with a quotation from one of his lectures, which truly captures the essence of the man:

My fifty years of teaching have been a continuous source of inspiration and joy to me. There is nothing more satisfying than to watch a young person's understanding broadening -- to see the look of pleasure on his face when a new idea strikes home. Whether the student be a freshman or a graduate student is not important; the growth of the mind is still there. I am, of course, especially interested in those who have done their doctorate theses with me, for I know them better than the others. Nearly all of them have been highly successful, either as teachers and researchers, or have achieved high places in chemical industry. But many of those who sat in my classes during their freshman year have remained my close friends over the years and have brought me great happiness.

Acknowledgments

We wish to acknowledge the assistance of John C. Bailar, III, M.D., Robert W. Parry, Stanley Kirschner, Theodore L. Brown, Ellen Handler, Trina Carter, and Diane Majors in the preparation of this article.

Literature Cited

1. Van Valkenburgh, H. B.; Bailar, Jr., J. C. *J. Am. Chem. Soc.* **1925**, *47*, 2134.
2. Kauffman, G. B. *J. Chem. Educ.* **1976**, *53*, 445.
3. Bailar, Jr., J. C. *Biog. Mem., Nat. Acad. Sci.* **1970**, *41*, 141.
4. Bailar, Jr., J C. *Studies in the Halogen Substituted Pinacols and the Possible Formation of Ketyl Radicals R₂COMgI*; Mack Printing Co.: Easton, PA, 1929.
5. Bailar, Jr., J. C. *J. Chem. Educ.* **1931**, *8*, 310.
6. Bailar, Jr., J. C. *Trans. Ill. State Acad. Sci.* **1931**, *23*, 307.
7. Bailar, Jr., J. C.; Auten, R. W. *J. Am. Chem. Soc.* **1934**, *56*, 774.
8. *LAS Newsletter, College of Liberal Arts and Sciences, University of Illinois at Urbana-Champaign*; Summer 1989, p 3.
9. Kauffman, G. B. *Alfred Werner: Founder of Coordination Chemistry*; Springer-Verlag: Berlin, Heidelberg, New York, 1966.
10. Walden, P. *Optische Umkehrerscheinungen (Waldensche Umkehrung)*; F. Vieweg: Braunschweig, 1919.
11. Werner, A. *Chem. Ber.* **1911**, *44*, 3279.
12. Werner, A.; Tschernoff, G. *Chem. Ber.* **1912**, *45*, 3294.
13. Boucher, L. J.; Kyuno, E.; Bailar, Jr., J. C. *J. Am. Chem. Soc.* **1964**, *86*, 3656.
14. Bailar, Jr., J. C. *Chem. Zvesti* **1965**, *19*, 153.
15. Bailar, Jr., J. C. *Rev. Pure Appl. Chem.* **1966**, *16*, 91.
16. Bailar, Jr., J. C. *Alfred Werner 1866-1919 Commemoration Volume*; Verlag Helvetica Chimica Acta: Basel, 1967; pp 82-92.
17. Bailar, Jr., J. C. *J. Chem. Educ.* **1989**, *66*, 537.
18. Bailar, Jr., J. C. *Inorg. Chem.* **1972**, *11* (5), frontispiece.
19. Corey, E. J.; Bailar, Jr., J. C. *J. Am. Chem. Soc.* **1959**, *81*, 2620.
20. Bailar, Jr., J. C., Ed.; *The Chemistry of the Coordination Compounds*; Reinhold: New York, 1956.
21. Kauffman, G. B.; Girolami, G. S.; Busch, D. H. *Coord. Chem. Rev.* **1993**, *128*, 1-48.

RECEIVED January 21, 1994

Figure 1. Portrait of Professor Kai Arne Jensen (1908–1992) drawn by the Danish artist Otto Christensen at the public defense of Professor Flemming Woldbye's thesis in 1969. (Reproduced with permission from reference 40. Copyright 1969 Det Berlingske Hus.)

Chapter 7

Kai Arne Jensen's Contribution to Coordination Chemistry

Hans Toftlund

Department of Chemistry, Odense University, DK 5230
Odense, Denmark

Kai Arne Jensen was born in Copenhagen in the year of 1908. In 1932 he got his degree in chemistry from the University of Copenhagen and in 1937 he compleated his doctorate at the University with a dissertation titled "On the Stereochemistry of Metals with Coordination Number Four" (in Danish). He became a lecturer in 1943, and in the period 1950-1978 he was professor at the University of Copenhagen. In the thesis Jensen confirmed the suggestion by AlfredWerner regarding the planar configuration of a series of Pt(II), Pd(II) and low spin Ni(II) complexes. He pionered the use of dipole moment measurements for stereochemical problems in coordination chemistry and was the first to prepare nickel(III) complexes. For generations Jensen was one of the most influential Danish chemists

The year 1993 marks the 100th anniversary of modern coordination chemistry as 1893 was the year Alfred Werner published his monumental article, "Beitrag zur Konstitution anorganischer Verbindungen" Z. anorg. Chem. **1893**, *3*, 267-330.

The year 1993 is also the 40th anniversary of the second in the series of well established International Conference on Coordination Chemistry. Kai Arne Jensen (Figure 1) had attended the first, held in an Imperial Chemical Industries (ICI) fundamental research laboratory at Welwyn in the U.K. in 1950, and although it was open it was attended almost entirely by participants from within the U.K. He realized that the field of coordination chemistry was so established that the creation of a series of international meetings was needed, and he proposed that the Danish Chemical Society organize a similar meeting. It was held in Copenhagen August 9-13, 1953 (*1*). Thus Jensen brought the conferences on coordination chemistry firmly into the international arena, where they continue to flourish. The last two surviving members of the organizing committee of the conference, Jannik Bjerrum and K. A. Jensen both died in 1992. It is natural to commemorate K. A. Jensen at the Werner Symposium as his early work was devoted directly to a rejection of the last pieces of criticism against Werner's fundamental thesis from 1893.

Jensen's publication list numbers about 250 papers, the majority of which were published in international journals. Before the Scandinavian journal, *Acta Chemica Scandinavica*, was founded in 1947 his favorite communication medium was the

0097–6156/94/0565–0083$08.00/0

Zeitschrift für anorganische und allgemeine Chemie, which was also used by Sophus Mads Jørgensen and Alfred Werner.

About 25 % of Jensen's papers deal with problems in coordination chemistry. An examination of the "Science Citation Index" shows that even today Jensen's production has an annual score of about 100 citations. 25 of these are to his early papers in *Zeitschrift für anorganische und allgemeine Chemie.* Considering the statistical fact that the citation rate of an average paper is 1.7 ,culminating a few years after the appearance, Jensen's unusually high score is a testimony to the importance of his work.

Jensen was a very talented researcher who used a scholarly approach to the problems that he wanted to solve. He mastered several physical chemical techniques and was the first to use some of them in coordination chemistry.

Biography (2)

Kai Arne Jensen was born in Copenhagen March 27 in the year 1908. In 1932 he received his degree in chemistry (*mag. scient.*) from the University of Copenhagen. He was immediately employed at the Inorganic Chemistry Laboratory at the Technical University, Copenhagen. The following year he became Professor Einar Biilmann's assistant at the Chemical Laboratory of the University of Copenhagen. In 1937 he completed his doctorate at the University with a dissertation titled "Om de Koordinativt Firegyldige Metallers Stereokemi" (3)(On the Stereochemistry of Metals with Coordination Number Four). He became a lecturer in 1943, and in 1950 he was made professor at the University of Copenhagen, a post that he held until his retirement in 1978.

In 1977 Jensen was awarded the Ørsted medal (4). He was a member of a long list of scientific societies. From 1952 to 1972 he served as the president of the IUPAC commission for inorganic chemical nomenclature.

Here I shall discuss only Jensen's contributions to coordination chemistry. However, from about 1938 his research gradually moved more and more into organic chemistry. For generations Jensen was one of the most influential Danish chemists. He was a splendid lecturer and wrote a successful system of textbooks (5 ,6).

Work included in Jensen's Thesis of 1937 (Figure 2)

During the early years of his scientific career, Jensen studied some analytical reagents used for the precipitation of metal ions. He found that thiosemicarbazide forms characteristic metal complexes with nickel (II)(27). This fact spurred him on to the study of the stereochemistry of nickel(II) complexes. As Biilmann's assistant, Jensen was supposed to work in organic chemistry. However, no restrictions were put on his choice of research area. Before Jensen came to Biilmann's laboratory, Kristian Højendahl had worked there for a short period and had constructed an apparatus for the measurement of dipole moments. The idea of solving stereochemical problems in coordination chemistry by using dipole moment measurements apparently did not come directly from Højendahl, but rather from the study of the literature (7).

In a pioneering work of 1935, Jensen introduced the measurement of dipole moments as a method for assigning the configurational isomers of [Pt $X_2(SR_2)_2$], where X is a halide ion and R are is an alkyl group (8). Contrary to Werners' assignment (9), Jensen proved that the α form is the *trans* isomer. The following year this work was extended to a series of platinum complexes of the form [Pt X_2Y_2], where Y is a tertiary phosphine, arsine or stibine (10).

At that time the question of whether Werner was correct or not in postulating a planar configuration for platinum(II) and palladium(II) complexes was still discussed. Jensen was able to show that some of the complexes have dipole moments close to

OM DE KOORDINATIVT FIREGYLDIGE METALLERS STEREOKEMI

AF

K. A. JENSEN

C. A. REITZELS FORLAG — KØBENHAVN
MCMXXXVII

Figure 2. The title page of K. A. Jensen's doctoral thesis of 1937, *On the Stereochemistry of Metals with Coordination Number Four* (Reproduced with permission from ref. 3. Copyright 1937, C. A. Reitzel).

zero, and therefore they must have a *trans* planar configuration. In some cases isomers were isolated which showed a very high dipole moment, consistent with a *cis* planar configuration (Figure 3) (*11*).

The *trans* isomers of the thioether complexes typically have dipole moments about 2.4 Debye units. Jensen attributed these dipole moments to peculiarities in the structures which preclude the possibility of a complete balancing of the dipole moment for groups in opposing positions (Figure 4).

If the geometry around the sulfur atoms is nonplanar, each group will have a local dipole moment. The direction of these moments can easily compensate each other in a *trans* conformation, so apparently all the alkyl groups lie on the same side of the plane containing the platinum, halide, and sulfur atoms (Figure 4) (*12*). Of course, another possibility is that the coordination geometry around the central platinum atom deviates from a perfect plane. From the magnitude of the dipole moment Jensen calculated an upper limit of 0.08 Å for the displacement of the platinum from the best plane through the four donor atoms Although in 1934 Cox *et al.* published a structure for *trans* -[Pt Cl$_2$(SMe$_2$)$_2$] based on single crystal X-ray-diffraction, showing the molecule to have D$_{2h}$ symmetry. Jensen was not convinced(*13*). Many years later, when X-ray diffraction had developed into a very accurate technique, he persuaded prof. Svend Erik Rasmussen to solve the structure of *trans* -[Pd Cl$_2$(SeEt$_2$)$_2$] (*14*).

The configuration was indeed confirmed, but the complex has a center of symmetry ruling out the possibilty of a finite dipole moment. There is, of course, always the possibility that the molecule adopts another conformation in solution.

Work Inspired by S. M. Jørgensen

Althougt Jørgensen was defeated by Werner, his experimental contribution to the foundation of coordination chemistry cannot be denied. Jensen was well acquainted with Jørgensen's work and often referred to it. Some of the compounds prepared by Jørgensen apparently have not found their final formulation in Werner's system.

In 1886 Jørgensen reported a strongly dichrotic compound with the composition [Pt Br$_3$(NH$_3$)(NH$_2$C$_2$H$_5$)] (*15*). This compound is formally a Pt(III) complex. Another Pt(III) compound, [Pt Cl(NH$_2$C$_2$H$_5$)$_4$]Cl$_2$, was reported by Wolffram in 1900 (*16*). About the time that Jensen wrote his thesis, the nature of Wolffram's red salt was discussed in the literature. Jensen realized that if it is a true Pt(III) complex, then it should have a magnetic moment so he asked Robert Asmussen to measure the magnetic susceptibility. The compound turned out to be diamagnetic so Jensen suggested a mixed valence composition with equal amounts of Pt(II) and Pt(IV) (*17*). It is an elegant proof but unfortunately not quite valid. In order to attain the appropriate coordination number for platinum, chloride bridges were postulated. However, a chloride bridge might very well serve as a path for antiferromagnetic coupling so the formulation as a Pt(III) complex cannot be ruled out. The X-ray crystal structure of Wolffram's red salt was solved in 1961(Figure 5) (*18*). It is indeed a mixed valence compound. This structural type has received much attention recently in connection with high temperature superconductors. Jensen seemed to have realized that X-ray diffraction is the obvious technique to use if one wants to solve structural problems in coordination chemistry.

Through a contact with Odd Hassel in Oslo he got access to a powder diffraction camera, and in 1936 he tried to solve the structure of another series of compounds, which apparently had a unusual oxidation states, namely the M$_2$Sb X$_6$ salts (*19*). He found a cubic structure with identical anions. Based on this result he proposed a genuine oxidation state of four for the antimony. He realized that this result might be wrong because the compounds were diamagnetic. The contradiction was resolved when Axel Tovborg Jensen and Svend Erik Rasmussen showed in 1955 that the double salts are in fact tetragonal. Some of the lines that Jensen quoted as singlets are in reality doublets (*20*). Later works by Peter Day have finally established these materials as Sb(III), Sb(IV) mixed valence compounds (*38*).

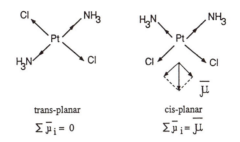

trans-planar

$\Sigma \bar{\mu}_i = 0$

cis-planar

$\Sigma \bar{\mu}_i = \bar{\mu}$

Figure 3. Addition of the bond moments for the two isomers of [Pt $Cl_2(NH_3)_2$].
The directions of the bond moments are represented by arrows pointing from +
to -. The lengths of the arrows are proportional to the magnitude of the bond
moments.

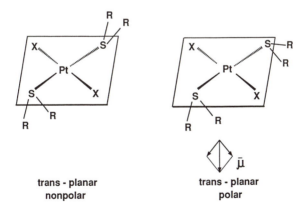

trans - planar
nonpolar

trans - planar
polar

Figure 4. Addition of the group moments of the thioether functions in two
different conformers of *trans*- [Pt $X_2(SR_2)_2$].

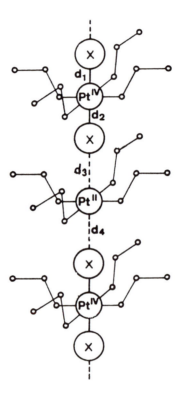

Figure 5. The X-ray crystal structure of Wolffram's red salt published by Craven
and Hall in 1961 (Reproduced with permission from ref. 18. Copyright 1961
International Union of Crystallography).

Another problem that Jensen wanted to solve by X-ray diffraction was the structures of the erythro and rhodo salts of chromium(III) (*21-22*). This turned out to be a very difficult problem. Because of some minor analytical inaccuracies these compounds were considered to be isomers. Werner suggested a "Valenzisomerie" (*23*). However, later it became clear that there is no fundamental difference between his "Hauptvalenzen und Nebenvalenzen" so another formulation had to be found. Jensen's suggestion that the erythro series have a μ-amido bridge was very clever. Amido bridged dinuclear complexes are known for cobalt(III) so it seems natural to suggest them for chromium(III) as well. Jensen showed that the powder diffraction diagrams for the two series were identical, which could be explained by a similar stereochemistry for NH_2 and OH. Much later the correct structures of the erythro and rhodo salts were shown to be aqua-μ-hydroxo and μ-oxo, respectively (*24*). The two series are therefore not isomers, but differ in their O/N ratio. Recent activity among Danish coordination chemists in the area of dinuclear chromium(III) complexes was thus probably initiated by Jensen.

Work inspired by Zeise

In Denmark there is a rich tradition for sulfur chemistry dating back to the early studies by William Christoffer Zeise (*25*). In 1832 Zeise discovered mercaptans and realized that the sulfur in these compounds "fills the place of oxygen in alcohol"(*26*). Except for the Ni(II) complex of the ligand ethylthiolate, prepared by P. Claesson in 1877, the coordination chemistry of this important class of compounds has been developed only in recent years. Jensen attacked this difficult topic very early. In 1934
he prepared an insoluble brown inner complex by the reaction of Ni(II) with 2-benzthiazolthiolate (*27*) . Later he prepared an extensive series of nickel-thiolato complexes. Based on the stöchiometry and magnetic susceptibility measurements, he concluded that these compounds consist of square planar low-spin $Ni(II)S_4$ units linked together in long chains with the sulfur atoms serving as bridges between the Ni(II) centers (*28*).

In a few cases it has been possible to isolate and crystallize nickel(II) thiolate complexes as oligomeric chains or rings. X-ray diffraction studies of these systems confirm the structure predicted by Jensen (Figure 6)(*29*) .

In 1829 Berzelius had found that a yellow salt is formed if an aqueous KCl solution is added to the residue from an evaporated solution of Na_2PtCl_6 in ethanol. In the years 1829 to 1830 Zeise made a series of careful studies leading to the discovery of the so-called Zeise's salt: $K[Pt\ Cl_3(C_2H_4)],H_2O$. He also analyzed the compound formed before the addition of the KCl : $[Pt\ Cl_2(C_2H_4)]_2$ called "Chloridum platinæ inflammabile" because of its flamability on heating (*30*). In the beginning of this century Biilmann had prepared a platinum(II) allyl alcohol complex, which was analogous to Zeise's salt (*39*). Jensen was the first to isolate platinum complexes of alkanedienes, in which one mole of the diene occupies two coordination sites. This work was not published. However, in 1949 he took up the project again and prepared the first platinum complex of cyclooctatetraene; $PtI_2(C_8H_8)$ (*31*). Again he used dipole moment measurement to prove that this compound is *cis* , and from cryoscopic data he could show that it is monomeric. As the most plausable structure he suggested one where the two double bonds in the tub form of cyclooctatetraene is bonded to platinum "with valence angle 90° perpendicular to two of the double bonds." This prediction has been confirmed for the corresponding Pd(II) compound by X-ray diffraction (*32*), but it was not so obvious at the time that it was proposed. This paper from 1953 accords with the simultaneously proposed molecular orbital structure of Zeise's salt, presented by Chatt and Duncanson at the conference in Copenhagen (Figure 7)(*1*). It is even more astonishing that Jensen actually had perceived the right structure of Zeises salt in 1937, but "pressure of other work prevented" the publication.

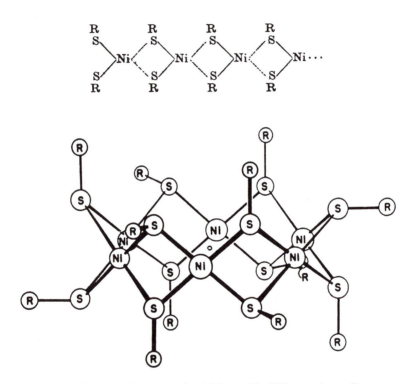

Figure 6. Suggested structure for thiolato nickel(II) complexes (Jensen, 1944) compared with a recent example of a reported structure of a cyclic hexameric thiolato-nickel(II) system (Reproduced from ref. 29).

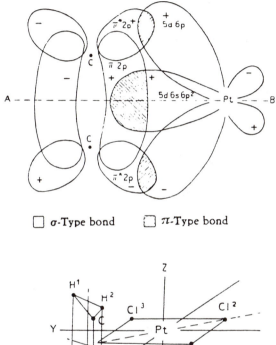

Figure 7. Two figures depicting the bonding and structure of Zeise's salt presented by Chatt and Duncanson at the Symposium on Co-ordination Chemistry, Copenhagen, 1953 (Reproduced with permission from ref. 1. Copyright 1954 The Danish Chemical Society).

Later Works

Jensen's synthesis and characterization of the first nickel(III) complexes was ahead of its time (*33*). In 1949 he discoved that [Ni $X_2(PR_3)_2$] can be oxidized to dark colored formally Ni(III) complexes [Ni $X_3(PR_3)_2$]. By the use of magnetic susceptibility measurements it was shown that they are genuine nickel(III) complexes. Measurements of the dipole moments were used to assign the structures. A value of about 2.5 D indicated a square pyramidal structure, but later a symmetric trigonal bipyramidal structure was chosen (Figure 8). This structure has been confirmed by X-ray diffraction for the corresponding Co(III) complex (*34*). The Co(III) complex has an unusual intermediate spin-state (S = 1). This has been rationalized by Christian Klixbüll Jørgensen, who also gave an assignment of the bands in the electronic spectrum (*35*).

Even though K.A. Jensen held a chair in organic chemistry from 1950, he was still interested in coordination chemistry. A long series of his papers dealt with thiosemicarbazides. At an early stage he observed that solutions of 1-monoaryl-substituted thiosemicarbazides and nickel(II) form very intensely colored solutions when they are exposed to air. In the 60's the nature of these complexes was further investigated. Jensen and his co-workers (Carl Theodor Pedersen and Klaus Bechgaard) showed that the ligands in these cases are oxidized to arylazothio-formamides (*36*). Consequently, these complexes may be formulated either as nickel(IV) complexes of thiosemicarbazides or as nickel(0) complexes of azothioformamides. However, ESR studies suggest that the most likely formulation is a nickel(II) center bonded to two radical anions. These ligands exhibit an extreme case of what Jørgensen calls "noninnocent behavior" (*37*).

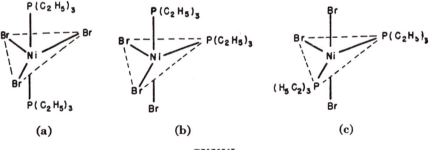

(a)   (b)   (c)

CXXVI

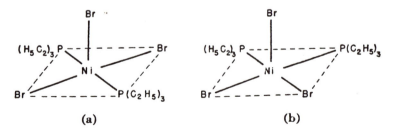

(a)   (b)

CXXVII

Figure 8. Possible structures of [NiBr$_3$(PEt$_3$)$_2$]. Jensen suggested the correct structure CXXVI(a), based on dipole moment measurements. (Reproduced with permission from reference 41. Copyright 1952 Prentice-Hall Inc.)

Jensen was an efficient transmitter of the tradition of sulfur chemistry in Denmark, and several young scientists were inspired by him. For example, Bechgaard became a leading figure in the successful search for organic metals and superconductors ("Bechgaard salts") via his deep insight into sulfur and selenium chemistry.

With the death of Kai Arne Jensen in Copenhagen on May 15, 1992, Denmark has lost one of its last polymaths, and the scientific world will miss him.

Literature Cited

1. *Proceedings of the symposium on Co-ordination Chemistry, Copenhagen, August 9-13 1953,* The Danish Chemical Society, Copenhagen, 1954.
2. Kjær, A. in *Dansk Biografisk Leksikon ;* Cedergren, S.,Ed.; Gyldendahl: København, 1981; Vol. 7; p. 302.
    Pedersen, C. T. *Dansk Kemi* **1992,** *73,* 27.
3. Jensen, K. A. *Om de Koordinativt Firegyldige Metallers Stereokemi ;* C. A. Reitzel: København, 1937.
4. *Dansk kemi* **1977,** *58,* 400.
5. Jensen, K. A. *Almen Kemi I-II ;* J. Gjellerup: København, 1957-59.
6. Jensen, K. A. *Organisk Kemi ;* J. Gjellerup: København, 1963-64.
7. Personal communication, K. A. Jensen, 1991.
8. Jensen, K. A. *Z. anorg. Chem.* **1935,** *225,* 97.
9. Werner, A. *Lehrbuch der Stereochemie ;* G. Fischer: Jena, 1904.
10. Jensen, K. A. *Z. anorg. Chem.* **1936,** *229,* 225.
11. Horn, G. W. ; Kumar, R.; Maverick, A. W.; Fronczek, F. R. ; Watkins, S. F. *Acta Cryst.* **1990,** *C46,* 135.
12. Kauffman, G. B.; Cowan, D. O. *Inorg. Synth.* **1960,** *6,* 211.
13. Cox, E. G.; Saenger, H.; Wardlaw, W. *J. Chem. Soc.* **1934,** 182.
14. Skakke, P. E.; Rasmussen, S. E. *Acta Chem. Scand.* **1970,** *24,* 2634 .
15. Jørgensen, S. M. *J. prakt. Chem.* **1886,** [2]*33,* 5236.
16. Wolffram, H. *Über äthylaminhaltige Platinbasen ,* Dissertation; Königsberg, 1900.
17. Jensen, K. A. *Z. anorg. allg. Chem.* **1936,** *229,* 252.
18. Craven, B. M.; Hall, D. *Acta Cryst.* **1961,** *14,* 475.
19. Jensen, K. A. *Z. anorg. allg. Chem.* **1937,** *232,* 193.
20. Tovborg Jensen, A.; Rasmussen, S. E. *Acta Chem. Scand.* **1955,** *9,* 304.
21. Jørgensen, S. M. *J. prakt. Chem.* **1882,** [2] *25,* 398.
22. Jensen, K. A. *Z. anorg. allg. Chem.* **1937.** *232,* 257.
23. Werner, A. *Neuere Anschauungen auf dem Gebiete der anorganischen Chemie* 2nd Ed.; F. Vieweg: Braunschweig, 1909; p. 290.
24. Wilmarth, W. K.; Graf, H.; Gustin, S. T. *J. Am. Chem. Soc.* **1956,** *78,* 2683.
25. Jensen, K. A. *Centaurus* **1989,** *34,* 324.
26. Zeise, W. C. *Pogg. Ann. Phys. Chem.* **1834,** *31,* 369.
27. Jensen, K. A. ; Rancke-Madsen, E. *Z. anorg. allg. Chem.* **1934,** *219,* 243.
28. Jensen, K. A. *Z. anorg. allg. Chem.* **1944,** *252,* 227.
29. Woodward, P.; Dahl, L. F.; Abel, E. W.; Crosse, B. C. *J. Am. Chem. Soc.* **1965,** *87,* 5251.
30. *Classics in Coordination Chemistry, Part 2: Selected Papers (1798-1899),* Kauffman, G. B., Ed.; Dover, New York , NY, 1976, and references therein.

31.   (a) Jensen, K. A. *Acta Chem. Scand.* **1953,** *7,* 866; (b) Jensen, K. A. *Acta Chem. Scand.* **1953,** *7,* 868.
32.   Goebel, C. V. *Diss. Abs.* **1967,** *B28,* 625.
33.   (a) Jensen, K. A.; Nygaard, B. *Acta Chem. Scand.* **1949,** *3,* 474 ; (b) Jensen, K. A.; Nygaard, B.; Pedersen, C. T. *Acta Chem. Scand.* **1963,** *17,* 1126.
34.   Enckevort, W. J. P. van; Hendriks, H. M.; Beurskens, P. T. *Cryst. Struct. Comm.* **1977,** *6,* 531.
35.   Jensen, K. A.; Jørgensen, C. K. *Acta Chem. Scand.* **1965,** *19,* 451.
36.   Jensen, K. A.; Bechgaard, K. ; Pedersen, C. T. *Acta Chem. Scand.* **1972,** *26,* 2913.
37.   Jørgensen, C. K. *Oxidation Numbers and Oxidation States* ; J. Springer Verlag: Berlin, 1969.
38.   Robin, M. B.; Day, P. *Adv. Inorg. Chem. Radiochem.* **1967,** *10,* 247.
39.   Biilmann, E. C. S. *Ber. deutsch. chem. Ges.* **1900,** *33,* 2196.
40.   *Berlingske Tidende,* Det Berlingske Hus, 1969.
41.   Martell, A. E.; Calvin, M. *Chemistry of the Metal Chelate Compounds;* Prentice-Hall, Inc.: Englewood Cliffs, NJ, 1952; p 321.

RECEIVED August 16, 1994

Jannik Bjerrum (1909–1992).

Chapter 8

Jannik Bjerrum (1909–1992)

His Early Years

Claus E. Schäffer

Department of Chemistry, H.C. Ørsted Institute, University
of Copenhagen, Universitetsparken 5, DK–2100 Copenhagen Ø, Denmark

Jannik Bjerrum made several contributions of wide scope in chemistry. The determination of stepwise complex formation in a constant ionic medium through measurement of the concentration of uncomplexed ligand made him world renowned. His book, *Metal Ammine Formation in Aqueous Solution,* is a classic. At first, he studied ammonia complexes of metal ions by determining the ammonia vapor pressure. Later, his introduction of the glass electrode for determining the concentrations of free Brønsted-base ligands was revolutionary. This technique became a standard one in inorganic and analytical chemistry as well as in biochemistry. The second contribution of wide scope that Bjerrum made was his discovery of the fact that complexation reactions that had been considered instantaneous could be slowed down in methanol at 200 K. However, here, the development of flow and relaxation techniques overtook him. This paper is mainly concerned with Bjerrum's life and with his pioneering work on the ammine complexes of copper(II), copper(I), and cobalt(III), but an account of the background and early history of this research and its legacy to general chemistry is also given.

There will be two contributions at this symposium to commemorate Jannik Bjerrum and his work and to try to point at certain historical perspectives that are closely connected with his activity through his lifetime. The present contribution is mainly concerned with Jannik himself, with his early papers on the copper-ammine complexes (*1-3*), and with his famous monograph, *Metal Ammine Formation in Aqueous Solution (4)*, while the other contribution, written by Christian Klixbüll Jørgensen, will cover the time from when he, Carl Johan Ballhausen, and a number of others including me were simultaneously associated with Jannik Bjerrum as students or postdocs. JB was at that time simultaneously a professor at the University of Copenhagen and Head of Chemistry Department A (Inorganic Chemistry) of the Technical University of Denmark. This was also the time when we became internationalized, first by having Geoffrey Wilkinson and his young postdoc Al Cotton as visitors for one year (1954) and then Arthur Adamson, Fred Basolo, and David Hume for the next year. Later many foreign scientists visited and worked in JB's laboratory: among them were

0097–6156/94/0565–0097$08.00/0

Mihály Beck, who also is a contributor to this symposium, Gilbert Haight, Clifford Hawkins, and Alan Sargeson in the sixties, and Norman Greenwood, Jack Halpern, Henry Taube, and Hideo Yamatera in the seventies. Jørgensen was employed at the laboratory until he left Denmark in 1959, and the year after that I moved with JB from the Technical University to the University of Copenhagen, whose chemistry, mathematics, and classical physics departments were obtaining new premises, named the H. C. Ørsted Institute. I was together with Jannik for more than forty years, until he stopped doing bench work in the laboratory at the age of eighty.

Jannik Bjerrum will be commemorated also in an "interview," which is planned to appear in *Coordination Chemistry Reviews*. Jørgensen and I shall therefore view Jannik and his work with our own eyes and, hopefully, with the eyes of many others who either knew him personally or had encountered his work.

Jannik Bjerrum's Life

As the eldest of four children and the only son, Jannik was perhaps a little spoiled. He had a lovable mother, who took good care of him and of the rest of the family. It was an academic family. His grandfather, also named Jannik Bjerrum, is still remembered in ophthalmology today, and his father, Niels Bjerrum, was the most versatile physical chemist of his generation (5) and simultaneously one of the founders of chemical physics and one of the pioneers in coordination chemistry after Werner (6). His father also had highly qualified academic friends, who came into their home or with whom Jannik went on boat trips in the summer (6). Thus Jannik's circumstances were in many respects ideal, particularly for a career in chemistry. However, Jannik's father was a person of eminence even beyond chemistry. The Danish community entrusted many important tasks to him, which required most of his attention. Moreover, his father was a strong personality who influenced the people and the society around him at a time when this was considered right and proper for a person of his stature. This situation did not always make life easy for Jannik. It was my impression from being together with the two men that the gentle-natured Jannik, even in his forties, did not feel completely at ease in his father's presence. It is never easy to have an exceptional father, and it was not only persons outside the family, who admired Niels, but Jannik's mother Ellen as well as Jannik himself also admired him very much. I almost feel guilty even now when referring to his father by his first name because he was in those days what is called in German *eine Respektsperson* (a person held in respect), and *Respektspersonen* were treated more formally at that time.

Jannik was also special, a quite different kind of character though; some persons would call him original. His world consisted exclusively of those things in which he was interested. His family was a large part of that; he had not only his wife, Grethe, but also seven children and 22 grandchildren, with 2 more on the way when he died. It is characteristic of the entire family that they have strong family ties, which never become bonds of slavery. They remain individualists, and more so than most other people. As far as Jannik was concerned, I think it is fair to say that science was given the highest priority throughout his life. Science was dear to him in an unusual way. It was my impression that he considered science sublime and most important of all for human life and society. His wife and children were also specially dear to him, and he gave them much attention in his special way. He liked children and young people, and while his approach to the former was almost always educational, he treated young people as his peers. In his later years his family meant more and more to him, and he enjoyed teaching his grandchildren more than he had ever enjoyed this kind of activity at the university.

As to Jannik's own interests outside of chemistry, I think history and botany had the highest rank. His relationship to the world of craftsmanship and even to its practitioners was insubstantial. When he was Pro-Vice-Chancellor of the University and attending one of those official dinners, his wife was once sitting with the Minister of Education of the Danish Worker's Party, the Social Democrats, and when this man learned that all the Bjerrum children were expected to have an academic education, he could not help expressing his opinion that this was unsound. However, for the Bjerrum family, it was a matter of course.

At the personal level Jannik was also different. Almost everyone would agree that he was charming. He liked to talk, and he was always pleasant in what he said and entertaining to listen to. He had such amusingly strong and rational opinions, often about issues that, at least for some of us, lay beyond reason. Also, he was unconventional in the subjects and the details that he would put forward at a lunch table. He had a characteristic, penetrating voice that he used loudly. So if you could not hear him in the lab, he was either not there at all, or he was sitting alone in his office.

He was spontaneous in what he said, but only rarely in what he did. Actually, at any stage of his life, he planned his future very carefully, and he always asked for advice when he had difficult decisions to make. He was highly organized and always finished necessary chores far ahead of time. People always figured that he was absent-minded. There was some truth to it. At any rate, it took me several years to learn to address him in such a fashion that *my* subject fully penetrated his attention. This was a useful ability to have acquired because he was omnipotent in those days in his role as a professor. However, though he liked his position, he had no wish to be distracted and absolutely no craving for exerting power, and he therefore left his responsibilities to his co-workers to a large extent and rarely interfered. When he did so and others resisted, he never bore a grudge. He did not affect conflicts, and a controversy rarely influenced his friendly relationships for more than a day. Few people disliked him. His spontaneity gave him an aura of innocence that was irresistible. Perhaps this was also the explanation for the fact that he was talked about as Jannik at a time when first names had hardly entered the Danish scene and when the use of titles had not left it at all. In spite of the Jannik, he was highly respected for his scientific importance, for his family background, and − at the time − merely because he was the professor. However, he never used his status to influence people at the personal level, and hardly at any level.

I remember him with most affection from the early years, that is, in the fifties when he was still a newly appointed professor. He liked to associate with all the young people. He had a natural gift for being encouraging, for letting one feel important. He saw to it that we were together with him on every international occasion, be it at excursions on Zealand with foreign visitors, drinking beer after a meeting, or at social events in connection with conferences abroad. He was always with us and introduced us to every foreign colleague from the literature as if this was an important occasion for the foreigner rather than for us. It made one feel fine, and, at the same time, it made one understand how international science was and had to be. Personally, I had the privilege of traveling with him many times. He needed a companion, and particularly under such circumstances when he was detached from daily duties, he was a good friend to whom you could get quite close. It could be a little tiring because he wanted things to happen all the time and he was able to eat constantly without getting fat. He had an ulcer, which he maltreated over more than a decade, and everyone who knew him also knew his ulcer.

All through his younger years Jannik had been given permission by his superiors to work independently, concentrating on his own projects (see the interview referred to above), and he repaid this good fortune with interest and interest on interest to his younger associates. At that time in Denmark there was hardly any education in research, and at least no formal one. Therefore it is not easy to set up an overall balance of the pluses and minuses of his policy of allowing his students a great degree of freedom, but it did create independent creativity, and I know of some who are deeply indebted to him for it.

Jannik also had good fortune in other respects. It is historically interesting that he almost had the study of stepwise complexation to himself for a whole decade, while the technique was still too laborious for everyone to acquire possession of it. Moreover, he came from one stronghold of pH understanding, Copenhagen with S. P. L. Sørensen, J. N. Brønsted, and N. Bjerrum, to the other, the Rockefeller Institute in New York with Leonor Michaelis (7), where he learned about Mac Innes' glass electrode (8), which he thereby was able to use before such electrodes became commercially available and before the time of electronic pH meters. It was when the Wheatstone bridge, precision potentiometers, and mirror galvanometers had a difficult time against the 10^7 ohm resistance of the homemade tubular glass electrode. With this equipment and in an environment of high humidity, Jannik developed the experimental basis for the book that became the bible of the subject. This book, which served as a Danish doctoral thesis at a time when almost every notable Danish scientist was an autodidact, had the special fate of appearing in a second edition (4) and becoming *This Week's Citation Classic* (9) more than 40 years after its first appearance. Moreover, the book has a special, historical attribute. Published in German-occupied Denmark, it was sent via Sweden to the United States in 1941, when it was allotted 13 columns in *Chemical Abstracts*.

Jannik ended his life a happy man who felt that he had accomplished what he wanted in his scientific as well as in his private life.

Inorganic Chemistry before 1930

In inorganic chemistry the classical methods of synthesis and analysis dominated the subject in the nineteenth century. However, Sophus Mads Jørgensen and several others with him had added microscopic investigations of crystals. This kind of investigation became a new, physical tool, which proved useful not only in identifying solids but also for relating complex ions of different central atoms to each other by their occurence in isomorphic crystals. An amusing example is $[RhCl(NH_3)_5]Cl_2$, whose structure, of course, was unknown at that time. Jørgensen named it purpureo rhodium chloride in spite of the fact that its color is yellow. It had already been realized beyond doubt, partly from considerations of isomorphism, that red $[CrCl(NH_3)_5]Cl_2$ had the *same* structure as purple $[CoCl(NH_3)_5]Cl_2$, and to express this result in a generalizing wording, Jørgensen simply named the chromium compound, purpureo chromium chloride. Here the color difference was minor, and I am sure everybody found Jørgensen's choice an obvious and useful nomenclature. However, Jørgensen went on, and when he had found that $[RhCl(NH_3)_5]Cl_2$ also belonged to this class, by analogy he named the yellow rhodium compound, purpureo rhodium chloride. It would seem to us today to have required a good deal of guts to name a yellow compound, purpureo. However, purpureo was the nomenclature at the time.

Around the turn of the century, electrochemical cells became another of those physical tools that crept into inorganic chemistry.

Werner's factual background for his three-dimensional coordination idea consisted to a large extent of Jørgensen's results from studies of inert complexes. With the eyes of today's chemist, an obvious consequence of Werner's coordination theory would be that complexation of an aqua ion almost had to take place in steps. Looking at the literature of this century's first 30 years teaches one otherwise. The problem was first addressed in detail by Niels Bjerrum, who studied the chromium(III) chloride system (10). He put forward the hypothesis − in the Werner language that has also become the language of today − that only the inner coordination sphere was important for the color of a complex. Then, after this hypothesis had contributed to allowing him to break with Arrhenius by suggesting that strong electrolytes were completely dissociated, he rationalized the whole complexation behavior of the Cr(III)-Cl system, including the isolation of the new ion, $[CrCl(OH_2)_5]^{2+}$, in salts. Later, again directed by his profound knowledge of physical chemistry, Niels Bjerrum studied the chromium(III)-thiocyanato system from the hexaaqua cation to the hexathiocyanato anion, thermodynamically as well as kinetically, isolated several of the intermediate complexes, and determined the stepwise complexity constants (*11*). He also presented a graph of the average number, $\bar{n}$, of ligands bound to the metal ion, and a series of graphs, the Bjerrum S-shaped curves, containing information about the degree of formation, α_k, of the individual complexes (containing k ligands) as a function of the concentration of free thiocyanate ion. This concentration was represented by its logarithm, which was used as the abscissa (semilogarithmic plot, *cf.* Figure 1c). This kind of mathematization was quite novel and could have been a major breakthrough, had the chemists of the time been more alert. However, Niels Bjerrum did not work his results out in enough detail (*11,12*) to make them appreciated.

The chemical basis for Niels Bjerrum's analysis had been obtained mainly by using classical methods in performing the chemistry itself, and it was difficult to think of other systems that could be studied in a similar way. The chromium(III) complexes were *labile enough* to be equilibrated, yet *inert enough* to be separated and isolated. Therefore there were several reasons why Niels Bjerrum's wonderful results from 1915 became an island in chemistry, isolated for some time. Regarding labile complexes, hardly anything happened before Jannik Bjerrum began his work. A few gross complexity constants had been determined early in the century with Bodländer (*13*) as the leading name, but otherwise quantitative knowledge was scarce in 1930.

Jannik Bjerrum's Early Work: The Copper(II)- and Copper(I)-Ammine Systems

Aged 21 and still a student at the University of Copenhagen, Jannik Bjerrum was given permission to work on his own in his father's laboratory at the Royal Veterinary and Agricultural School. This led to two long papers that were published in the series from the Royal Danish Academy of Sciences and Letters (*1,2*). Another paper in the same series, also as far as chemistry is concerned (*3*), was the result of JB being a visitor at J.N.Brønsted's laboratory at the University and at J.A.Christiansen's laboratory at the Technical University. All three papers were written in German and were never quoted much. Their subject was not fashionable when they appeared, and when it became fashionable more than 10 years later, their more general aspects had already been included in the monograph (*4*). In fact, these three papers form another island in chemistry, isolated in time. There was little activity in the area of stepwise complex formation until the monograph (*4*) had become known. In the next three subsections, I shall discuss the first three papers: "Investigations of Copper-Ammine Complexes", one after the other.

I. Determination of the Complexity Constants of the Ammine Cupric Ions by Measurements of the Vapor Pressure of Ammonia and by Determination of the Solubility of Basic Cupric Nitrate (Gerhardtite)(*1*). JB chose to work with copper nitrate because nitrate did not seem to complex with copper. He observed that a basic salt, Gerhardtite, $Cu(NO_3)_2 \cdot 3Cu(OH)_2$, precipitated when ammonia was added to copper nitrate. I think that at first he considered this a nuisance, but it became his good fortune for several reasons. He had the idea of adding ammonium nitrate, which by making the solution more acidic solved his immediate problem of preventing the basic salt from precipitating. The ammonium nitrate provided additional advantages. It favored ammonia in its competition with hydroxide ion so much that for all practical purposes only ammonia complexes were formed. Finally, since JB had chosen, in order to have specified conditions, to use 2 M ammonium nitrate throughout, he simultaneously obtained a *medium* in which the *concentration mass action law* was likely to hold in spite of the fact that the equilibrium expressions contained ions. Actually, this idea of using ammonium salts in connection with the study of ammonia complexation became the standard procedure in subsequent work, and, much more importantly, it focused attention upon the general idea of using a medium of high salt concentration, which I believe goes back to Bodländer and which had been elaborated by Brønsted. It now became general practice. Therefore, from then on, complexation in aqueous solution was no longer studied in water or in water with a known ionic strength (below 0.1 M) but in a medium consisting of a high concentration of some salt chosen so that its cations and anions were likely to have *no mass action effect* on the equilibria studied, that is, they were unable to form complexes with any of the species present in the complex equilibria.

JB now addressed quantitatively the famous copper-ammonia complexation reaction whose history goes back to Andreas Libavius' observations in the 16th century. In Table I JB's numerical results, valid for the medium, 2 M ammonium nitrate, and at 291 K, are given. Below the array of numbers, the general step constant, K_n, has been defined and denoted *(14)* in the way that has become international, partly through the influence of his book *(4)* and partly through that of *Stability Constants,* whose first edition *(15)* arose out of JB's private compilation. Table I, containing JB's first set of data, will be used as a running example through the rest of this paper.

Table I. Data for the Copper(II)-Ammine System

I	II	III	IV	V	VI	VII	VIII
n	$10^{-3}K_n$	$\log(K_n/M^{-1})$	K_n/K_{n+1}	K_n'/K_{n+1}'	$T_{n,n+1}$	$S_{n,n+1}$	$L_{n,n+1}$
1	20.5	4.31					
			4.39	8/3	0.64	0.43	0.22
2	4.67	3.67					
			4.25	9/4	0.63	0.35	0.28
3	1.098	3.04					
			5.46	8/3	0.74	0.43	0.31
4	0.2012	2.30					
5	0.000345	−0.46					

$$K_n = \frac{[ML_n]}{[ML_{n-1}][L]}$$

Column II contains the results for the step constants that came from JB's analysis of data.

The major qualitative conclusion was that complexation takes place in consecutive steps. At first the aqua ion takes up one ligand, then the next one, and so on. It is a stepwise building-up process from the point of view of the ligand, but it is still today not known if uncomplexed copper(II) in aqueous solution is a tetraaqua, a pentaaqua, or a hexaaqua ion. In fact, it is even doubtful that a clear distinction can be made. Similar remarks apply to the number of water molecules in the complexes. Thermodynamically, this is a non-issue because the mass action of water (water activity) is effectively constant in 2 M ammonium nitrate. At any rate, the experiments clearly revealed the stepwise character of the formation of the tetraammine-copper(II) complex. Moreover, it was now obvious that this system could no longer be described fully by using only one complexity constant but that it required one constant for each of its steps. The quantitatively new result was the numerical values for the set of step constants.

The copper(II)-ammine system was the first labile metal-ligand system that had been studied in this detail and deprived of its secret of being stepwise in character. This was known to be the case also for the inert chromium(III)-thiocyanato system

(*11,12*), and it was therefore likely that it was a general property of this kind of system. As we all know today, this turned out to be the case.

The results of Table I allowed a deeper analysis. Of course, the larger the step constants, the more dramatic the underlying chemistry. Therefore the most noteworthy of column II is the fact that the first four step constants are pronouncedly larger than the fifth one. JB expressed this by introducing a new terminology. He said that the characteristic coordination number N and maximum coordination number N' for copper(II) are equal to 4 and 5, respectively.

Because of its special thermodynamic stability, the tetraammine system will be discussed separately. Instead of using the most direct and conventional thermodynamic modeling in terms of its four step constants (column II), it can be alternatively modeled in terms of the geometric average of these constants, K_{av}, combined with three ratios of step constants, for instance those of consecutive constants (column IV). K_{av} is simply related to the gross constant β_4 as follows:

$$K_{av}^{\,4} = K_1 K_2 K_3 K_4 = \beta_4 = 21.0 \cdot 10^{12} \ M^{-4} \tag{1}$$

$$K_{av} = 2.14 \cdot 10^3 \ M^{-1} \tag{2}$$

Equation 1 shows that describing the tetraammine system solely by K_{av} is equivalent to describing it solely by β_4. Thus one might say that all JB's fundamentally new results about the tetraammine system are concentrated in the step constant ratios, which contain information about the spreading of the five species from the aqua ion to the tetraammine over the concentration scale of uncomplexed ammonia. We shall be throwing much more light on this issue later in this paper. Here we limit ourselves to the statement that the value for K_{av} of equation 2 implies that the system is formed in the neighborhood of $[NH_3] = K_{av}^{-1} = 4.7 \cdot 10^{-4} \ M$ (*cf.* Figure 1a). Moreover, including three independent step constant ratios means that *all* the available information about the spreading mentioned is included.

JB considered the logarithm of two consecutive step constants, $\log(K_n/K_{n+1})$ (column III), which he named (*4*) "total effects", $T_{n,n+1}$ (column VI). This leads to a discussion of affinities. The major quantity here is the average standard affinity, $RT\ln K_{av}$, for taking up one mole of ammonia. Since $RT\ln 10 = 5.574 \ kJ \cdot mol^{-1}$ at 291 K, one obtains $RT\ln K_{av} = 18.6 \ kJ \cdot mol^{-1}$. This free-energy quantity, which is the arithmetic average of the standard affinities of the four step reactions (*cf.* equation 1), is the *primary* piece of chemical information about the formation of $Cu(NH_3)_4^{2+}$ from the aqua ion. Its magnitude is of the order of magnitude of the energy of a hydrogen bond. When it is small compared with energies of normal chemical bonds, one has to remember not only that it is not an enthalpy but a standard free-energy quantity but also that it refers to some sort of a *substitution* of ammonia for water in the inner coordination sphere. In this sense all the step constants are *difference* quantities when referred to the microscopic situation.

It was mentioned above that K_{av}, together with the three K_n/K_{n+1} ratios, form a complete set of equilibrium-constant observables for the tetraammine system. Likewise, $\log(K_{av}/M^{-1})$, together with the three "total effects", $T_{n,n+1}$, form a complete set of standard affinity observables. If the three "total effects" of column VI are multiplied by $RT\ln 10$, the resulting standard affinities are 3.6, 3.5, and 4.1 $kJ \cdot mol^{-1}$, respectively. It is notable that these three quantities are close together, with an average of 3.7 $kJ \cdot mol^{-1}$. This average quantity might be taken as the *secondary* piece of factual information about the tetraammine-copper(II) system, since it is rather clear that this quantity, together with $\log(K_{av}/M^{-1})$, will reproduce the data quite well. However, this is not the customary way of analyzing the data. Each of the three quantities that were averaged is the differential standard affinity of two consecutive step reactions, or, equivalently, the standard affinity of the comproportionation (reverse disproportionation) reaction producing the intermediate complex,

$Cu(OH_2)_{4-n}(NH_3)_n^{2+}$ ($n = 1, 2$, and 3). Again referring to Table I, the first two of the consecutive step reactions have been used to illustrate this in equations 3, 4, and 5:

$$Cu^{2+} + NH_3 = Cu(NH_3)^{2+} \qquad\qquad K_1 \qquad\qquad (3)$$

$$Cu(NH_3)^{2+} + NH_3 = Cu(NH_3)_2^{2+} \qquad\qquad K_2 \qquad\qquad (4)$$

$$Cu^{2+} + Cu(NH_3)_2^{2+} = 2\ Cu(NH_3)^{2+} \qquad\qquad K_1/K_2 \qquad\qquad (5)$$

JB's "total effects" were written as a sum of two terms, called the "*statistical effect*" (column VII) and the "*ligand* effect" (column VIII), where the latter one in general could be split into an "*electrostatic* effect" and a "*residual* effect" (called the "*rest* effect"). For the step constants of consecutive steps, this may be written:

$$\log(\frac{K_i}{K_{i+1}}) = T_{i,i+1} = S_{i,i+1} + L_{i,i+1} = S_{i,i+1} + E_{i,i+1} + R_{i,i+1} =$$
$$\log(\frac{K_i'}{K_{i+1}'}) + E_{i,i+1} + R_{i,i+1}; \qquad (6)$$

where the primed constants are the statistical step constants. They are purely theoretical quantities, calculated on two assumptions:

1) the N coordination positions are equally probable, that is, they remain statistically equivalent during the stepwise complex formation.

2) The average step constant for these equivalent steps is that known from experiment.

Thus, in this statistical model, it is assumed that the free-energy discrimination between the five species of the tetraammine system does not arise from an energetic difference between the binding of the two kinds of ligand (here ammonia and water) but rather from the standard entropy that depends on the number of ways, $\binom{N}{i}$, the individual species, $Cu(OH_2)_{N-i}(NH_3)_i^{2+}$, can be built up from Cu^{2+}, OH_2, and NH_3. According to this model that provides the four statistical constants, K_i' ($i = 1,2,3$, and 4), corresponding to the five generalized gross constants β_i' ($i = 0,1,2,3$, and 4), the following expression is perhaps the simplest one:

$$\beta_i' = \binom{N}{i} \cdot (K_{av})^i \qquad (i = 0,1,...N) \qquad (7)$$

implying (*cf.* equation 1)

$$K_i' = K_{av} \cdot (N-i+1)/i \qquad (i = 1,2,...N) \qquad (8)$$

We shall call the coefficient to K_{av} in equation 8 *the statistical factor*. From this equation the following symmetry relationship is derived

$$K_{av}^2 = K_i' \cdot K_{N-i+1}' \qquad (i = 1,2,...N) \qquad (9)$$

The statistical model is an oversimplified model for part of the inner-sphere standard entropy, but it is parameter-free as far as the ratios between the statistical constants, K_i'/K_j', are concerned. Thus the numbers in columns V and VII arise from this purely mathematical model. The numerical values of K_i' are for the tetraammine-copper(II) system K_i'($i = 1,2,3$, and 4) = 8560, 3210, 1427, and 535 M^{-1}, respectively, whose exact ratios are to be found in column V. This simple *one-parameter model* turned out to account for most of the deviations of the experimental step constants from their geometric average, K_{av}, which is the only parameter of the model.

The electrostatic effects, $E_{i,j}$, referring to complexes with i and j ligands, are per definition absent here because the ligands are uncharged. Then the "ligand effect" and the "residual effect" make up the same concept. This has been expressed in equation 10:

$$E_{i,i+1} = 0; \quad L_{i,i+1} = R_{i,i+1} = \log(\frac{K_i}{K_{i+1}}) - \log(\frac{K_i'}{K_{i+1}'}) \tag{10}$$

However, the story is not quite finished at this point. It is also possible to combine the statistical model with the entire set of data. At first this will be done by direct reference to Table I. The "total effects" (column VI) and their associated standard affinities for the three comproportionation reactions of the type of equation 5 were discussed above. These experimental standard affinities, $RT\ln10 \cdot T_{n,n+1}$, may be split into a statistical set $RT\ln10 \cdot S_{n,n+1}$ and a residual set $RT\ln10 \cdot R_{n,n+1} = RT\ln10 \cdot L_{n,n+1}$. The statistical set, which is derived solely from the parameter-free statistical model and as such is a set of purely mathematical quantities, has the values 2.4, 2.0, and 2.4 $kJ \cdot mol^{-1}$, while the residual set has the values 1.2, 1.5, and 1.7 $kJ \cdot mol^{-1}$ with a set average of 1.5 $kJ \cdot mol^{-1}$. This average may be taken as the *secondary* piece of real, chemical information about the stepwise formation of the tetraammine system in the data reduction by the statistical model, when K_{av} is taken as the *primary* quantity. This small quantity, 1.5 $kJ\,mol^{-1}$, is the average statistically-corrected standard affinity for the three consecutive comproportionation reactions of the type of equation 5. Because the statistical correction has been made, it is believed to represent a more *chemical* quantification of the system's preference to *unlike* ligands in the coordination sphere. This is probably why it was associated with the term "ligand effect". However, the preference might equally well be rationalized as a central ion property: ammonia is a better donor than water, and therefore the central ion already complexed with ammonia will have reduced accepter properties.

The *tertiary* piece of chemical information is that the quantities of column VIII of Table I increase. This is probably significant for the copper(II) system, but JB himself (4) expresses some doubt as to whether his results for other central ions are accurate enough for similar conclusions to be reliable. Even the tetraammine-copper(II) system is rather well described by the use of only two chemical parameters (K_{av} together with the average of $RT\ln10 \cdot T_{n,n+1}$) or, perhaps better, by adding the parameter-free statistical description by replacing $RT\ln10 \cdot T_{n,n+1}$ by $RT\ln10 \cdot L_{n,n+1}$.

If the entire set of residual affinities (column VIII) is taken together with $\log(K_{av}/M^{-1})$, this is just a parameter transformation relative to a description in terms of the four consecutive step constants. An alternative method of performing an equivalent transformation has become rather common: a statistical correction of the experimental step constants by division by the statistical factors. For the tetraammine system these factors are by equation 8 equal to 4, 3/2, 2/3, and 1/4, respectively, giving the statistically corrected constants (in M^{-1}) of 5125, 3113, 1647, and 805, whose logarithms have differences that are the "ligand effects" (column VIII) or "residual effects" (*cf.* equations 1 and 2).

It is likely that JB's fundamentally new results were in 1930 looked upon with scepticism, if not suspicion, because they had been obtained by such an intangible, mathematical analysis of data. Of course, the mathematics is the same as that involved in obtaining the acidity constants of the tetraprotonic pyrophosphoric acid, $H_4P_2O_7$, whose step constants are much more different, making the relation between experiment and theory more direct and transparent. Moreover, this system resembles the phosphoric acid system where the steps in the complexation (protonation) of the phosphate ion take place in quite distinct parts of the pH scale.

Because of the scepticism toward mathematization, which, I am sure, JB at the bottom of his heart shared with his contemporaries, he was probably pleased that he could support his results with more readily understood results from solubility measurements. These will be touched upon presently. Also in publication II (2), discussed below, additional evidence was provided for the stepwise character of complex formation.

The condition for equilibrium between the copper-ammine solution and the solid phase, $Cu(NO_3)_2 \cdot 3Cu(OH)_2$, can be formulated as the expression, $[Cu^{2+}][NH_3]^{3/2}$, being a constant. This expression was shown to be quantitatively consistent with the step complexity constants found from the analysis of the vapor pressure data over a wide range of concentrations of free ammonia.

II. The Complexity Constants of the Pentaammine Cupric Complex and the Absorption Spectra of the Ammine Cupric Ions (2). This publication, which was based upon measurements of absorption spectra using a visual comparison König-Martens spectrometer, is interesting in the present context for four reasons.

1. It shows that each step of complex formation from the aqua ion to the tetraamminecopper(II) ion is accompanied by a blue-shift of the absorption smaximum of 1200 cm^{-1}, while the formation of the pentaammine complex gives a red-shift from the tetraammine complex, also of approximately 1200 cm^{-1}. The maximum molar absorptivity similarly increases from 10 for the aqua ion with steps of 10 all the way up to the tetraammine and then a step of 30 to the pentaammine complex.

2. In the interval between $[NH_3] = 0.5$ and $10\ M$, all absorption spectra can be written as a linear combination of those of the tetraammine and the pentaammine complex showing that only these two complexes exist in this interval of concentrations of free ammonia.

3. The half-width of the experimental absorption curves is invariably larger than that of the calculated curves for the individual complexes. This is additional evidence that the mathematical analyses of vapor pressure data as well as of absorption data are realistic in supporting the existence of monoammine, diammine, triammine, tetraammine, and pentaammine (14).

4. The complexity constant for formation of the pentaammine from the tetraammine is seven thousand times smaller than the geometric average, K_{av}, of the first four step constants. This is an example of how quantitative data, by their irregularity, can sometimes provide interesting new qualitative results. In this case, of course, the qualitative interpretation is that the first four ammonia molecules end up binding to copper(II) equivalently, probably in a square arrangement, while the fifth ammonia binds quite differently, probably perpendicular to the plane of the other four.

As one often sees in the history of chemistry, observations of scientists in their early years influence their work of much later years. There is no doubt that JB's early interest in the colors of complexes and, in particular, in the relationship between the colors and the structures became the basis for his inspiration of his entire laboratory almost 25 years later (see here Christian Klixbüll Jørgensen's contribution about Jannik Bjerrum at this symposium).

III. Determination of the Complexity Constants of the Ammine Cuprous Ions by Electrochemical Measurements and of the Equilibrium between Cuprous and Cupric Ammonia Complexes in the Presence of Copper Metal (3). Every chemist knows that a solution of copper(II)-aqua ions in equilibrium with copper metal contains virtually no copper(I). The qualitative observation which is the basis for the publication that will now be discussed is that if ammonia is gradually added to this equilibrium, established in $2\ M$ ammonium nitrate, copper metal will gradually dissolve, the blue color of copper(II)-ammines will fade, and, eventually, the solution will become colorless.

By using a copper amalgam electrode, which in effect is a copper(I)- and copper(II)-aqua ion electrode, and a pure mercury electrode, which is a redox electrode, JB remeasured and confirmed the gross complexity constant β_4 and the fifth consecutive constant K_5 for the copper(II)-ammine system and provided, in addition to a number of standard redox potentials, the two step constants of the copper(I) system. These were found to be $8.6 \cdot 10^5 \, M^{-1}$ and $8.6 \cdot 10^4 \, M^{-1}$. These constants are more different than those of the statistical model because they have a ratio of 10.0 as compared with the statistical ratio of 4 (*cf.* equation 8). A data transformation of the two constants is parametrization by their geometric average of $2.72 \cdot 10^5 \, M^{-1}$ (average standard affinity per ammonia of 30.3 kJ mol^{-1} at 291 K) plus a term referring to the comproportionation constant analogous to that of equation 5. The most direct expression for this term is the "total effect", $T_{1,2} = 1.0$, or the associated standard free energy term $RT\ln 10 \cdot T_{1,2} = 5.6 \, kJ \cdot mol^{-1}$. The statistically corrected standard free energy term $RT\ln 10 \cdot L_{1,2} = RT\ln 10 \cdot R_{1,2} = 2.2 \, kJ \cdot mol^{-1}$ (*cf.* the above discussion of the tetraammine-copper(II) system, in particular, equations 5 and 6). It is seen that in the copper(I)-ammine system the statistical model is able to account for 60% of the experimental spreading of the three species on the log([NH$_3$]/*M*) scale.

Most systems that have been studied possess a spreading that is wider than the statistical one or, in other words, show a positive *residual effect*. It is therefore noteworthy that the silver(I)-ammine system (*4*), which, in regard to coordination numbers (*17*) and characterizing d^{10} configuration, is analogous to the copper(I) system, has a *negative residual effect*. This unexpectedly low occurrence of the silver(I)-monoammine complex − only 20% at the maximum − is not understood. However, in view of the very small standard free energy differences that we are concerned with, it will probably be a while before we come to understand the subtle difference between copper(I) and silver(I).

In his later years JB became interested in complexes with very small complexity constants. This is a tricky business because under the circumstances where such complexes are formed, a constant medium to make a concentration mass action law applicable cannot be maintained. Nevertheless, from an inorganic chemical point of view, the qualitative question of which species are formed under specified conditions is especially interesting, whereas the question of whether their stability can be described by good constant constants is only of secondary importance. For this kind of investigation, fingerprint methods are particularly suited, and JB returned to his beloved absorption spectra and provided strong evidence for the formation of a triammine complex in strongly ammoniacal solutions in the copper(I) (*17*) as well as in the silver(I) system (*19*).

One of Jannik Bjerrum's last papers (*20*) was coauthored by his youngest son, Morten and concerned JB's father's first system (*10*), the extremely weakly complex chromium(III)-chloride system.

The Concepts of Stepwise Complex Formation, Its Internationalized Notation, and Its Influence on Contemporary General Chemistry

In the following it is assumed that the central system takes up ligands of one particular kind, consecutively, while the complexed system remains mononuclear throughout the entire complexation process (*4*). The central system need not be a metal ion. Other examples are a Lewis acid taking up bases or a base taking up protons.

For this kind of system mass action conditions are such that all variable ratios between concentrations of all species present (*14*) depend on only one variable: the concentration, [L], of free, that is, uncomplexed ligand (*4*). The stoichiometric concentrations of metal ion, C_M, and of ligand, C_L, with which the experimentalist begins, can be expressed:

$$C_M \equiv \sum_{i=0}^{i=N} [ML_i] = \sum_{i=0}^{i=N} \beta_i [M][L]^i \tag{11}$$

$$C_L \equiv [L] + \sum_{i=0}^{i=N} i[ML_i] = [L] + \sum_{i=0}^{i=N} i\beta_i[M][L]^i \qquad (12)$$

where, by definition, $\beta_0 \equiv 1$, and where the product complexity constants, β_i, are given by

$$\beta_i = \frac{[ML_i]}{[M][L]^i} = K_1 K_2 K_3 \ldots K_i \qquad (13)$$

There is a one-to-one relationship between the concentration of uncomplexed ligand [L] and the average number of bound ligands, $\bar{n}$. This is a consequence of the mass action equations. The monotonic function, $\bar{n}([L])$, is called the formation function of the system, and for a fixed value of this function, the ratios of concentrations of all species (14) are fixed (4). The formation function has the form

$$\bar{n} \equiv \frac{C_L - [L]}{C_M} = \frac{\sum_{i=0}^{i=N} i\beta_i[L]^i}{\sum_{i=0}^{i=N} \beta_i[L]^i} \qquad (14)$$

where, after the introduction of the mass action expressions, the fraction has been reduced by the concentration, [M], of free metal ion.

Solutions of the same complex system having the same $\bar{n}$, and thereby the same [L], are called corresponding solutions (18). They obey Beer's law in terms of stoichiometric concentrations in the spectral range where only the complex species including the aqua ion contribute to the absorption.

A graph, $\bar{n}$ versus $\log([L]/M)$, has an S-shape when the step constants are close to those of the statistical model. The functional value, $\bar{n}$, then moves asymptotically either toward the characteristic or the maximum coordination number for large values of [L] (Figure 1a). The graph, $\bar{n}$ versus p[L], has, of course, the shape of the mirror image of an S.

It is the focusing upon the intermediate species that is the idea of the stepwise complexation. Because the absolute concentrations of these species depend on C_M as well as on [L], it is the fractional distribution of C_M upon the concentrations of the different ML_n species that is of special interest. These quantities are called the degrees of formation, α_n, of the individual complexes. They are given by

$$\alpha_n = \frac{[ML_n]}{C_M} = \frac{\beta_n[L]^n}{\sum_{i=0}^{i=N} \beta_i[L]^i} \qquad (15)$$

and the meaning of the sum rule

$$\sum_{i=0}^{i=N} \alpha_i = 1 \qquad (16)$$

is immediately understandable by the mole-fraction character of the alphas. By using the expression 14, another sum rule

$$\sum_{i=0}^{i=N} i\alpha_i = \bar{n} \tag{17}$$

can be derived.

Introduction of the alphas into the expression for $\bar{n}$ gave the most beautiful general expressions of the entire mathematization process. Equation 18 was derived by JB in his monograph (4), and he always talked about it as the extended Bodländer formula because Bodländer (16) had derived it for the special case of $n = 0$. This formula is

$$-\frac{d(\log \alpha_n)}{d(\log[L])} = \bar{n} - n \tag{18}$$

It should be noted that n is an integer, whereas $\bar{n}$, of course, varies continuously between zero and the maximum coordination number N' (cf. Figure 1a). If the differential quotients of equation 18 are plotted versus $\log([L]/M)$, an array of graphs is obtained, the components of which are congruent to the graph, $\bar{n}$ versus $\log([L]/M)$. The meaning of equation 18 is that if the degree of formation of any complex species is known as a function of [L], then the entire system is known.

The concepts and the notation that have now been reviewed make up the legacy of this research to general chemistry. In addition, certain graphical representations have become standard ones, and they will be presented now.

Figure 1a contains three graphs that all represent $\bar{n}$. The least steep graph is the "experimental" formation function that is calculated by using equation 14 together with the step constants of Table I, valid for the copper(II)-ammine system. This graph has two inflection points and $\bar{n} > 4$ for high values of $\log([NH_3]/M)$. It includes the information, K_5, about the weakly bound pentaamminecopper(II) complex.

The graph that is steepest at the common inflection point applies to the one-step complexation reaction. This graph is obtained by assuming that the only species present are the free copper ion and the tetraammine complex, with the β_4 value as implied by Table I. In this case one has the expressions:

$$\alpha_4 = \frac{[ML_4]}{[M] + [ML_4]} = \frac{\beta_4[L]^4}{1 + \beta_4[L]^4} \tag{19}$$

$$\bar{n} = \frac{4\beta_4[L]^4}{1 + \beta_4[L]^4} \tag{20}$$

This graph is an illustration of the consequence of the old view on complex formation that the species of highest coordination number − in the copper(II)-ammonia system, the *characteristic* coordination number, which is 4 − is formed in one step.

The graph of intermediate slope illustrates the formation of the system that has the β_4 value of the copper(II)-ammine system but a statistical distribution of all complex species, N being assumed equal to four.

In order to comprehend the contents of the three model graphs, it should be realized that only β_4 is required in order to draw the two model-theoretical graphs (the two steepest ones): in the steepest graph β_4 is the only complexity constant, and in the intermediate one, the other β values follow from β_4 by the probability considerations (cf. equation 8). The fact that the experimental graph is the least steep one means that nature has spread the complex species over a range of [L] that is larger than the one expected on statistical grounds (positive *ligand effect*). In other words, in addition to the statistical spreading, there is also a chemical one, which, in this case, may be phenomenologically characterized either as a symbiotic effect of unlike ligands or an antisymbiotic effect of like ligands. In this particular chemical system the cause of this

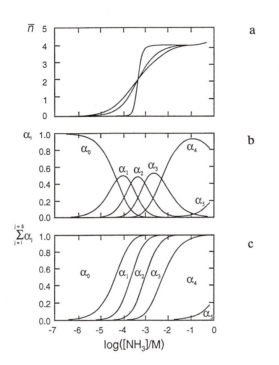

Figure 1. (a) Three versions of the graph of the formation function, $\bar{n}$, for the tetraammine/pentaamminecopper(II) system. The one-step graph ($N = 4$), the statistical graph ($N = 4$), and the experimental graph ($N = 5$) have decreasing slopes at the common inflection point. (b) Relative distribution of the six species from the copper(II)-aqua ion to the pentaamminecopper(II) ion. (c) "Experimental" delimitation graphs for the pentaamminecopper(II) system (*cf.* equation 21). Note that the five graphs of this figure add up to the experimental graph of Figure 1a.

effect is probably to be found mainly with the central ion because ammonia and water are expected to be similar in their bonding properties, whereas in a case like phosphine and water the ligands are more likely to be responsible for the deviation from statistics.

Figure 1b shows the distribution of species, calculated from K_1, K_2, K_3, K_4, and K_5 of the copper(II)-ammine system with increasing $\log([L]/M)$ (Table I). At low [L], the degree of formation of the aqua ion, α_0, begins to decrease from unity, and until high [L], α_4 is almost the mirror image of α_0. Only intermediate complexes have degrees of formation that increase, go through a maximum, and then decrease again. The crossing of graphs, $\alpha_{n-1} = \alpha_n$, occurs approximately at $\log([L]/M) = -\log(K_n/M^{-1})$. Had the distribution been statistical with $N = 4$, this set of graphs would have been symmetrical about the vertical line $\log([L]/M) = -\log(K_{av}/M^{-1})$, α_2 would have had its maximum value of 24/64 on this line, and α_1 and α_3 would have had their maxima of 27/64 at the abscissa $-\log(K_{av}/M^{-1}) - \log(3)$ and $-\log(K_{av}/M^{-1}) + \log(3)$, respectively. The experimental distribution is seen to be rather close to these values. However, the intermediate species have slightly higher values for their maxima and slightly larger distances between these, in accordance with the fact that the spreading found experimentally is wider than the statistical one. It should be noted that the inclusion of α_5 in the figure does not influence the comparison with the statistical considerations based upon $N = 4$ because K_5 is so much smaller than the other constants.

Figure 1c gives the five graphs that divide the presentation area into segments belonging to the six metal-containing species (11,12,14). From left to right these five Bjerrum S-graphs are defined by

$$\alpha_1+\alpha_2+\alpha_3+\alpha_4+\alpha_5; \quad \alpha_2+\alpha_3+\alpha_4+\alpha_5; \quad \alpha_3+\alpha_4+\alpha_5; \quad \alpha_4+\alpha_5; \quad \alpha_5;$$
$$\text{all depicted versus } \log([L]/M) \tag{21}$$

so that the sections of the ordinates delimited by the graphs for a given value of the abscissa are the degrees of formation (fractions of C_M) of the appropriate complexes (4,11,14). The inflection point of a graph occurs approximately at $\log([L]/M) = -\log(K_i/M^{-1})$, and the graphs are more widely separated than expected according to the statistical considerations. Moreover, by comparison of equations (17) and (21), it is seen that the functions represented by the five graphs of Figure 1c add up to the formation function represented by the experimental graph, $\bar{n}$ versus $\log([L]/M)$, of Figure 1a.

The Monograph's Other Contributions to Inorganic Chemistry

The stepwise complexation of a number of metal ions with ammonia and ethylenediamine was studied under conditions standardized by a constant ionic medium. In addition to the quantifications embodied in these investigations and the thorough general-chemistry review that was already discussed above, the monograph (4) also contains the following qualitative results, valid for ammine and amine complexes and of general interest to inorganic chemistry.

Copper(I), silver(I), and mercury(II) have in common that their *characteristic* coordination number is two. Only at much higher $[NH_3]$ will copper(I) and silver(I) take up one (17,19) and mercury(II) take up two extra ammonia molecules. Copper(II), zinc(II), and cadmium(II) have the *characteristic* coordination numbers of four and the maximum ones of five, possibly six, and six, respectively. Manganese(II), iron(II), cobalt(II), and nickel(II) are increasingly complexing and have the characteristic and the maximum coordination number of six. Zinc(II) forms complexes with ammonia with an affinity between those of cobalt(II) and nickel(II), while copper(II) forms complexes whose formation constants are almost two orders of magnitude greater per NH_3. Magnesium, and particularly calcium and lithium, form much weaker ammine complexes so that their coordination numbers cannot be determined. Having summarized these results, one has to add that the characteristic coordination number of a given metal ion may vary with the ligand. For example, in a series of later papers, JB found that phosphine and arsine ligands behaved quite differently from ammonia in this regard.

JB's studies of equilibria within the inert cobalt(III) systems with ammonia and ethylenediamine are of a more special nature. However, since they belong to the most original parts of his book (4), a discussion of them will not be completely omitted here. They can be divided into two categories: step complexation and gross complexation. These will be discussed in turn.

The observation that charcoal catalyzes the basic aquation of hexaammine-cobalt(III) had been noted in literature, and JB observed that this catalysis was accelerated in the presence of cobalt(II). These observations were the basis for the equilibrium studies that now will be described.

If ammonia is added to the pentaammineaquacobalt(III) ion, an "instantaneous" acid-base equilibrium is established:

$$A: \quad [Co(OH_2)(NH_3)_5]^{3+} + NH_3 \rightleftarrows [Co(OH)(NH_3)_5]^{2+} + NH_4^+ \qquad (22)$$

However, this equilibrium is only a partial one. If charcoal is added, a more complete equilibrium also involving the scheme B can be established:

$$B: \quad [Co(OH)(NH_3)_5]^{2+} + NH_4^+ = [Co(NH_3)_6]^{3+} + H_2O \qquad (23)$$

It is immediately obvious from equation 23 that this equilibrium is mass-action independent of $[NH_3]$. This is another way of remarking the general fact that a distinction between ammonia and hydroxide complexation cannot be made on the basis of pH measurements, when $[NH_4^+]$ is constant.

JB determined the equilibrium constants for the reactions A and B as expressed by $\log K_A = 2.75$ and $\log(K_B/M^{-1}) = 1.66$, valid for 2 M ammonium nitrate. For K_A, glass electrode measurements were used; for K_B, spectrophotometric measurement were used. Addition of the two reactions gives the sixth of the consecutive step complexation reactions of the cobalt(III)-ammine system, for which one accordingly has $\log(K_6/M^{-1}) = 4.41$.

The gross complexity constant of the cobalt(III)-ammine system was determined by redox electrode measurements. JB found that for a solution containing Co^{II} and Co^{III} under conditions at which the hexaamminecobalt(III) would be the completely dominant Co^{III} species at equilibrium, his platinum redox electrode had a potential that followed the expression:

$$E_{redox} = E^0_{redox} - (RT/F)\ln\alpha_6 \qquad (24)$$

where α_6 is the degree of formation of the hexaamminecobalt(II) ion. This observation shows that a redox equilibrium is established at the electrode, and from the value, $+0.035$ volt, found for E^0_{redox}, combined with the value, $+1.844$ volt, known to apply to the Co^{II}/Co^{III} aqua ion system, this result allowed $\log(\beta_6/M^{-6})$ for the cobalt(III)-ammine system to be calculated as 35.2. This is a large constant compared with those of the divalent metal ions.

The two Co^{III} constants, K_6 and β_6, agree well with each other by giving rise to an average "*ligand* effect" of $(35.2/6 - 4.41 - \log 6)/2.5 = 0.3$ (*cf.* Table I and the copper(I)-ammine discussion).

Reaction Velocities of "Instantaneous" Complexation Reactions

In 1952, that is, before the time of flow and relaxation methods, JB discovered that the reaction between iron(III) ions and thiocyanate ions, a reaction that is famous in analytical chemistry for its intensive, blood-red color, can be slowed down practically to a standstill if the reaction is performed in cold methanol. This method of making "instantaneous" reactions measurable turned out to apply quite generally not only to complexation reactions but also to certain "fast" redox reactions, for example, the reaction between copper(II) and cyanide ion. JB's observation was only followed up to a limited extent (*21*) though a number of novel results were obtained by the cooling method. There is not much doubt that this method would sooner or later have considerably influenced the subject of inorganic chemistry, had it not been because flow and relaxation methods, which were invented so soon after, gave similar information.

Conclusion

Those who knew Jannik Bjerrum will remember him clearly for the rest of their lives. He was really a character. He was friendly and entertaining and completely devoted to chemistry. As to his work, it is strange that one of his most original contributions, that of making the rates of fast complexation reactions measurable, will probably remain only a historical curiosity. He always regretted that circumstances did not allow him to follow up his pioneering results in this area. On the other hand, Jannik Bjerrum was lucky and successful with regard to his contribution to the area of stepwise complex formation, and it will be difficult to view this area in historical perspective without remembering him as the master of the subject.

It is strange how completely this area of chemistry has left the forefront of today's science and yet how it retains a well-established position, almost as one of the foundations of chemistry, methodologically as well as conceptually, far beyond its original narrower focus.

Literature Cited

1. Bjerrum, J. *Mat. Fys. Medd. Dan. Vid. Selsk.* **1931**, *11*, 1-58.
2. Bjerrum, J. *Mat. Fys. Medd. Dan. Vid. Selsk.* **1932**, *11*, 1-64.
3. Bjerrum, J. *Mat. Fys. Medd. Dan. Vid. Selsk.* **1934**, *12*, 1-67.
4. Bjerrum, J. *Metal Ammine Formation in Aqueous Solution: Theory of the Reversible Step Reactions*; P. Haase and Son: Copenhagen, 1941 (1st ed.) and 1957 (2nd ed.). A copy of this monograph is still available from the present writer. Reviewed in *Chem. Abstr.* **1941**, *35*, 6527-6534.
5. Guggenheim, E. A. *Proc. Chem. Soc.* **1960**, 104-114.
6. Kauffman, G. B. *J. Chem. Educ.* **1980**, *57*, 779-782; 863-867.
7. Michaelis, L. *Die Wasserstoffionenkonzentration. Teil I. Die theoretischen Grundlagen;* Julius Springer: Berlin, 1922.
   Michaelis, L. *Oxydations-Reduktions-Potentiale. Zweiter Teil der "Wasserstoffionenkonzentration,"* 2nd Ed.; Julius Springer: Berlin, 1933.
8. MacInnes, D. A.; Belcher, D. *Ind. Eng. Chem. Anal. Ed.* **1933**, *5*, 199-200.
9. Bjerrum, J. *Current Contents* **1982**, *32*, 20.
10. Bjerrum, N. *Z. phys. Chem.* **1907**, *59*, 336-383; 581-604; *Z. anorg. Chem.* **1909**, *63*, 140-150.
11. Bjerrum, N. *Z. anorg. Chem.* **1921**, *118*, 131-164; **1921**, *119*, 39-53; 54-68; 179-201.

12. Bjerrum, N. *Selected Papers (in English translation)*; E. Munksgaard: Copenhagen, 1949. This book contains references to earlier papers in Danish and a partial translation of refs. 8 and 9 into English.
13. Bodländer, G.; Fittig, R. *Z. phys. Chem.* **1902**, *39*, 597-612.
14. In the present context n usually refers to the number of ligands in the complex ion under consideration, regardless of whether isomers exist or not. The reason for this is that if fingerprint methods for the determination of the concentrations of individual isomers are absent, then it is always the sum of the concentrations of isomers, which can be calculated from any set of data. One may say that isomers have a common mass action effect in these cases. On the other hand, the mass action equilibria between isomers are, of course, independent of [L] as well as of C_M.
15. Bjerrum, J.; Schwarzenbach, G.; Sillén, L. G. *Stability Constants*; Special Publication No. 7; The Chemical Society: London, 1958.
16. Bodländer, G. *Festschrift für R. Dedekind*; Friedrich Vieweg und Sohn: Braunschweig, 1901.
17. Bjerrum, J. *Acta Chem. Scand., Ser. A* **1986**, *A40*, 233-235.
18. Bjerrum, J. *Mat. Fys. Medd. Dan. Vid. Selsk.* **1942**, *21*, 1-22.
19. Bjerrum, J. *Acta Chem. Scand., Ser. A* **1986**, *A40*, 392-395.
20. Bjerrum, M. J.; Bjerrum, J. *Acta Chem. Scand., Ser. A* **1990**, *44*, 358-363.
21. Bjerrum, J.; Poulsen, K. G. *Nature* **1952**, *169*, 463-464.

RECEIVED January 21, 1994

*"Jannik's Tea Table." From left to right: Flemming Woldbye,
Carl Johan Ballhausen, Jannik Bjerrum, Arthur W. Adamson,
Edmund Rancke-Madsen, Niels Hofman-Bang, Ingeborg Poulsen,
Knud Georg Poulsen, Claus Erik Schäffer, and
Christian Klixbüll Jørgensen. (Photo courtesy of Fred Basolo.)*

Chapter 9

Jannik Bjerrum's Later Life—Turning Toward Chemical Physics

Personal Recollections of a Grateful Student

Christian K. Jørgensen

Section of Chemistry, University of Geneva, 30 Quai Ansermet, CH 1211 Geneva 4, Switzerland

An interest since 1934 in absorption spectra of copper(II) and nickel(II) complexes as "fingerprint" identification of each species in solution and the awakening awareness of new quantum physical interpretations of excited states (until 1956 by the fragile hypothesis of perturbation by the tiny nonspherical portion of the Madelung potential due to the surrounding electric charges) involved Jannik Bjerrum somewhat remotely during his later years in the (rapidly modifying) frame of "ligand field" theory and the Angular Overlap Model elaborated by Claus Schäffer.After his retirement in 1979, Jannik returned to the old problem of very small formation constants and the relationships between collective properties in physical chemistry and in modern chemical physics.

Jannik Bjerrum received a thorough training (both in his home laboratory and at the University of Copenhagen) in classical chemistry,including qualitative and quantitative analysis (1). He developed new synthetic methods and worked on ammonia and ethylenediamine complexes (the favorite ligands (2,3) of Sophus M. Jørgensen) of copper(II) and copper(I) since 1931, followed by the elaboration of the glass-electrode method of studying complexes of Brønsted bases with ions of metallic elements (4,5) (the two major problems being competing formation of monomeric and/or polymeric hydroxo complexes and the slow or even irreproducible kinetics) such as colored nickel(II) and colorless zinc(II), cadmium(II), and nitrogen-bonded complexes.

There is no doubt that Jannik Bjerrum obtained much inspiration from the kinetic and equilibrium studies of his father Niels Bjerrum (1879-1958), the physical chemist (6) best known for his work on chromium(III) complexes. Jannik obtained a comparable success, studying octahedral cobalt(III) complexes of the N_6 and N_5O type (later extended to N_4O_2 but not beyond S.M.Jørgensen's type N_3O_3 prepared via [$(O_2NO)_3Co(NH_3)_3$] crystals). Jannik's great innovation of using activated charcoal as a catalyst for cobalt(III) equilibria also showed its intrinsic limitations by inducing

0097–6156/94/0565–0117$08.00/0

precipitation of black hydroxyoxides that resemble the mineral (known for its incorporation of many trace metals) heterogenite.

Jannik's concept of the formation curve is a tool for astutely analyzing the collective (average) properties of a mixture of a given M ion's complexes in solution that attains its purpose by the choice of highly variant total M and total L concentrations. Jannik also used metallic electrodes to obtain free [M] via Nernst's law, but few elements are "soft" enough, although they may work in amalgams. In 1946 he suggested the principle of corresponding solutions, which states that (exclusively monomeric) complexes have identical spectra (per mole) for the same n. Because a monochromatic mercury atomic line at 436 nm was available, he studied $CuCl_4^{2-}$, the only constituent absorbing there (in an electron transfer band), and he used his (4,5) scheme "upside-down" to deduce the relative amounts of n = 0,1,2,and 3 copper(II) chloro complexes. With modern electronic recording spectrophotometers, this idea is versatile; I used it to show that K_4 for $PdCl_4^{2-}$ is 6 $\underline{M}^{-1}$ rather than of order 100 $\underline{M}^{-1}$ as reported.

If the determination of formation constants is an ultimate goal, the optimal data would be sets of, $\underline{e.g.}$, 10 or 20 narrow signals from each species in transparent systems, with intensities (above background) proportional to concentrations and also mutually non-interfering. Raman spectra were the closest to fit this demand in 1950-60, but A.W.Langseth (1895-1961) at the University of Copenhagen pointed out that a broad background marred aqueous solutions and that even organic solvents or low-melting salts required exceptional precautions in order to be sufficiently limpid. (These difficulties have been almost completely removed by modern laser equipment).

Hence Jannik turned to absorption spectra in the visible, which he had already studied with primitive spectrophotometers (4) of copper(II) and of nickel(II) aqua-ammonia complexes. Had he been more interested in the narrow absorption bands of trivalent lanthanides (7,8) having electron configurations $4f^2$ of praseodymium(III) through $4f^{13}$ of ytterbium(III), he might have turned toward their complexes (mainly bonded to oxygen atoms in solvents or in, usually multidentate anions). This approach only gave an intended (9) demonstration of the exchange of methanol with (about 30 times more strongly bound H_2O) on $4f^3$ neodymium(III), but the spectral changes (in particular of the exceedingly narrow band moving from 427.3 nm in water in direction of 430 nm) are really due more to Nd(III) in anhydrous methanol binding chloride, $ClNd(CH_3OH)_n^{2+}$, subsequently releasing chloride by addition of small amounts of water, providing mixed complexes $(H_2O)_x Nd(CH_3OH)_y^{3+}$. Above 20 volume procent (10 molar) water, it remains a unsettled question whether vanishing y is accompanied by x = 9 (as in several crystals) or, in part, 8. Hydrochloric acid (10,11) has to be above 5 $\underline{M}$ before $ClNd(OH_2)_2^{2+}$ can be detected. Related studies (12-14) of cobalt(II) and nickel(II) nitrate and various slowly reacting chromium(III) salts in ethanol with low water content unexpectedly showed that agreement with the mass-action law (in concentration units) is the exception rather than the rule in such solutions, complicated by the much stronger affinity to nitrate, as compared with pure water. Something valuable emerged from this confusion; it showed that relying on "neutral salt media" to keep the mass-action law approximately valid (without aberrant variations of "activity coefficients") is barely conceivable for water concentrations below 10 volume percent. It was shown ([9],[10],[12]) that these anomalies are not essentially due to lower dielectric constants of the bulk solvent; the relative binding of water compared to nitrate or chloride anions is "chemical" in origin.

Jannik Bjerrum and I (having different, but complementary,distinct motivations) sought a convincing explanation for absorption bands (i.e.,colors) of complexes containing a partially filled 3d-, 4d-, or 5d-shell (the corresponding $4f^q$ problem can be reduced ([7]) to the distribution of J-levels of an isolated Ln^{3+} ion, each of their $(2J+1)$ states differing very slightly in energy because of the ligating atoms marginally perturbing ([15]-[19]) the spherical symmetry). In 1952 the prevailing rationalization of chemical bonding in d^q complexes was (rather unconvincingly) based on hybridization theory ([20]-[23]) incorporating the Lewis paradigm (from 1916) of electron-pair bonds.

One afternoon in late fall 1952 Jannik's student Carl Johan Ballhausen and I encountered a paper ([24]) by the late Friedrich Ilse and by Hermann Hartmann,who ascribed the purple color of conceptually simple $3d^1$ systems, such as $Ti(OH_2)_6^{3+}$, to the energy difference between high-lying (4 states) and the ground state (6 almost coincident states) due to the (very small) nonspherical deviation $U(x,y,z)$ from the (huge) spherical part $U_0(r)$ of the Madelung potential (having the titanium nucleus at $(0,0,0)$ and $r^2 = x^2 + y^2 + z^2$). However much such an "electrostatic crystal field" got described in 10 or 20 books (the most instructive perhaps being the one ([25]) by the late John S. Griffith), this description collapsed at the Solvay meeting on Chemistry at Brussels (May 1956) and had actually reduced itself ad absurdum by Hermann Hartmann and Hans Ludwig Schläfer's paper ([26]) in 1954 showing that the "crystal field parameter" (only one for a regular octahedral d-group MX_6 with six identical M-X distances R) is the same (within a factor of 1.1) for $3d^3$ chromium(III) and $3d^2$ vanadium(III) complexes of oxide, hydroxide, oxalate, malonate, and aqua ligands (usually slightly smaller for anion ligands than for H_2O) and invariably larger ([27]) for ammonia. A further difficulty was that noncubic perturbations in MX_5L, trans- and cis-MX_4L_2, quadratic MX_4, and linear XMX should have been much more prominent and dependent on R^{-3} in contrast to R^{-5} for MX_6 with all nuclei on Cartesian axes.

During conversations Jannik emphasized (and cautiously elaborated by textbook arguments)that formation constants K_n are related ([28]) to to $RT(ln \; K_n)$, replacing one or several aqua ligands with the n'th instance of the ligand considered. Actually, in his thesis ([4]) he discussed the far larger free energy decrease of a gaseous ion M^{z+} when immersed in water and binding N aqua ligands (including quite significant "second-sphere" stabilization). This was transgressing the very powerful taboo in pre-1950 physical chemistry that numerical differences of free energy for transferring a set of species from one phase (e.g., a vacuum, or very low pressure of helium) to another (e.g., a solvent) are dogmatically devoid of meaning, unless the set of species is electrically neutral (as are Al^{3+} and three Cl^- but not any single ion). This initiated a reevaluation of the Debye-Hückel concept of single-ion activities as a function of concentration and almost resulted in the abandonment of the Arrhenius paradigm by a few critical readers. Niels Bjerrum's last paper ([29])discussed single-ion activities, a typical Danish scruple.

A major reason for the clash of classical physical chemistry with
the concepts of chemical physics was the series limits in the 1895
Rydberg formula (30) providing precise ionization energies I_n [i.e.,
the energy difference between the ground state of M^{n+} and the ground
state of $M^{(n-1)+}$] for many monatomic gaseous ions and atoms (n = 1).
Admittedly, the huge I_n are differences of <u>energy</u> rather than of <u>free</u>
energy, but it is also true that the separation between concepts (of
major concern to physical chemistry) such as enthalpy and free energy
usually are several orders of magnitude smaller than I_n. As Planck
stated concerning adversaries of new paradigms, the persons finding
it sacrilegious to define the chemical ionization energy for the

reaction $\frac{1}{2}$ H_2 (g) + H_2O (aq) = H_3O^+ (aq) + e^- (g)

slowly left the arena. One day in 1957 Jannik forgot that on p. 79 of
his thesis (4) the hydration energy of a gaseous proton is close to
10.5 eV. A review by Rosseinsky (31) estimated the value as 11.2 eV,
and slowly, this concept became less objectionable (28,30,32), and a
IUPAC report (33) gave a constant (4.42 V compared to 4.5 V by
Rosseinsky) that one should add to a standard oxidation potential E^0
in aqueous solution (4.6 V in CH_3CN) to obtain I_{chem} = (E^0 + 4.42) eV,
easy to compare with I_n of gaseous ions (or, for that matter, with
photoelectron I_b values of gaseous molecules). I_{chem} for removal of
two electrons, <u>e.g.</u>, Tl(I)→Tl(III) are twice this expression and for
aqua ions 2(+1.25 + 4.42) = 11.34 eV, because E^0 for thallium(I) aqua
ions is +1.25 V relative to the standard hydrogen electrode.

      In Copenhagen an early interest in "crystal field" theory centered
around the stabilization of partly filled shells d^q in octahedral
complexes. Jannik, being fully aware of the dependence of observed
constants on (perhaps minute) differences of energy, wrote a short (34)
note and presented a paper (35) in Amsterdam comparing 3d-group M(II)
K_n values with a linear interpolation between Ca(II) and Zn(II) and
rationalizing their variation by the one-electron energy difference
(E_1-E_2) = $\mathbf{\Delta}$ between the antibonding (x^2-y^2) and $(3z^2-r^2)$ orbitals (at
identical energy in regular octahedral MX_6) and orbitals (xy), (xz),
and (yz) at a definite lower energy. As later elaborated (36,37) the
agreement is plausible, but a controversy with R.J.P.Williams arose
concerning the usually much higher K_n values of copper(II) complexes
compared (38,39) to nickel(II) and zinc(II) (with a few near (35)
exceptions for the ligands ethylenediaminetetraacetate and 1,10-
phenanthroline). According to Williams, the covalent bonding is much
more pronounced in the rather oxidizing Cu(II), but this comparison
is unfortunately muddled by the consequence of the Jahn-Teller effect
that (x^2-y^2) contains one strongly antibonding electron in quadratic
or distorted octahedral complexes, but the $(3z^2-r^2)$ orbital two weakly
antibonding electrons. This line of thought counts the number (<u>a</u>) of
antibonding electrons (16,40) representing a stabilization, under
equal circumstances proportional to (2q-5<u>a</u>) in MX_6. The idea remains
viable as a rationalization of differences, when comparing with q= 0,
5 (high-spin S= 5/2), and 10. It may be added that the electrostatic
version of "crystal field" stabilization (34,35) was based on a paper
(41) published in 1952 by two solid-state physicists at the Philips
laboratory in Eindhoven. However, these authors were more interested
in differential changes of ionic radii (being somewhat shorter for d^q
electron densities not being spherically symmetric, in contrast to

d^0, high-spin d^5, and d^{10}). In all of these cases, the slopes between
reference closed-shells Ca(II) and Zn(II) or Sc(III) and Ga(III) are
not readily amenable to a general hypothesis (39,42), and today,
"ligand field" effects seem to be a minor, superposed contribution.
With time the latter quantities descended to a status similar to
solvent and substituent inductive pK changes of acids.

Being more impressed by extensive calculations and applied
mathematics than I, in 1956-1960 Jannik established a modus vivendi
with the nonspherical electrostatic paradigm as the conceptual picture
of d-group energy levels. Then, a constructive alternative was slowly
established by Claus E, Schäffer (16,40,43) as the angular overlap
(A.O.M.) model of one-electron energy differences between the five d-
like orbitals (derived from ideas behind the extended Hückel treatment)
An important argument was the verified comparison (40) of the π-anti-
bonding effect (OH^-,F^-,Cl^-,Br^-,I^-) and the π-bonding effect in CN^-,
pyridine, 2,2'-bipyridine, (likely to be larger (44) for CO) with the
σ-antibonding effect of ammonia ("one-lone-pair-ligands" being called
L) in chromium(III) ML_6, ML_5X, and trans- and cis-ML_4X_2 (40,45).

In addition to the one-electron energy parameters, a fairly
complicated system of interelectronic repulsion (Slater-Condon-Shortley
parameters) introduced in 1954 by Tanabe and Sugano (46) and elaborated
by Claus E.Schäffer (27,47,48) agreed with expectations of the (not
overwhelming) extent of covalent bonding induced from the nephelauxetic
series of central atoms (in a definite oxidation state) and of ligands.
The most appealing rationalization of the nephelauxetic (cloud-
expanding effect (49) is a superposition of the moderate expansion
of the 3d (or 4d or 5d) radialfunction (due to the corresponding
central field adapting to an efficient M charge below the integer z
indicated by the oxidation state) and the more "chemical" effect of
delocalization of the antibonding molecular orbitals acquiring nodes
between the M and X nuclei (or even between (27) two adjacent X nuclei)

The work in Copenhagen begun in 1953 is a prime example of the
inductive approach to selected areas of theoretical chemistry, unlike
the more fashionable deductive method inherited from physics via
ancient geometry, Descartes, Newton, Maxwell, Einstein, and Dirac.
This does not imply that applied group theory (for which I am grateful
for the fruitful collaboration with Claus Schäffer since 1956) was not
a major component in answering the question "why are d^q complexes so
frequently colored ?".However,if,comparing atomic spectra, the closely
related question of 4f^q energy levels, and positions, band shapes, and
intensities of d^q systems in solution and (frequently syncrystallized
as in ruby) in solids (45) is not (in practice) a systematic derivation
from the nonrelativistic Schrödinger equation (30), any more than one
can argue that the weather forecast or the results of roulette in a
gambling casino are, strictly speaking, Newtonian mechanics.

Chemistry Department A of the Technical University of Denmark
(T.U.) occupied the major part of the Farimagsgade wing of the Liebig-
style laboratory building (colloquially called "Sølvtorvet" [Silver
Market] after the adjacent plaza) until the T.U. part moved to
Lundtofte in a field 17 km north of Copenhagen. In 1962 Jannik moved
(with his smaller staff appointed by the university) to the new
buildings for chemistry, nonnuclear physics, and mathematics in the
five-pronged E-shaped "H.C.Ørsted Institute" at Universitetsparken 5.
During the time when I studied and worked at the "Sølvtorvet" building
(1950-1958) the organization was rather unique and slightly unreal and

surrealistic. About a dozen people, including Jannik much of the time, spent one to several hours drinking tea every afternoon (I feel that any reference to a painting by Leonardo da Vinci would be rather tasteless, but the scene resembled it) and talking about chemistry, some mathematics, and very little internal politics (Jannik, the highest-ranking representative of the Danish King, was present,taking his ulcer medication). They were the "inner group" (with a small minority of real T.U. people) trying its best to manage the unwieldy output of students, first and second year stud.polyt. (chemical engineering) about a hundred each year, including 1.4 university stud.mag.(aspirants to become tea drinkers). At any time, our laboratory might burn or blow up, but never have a student revolt(it was before 1968). The closest analogy is a dozen Oxford college fellows trying to form a civilizing Langmuir-Blodgett film on the affluent natives of Zanzibar (destined to do more managing than chemical research and reading).

Jannik's pedagogic influence was not only the tacit (not alluding to me) example of how to behave when practicing the most fascinating of all professions. He had a great influence by alloting the final tasks before obtaining the medieval degree of magister scientiarum (meaning that one is a full-fledged university teacher).

I was a singularity inside a tiny minority of tea drinkers (1). I knew every comma in the Royal Resolution about how to become a candidatus magisterii (it saved from 43 afternoons of organic lab synthesis). This species (and with chemistry as major subject, only marginally less scarce than unicorns) is expected to become a Gymnasium (secondary school) teacher in his speciality after having then passed a month of practical pedagogics (which I never dared to try). For the first three years, such students studied nearly equal amounts of mathematics, astronomy, and chemistry (disregarding their specialization, except for practical laboratory work some afternoons). I literally shopped around (as the university advised) several days in August 1950, looking at: theoretical physics; atomic spectra (with my old friend, Professor Ebbe Rasmussen); astrophysics; and finally inorganic (etc.) chemistry. Arriving at this scenario of King Arthur and his Round Table (even Arthur Adamson from Los Angeles arrived later and learned his languages nos. 21 to 29 in Copenhagen), I stayed. In May 1951, I presented a seminar, "Differences and Some Similarities between Lanthanides and Transthorium Elements" (that the Royal Resolution tolerated during any year but scheduled for me in 1954). A few perspicacious colleagues realized that there would be minor problems one day.

Apparently Jannik was satisfied with a rationalization of the d-group colors now developing, but the arguments by Racah (50) about $4f^q$ were adapted by Claus Schäffer to d^q configurations both for spherical and octahedral symmetry. It should be remembered(51,52) that a pure configuration d^q or f^q has many admirable properties (like a polyhedron with many different faces) but that the Schrödinger many-electron solutions normally contain several percent (30) of each of a few different configurations possessing squared amplitudes mixed with the preponderant (49) electron configuration (evaluated by Hartree-Fock technique) determining the J-values of the lower-lying energy levels. Very diplomatically Jannik called all rationalizations of observed excited states "ligand field" theory, disregarding the fact that one should then add a year (1955; 1959;1965; 1971; 1986) etc.(?) as when speaking about a developing airplane type.

As the actresses in the "salon" said to Louis XIV in the movie "Versailles" by Sacha Guitry, "We were the Golden Age", the tea drinkers are not overstepping their qualifications by saying something similar. I have reached the conclusion that I remained an inorganic chemist due to a meeting in September 1950 with my hierarchical superior, Lene Rasmussen (Magister 1953; Lecturer at the Ife University, Ibadan,in Nigeria,1964 and later back at Aarhus University) in the qualitative analysis course (this would correspond to General Chemistry in United States because there were 55 elements to separate) during both first year semesters. I was later appointed sub-instructor there, and for a few years Amanuensis II. Jannik required Lene to write a short book on unusual oxidation states (for the Magister degree), and she and I decided what was "unusual". We both recognized many proliferating mistakes in literature but also the adamant concept that facts are stubborn (as one living in St.Petersburg once wrote).

Since my period of photoelectron spectra of inorganic solids (1970-1979), which I spent reshuffling and wrinkling quantum chemistry quite a bit, I now realize that I have recycled (or regressed) to the role of a part-time atomic spectroscopist as well as astrophysicist because I was invited to write reviews for Comments on Inorganic Chemistry (30), edited by Fred Basolo, and for Comments on Astrophysics (53), edited by Virginia Trimble. In 1992, at the XXIX.ICCC in Lausanne where I received a free sample (909 pages) of Inorganica Chimica Acta, I suddenly felt as if I had stumbled over the exit from the Elverhøj ("The Fairy Hill") where I had lived as a chemist for 40 years. -- But what an experience !

When Jannik retired in 1979, he returned to his earlier passions in the laboratory. Just as Lene and I had determined the largest formation constants of any ammonia or ethylenediamine complexes (54), of palladium(II), Jannik determined the smallest constants: iron(III) in $FeCl_4^-$ and various $FeCl_n(OH_2)_x^{(3-n)+}$ (55), $CuCl_n(OH_2)_x^{(2-n)+}$ (56); $NiCl(OH_2)_5^+$ (57), octahedral $ClCr(OH_2)_5^{2+}$, $Cl_2Cr(OH_2)_4^+$, and probably $Cl_3Cr(OH_2)_3$ in 12.5 $\underline{M}$ LiCl (58); $BrNi(OH_2)_5^+$ with K_1 = 0.005 M^{-1}(59). and not far (60) from $K_1 \sim 0.003$, $K_2 \sim 0.00035$, and $K_3 \sim 4 \cdot 10^{-5}$ for $Br_nCr(OH_2)_{6-n}^{(3-n)+}$. Jannik (61)reviewed octahedral $(H_2O)_5Co(NCS)^+$ and $(H_2O)_4Co(NCS)_2$ as well as tetrahedral $Co(NCS)_4^{2-}$ and $(H_2O)Co(NCS)_3^-$ and compared them with nickel(II) thiocyanates in various solvents.

These results are important for the spectra of inner-sphere complexes with a number of aqua and anionic ligands, whereas I submit respectfully that the arguments about definite activity coefficients (62) are rather circular. Jannik's last publications may illustrate the pragmatic fact that spectra in the visible region are much more uniquely characteristic for the first coordination sphere, as Niels Bjerrum told Werner in 1909 for cases like $[ClCr(OH_2)_5]^{2+}$ as compared to ion-pairs. Disregarding undoubted "salt effects" (12,63) this is a major distinction between chemical physics and physical chemistry.

Acknowledgments

I am grateful to my early student comrade Claus E.Schäffer, for most valuable advice, constructive criticism, and continued discussions about applied group theory.
I am also grateful to Professor Fred Basolo for his friendly conversations, since he first visited Jannik and Copenhagen and for favoring inductive discussions of significant details of absorption spectra of complexes rather than relying on ephemeral fashion in computer chemistry.
I finally wish to thank Professor George B. Kauffman for his kind invitation to participate in the Denver symposium and for his careful and extended effort to edit and ameliorate my papers.

Literature Cited

1. Kauffman, G. B. J. Chem. Educ. 1985, 62, 1002.
2. Kauffman, G. B. J. Chem. Educ. 1959, 36, 521.
3. Kauffman, G. B. Chymia 1960, 6, 180.
4. Bjerrum, J. Metal Ammine Formation in Aqueous Solution, 2nd ed.; P. Haase: Copenhagen, 1957.
5. Bjerrum, J. Chem. Rev. 1950, 46, 381.
6. Kauffman, G.B. J.Chem. Educ. 1980, 57, 779, 863.
7. Carnall, W.T.; Fields, P.R.; Rajnak, K. J.Chem.Phys. 1968, 49, 4412, 4424, 4443, 4447, 4450.
8. Reisfeld, R.; Jørgensen, C.K. Lasers and Excited States of Rare Earths; Springer: Berlin and New York, 1977.
9. Bjerrum, J.; Jørgensen, C.K. Acta Chem. Scand. 1953, 7, 951.
10. Jørgensen, C.K. Mat. fys. Medd. Dan. Vid. Selsk. (Copenhagen) 1956, 30, No. 22.
11. Malkova, T.V.; Shutova ,G. A.; Yatsimirskii, K. B. Russ. J. Inorg. Chem. 1964, 9, 993.
12. Jørgensen, C. K. Acta Chem. Scand. 1954, 8, 175.
13. Katzin, L.I.; Gebert, E. Nature 1955, 175, 425.
14. Jørgensen, C.K.; Bjerrum, J. Nature 1955, 175, 426.
15. Jørgensen, C. K.; Pappalardo, R.; Schmidtke, H.H. J.Chem. Phys. 1963, 39, 1422.
16. Schäffer, C. E.; Jørgensen, C.K. Mol. Phys. 1965, 9, 401.
17. Jørgensen, C. K. Chem. Phys. Lett. 1967, 1, 11.
18. Jørgensen, C. K. Chem. Phys. Lett. 1982, 87, 320.
19. Jørgensen, C. K.; Faucher, M.; Garcia, D. Chem. Phys. Lett. 1986, 128, 250.
20. Taube, H. Chem. Rev. 1952, 50, 69.
21. Jørgensen, C. K. Chimia 1971, 25, 109, 213.
22. Jørgensen, C. K. Chimia 1984, 38, 75.
23. Jørgensen, C. K. Top. Current Chem. 1984, 124, 1.
24. Ilse, F. E.; Hartmann, H. Z.Physik. Chem. (Leipzig) 1951, 197,239.
25. Griffith, J. S. Theory of Transition-metal Ions; Cambridge University Press: Cambridge, 1961.
26. Hartmann, H.; Schläfer, H. L.Angew. Chem. 1954, 66, 768.
27. Jørgensen, C.K. Modern Aspects of Ligand Field Theory; North-Holland: Amsterdam, 1971.
28. Jørgensen, C. K. Top. Current Chem. 1975, 56, 1.
29. Bjerrum, N. Acta Chem. Scand. 1958, 12, 945.
30. Jørgensen, C. K. Comments Inorg. Chem. 1991, 12, 139.
31. Rosseinsky, D. R. Chem. Rev. 1965, 65, 467.

32. Jørgensen, C. K. Comments Inorg. Chem. 1981, 1, 123.
33. Trassati, S. Pure Appl. Chem. 1986, 58, 955.
34. Jørgensen, C. K.; Bjerrum,J. Acta Chem. Scand. 1955, 9, 180.
35. Bjerrum, J.; Jørgensen, C. K. Rec. Trav. Chim. (Netherlands) 1956, 75, 658.
36. George, P.; McClure, D.S. Progress Inorg. Chem. 1959, 1, 382.
37. Dunitz, J. D.; Orgel, L.E. Adv. Inorg. Chem. Radiochem. 1960, 2,1.
38. Williams, R. J. P. Discus$. Faraday Soc. 1958, 26, 123, 180, 186.
39. Jørgensen, C. K. Chimia 1974, 28, 6.
40. Glerup, J.; Mønsted, O.; Schäffer, C.E. Inorg.Chem. 1976, 15,1399.
41. Santen, J. H. van; Wieringen, J. S. van Rec. Trav. Chim. (Netherlands) 1952, 71, 420.
42. Jørgensen, C. K. Chimia 1992, 46, 444.
43. Schäffer, C. E.; Jørgensen, C. K. Mat. fys. Medd. Dan. Vid. Selsk. (Copenhagen) 1965, 34, No. 13.
44. Jørgensen, C. K. Chem. Phys. Lett. 1988, 153, 185.
45. Reisfeld, R.; Jørgensen, C. K. Struct. Bonding 1988,69, 63.
46. Tanabe, Y.; Sugano, S. J.Phys. Soc. Japan 1954, 9, 753, 766.
47. Schäffer, C. E.; Jørgensen, C. K. J.Inorg. Nucl. Chem. 1958, 8, 143.
48. Jørgensen, C. K. Progress Inorg. Chem. 1962, 4, 73.
49. Jørgensen, C. K. Oxidation Numbers and Oxidation States; Springer: Berlin and New York; 1969.
50. Racah, G. Phys. Rev. 1949, 76, 1352.
51. Brorson, M.; Schäffer, C. E. Inorg. Chem. 1988, 27, 2522.
52. Bendix, J.; Brorson, M.; Schäffer, C. E. Inorg. Chem. 1993, 32, 2838.
53. Jørgensen, C. K. Comments on Astrophysics 1993, 17, 49.
54. Rasmussen, L.; Jørgensen, C. K. Acta Chem. Scand. 1968, 22, 2313.
55. Bjerrum, J.; Lukes, I. Acta Chem. Scand. 1986, A 40, 31.
56. Bjerrum, J. Acta Chem. Scand. 1987, A 41, 328.
57. Bjerrum, J. Acta Chem. Scand. 1988, A 42, 714.
58. Bjerrum, M. J.; Bjerrum, J. Acta Chem. Scand. 1990, 44, 358.
59. Bjerrum, J. Acta Chem. Scand. 1990, 44, 401.
60. Bjerrum, M. J.; Bjerrum, J. Acta Chem. Scand. 1991, 45, 23.
61. Bjerrum, J. Coord. Chem. Rev. 1990, 100, 105.
62. Bjerrum, J. Coord, Chem. Rev. 1989, 94, 1.
63. Jørgensen, C. K.; Parthasarathy, V. Acta Chem. Scand. 1978, A 32, 957.

RECEIVED March 28, 1994

David P. Mellor (1903–1980).

Frank P. Dwyer (1910–1962).

Sir Ronald S. Nyholm (1917–1971).

Chapter 10

The Contributions of David P. Mellor, Frank P. Dwyer, and Ronald S. Nyholm to Coordination Chemistry

Stanley E. Livingstone

School of Chemistry, University of New South Wales, Kensington, NSW 2033, Australia

David P. Mellor (1903-1980), Frank P. Dwyer (1910-1962) and Sir Ronald S. Nyholm (1917-1971) were three Australians who were largely responsible for the great interest in coordination chemistry in Australia and for Australia's reputation in this field. All three had close contacts with Sydney University (SU), the University of New South Wales (UNSW), and its precursor Sydney Technical College (STC). Mellor was Reader at SU and Professor of Chemistry at UNSW (1956-1969). Dwyer was a graduate of SU, Head Teacher at STC, Senior Lecturer at SU, and Professor of Biological Inorganic Chemistry at the Australian National University (1961-1962). Nyholm was a graduate of SU, Teacher at STC, Associate Professor at UNSW, and Professor of Chemistry at University College London (1955-1971).

It seems that the earliest research in coordination chemistry performed in Australia was by Masson and Steele at the University of Melbourne in 1898. Sir David Orme Masson was Professor of Chemistry in Melbourne (1886-1924). Bertram Steele later worked with Sir William Ramsay at University College, London and became Foundation Professor of Chemistry at the University of Queensland, Brisbane. Their paper, "The blue salt of Fehling's solution and other cuprotartrates"[1], reported that the deep blue color of Fehling's solution was due to an anionic copper species. This work was not extended further until 1920 when Masson put his research students Packer and Wark to work on cuprotartrates; they confirmed Masson and Steele's results and isolated several cuprotartrates[2]. Ian William (later Sir Ian) Wark extended this work further in his PhD studies at University College, London[3,4]. John Packer subsequently became Professor of Chemistry at Canterbury University College, Christchurch, New Zealand in 1944.

Early Coordination Chemistry in Sydney

Coordination chemistry in Sydney had its beginning when Eustace Ebenezer Turner (1893-1966) was appointed Lecturer in Chemistry at the University of Sydney in 1919. During the First World War Turner had been research assistant to Professor William

0097–6156/94/0565–0127$08.00/0

Jackson Pope at Cambridge, working on the synthesis of arsines as potential war gases. In Sydney Turner began a collaborative research effort with George Burrows (1888- 1950), who had been appointed to the staff at the University of Sydney in 1919. This work was published in six papers in the *Journal of the Chemical Society* (4) and included one paper on metal-arsine complexes(5). In 1921 Turner returned to England, where he later became Professor of Chemistry at Bedford College, London and a Fellow of the Royal Society.

Burrows continued research on coordination chemistry and between 1928 and 1938 published 15 papers, mostly on metal-arsine complexes(4). His Liversidge Lecture to the Royal Society of New South Wales, titled "Organic arsenicals in peace and war"(6), was an excellent review of arsines and metal-arsine complexes. Burrows was to have a marked effect on subsequent research on coordination chemistry in Australia. Mellor has stated(7): "Burrows lectured on coordination compounds to advanced classes, instituted laboratory courses on the subject, and before long began to attract staff and students to his chosen field. I think that it is fair to say that to Burrows, more than any other man, we owe the formation, at the University of Sydney, of the first school of coordination chemistry in this country [Australia]." It was in this atmosphere that Mellor, Dwyer, and Nyholm began their research careers and it was one of the points from which metal-arsine chemistry emanated.

Mellor

David Paver Mellor was born in Launceston, Tasmania and graduated from the University of Tasmania in 1924. He was appointed Lecturer in Chemistry at the University of Sydney in 1929 and promoted to Reader in Inorganic Chemistry in 1946. His lifelong interest was in the structure of inorganic compounds and metal complexes. His early interest was in X-ray crystallography and his first three publications — all with his research student Frank Dwyer — were in this area: *viz.*, "The crystal structure of indium," "The occurrence of β-cristabolite in Australian opals," and "An X-ray study of opals"(8).

In the 1930's and 1940's Mellor and his 15 or so research students worked on metal complexes, using a variety of physical techniques, including refractive index, X-ray diffraction, magnetic susceptibility, and thermodynamic stability determinations (for a list of publications, see ref. 4). Indeed, Mellor was one of the earliest coordination chemists to study X-ray crystal structures, magnetic moments, and stability constants of metal complexes. In 1938 Mellor went to work with Linus Pauling at the California Institute of Technology. He returned to Sydney full of enthusiasm for the ideas which had just been published in Pauling's book *The Nature of the Chemical Bond*, one of the most influential chemical books of this century. He introduced Pauling's concepts into both his teaching and research, and his subsequent publications show Pauling's influence, especially in relation to magnetochemistry.

A five-part series "Magnetic studies of coordination compounds" followed(9-11). In two papers of this series Mellor and Craig reported the magnetic moments of 37 nickel and 21 cobalt complexes with the chromophores: O_4, O_2N_2, N_4, N_2S_2, O_2S_2, and S_4. The authors attempted to explain the preference for high- or low-spin magnetic behavior on the nature of the chromophore, *i.e.*, the donor atoms. This was an important concept, since it anticipated Chatt's idea of (a) and (b) class metals(12) and Pearson's "hard" and "soft" ligands(13). David Craig did his MSc research under Mellor's supervision – PhD degrees were not awarded by Australian universities until after the Second World War – and later occupied Chairs of Physical Chemistry at the University of Sydney, University College, London, and the Australian National University, Canberra, and he became a Fellow of the Royal Society and of the Australian Academy of Sciences. He was later, together with L.E. Sutton, R.S.

Nyholm and A. Maccoll, to influence the theoretical understanding of metal-ion complex chemistry, especially the role of π-bonding.

In three papers, "The significance of large angle distortions in relation to the stereochemistry and magnetic properties of quadricovalent metals," "The stereochemistry of some metallic complexes with special reference to their magnetic properties and the Cotton effect," and "A study of the magnetic behaviour of complexes containing the platinum metals," Mellor did much to clear up the state of knowledge of the magnetic behaviour of transition metals(*14*). His review, "The stereochemistry of square complexes"(*15*), was a standard reference work. For this and his other work on metal complexes Mellor was awarded the DSc degree by the University of Tasmania.

Arguably Mellor's most significant contribution to coordination chemistry was the establishment of the stability sequence for bivalent metals in complexes with oxygen and nitrogen donors: *viz.*, Pd > Cu > Ni > Co > Zn > Cd > Fe > Mn > Mg(*16*). This work was published before the classic work of Irving and Williams (*17*) who showed that the sequence for the first transition series is: Mn < Fe < Co < Ni < Cu > Zn. Mellor published a few more papers on X-ray structure determinations, including one on Zeise's salt, $K[PtCl_3(C_2H_4)].H_2O$, showing that the C-C axis of the ethylene ligand is perpendicular to the plane containing the $PtCl_3$ grouping(*18*).

In 1955 Mellor was appointed Professor of Chemistry at the University of New South Wales. Unfortunately, he did no more experimental research, but he was the author of four books: *The Rôle of Science and Industry in Australia in the War of 1939-1945*(*19*), *Chelating Agents and Metal Chelates* (*20*),*The Evolution of the Atomic Theory* (*21*), and *Chemistry of Chelation and Chelating Agents* (*22*).

Dwyer

Francis Patrick Dwyer was born in Nelson's Plains, a small town in New South Wales, and he graduated from Sydney University in 1930. For his MSc degree, under Mellor's supervision, he worked on glyoxime complexes of palladium, and he isolated *cis* and *trans* isomers of bis(benzylmethylglyoximato)palladium(II)(*23*); this was an important contribution, since the stereochemistry of four-coordinate Pd(II) and Pt(II) had not been unequivocally established. In 1934 Dwyer was appointed Head Teacher of Inorganic Chemistry at Sydney Technical College, and here he carried out an extensive investigation of the reactions of diazoamino compounds with metal salts to produce highly colored "lakes", which are loosely bound complexes chemisorbed at the surface of metal hydroxides. Using these reactions, Dwyer devised a number of microanalytical tests for various metal ions; this research was reported in 18 papers in *Journal of the Society of Chemical Industry* , the *Journal .and Proceedings of the Australian Chemical. Institute,* tand the *Journal and Proceedings of the Royal Society of New South Wales (4)*. For this work he was awarded the DSc degree by the University of Sydney in 1946.

Ronald Sydney Nyholm was born in Broken Hill, a large, isolated mining town in New South Wales, 1200 km west of Sydney. He graduated from Sydney University in 1938 and in 1940 was appointed Teacher of Chemistry at Sydney Technical College. Dwyer, with his experimental skills, and Nyholm with his experience with arsines gained from Burrows, developed a good working relationship. In a 10-part series, "The chemistry of bivalent and trivalent rhodium"(*24,25*), they reported a number of Rh(III) and Rh(II) complexes of arsines, pyridine, dimethylglyoxime, and dimethyl sulfide. The Rh(II) complexes included the halogen-bridged dimeric complexes $[RhX_2(Ph_2MeAs)_3]_2$ (X = Cl, Br, I); the presence of Rh(II) in these diamagnetic complexes was later questioned, since Rh(II) would be expected to be paramagnetic, and a reinvestigation by Nyholm in 1960(*26*) showed that at least some of these compounds are hydrido complexes of Rh(III). However, this should not be taken as criticism of their early work, since hydrides of transition metals were unknown at that time. It is of interest that hydrido complexes of rhodium have been extensively used as catalysts in the organic chemical industry. In a 4-part series, "The chemistry of bivalent and trivalent iridium"(27,28), they reported similar arsine complexes of Ir(II) and Ir(III).

Dwyer and Nyholm reported complexes of Ru(III), Ru(II), Os(III), and Os(II) with Ph$_2$MeAs(29,30). The stability constant of [Fe(phen)$_3$]2+ and the redox potentials of the systems [Ru(phen)$_3$]3+/Ru(phen)$_3$]2+, [OsBr$_6$]2-/[OsBr$_6$]3-, and [OsCl$_6$]2-/[OsCl$_6$]3- were determined(31-33).

This brought to a close the collaboration of Dwyer and Nyholm at Sydney Technical College. It had established a tradition for research in coordination chemistry within the Chemistry Department, which a few years later became the School of Chemistry in the University of New South Wales.

Dwyer did further work on redox potentials of the Ru(III)/Ru(II) and Os(III)/Os(II) systems containing the ligand 2,2'-bipyridyl, showing that the potential is influenced by the charge on the ion and the degree of chelation(34) At Sydney University Dwyer collaborated with Frank Lions, Reader in Organic Chemistry, designing and synthesizing the sexadentate ligand, 1,8-bis(salicylideneamino)-3,6-dithiaoctane (HONSSNOH) and resolving the Co(III) complex [Co(ONSSNO)]I, which gave a molecular rotation of ± 50,160°; this was the highest molar rotation recorded up to that time(35) . Cobalt complexes of other sexadentates were prepared and optically resolved (36). Dwyer developed further his skill in resolving metal complexes into their optical isomers and carried out many such resolutions(7). The major thrust of this work was the elegant design of the sexadentate ligand to give a specific configuration about the metal ion . It was one of the first examples of such design of a ligand.

Using chiral molecules, Dwyer discovered a phenomenon to which he gave the name "configurational activity". Briefly, this means that the activity coefficients of optical isomers are different in an asymmetric environment provided by another optically active ion(7,34) He predicted that configurational activity should be of biological importance, since the rates of many enzyme processes should be changed by the addition of inert optically active complex ions(38). This work provided a better theoretical framework for the Pfeiffer effect and subsequently influenced the theory of chiral interactions.

Dwyer became interested in bioinorganic chemistry and reported on the toxicity to mice of some complexes such as the d - and l - isomers of [Co(en)$_3$]3+ and [Os(bipy)$_3$]2+ (39) . He and his colleagues showed that these ions are choline esterase inhibitors, and this led to his appointment in 1957 as Reader in Biological Inorganic Chemistry at the Australian National University, Canberra, where he became Professor in 1961. Here he studied the curare-like poison effect of metal complexes such as [Ru(terpy)$_2$](ClO$_4$)$_2$. Of greater interest and practicality were the studies made by Dwyer and his colleagues on the effects of similar complexes onl microorganisms: e.g., tris(3,4,7,8-tetramethyl-1,10-phenanthroline)nickell(II) acetate proved especially effective in the prevention of infection due to Staphylococcus aureus in newborn babies and in surgical wounds(7,37). In his last paper published before his death Dwyer described a novel method for resolving amino acids by means of a metal complex; this method has the advantage of yielding both isomers in pure form(40). It was also in this period that conformational analysis of coordination complexes began in a serious way. Dwyer and his associates were major contributors to the development of the understanding of steric effects and their influence on the structures of metal complexes.

Dwyer died suddenly of a heart attack in 1962, aged 51, at the peak of his resaearch activity. He was the author of 160 publications, a complete list of which is given in Sutherland's biography(41).

Nyholm

In 1947 Nyholm was awarded an ICI Fellowship to work with Professor Ingold at University College, London and was awarded his PhD degree in 1950. He was appointed to the staff at University College and worked with the diarsine *o*-phenylenebisdimethylarsine (I; As-As), which had been reported by Joseph Chatt in 1938 (*42*). Nyholm was able to show that this diarsine is effective in stabilizing high - *e.g.*, Ni(III) and Ni(IV) – and low – *e.g.*, Fe(0), Cr(0), Mo(0), W(0), and Fe(I) – oxidation states. Table I lists the diarsine

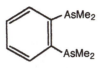

(I)

complexes reported by Nyholm and his co-workers during the period 1950-1961. The stabilization of unusual oxidation states by the diarsine (I) prompted considerable theoretical discussion (*43,44*) and a search for other ligands that might behave similarly.

Another aspect of Nyholm's extensive study of metal complexes of the diarsine (I) was the occurrence of unusually high coordination numbers, *e.g.*, six for Pd(II) and Pt(II) and eight for Ti(IV) and V(IV). Nyholm's interest in these high coordination numbers extended to the stereochemistry involved, and his research group established the stereochemistry of some of these complexes by X-ray crystallography. The compounds $[MI_2(As-As)_2]$ (M = Pd, Pt) were shown to be distorted octahedral with the four arsenic atoms in the plane of the square and the two iodine atoms in the apical positions with unusually long bonds(*45*), while the complex $[TiCl_4(As-As)_2]$ was shown to be dodecahedral (*46*).

In 1952 Nyholm returned to the University of New South Wales (formerly Sydney Technical College), where his effect on the Inorganic Chemistry Department was immediate and profound. He brought back fresh ideas of what he called the "Renaissance of inorganic chemistry" and stimulated a greater enthusiasm for research in coordination chemistry. He had become interested in magnetochemistry and extended his work in that area while in Sydney; Craig(*47*) has stated that "Nyholm advanced the method of measurement of magnetic moments itself." He was promoted to Associate Professor in 1953 and received his DSc degree (London) in 1954. Early in 1955 he left Australia to accept a Chair of Chemistry at University College, London. It has been said: "He had spent only three years in Sydney, yet in that time he had taken coordination chemistry in Australia out of one rut and put it in another"(*4*)· During that period he published four important reviews, *viz.*, "Stereochemistry and valence states of nickel"(*48*), "Magnetism and inorganic chemistry"(*49*), "The stereochemistry of complex compounds"(*50*), and "Magnetism and stereochemistry"(*51*).

At University College Nyholm soon built a flourishing research group which attracted PhD and postdoctoral students from Britain, the USA, Australia, India, New Zealand, and elsewhere. The scope of his research widened and included further work on arsine and phosphine metal complexes; magnetic, infrared, and nmr studies of metal complexes; reactivity of coordinated ligands; complexes containing metal-metal bonds; compounds containing a platinum-carbon bond; carbonyl, olefin, cyclopentadienyl, and perfluoro complexes; alkali metal complexes; complexes with unsaturated phosphines and arsines as ligands; the stabilization of Fe(IV); and oxidative addition reactions.

Table I. Metal Complexes of o-Phenylenebisdimethylarsine (As-As)

Complex	Metal	Complex	Metal
$[M^0(As\text{-}As)_2]$	Ni	$[M^{II}X_2(As\text{-}As)(CO)_3]$	Mo
$[M^0(As\text{-}As)(CO)_2]$	Ni	$[M^{II}X(As\text{-}As)(CO)_2]^+$	Mo
$[M^0(As\text{-}As)(CO)_3]$	Fe	$[M^{III}(As\text{-}As)_2]^{3+}$	Au
$[M^0(As\text{-}As)_2\,CO]$	Fe	$[M^{III}X_3(As\text{-}As)]$	Ni
$[M^0(As\text{-}As)(CO)_4]$	Cr, Mo, W	$[M^{III}Cl_2(As\text{-}As)H_2O]^+$	Mn
$[M^0(As\text{-}As)_2(CO)_2]$	Cr, Mo, W	$[M^{III}I(As\text{-}As)_2]^{2+}$	Au
$[M^I(As\text{-}As)_2]^+$	Cu, Ag	$[M^{III}X_4(As\text{-}As)]^-$	Cr, Ru
$[M^II(As\text{-}As)(CO)_2]$	Fe	$[M^{III}X_3(As\text{-}As)H_2O]$	Cr
$[M^{II}X_2(As\text{-}As)]$	Ni, Zn, Cd, Hg	$[M^{III}X_2(As\text{-}As)_2]^+$	Cr, Tc, Re,
$[M^{II}(As\text{-}As)_2]^{2+}$	Ni, Pd, Pt		Fe, Ru, Os,
			Co, Rh
$[M^{II}\,X(As\text{-}As)_2]^+$	Ni, Pd, Pt		Ni, Au
$[M^{II}\,ClO_4(As\text{-}As)_2]^+$	Ni	$[M^{IV}X_4(As\text{-}As)]$	Mo
$[M^{II}X_2(As\text{-}As)(CO)_2]$	Fe	$[M^{IV}X_2(As\text{-}As)_2]^{2+}$	Fe, Os, Ni,
			Pd, Pt
$[M^{II}\,X_2(As\text{-}As)_2]$	Mn, Tc, Re, Fe		
	Ru, Os, Co, Ni,	$[M^{IV}Cl_4(As\text{-}As)_2]$	Ti, Zr, Hf,
	Pd, Pt		V
		$[M^{V}Cl_4(As\text{-}As)_2]^+$	Tc, Re

Particular mention should be made of complexes containing metal-metal bonds, a subject on which Nyholm and his collaborators published many papers. Indeed, Nyholm could be regarded as one of the leading experts in this field, as evidenced by his review "Metal-metal bonds in transition metal complexes"(52)·

Nyholm also became active in the area of chemical education and took a leading part in launching the journal *Education in Chemistry* and the large work *Comprehensive Inorganic Chemistry* (53) . He was President of The Chemical Society (1968-1970) and during his term as President saw the amalgamation of the society with the Faraday Society, the Royal Institute of Chemistry, and the Society for Chemical Industry. He received many honors: Corday Morgan Medal (1952); H. G. Smith Medal of the Royal Australian Chemical Institute (1955); Fellowship of the Royal Society (1958); Royal Medal of the Royal Society of New South Wales (1959); Tilden Lecturer of the Chemical Society (1961); first Dwyer Memorial Lecturer (1963); Liversidge Lecturer of the Chemical Society (1967); Gold Medal of the Italian Chemical Society (1968); 'Sigillum Magnum' Medal of the University of Bologna (1969); and Honorary DSc, University of New South Wales (1969). He was knighted in 1967. His publications numbered 278 ; a full list is given in ref. *47*. He was killed in a car accident near Cambridge, England in 1971, aged 54. Like Dwyer, he was at the height of his career.

Conclusion

As a tribute to three great Australian chemists, in 1972 three lecture theaters in the School of Chemistry at the University of New South Wales were named, respectively, Dwyer, Nyholm, and Mellor.

Literature Cited

1. Masson, O; Steele, B. D. *J. Chem. Soc.* **1899**, *75*, 7251

2. Packer, J; Wark, I. W. *J. Chem. Soc.* **1921**, *119*, 1248.

3. Wark, I. W. *J. Chem. Soc.* **1923**, *123*, 1815; **1924**, *125*, 2994.

4. Baker, A. T.; Livingstone, S. E. *Polyhedron* **1985**, *4*, 1337.

5. Burrows, G. J.; Turner, E. E. *J. Chem. Soc.* **1921**, *119*, 1448.

6. Burrows, G. J. *J. Proc. Roy. Soc.* N.S.W. **1940**, *74*, M1.

7. Mellor, D. P. *Proc. Roy. Aust. Chem. Inst.* **1970**, *37*, 199.

8. Dwyer, F. P.; Mellor, D. P. *J. Proc. Roy. Soc. N.S.W.* **1932**, *66*, 234, 378; **1934**, *68*, 47.

9. Mellor, D. P.; Goldacre, R. J. *J. Proc. Roy. Soc. N.S.W.* **1939**, *73*, 233.

10. Mellor, D. P.; Lockwood, W. H. *J. Proc. Roy. Soc. N.S.W.* **1940**, *74*, 141.

11. Mellor, D. P.; Craig, D. P. *J. Proc. Roy. Soc. N.S.W.* **1940**, *74*, 475, 495; **1941**, *75*, 27.

12. Ahrland, S.; Chatt, J.; Davies, N. R. *Quart. Rev. Chem. Soc.* **1958**, *12*, 265.

13. Pearson, R. G. *J. Am. Chem. Soc.* **1963**, *85*, 3533.

14. Mellor, D. P. *J. Proc. Roy. Soc. N.S.W.* **1940**, *74*, 107; **1941**, *75*, 157; **1943**, *77*, 145.

15. Mellor, D. P. *Chem. Rev.* **1943**, *33*, 137.

16. Mellor, D. P.; Maley, L. *Nature (London)* **1947**, *159*, 370; **1948**, *161*, 436.

17. Irving, H.; Williams, R. J. P., *Nature (London)* **1948**, *162*, 746; *J. Chem. Soc.* **1953**, 3192.

18. Wunderlich, J. A.; Mellor, D. P. *Acta Cryst.* **1954**, 7, 130.

19. Mellor, D. P. *The Role of Science and Industry*, Series 4, in *Australia in the War of 1939-45* ; Australian War Memorial: Canberra, A.C.T., 1958; Vol. 5.

20. *Chelating Agents and Metal Chelates*; Dwyer, F. P.; Mellor, D. P., Eds.; Academic Press: New York, N.Y., 1964.

21. Mellor, D. P. *The Evolution of the Atomic Theory*; Elsevier: Amsterdam, 1971.

22. Mellor, D. P. Chemistry of Chelation and Metal chelates. In *The Chelation of Heavy Metals*; Levine, W. C., Ed.; *International Encyclopaedia of Pharmacology and Therapeutics* ; Section 20, Pergamon Press: Oxford, 1979.

23. Dwyer, F. P.; Mellor, D. P. *J. Am. Chem. Soc.* **1934**, *56*, 1551; **1935**, *57*, 605; *J. Proc. Roy. Soc.* **1934**, *68* ,107.

24. Dwyer, F. P.; Nyholm, R. S. *J. Proc. Roy. Soc. N.S.W.* **1941**, *75*, 122, 127, 140; **1942**, *76*, 129, 133, 275; **1944**, *78*, 67, 266; **1945**, *79*, 126.

25. Dwyer, F. P.; Nyholm, R. S.; Rogers, L. E. *J. Proc. Roy. Soc. N.S.W.* **1947**, *81*, 267.

26. Lewis, J.; Nyholm, R. S.; Reddy, G. K. *Chem. Ind.* **1960**, 1386.

27. Dwyer, F. P.; Nyholm, R. S. *J. Proc. Roy. Soc. N.S.W.* **1943**, *77*, 116; **1945**, *79*, 121.

28. Dwyer, F. P.; McKenzie, H. A.; Nyholm, R. S. *J. Proc. Roy. Soc. N.S.W*, **1944**, *78*, 260; **1947**, *81*, 261.

29. Dwyer, F. P.; Humpoletz, J. E.; Nyholm, R. S. *J. Proc. Roy. Soc. N.S.W.* **1946**, *80*, 217.

30. Dwyer, F. P.; Nyholm, R. S.; Tyson, B. T. *J. Proc. Roy. Soc. N.S.W.* **1947**, *81*, 272.

31. Dwyer, F. P.; Nyholm, R. S. *J. Proc. Roy. Soc. N.S.W.* **1946**, *80*, 28.

32. Dwyer, F. P.; McKenzie, H. A.; Nyholm, R. S. *J. Proc. Roy. Soc. N.S.W..* **1946**, *80*, 183.

33. Dwyer, F. P.; Humpoletz, J. E.; Nyholm, R. S. *J. Proc. Roy. Soc. N.S.W.* **1946**, *80*, 212, 242.

34. Buckingham, D. A.; Sargeson, A. M. In *Chelating Agents and Metal Chelates*; Dwyer, F. P.; Mellor, D. P., Eds.; Academic Press: New York, N.Y., 1964; pp 272, 197.

35. Dwyer, F. P.; Lions, F. *J. Am. Chem. Soc.* **1947**, *69*, 2917; **1950**, *72*, 1546

36. Dwyer, F. P.; Gill, N. S.; Gyarfas, E. C.; Lions, F. *J. Am. Chem. Soc.* **1950**, *72*, 5037.

37. Shulman, A.; Dwyer, F. P. Metal Chelates in Biological Systems. In *Chelating Agents and Metal Chelates*; Dwyer, F. P.; Mellor, D. P., Eds.; Academic Press: New York, N.Y., 1964; p 383.

38. Dwyer, F. P.; Gyarfas, E. C.; O'Dwyer, M. F. *Nature (London)* **1951**, *167*, 1036; Dwyer, F. P.; Gyarfas, E. C. *Nature (London)* **1951**, *168*, 29.

39. Dwyer, F. P.; Gyarfas, E. C.; Rogers, W. P.; Koch, J. H. *Nature (London)* **1952**, *170*, 190.

40. Dwyer, F. P.; Halpern, J. *Nature (London)* **1962**, *196*, 270.

41. Sutherland, K. L. Biographical Memoir, *Australian Academy of Sciences Year Book*, Canberra, A.C.T., 1963; p 32.

42. Chatt, J.; Mann, F. G. *J. Chem. Soc.* **1939**, 1622.

43. Chatt, J. *J. Inorg. Nuclear Chem.* **1958**, *8*, 515.

44. Orgel, L. E. *International Conference on Coordination Chemistry London*: Chemical Society: London, 1959; p 93; *An Introduction to Transition Metal Chemistry : Ligand-Field Theory ;* Methuen: London, 1960, p 132.

45. Harris, C. M.; Livingstone, S. E. *Rev. Pure App. Chem (Australia)* **1962**, *12*, 16.

46. Clark, R. J. H.; Lewis, J.; Nyholm, R. S.; Pauling, P.; Robertson, G. B. *Nature (London)* **1961**, *192*, 222.

47. Craig, D. P. *Biographical Memoirs of Fellows of the Royal Society* **1972**, *18*, 445.

48. Nyholm, R. S. *Chem. Rev.* **1953**, *53*, 263.

49. Nyholm, R. S. *Quart. Rev. Chem. Soc. (London)* **1953**, *7*, 263.

50. Nyholm, R. S. *Progr. Stereochem.* **1954**, *1*, 322.

51. Nyholm, R. S. *J. Proc. Roy. Soc. N.S.W.* **1955**, *89*, 8.
52. Lewis, J.; Nyholm, R. S. *Sci. Progr.* **1964**, *52*, 557.
53. *Comprehensive Inorganic Chemistry*; Bailar, J. C.; Emeléus, H. J.; Nyholm, R. S.; Trotman-Dickenson, A. F., Eds.; Pergamon Press : Oxford, 1973.

RECEIVED February 18, 1994

Yuji Shibata (1882–1980) in 1942.

Ryutaro Tsuchida (1903–1962) in 1958.

Chapter 11

History of Coordination Chemistry in Japan During the Period 1910 to the 1960s

Kazuo Yamasaki

Department of Chemistry, Nagoya University, Nagoya 464, Japan

The history of coordination chemistry in Japan is briefly presented. Yuji Shibata, founder of coordination chemistry in Japan studied extensively the absorption spectra of complexes of various metals from 1915 to 1917 after returning from Europe. His researches also included the spectro- chemical detection of complex formation in solution, coagulation of arsenic sulfide sols by complex cations, and catalytic oxidation and reduction by metal complexes in solution. Ryutaro Tsuchida published the "spectrochemical series" in 1938 based on the results of his measurements of absorption spectra of cobalt complexes. One of the most remarkable results after World War II is the determination of absolute configurations of cobalt complexes using X-rays in 1954 by Y. Saito and his coworkers.

In this review the history of coordination chemistry in Japan is briefly presented beginning with Yuji Shibata(1882-1980), the only Japanese co-worker of Alfred Werner's. The period covered is from 1910 to the 1960s, about 50 years.

Alfred Werner's first paper on the coordination theory (*1*) was published in 1893 and was introduced to Japan in an abridged form in 1897 by Riko Majima (1874-1962), who was a postgraduate student in the College of Science, Tokyo Imperial University. Majima, who later became the father of organic chemistry in Japan, introduced this paper and three subsequent papers of Werner's in the *Tokyo Kagaku Kaishi (Journal of the Tokyo Chemical Society)* under the title of "Theory of Molecular Compounds" (2). If read today, Majima's articles are difficult to under- stand even for us who know the coordination theory, indicating that Werner's ideas were not easily understandable at that time.

0097–6156/94/0565–0137$08.00/0

Research During the Period 1910-1930s

Spectrochemical Works of Shibata and his Co-workers.   More than ten years passed before Yuji Shibata carried out the first research on coordination chemistry in 1910.   Shibata was born in Tokyo as the second son of Shokei Shibata, an eminent pharmacologist who had studied organic chemistry in Berlin under August Wilhelm von Hofmann during the period 1870-1874. Yuji Shibata graduated from the Department of Chemistry, College of Science, Tokyo Imperial University in 1907. He first studied organic chemistry and published three papers related to the Grignard reagents (*3-5*). However, he was asked by Joji Sakurai (1858-1939), Chairman of the Department of Chemistry to teach inorganic chemistry because Tamemasa Haga, the professor of inorganic chemistry, was seriously ill, and the university urgently needed a candidate for the future professor of inorganic chemistry. Shibata accepted this proposal and decided to study inorganic chemistry in Europe.   Majima, who was Shibata's senior by 10 years   and who was staying in Europe at   that time, recommended Werner in Zürich, a choice that was to have great consequences. If Shibata had continued his organic chemistry studies or had studied under some other chemists in Europe, coordination chemistry in Japan might have had an entirely different aspect than the present one.

When Shibata arrived in Zürich in 1910, it was rumored that Werner might move to a German university,   and Shibata decided to study first under Werner's teacher, Arthur Hantzsch (1857-1935) in Leipzig. There Shibata studied the absorption spectra of cobalt (II) salts with a small quartz spectrograph. He found that the color of a solution of a cobalt salt with   coordination number six was red, while that of a solution of a cobalt salt with coordination number less than six was blue (*6*). After staying one year in Leipzig, Shibata moved to Zürich in 1911 to study under Werner. In Zürich he prepared *cis*-$[Co(NH_3)_2(en)_2]X_3$ and succeeded in resolving it into optical isomers. This was the first example of the resolution of a $[Cob_2(AA)_2]$- type complex, AA and b being didentate and monodentate ligands, respectively (*7*).

In 1912 Shibata moved to Paris to study under Georges Urbain (1872-1938). He intended to study the rare earth elements, but Urbain advised him not to do so because such study required tedious fractional crystallization, which was not suitable for a foreign chemist with only limited time to spend. Instead, Urbain suggested that Shibata carry out absorption spectrographic studies of cobalt complexes. Fortunately, Shibata was able to use the newly obtained medium-sized quartz spectrograph of Adam Hilger, type E2,   and he carried out   absorption measurements of cobalt-ammine complexes (*8*). In Urbains's laboratory Shibata also learned from Jacques Bardet the technique of emission spectrographic analysis , which Shibata later used to analyze the rare earth minerals found in Japan.

Shibata returned to Japan in 1913, one year before the outbreak of the First World War, and he was appointed associate professor at his alma mater. Fortunately, he was able to purchase the same medium-sized quartz spectrograph, E2, through the special favor of Joji Sakurai, the Department Chairman,   and he actively began spectrochemical studies of metal complexes.   This spectrograph was in use for more than fifty years in the Department of Chemistry, Tokyo University. The present author also used this quartz spectrograph in 1939 for his doctoral work on the absorption spectra of 2,2'-bipyridine complexes (*9-12*).

During the period 1914-17 Shibata measured the visible and ultraviolet absorption spectra of more than 120 metal complexes, including not only colored complexes of such as cobalt, chromium, nickel, and copper but also those of colorless metals like silver, zinc, and mercury. He published the results in French in the *Journal of the College of Science of Tokyo Imperial University* during the period 1915-1917 (*13-15*). These were the world's pioneering works on the absorption spectra of metal complexes, and Shibata was quickly recognized as the foremost specialist in this field. As is well known, absorption spectra were then measured by the so-called Hartley-Baly method, using a Baly tube and photographic plates. It was a semiquantitative method, *i.e.,* the wavelength of the absorption maximum was accurately measured, but the absorption intensity was only qualitative.

At almost the same time as Shibata's research Luther and Nikolopulos (*16*) studied the visible absorption spectra of cobalt-ammines in 1913. Werner himself also began the spectrochemical study of metal complexes (*17*), but his fatal disease had already begun, and furthermore World War I interrupted the academic activities of European chemists. These were the main reasons why no coordination chemist in Europe carried out detailed spectrochemical studies of metal complexes.

In 1921 Shibata devised a spectrochemical method for detecting complex formation in solution (*18,19*). The so-called method of continuous variation, reported by Paul Job (*20*) in 1925, involves the same principle as Shibata's method. By using this method Shibata, with his student Toshi Inouye, studied the complex formation between $HgCl_2$ and chlorides of other metals.

Works other than Spectrochemical Studies. Shibata extended his study to coagulation studies of negatively charged arsenic sulfide sols by complex cations, and he confirmed the extension of Freundlich's relationship between the coagulation concentration and the cation valency to higher valencies (*21*). This method was used to determine the ionic charge of a complex cation, together with the conductometric method developed earlier by Miolati and Werner (*22,23*). A similar experiment for the determination of anionic valency was later carried out in 1953 by R. Tsuchida, Akitsugu Nakahara and Kazuo Nakamoto using positively charged $Fe(OH)_3$ sols (*24*).

In 1917 Shibata and his student Toshio Maruki (*25*) proposed the *cis*-structure of the two ammine groups of Erdmann's salt, $NH_4[Co(NO_2)_4(NH_3)_2]$, based on the fact that its derivative $X[Co(C_2O_4)(NO_2)_2(NH_3)_2]$ was resolvable into optical antipodes. This result was confirmed by William Thomas (*26*), while E.H.Riesenfeld and R.Klement (*27*) in 1922 and Bhabes Chandra Ray (*28*) in 1937, by denying the above-mentioned optical resolution, claimed the salt possesses the *trans*-structure. This problem of the structure of Erdmann's salt was finally solved in 1957 by Yoshimichi Komiyama, who determined its *trans*-structure by X-ray crystal analysis (*29*). The reason for such confusion may have been the difficulty in measuring the small rotation angles of a colored complex solution by the visual method. The corresponding *cis*-isomer, K[Co(NO$_2$)$_4$(NH$_3$)$_2$], which has been a long standing missing link in a series of nitroammine complexes, [Co(NO$_2$)$_x$(NH$_3$)$_{6-x}$]$^{3-x}$ (x=1-5), has recently been isolated and its crystal structure determined by Takashi Fujiwara *et al.* (*30*).

Shibata's research field was further extended to the study of the catalytic property of cobalt complexes. In 1918 he and his elder brother, Keita Shibata (1877-1949), professor of phytochemistry and plant physiology at Tokyo Imperial University, found that myricetin, a kind of flavonol, is easily oxidized in solution by cobalt complexes at room temperature, and they published the results in 1920 (*31*).

This study was continued for more than 20 years both in the laboratories of phytochemistry and inorganic chemistry with many co-workers, including Ryutaro Tsuchida and the present author. The primary experimental technique used was the measurement of oxygen absorption by a Warburg manometer; pyrogallol and other oxidizable substances were used as the substrate. The most active compounds were cobalt complexes containing anionic ligands such as [CoCl(NH$_3$)$_5$]X$_2$, whereas complexes such as [Co(NH$_3$)$_6$]X$_3$ were inactive. In addition to cobalt some complexes of copper and nickel were active, but those of chromium were inactive. The catalytic oxidation was studied mainly by oxygen absorption alone because the dark colors of the oxidized solution disturbed the absorption spectral measurements. Thus the details of reaction processes were not easily elucidated. The so-called asymmetric oxidation of optically active substrates such as *l*-dioxyphenylalanine(*l*-dopa) and *d*-catechin by optically active cobalt complexes such as [CoCl(NH$_3$)(en)$_2$]X$_2$ was also studied with some positive results (*32,33*).

As the oxidation mechanism the Shibata brothers proposed the activation of water molecules which were replaced by the aquation reaction of anionic ligands in a complex ion. They summarized the results of their early 16 papers, and in 1936 they published a book *Katalytische Wirkungen der Metallkomplexverbindungen* (*34*). This book was not frequently cited in the chemical literatures of English- speaking countries probably because it was written in German, and the publication was close to the outbreak of World War II. In 1974, however, 40 years after the publication of this book, Eastman Kodak Co. requested a US patent on the use of a cobalt complex such as [Co(NH$_3$)$_6$]Cl$_3$ as an amplifier in the development of color films (*35*). Although no reference is made to the works of Shibata, it is supposed that the chemists at Eastman Kodak Co. studied the works of Shibata and co-workers.

On reaching the retirement age of sixty of the Tokyo Imperial University in 1942 Yuji Shibata moved to the newly established Nagoya University as the Dean of the Science Faculty. Thus the research on catalytic oxidation was suspended, although this research was his favorite topic. His work on the catalytic action of metal complexes was further extended by his co-workers to the catalytic reduction in hydrogen, which had been initiated in 1939 by Shibata himself. Among the results obtained, the formation of [Co(CN)$_5$H]$^{3-}$, a strong reducing agent, reported by Masaakira Iguchi(*36*) in 1942, was remarkable, and its structure has been extensively discussed (*37*).

Works of Other Researchers. Before 1930 studies on coordination chemistry in Japan were almost exclusively concentrated in Yuji Shibata's laboratory, and only a few reports from other laboratories were published. For instance, Satoyasu Iimori (1885-1982), who later became the founder of radiochemistry in Japan, studied the replacement of CN- groups in the hexacyanoferrate(III) ion with water in 1915 (*38*) and photochemical reactions of cyano complexes of platinum and nickel and photochemical cells in 1918 (*39*).

Furthermore, measurement of formation constants of nickel, cadmium, and zinc cyano complexes by Koichi Masaki were reported in 1931 (*40-43*). Another interesting paper (*44*) on the effects of chemical bonding on the X-ray absorption edges using metal complexes was reported in 1927 by Shin-ichi Aoyama, Kenjiro Kimura, and Yosio Nisina, who were staying in Bohr's research institute in Copenhagen.

In 1930 the Department of Chemistry was established in Osaka University, and Ryutaro Tsuchida(1903-1962) was appointed the professor of inorganic chemistry. Also Taku Uemura(1893-1980) began the research on coordination chemistry at the Tokyo Institute of Technology at nearly the same time. Both men were former co-workers of Yuji Shibata's. Thus young coordination chemists graduated from these newly established laboratories, and the numbers of published research papers gradually increased.

Research During the Period 1930-1945

Because the number of researches in this period increased considerably, only the main works will be mentioned.

Spectrochemical Studies. In the newly established laboratory Tsuchida began to measure quantitatively the absorption spectra of cobalt complexes, first by remeasuring the absorption spectra reported by Shibata and then preparing new complexes with various ligands. The results were summarized in the shift rules of absorption bands with the replacement of ligands, *i.e.*, the spectrochemical series, which was reported first in 1938 (*45*) and was refined in 1955 (*46*). Its importance was recognized after World War II by its relation to the crystal field theory. In this field of research important works of physicists such as Yukito Tanabe and Satoru Sugano after the war deserve mention (*47*).

In 1938 Tsuchida devised a method for measuring the absorption spectra of crystals by combining a microscope and a spectrograph (*48*). This method was later used by him and his co-workers for measuring the dichroism of crystals (*49*). In 1939 Tsuchida further proposed a theory that the electron pairs used for coordinate bonds and the unshared electron pairs of the central metal atom are distributed in a symmetrical way around the central metal (*50,51*). This idea antedated the similar theory proposed in 1940 by Sidgwick and Powell (*52*), but Tsuchida's works were not known to foreign chemists because of war.

Preparation of New Compounds. In 1938 Tokuichi Tsumaki prepared a cobalt complex with salicylaldehyde-ethylenediimine and found that it combined reversibly with oxygen (*53*). This compound provided the impetus for the detailed studies during the war by Melvin Calvin and his co-workers on oxygen adducts of cobalt complexes (*54*). Tsuchida and Kobayashi prepared a cobalt complex of dimethylglyoxime, a nonelectrolyte, and to prove its structure by its resolvability or nonresolvability, they devised a method called asymmetric adsorption on optically active quartz powder (*55*). This method was improved in 1970 by Yuzo Yoshikawa and Kazuo Yamasaki using Sephadex ion exchanger for the complete chromatographic resolution of metal complexes (*56,57*).

Research Activities after World War II

After the war, which caused tremendous damage all over Japan, the educational systems were reformed, and many new universities were established, which were smaller in size than the earlier universities. With the reconstruction of Japan research in coordination chemistry has become an active field and has been extended to non-Werner type complexes and organometallic chemistry. Only a few postwar researches will be mentioned here.

One of the most remarkable results in the field of structural studies is the determination of absolute configurations of the optically active cobalt complex ion $[Co(en)_3)]^{3+}$ by Yoshihiko Saito *et al.* in 1954, using the anomalous scattering of X-rays (*58,59*). This work has expanded enormously, and absolute configurations of about 150 complexes are known as of the end of the 1970s, 80 of which have been determined in Japan (*60*). Another work to be mentioned is the new synthetic method for cobalt (III) complexes starting from $[Co(CO_3)_3]^{3-}$ ion, which was devised by Muraji Shibata of Kanazawa University in 1964 (*61*). This method has been much used to synthesize many new cobalt complexes (*62*).

Finally, it may be appropriate to describe here the national meeting of coordination chemists. The first symposium on coordination chemistry was held in 1942, and eight papers were presented. After World War II it was reorganized, and the first meeting was held in 1952. It continues to the present day. About 700 chemists attended in 1993. In 1970 Japanese coordination chemists organized a small society, the Japan Society of Coordination Chemistry. The above mentioned symposium is held by this Society jointly with the Chemical Society of Japan. In 1961, when the Sixth International Conference on Coordination Chemistry (VI ICCC) was held in Detroit, Michigan, U.S.A., several Japanese coordination chemists were invited, which strongly influenced their research. Ryutaro Tsuchida was invited, but his health was not good enough to permit him to go abroad. In 1962 he died of stomach cancer at the age of 59.

Five years later, in 1967, the 10th ICCC was held in Japan at Tokyo and Nikko with Yuji Shibata as the President and the present author as the General Secretary of the Organizing Committee, respectively. Many foreign coordination chemists participated in this ICCC, including John C. Bailar, Jr., Fred Basolo, Ronald

S. Nyholm, G. Wilke, K. B. Yatsimirskii, Lars Gunnar Sillén, Jannik Bjerrum, and other eminent coordination chemists. This conference made a strong impact on the Japanese coordination chemists, especially on the younger ones. Also, foreign chemists had the opportunity to recognize the state of coordination chemistry research in Japan.

After working in Tokyo University for 29 years (1913-1942), devoting the latter half of his life to the establishment of Nagoya University (1942-1948) and Tokyo Metropolitan University (1949-1957) in the difficult years during and after World War II and serving as the President of the Japan Academy for eight years(1962-1970), Yuji Shibata, the founder of coordination chemistry in Japan calmly passed away on the morning of January 28th, 1980, his 98th birthday.

Acknowledgment. The present author is grateful to the Editor , Professor George B. Kauffman for his advice and linguistic revision of the manuscript.

Literature Cited
Papers marked with * are in Japanese.
1. Werner, A. *Z. anorg. Chem.* **1893,** *3*, 267.
   For a discussion and an annotated English translation see Kauffman, G .B. *Classics in Coordination Chemistry, Part 1: The Selected Papers of Alfred Werner*; Dover: New York,1968; pp 5-88.
2.* Majima, R. *Tokyo Kagaku Kaishi* **1898,** *19*, 233.
3. Shibata,Y. *J. Chem. Soc.* **1909,** *95*, 1449.
4. Shibata,Y. *Ber.* **1910,** *43*, 2619.
5. Shibata, Y. *J. Chem. Soc.* **1910,** *97* , 1239.
6. Hantzsch, A.; Shibata, Y. *Z. anorg. Chem.* **1912,** *73, 309.*
7. Werner, A.; Shibata, Y. *Ber.* **1912,** *45*, 3287.
8. Urbain, G.; Shibata, Y. *Compt. rend.* **1913,** *157, 594.*
9. Yamasaki, K. *Bull. Chem. Soc. Japan* **1937,***12, 390.*
10. Yamasaki, K. *Bull. Chem. Soc. Japan* **1938,** *13, 538.*
11. Yamasaki, K. *Bull. Chem. Soc. Japan* **1939,** *14, 130.*
12. Yamasaki, K. *Bull. Chem. Soc. Japan* **1939,** *14*, 461.
13. Shibata Y. *J. Coll. Sci. Imp. Univ. Tokyo* **1915,** *37*, Art.2, 1.
14. Shibata, Y. *J. Coll. Sci. Imp. Univ. Tokyo* **1916,** *37*, Art.8, 1.
15. Shibata, Y. *J. Coll. Sci. Imp. Univ. Tokyo* **1917,** *41*, Art.6, 1.
16. Luther, R.; Nikolopulos, A. *Z. physik. Chem.* **1913,** *82*, 361.
17. In his letter addressed to Shibata with the date of Dec. 28, 1913
   Werner wrote, "Thank you very much for your friendly congratulations for the Nobel prize. I am very pleased to learn that you still remain faithfully in metal-ammines. We have also investigations on the absorption spectra going on, which, however, shall never disturb your work. I wish you best results" (translated from the German). Private communication of Shibata to Yamasaki.
18. Shibata, Y.; Inoue, T.; Nakatsuka, Y. *Japanese J. Chem.* **1922,** *1*,1.
19. Shibata, Y.; Inoue, T. *Japanese J. Chem.* **1926,** *2*, 109.

20. Job, P. *Compt. rend.* **1925,** *180,* 928.
21.*Matsuno, K. *Tokyo Kagaku Kaishi* **1918,** *39,* 908.
22. Werner, A.; Miolati, A. *Z. physik. Chem.* **1893,** *12,* 35.
    For discussions and English translations see Kauffman, *op.cit.,* pp 89-139.
23. Werner, A.; Miolati, A. *Z. physik. Chem.* **1894,** *14,* 506.
24.*Nakahara, A;Nakamoto, K; Tsuchida, R. *Nippon Kagaku Kaishi* **1953,** *74,*
    488.
25. Shibata, Y.; Maruki, T. *J. Coll. Sci. Imp. Univ. Tokyo* **1917,** *41,* Art.2, 1.
26. Thomas, W. *J. Chem. Soc.* **1923,** *123,* 617.
27. Riesenfeld, E.H.; Klement, R. *Z. anorg. Chem.* **1922,** *124,* 1.
28. Ray, B.C. *Indian Chem. Soc.* **1937,** *14,* 440.
29. Komiyama, Y. *Bull. Chem. Soc. Japan* **1957,** *30,* 13
30. Fujiwara, T.;Fuyuhiro, A.;Yamanari, K.; Kaizaki, S. *Chem. Lett.* **1990,** 1679.
31.*Shibata, Y.;Shibata, K. *Tokyo Kagaku Kaishi* **1920,** *41,* 35.
32. Shibata, Y.; Tsuchida, R. *Bull. Chem. Soc. Japan* **1929,** *4,* 142.
33. Shibata, Y.; Tsuchida, R. *Bull. Chem. Soc. Japan* **1931,** *6,* 210.
34. *Katalytische Wirkungen der Metallkomplexverbindungen*;
    Shibata, K.;Shibata, Y. Eds.; The Iwata Institute of Biochemistry:
    Publication No.2; Tokyo, 1936.
35. Eastman Kodak Co., *U. S. Patent,* 3841873 (Oct.15,1974); *CA,* **1975,**37293.
36.*Iguchi, M. *Nippon Kagaku Kaishi* **1942,** *63,* 634.
37. King, N.K.; Winfield, M.E. *J. Am. Chem. Soc.* **1961,** *83,* 3366.
38.*Iimori, S. *Tokyo Kagaku Kaishi* **1915,** *36,* 150.
39.*Iimori, S. *Tokyo Kagaku Kaishi* **1917,** *38,* 507.
40. Masaki, K. *Bull. Chem. Soc. Japan* **1929,** *4,* 190.
41. Masaki, K. *Bull. Chem. Soc. Japan* **1931,** *6,* 60.
42. Masaki, K. *Bull. Chem. Soc. Japan* **1931,** *6,* 89.
43. Masaki, K. *Bull. Chem. Soc. Japan* **1931,** *6,* 233.
44. Aoyama, S.; Kimura, K.;Nisina, Y. *Z. Physik* **1927,** *44,* 810.
45. Tsuchida, R. *Bull. Chem. Soc. Japan* **1938,** *13,* 388.
46. Shimura, Y.; Tsuchida, R. *Bull. Chem. Soc. Japan* **1956,** *29,* 311.
47. Tanabe, Y.; Sugano, S. *J. Phys. Soc. Japan* **1954,** *9,* 753.
48. Tsuchida, R.;Kobayashi, M. *Bull. Chem. Soc. Japan* **1938,** *13,* 619.
49. Yamada, S.; Tsuchida, R. *Bull. Chem. Soc. Japan* **1952,** *25,* 127.
    Subsequent papers are published in the same journal through 1960.
50. Tsuchida, R. *Bull. Chem. Soc. Japan* **1939,** *14,* 101.
51. Tsuchida, R.;Kobayashi, M.; Kuroya, H. *Rev. Phys. Chem. Japan*
    1939, *13,* 151.
52. Sidgwick, N. V.;Powell, C. F. *Proc. Roy. Soc.* **1940,** *A176,* 153.
53. Tsumaki, T. *Bull. Chem. Soc. Japan* **1938,** *13,* 252.
54. Calvin, M.; Bailes, R. H.; Wilmarth, W. K. *J. Am. Chem. Soc.* **1946,**
    68, 2254.
55.*Tsuchida, R.; Kobayashi, M.; Nakamura, A. *Nippon Kagaku
    Kaishi* **1935,** *56,* 1339.

56. Yoshikawa, Y.;Yamasaki, K. *Inorg. Nucl. Chem. Lett.* **1968,** *4,* 697.
57. Yoshikawa,Y.; Yamasaki, K. *Coord. Chem. Rev.* **1979,** *28,* 205.
58. Saito, Y.; Nakatsu, K.; Shiro, M.; Kuroya, H. *Acta Cryst.* **1954,** *7,* 636.
59. Saito, Y.; Nakatsu, K.; Shiro, M.; Kuroya, H. *Acta Cryst.* **1955,** *8,* 729.
60. Saito, Y. *Topics in Stereochemistry*; Eliel, E. W.; Allinger, N. L. Eds.; John Wiley: New York, 1978, Vol.10; pp 95-174.
61. Shibata, M.; Kyuno, E.; Mori, M. *Inorg. Chem.* **1964,** *3,* 1573.
62. Shibata, M. *Modern Syntheses of Cobalt (III) Complexes; Topics in Current Chemistry*; Springer-Verlag: Berlin, 1983, Vol. 110, pp 1-118.

RECEIVED May 11, 1994

SPECIALIZED ASPECTS

Chapter 12

The Compleat Coordination Chemistry
What a Difference a Century Makes!

Daryle H. Busch

Department of Chemistry, University of Kansas, Lawrence, KS 66045

Alfred Werner's coordination chemistry was remarkably broad, effectively encompassing most of the molecularly discrete compounds of the metallic elements, and the field has expanded enormously during its century-long rule. Today coordination chemistry is recognized to involve all manner of interactions between separately recognizable atomic and moelcular entities, of all possible charge variations. The coordination entity persists as the unifying concept, and the principles of the field are perceived as underlying much of the mutual organizing that occurs between unlike and even like molecular and atomic species. The common structural principles of molecular organization underlie all of those interactions that lead to the formation of a coordination entity in the various realms of receptor/receptee combinations, be it metal ion/ligand, host/guest, or intermolecular hydrogen bonding and/or stacking. Complementarity and constraint are key considerations, and their interplay is of major concern in molecular design. The presence within a molecule of multiple receptors of varied kinds has opened new vistas for chemical research and new promises for technology.

Coordination chemistry emerged in the publications of Alfred Werner in order to explain chemical substances that were, at that early time, viewed as "complex compounds" because they failed to conform to the contemporary valency rules (1, 2). The major purpose of this essay is to focus on the realization that coordination chemistry, the seminal, but highly augmented, legacy to science of Alfred Werner, is foundational to the understanding of the global issue of the organization of molecules in whatever sample of matter such relationships may occur, be it natural or synthetic. Werner's contribution included foundations for the understanding of the molecularly discrete compounds of the metallic elements, but the breadth of his view was evident in the statement "Almost all compounds of the first order (saturated hydrocarbons form the sole exception) possess the property of combining with other compounds of like nature"*(2a)*. The early evolution of coordination chemistry focused on compounds of the metals, and thus the alkali metal ions were essentially excluded from consideration.

During the greater part of its first hundred years, coordination chemistry focused on the concept of a monatomic, cationic, central atom bonded to Lewis bases as ligands. It has been a *cation core-focused model*. The cation core-focus and emphasis on metal ions as central atoms provide the point of departure from which to view the expansion of coordination chemistry into its current natural and compleat form. The discussion in this and the following sections will attempt to justify this concept of completeness that derives from pulling the totality of the field together. While the scope of coordination chemistry is now essentially complete, the fundamental understanding of the underlying principles continues to unfold.

The *coordination entity* or *complex* has come to be recognized as the unifying concept of coordination chemistry. The traditional coordination entity has each of the following attributes: a central (metal) atom, a number of ligands, a coordination number, and a polyhedral structure. The *coordination polyhedron* often occupies center stage as the basis for structural understanding.

Coordination chemistry uniquely provides and effectively maintains a stewardship over certain parts of the conceptual foundations and chemical content of chemistry. Because coordination chemistry was the home of the chemistry of transition metal compounds, certain conceptual topics were largely developed by the followers of that discipline; for example, stereochemistry of higher coordination numbers (5 and above), bonding and spectroscopy in systems having d-orbitals, and mechanisms of reactions of the metallic elements. The principles are so basic that they have immediate application as undreamed-of new substances serendipitously appear in chemistry (*e.g.*, dihydrogen complexes, metal derivatives of fullerenes, and metal-containing liquid crystals). Coordination chemistry is a field that spawns fields: transition metal organometallic chemistry, homogeneous catalysis, bioinorganic chemistry; and it provides a foundation for other burgeoning fields: solid state chemistry; extended and mesoscopic materials; photonic materials; models for solid surfaces; separations science; and molecular electronics, machines and devices.

Those trained as traditional coordination chemists who read this might share the view that it was a truly remarkable experience to watch exceptionally creative organic chemists discover chemistry beyond the molecule. The experience is illustrative of the compartmentalization of knowledge in today's world. It takes no challenging analysis to conclude that these new aspects of the compleat coordination chemistry constitute a replay, in a different realm, of the thesis of Werner's work of an even hundred years ago. The greatest importance of these new developments is the completion of the realm of coordination chemistry. The work begun by an organic chemist, Alfred Werner, in the preceding century has been completed by modern organic chemists *(3-5)*. The total result is a fundamental structural perspective on the intermolecular interactions between all manner of molecular partners, an extension of the notion of a coordination entity to every kind of interacting pair, and the incorporation of all kinds of intermolecular forces into coordination chemistry.

Expansion of the Boundaries of Coordination Chemistry

The compleat coordination chemistry exists today because all of the binding interactions that yield distinct molecular species, *i.e.*, *coordination entities,* by the union of two or more lesser molecular species, *i.e.*, complex formations, can now be included within the expanded field. Quoting from Lehn *(3)*, "the chemistry of artificial receptor molecules may be considered a generalized coordination chemistry, not limited to transition metal ions but extending to all types of substrates (receptees): cationic, anionic, or neutral species of organic, inorganic, or biological nature."

At this point, please bear with the author and consider a suggestion with regard to nomenclature. The word *receptor* is a most general term that encompasses both ligands and hosts, and its use is to be strongly encouraged. In contradistinction, the word "substrate" is not a suitable conjugate term for truly wide-ranging applications. Substrate is appropriate when the receptor is part of an enzyme; however, the term already has two meanings (also the underlying or supporting substance in certain materials usages). The obvious conjugate is *receptee*, and it is recommended that this term be used *(6)*. This new term is not otherwise burdened with context, and its conjugate relationship is obvious.

The perspectives provided by Lehn's supramolecular chemistry *(3,7)* and Cram's host/guest complexation *(4)* do indeed broaden the realm of coordination chemistry, but the focus still remains on a molecular *coordination entity.* On the other hand, the coordination polyhedron has lost its pivotal position in the broad definition of coordination chemistry. Furthermore, all manner of intermolecular interactions and interacting pairs are included, and the forces included range from van der Waals and subtle hydrophobic interactions through strong covalent bonds. Coordination chemistry demands only that *the molecular entities that unite to form the complex still be recognizable substructures within the complex (6).* It is particularly instructive, at this point, to examine examples of coordination entities formed by various modes of interaction that were not recognized in traditional coordination chemistry.

For many years, only a few examples of reasonably well-characterized complexes were known for the alkali metal ions, especially the larger ones. This situation prevailed until those previously ignored elements were brought into coordination chemistry by the discovery of the ionophores *(8)* and by the demonstration that cyclic polyglycol ethers and cyclic polyglycol-like ethers of appropriate ring size can bind to alkali metal ions *(9-11)*. Optimized alkali metal and alkaline earth metal ion ligands were developed using macrobicyclic "cryptate" ligands *(7,12)* (Figure 1) and increasingly rigid ligands *(4)*. In the second and third major developments, variants of the cryptates yielded early examples of receptors for such complex cationic species as NH_4^+ and $^+NH_3(CH_2)_nNH_3^+$ (Figure 2) *(14-16)* and for such anions as halides *(17,18)*, carboxylates, and phosphates *(19-21)*.

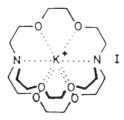

Figure 1.   Cryptate complex of K$^+$ (Reproduced with permisison from reference 13. Copyright 1985, John Wiley & Sons, Inc).

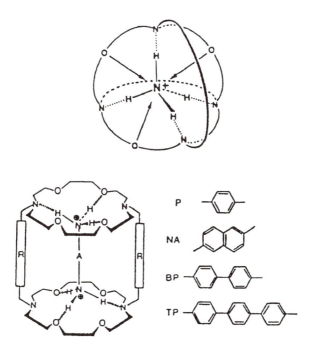

Figure 2.   Complexes of ammonium ions with hydrogen bonding receptors (Reproduced with permisison from reference 3. Copyright 1988, Kluwer.).

Still more complicated receptees, for example, $[Co(NH_3)_6]^{3+}$, $[Co(NH_2CH_2CH_2NH_2)_3]^{3+}$, $[Co(1,3,6,8,10,13,16,19\text{-octaazabicyclo}[6.6.6]\text{eicosane})]^{3+}$, (in lasalocid A) (22,23), and ferri- and ferrocyanide (with macrocyclic polyammonium hosts) (24,25), have been incorporated as central moieties in coordination entities. The use of complexes as receptors or receptees is, in principle, indefinitely extendable. Indeed, NH_3 is a complex molecule that is a ligand in the hexaammine complex that is the receptee in the lasalocid complex. Placement of $[Co(NH_3)_6](LAS)_3$ in a membrane, as receptor, is very like a fourth level of complexation. The receptor for $^+NH_3\text{-R-}NH_3^+$ (Figure 2) is illustrative of ditopic compartmental receptors (14-16). In principle, there is no limit to the number and relative orientations of the receptee sites of polytopic compartmental ligands.

The fourth expansion of the boundaries of coordination chemistry includes neutral molecules, generally held by hydrophobic interactions, and the founding receptor molecules are the cyclodextrins and the cyclophanes. The cyclodextrins (26) are cyclic oligosaccharides consisting of six or more α-1,4-linked D-glucose units (Figure 3). These molecules are roughly the shapes of truncated cones, and they have hydrophilic exteriors and hydrophobic cavities. In aqueous solutions, various poorly solvated species retreat to these cavities, forming complexes.

The cyclophanes of Tabushi (29,30) and Murakami (31) are especially suited for hydrophobicly binding aromatic molecules in aqueous solutions (Figure 4). With cationic cyclophane hosts, both hydrophobic and electrostatic forces come into play, and not only aromatic molecules but their anionic derivatives are also bound (32-34).

Finally, the base-pairing that occurs naturally in genetic materials has been incorporated into small molecule complexation (35-39). These developments have brought specific hydrogen bonding patterns into coordination chemistry. Specific hydrogen bonding has often been augmented by stacking interactions between aromatic groups in studies on base-pair emulative receptors. Rebek's genetic base receptor (40-42) and its complex with adenine are shown in Figure 5, along with Hamilton's barbiturate receptor (43-46), the latter being derived from a macrocyclic parent structure.

Molecular Organization

The most general function of coordination chemistry is to organize molecules (47). In the simplest possible case, the binding of water molecules to a metal ion, e.g., nickel(II), in aqueous solution, the ligands (water molecules) are placed in a totally organized state when compared to their usual condition as free solvent molecules. It is, in fact, true that the study of coordination chemistry has revealed a stream of generalizations, each of which is a manifestation of molecular organization. The abilities of receptors and receptees to recognize each other and the strength of their interactions depend on a few simple structural factors that underlie the phenomena of molecular organization.

The structural factors that are fundamental to molecular organization are size, shape, electronic relationships, topology, and rigidity. The first three (size, shape, and electronic) combine to constitute the phenomenon of receptor/receptee

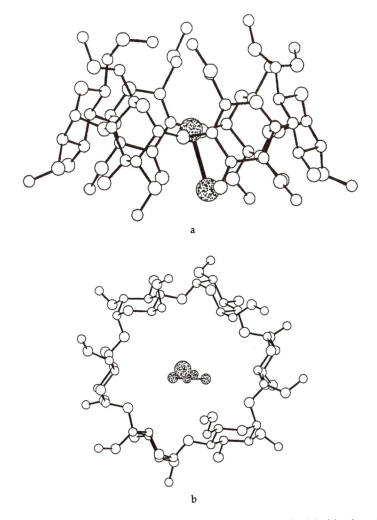

a

b

Figure 3.   The iodine complex of dimethyl-α-cyclodextrin: (a) side view; (b) view from above opening (Reproduced with permisison from reference 26. Copyright 1991, Academic Press.).

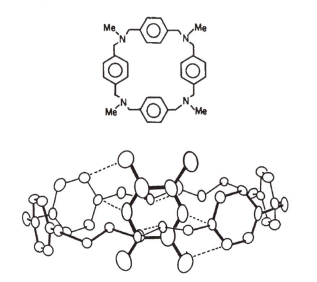

Figure 4.   Cyclophane complex with a neutral guest: top, host molecule;
             bottom, structure of complex (Reproduced with permisison from
             reference 13. Copyright 1985, John Wiley & Sons, Inc).

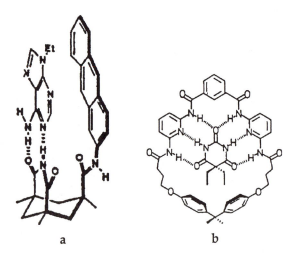

Figure 5.   Specific hydrogen bonding receptors: a) Adenine complex; b)
             barbiturate complex (a Reproduced with permisison from
             reference 37. Copyright 1990 VCH; b Reproduced from reference
             45. Copyright 1988, American Chemical Society).

complementarity. Complexation will occur when complementarity is adequate. The fit of metal ions into low energy conformations of macrocycles, a subject that has been much discussed, exemplifies complementarity. Furthermore, the necessity for a favorable binding force requires an additional complementarity, which may be exemplified by charges of opposite sign or by the hard and soft donor and acceptor atoms. Rebek *(36)* states, "the principle of molecular recognition: identification is most effective with surfaces of complementary shape, size, and functionality." Expanding on this and Lehn's words*(3)*, *complementarity is a congruence of shape and size factors and energetic or electronic compatibility between receptor and receptee, host and guest, or central atom and ligand.* Additional constraints represented by topological and rigidity considerations may contribute greatly to the strength of the receptor/receptee interaction. Thus topological and rigidity constraints are the design factors available for arbitrarily enhancing affinity (Figure 6).

It is particularly instructive to see how classic developments in coordination chemistry emphasize the structural factors that define molecular organization. For polydentate ligands, the role of molecular organization is strikingly evident in the various so-called "effects" that have been found to give stronger metal complexes: the chelate effect, the macrocyclic effect, the cryptate effect, and pre-organization or multiple-juxtapositional fixedness.

The chelate effect was the first major phenomenon that attributed enhanced stability of a complex to a general structural characteristic of a ligand—the ability to form a chelate ring. For complementary ligands, increasing the number of chelate rings increases the stability of complexes *(48)*. This phenomenon is of topological origin. If a donor atom of a potentially chelating ligand is not attached to a metal ion while other donor(s) of the same ligand are attached, then the free donor atom is held close to the metal ion, as if it were present in very high concentration. The kinetic manifestation of the chelate effect occurs as an abnormally rapid rate of binding of the free ligating atom of a chelate ligand, whereas the rate of dissociation of the terminal donor atom may be quite similar to that of a corresponding monodentate group *(49)*. Thus the chelate effect arises from topological relationships and is manifested in an abnormally rapid rate of binding as chelate rings are closed.

The macrocyclic effect produces increases of two to four orders of magnitude in the stabilities of complexes over those of comparably complementary acyclic chelating ligands *(50-52)*. This is again a reflection of increased topological constraint. In order to remove a macrocycle from a metal ion it is not possible to start at one end of the ligand and dissociate a terminal donor since rings have no ends. Instead, the ring must fold or contort in some other way in order to increase a metal-donor bond distance. Like the chelate effect, the macrocyclic effect also has a topological origin, but the macrocyclic effect is characterized by both slow rates of binding and dissociation, and the rate of dissociation is more greatly diminished *(50-52)*. The cryptate effect, which provides additional orders of magnitude of stabilization of complexes, arises from a still more constrained topology. These relationships are summarized in Figure 7.

Figure 8 shows that the effect of increasing the rigidity of the ligand framework on the labilities of transition metal complexes with amine ligands *(49,*

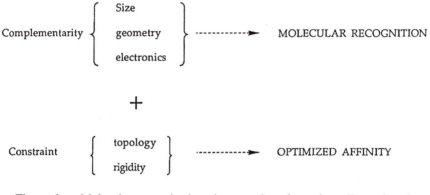

Figure 6. Molecular organization in complex formation (Reproduced reference 6. Copyright 1993, American Chemical Society).

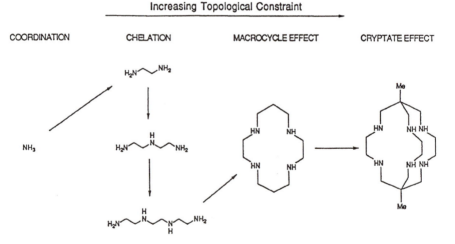

Figure 7. Topology and the chelate, macrocycle, and cryptate effects (Reproduced with permisison from reference 53. Copyright Battelle, 1993).

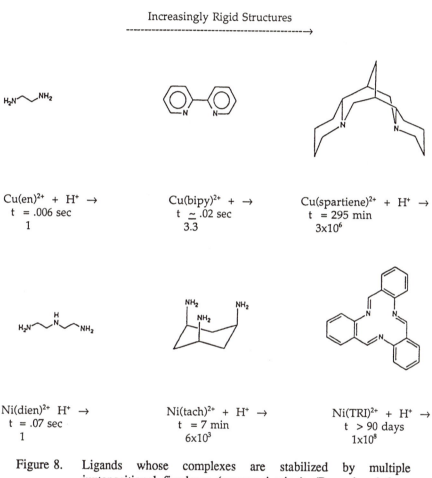

Figure 8.   Ligands whose complexes are stabilized by multiple juxtapositional fixedness (preorganization). (Reproduced from reference 6. Copyright 1993, American Chemical Society).

54-59) can be as great as 10^7 to 10^8, an observation that was labeled *multiple juxtapositional fixedness (59-60)*. A similar dramatization of the benefit of increasingly rigid or *preorganized (61)* structures is shown in Figure 9, where estimated free energies of binding to a metal ion are given for families of cyclic ethers having differing levels of flexibility *(4)*. The rigid single-ring spherand ligand binds the lithium cation more strongly than does the topologically more constrained, bicyclic, but flexible, cryptand.

The recurrence of the word *constraint* is central to this analysis of the structural factors contributing to the affinities of the bound pairs in coordination entities. *Size* and *shape* lead to *optimized complementarity* when the receptor and receptee enjoy the best fit; in terms of classic stereochemistry, nothing more can be done to enhance the binding affinity when this optimized mutual compatibility has been achieved. However, the addition of *topological* and *flexibility constraints* can *enhance affinity* a great deal more as long as their addition does not interfere with complementarity. These conclusions elaborate slightly on Cram's statement, "Just as preorganization is the central determinant of binding power, complementarity is the central determinant of structural recognition" *(4)*. The relationships among the structural factors of molecular organization *(6, 47)* are summarized in Figure 6.

Multiple Varied Receptors Within a Single Coordination Compound

In the simplest case, ditopic, tritopic, ... polytopic receptors would repeat identical receptor sites along a chain, sheet, or 3-dimensional matrix after the fashion of functionalized polymers, especially resins. Compartmental ligands continue to be of great interest for such functions as receptee separations, conducting polymers, ferromagnetically coupled molecules, and the like, but far more intricate levels of molecular organization now exist among the ambitions of coordination chemists. The inventory of receptor sites includes the following, and examples of any or all might be incorporated into the design of a single multireceptor supramolecular system: macrocycles, macrobicycles,..., macropolycycles, fixed but open cleft structures, and flat platforms, whose binding is based on (a) donor atoms, (b) hydrogen bonds, (c) charged groups, (d) hydrophobic interactions, and (e) stacking interactions. Many examples exist of the pairing of disparate receptors. A few early examples are crown ether face-to-face with a porphyrin *(62)* (Figure 10) to provide an alkali metal site near a transition metal site; porphyrin appended to a cyclodextrin *(64)* (Figure 11); a picnic basket porphyrin *(65)* (Figure 12); and a vaulted cyclidene *(66-69)* (Figure 13) to locate an oxidizing center near an receptor site for an organic molecule; and Rebek hydrogen bonding receptors on porphyrins *(70)* (Figure 14) to bind complementary molecules near metal site.

Enzymes, especially metalloenzymes, offer a compelling challenge to those who would like to apply multiple receptor site systems. An example of a possible set of receptors follows: (a) metal ion site, (b) substrate binding site, (c) environment at the active site, (d) nucleophile or electrophile as cofactor, (e) electron transfer system, and (f) proton transfer system.

The greatest importance of coordination chemistry in the future will almost certainly be in bringing higher levels of molecular organization into the design of

Increasingly Rigid Structures
--→

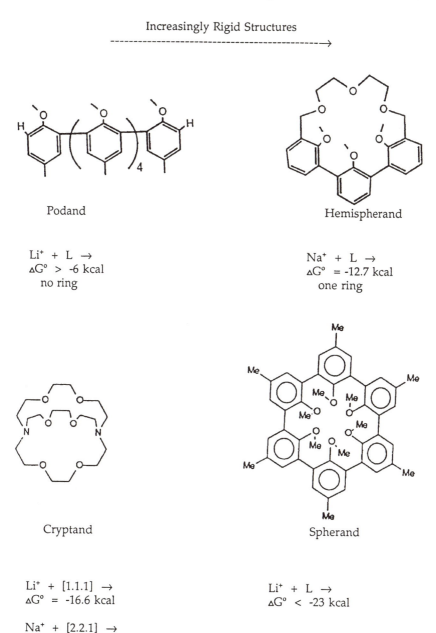

Podand

Li$^+$ + L →
ΔG° > -6 kcal
no ring

Hemispherand

Na$^+$ + L →
ΔG° = -12.7 kcal
one ring

Cryptand

Li$^+$ + [1.1.1] →
ΔG° = -16.6 kcal

Na$^+$ + [2.2.1] →
ΔG° = -17.7 kcal
two fused rings

Spherand

Li$^+$ + L →
ΔG° < -23 kcal

one rigid ring

Figure 9.   Preorganized ligands (multiple juxtapositional fixedness) (Reproduced from reference 6. Copyright 1993, American Chemical Society).

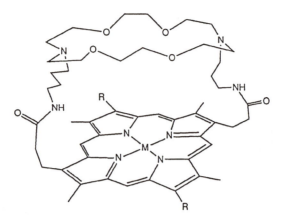

Figure 10. Multiple-varied receptors: face-to-face porphyrin/crown ether.
(Reproduced with permission from reference 69. Copyright 1989,
Kluwer).

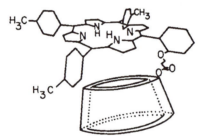

Figure 11. Multiple-varied receptors: porphyrin/cyclodextrin (Reproduced
with permission from reference 63. Copyright 1985, John Wiley
& Sons, Inc.).

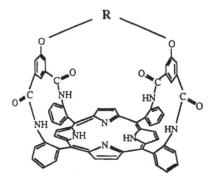

Figure 12. Multiple-varied receptors: macrobicycle with sites for metal ion
and substrate (Reproduced from reference 65. Copyright 1990,
American Chemical Society).

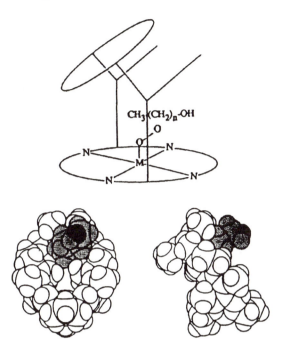

Figure 13. Multiple-varied receptors--macrobicyclic cyclidene with sites for metal ion and substrate: top, sketch of ternary complex; right, side view of substrate complex; left, view into cavity that contains substrate. (Reproduced with permission from reference 69. Copyright 1989, Kluwer).

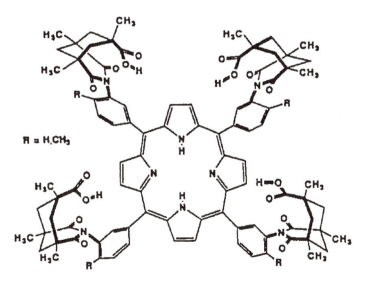

Figure 14. Multiple-varied receptors: porphyrin/Rebek receptor. (Reproduced with permission from reference 36. Copyright 1988, VCH).

molecules and complicated molecular systems. The key intellectual tools for the design of remarkably complicated, highly ordered molecular systems exist today, and these concepts provide reason for great excitement for the present and following generations of chemists.

Acknowledgment

The perspective championed here is the result of extended periods of research in coordination chemistry under the sponsorship of the National Science Foundation and the National Institutes of Health.

Literature Cited

1.   Werner, A. *Z. Anorg. Chem.* **1983**, *3*, 267. For a discussion and annotated English translation see Kauffman, G. B. Classics in Coordination Chemistry, Part 1: The Selected Papers of Alfred Werner; Dover: New York, 1968; pp 5-88.

2.   Werner, A. *New Ideas on Inorganic Chemistry*; Hedley, E. P., Transl.; Longmans, Green: London, 1911; (a) p 23.

3.   Lehn, J.-M. *J. Inclusion Phen.* **1988**, *6*, 353.

4.   Cram, D. J. *J. Inclusion Phen.* **1988**, *6*, 397.

5.   Pedersen, C. J. *J. Inclusion Phen.* **1988**, *6*, 337.

6.   Busch, D. H. *Chem. Rev.* **1993**, *93*, 847.

7.   Lehn, J.-M. In *Perspectives in Coordination Chemistry*; Williams, A. F.; Floriani, C.; Merbach, A. E., Eds.; Verlag Helvetica Chimica Acta: Basel; VCH: Weinheim, 1992; p 447 ff.

8.   Moore, C.; Pressman, B. C. *Biochem. Biophys. Res. Commun.* **1964**, 562. Pressman, B. C. *Ann. Rev. Biochem.* **1976**, *45*, 501.

9.   Pedersen, C. J. *J. Am. Chem. Soc.* **1967**, *89*, 2495.

10. Pedersen, C. J. *J. Am. Chem. Soc.* **1967**, *89*, 7017.

11. Pedersen, C. J.; Fresndorf, H. K. *Angew. Chem., Intl. Ed.* **1972**, *11*, 16.

12. Lehn, J.-M.; Sauvage, J. P. *J. Am. Chem. Soc.* **1973**, *97*, 6700.

13. Meade, T. J.; Busch, D. H. *Prog. Inorg. Chem.* **1985**, *33*, 59.

14. Graf, E.; Lehn, J.-M.; LeMoigne, J. *J. Am. Chem. Soc.* **1982**, *104*, 1672.

15. Kotzyba-Hibert, F.; Lehn, J.-M.; Vierling, P. *Tetrahedron Lett.* **1980**, 941.

16. Pascard, C.; Riche, C.; Cesario, M.; Kotzyba-Hibert, F.; Lehn, J.-M. *J. Chem. Soc., Chem. Commun.* **1982**, 557.

17. Simmons, H. E.; Park, C. H. *J. Am. Chem. Soc.*, **1968**, *90*, 2428, 2931.

18. Hoseini, M. W.; Lehn, J.-M. *J. Am. Chem. Soc.* **1982**, *104*, 3525.

19. Lehn, J.-M; Sonveaux, E.; Willard, A. K. *J. Am. Chem. Soc.* **1978**, *100*, 4914.

20. Dietrich, B.; Guilhem, J.; Lehn, J.-M.; Pascard, C.; Sonveaux, E. *Helv. Chim. Acta* **1984**, *67*, 91.

21. Dietrich, B.; Fyles, D. L.; Fyles, T. M.; Lehn, J.-M. *Helv. Chim. Acta* **1979**, *62*, 2763.

22. Chia, P. S. K.; Lindoy, L. F.; Walker, G. W.; Everett, G. W. *J. Am. Chem. Soc.* **1991**, *113*, 2533.

23. Alston, D. R.; Siawin, A. M. Z.; Stoddart, J. F.; Williams, D. J.; Zarzycki, R. *Angew. Chem., Int. Ed. Engl.* **1987**, *26*, 693.

24. Gross, P. F.; Hosseini, M. W.; Lehn, J.-M; Sessions, R. B. *J. Chem. Soc., Chem. Commun.* **1981**, 1067.

25. Manfrin, M. F.; Moggi, L.; Castelvetro, V.; Balzani, V.; Hosseini, M. W.; Lehn, J.-M. *J. Am. Chem., Soc.* **1985**, *107*, 6888.

26. Harata, K. In *Inclusion Compounds;* Atwood, J. L.; Davies, J. E. D.; MacNicol, D. D.,EdsAcademic Press: London, 1991; Vol. 5, Chap. 9.

27. Saenger, W. In *Inclusion Compounds* Atwood, J. L.; Davies, J. E. D.; MacNicol, D. D.,Eds.; Academic Press: London, 1984; Vol. 2, Chap. 8.

28. *Cyclodextrins and Their Industrial Uses;* Duchene, D.; Ed.; Editions de Sante: Paris, 1987.

29. Tabushi, I.; Kuroda, Y.; Kimura, Y. *Tetrahedron Lett.,* **1976**, *37*, 3327.

30. Tabushi, I.; Yamamura, K. *Topics in Current Chem.* **1983**, *113*, 145.

31. Murakami, Y. *Topics in Current Chem.* **1983**, *115*, 107.

32. Diederich F.; Griebel, D. *J. Am. Chem. Soc.* **1984**, *106*, 8037.

33. Diederich, F.; Dick, K.; Griebel, D. *Chem. Ber.* **1985**, *118*, 3588.

34. Dieterich, F. *Angew. Chem., Int. Ed., Engl.* **1988**, *27*, 362.

35. Hamilton, A. D.; Pant, N.; Muehldorf, A. *Pure & Appl. Chem.* **1988**, *60*, 533.

36. Rebek, Jr., J. *Environmental Influences and Recognition in Enzyme Chemistry;* VCH Publishers: New York, 1988; Chap. 8.

37. Rebek, Jr., J. *Angew. Chem., Int. Ed. Engl.* **1990**, *29*, 245.

38. Rebek, Jr., J. *Accounts Chem. Res.* **1990**, *33*, 399.

39. Jorgensen, W. L.; Severance, D. L. *J. Am. Chem. Soc.* **1991**, *113*, 209.

40. Tjivikua,T.; Deslongchamps, G.; Rebek, Jr., J. *J. Am. Chem. Soc.* **1990**, *112*, 8408.

41. Nowick, J. S.; Ballester, P.; Ebmeyer, F.; Rebek, Jr., J. *J. Am. Chem. Soc.* **1990**, *112*, 8902.

42. Jeong, K. S.; Jivkua, T.; Meuhldorf, A.; Famujlok, M.; Rebek, Jr., J. *J. Am. Chem. Soc.* **1991**, *113*, 201.

43. Hamilton, A. D.; Van Engen, D. *J. Am. Chem. Soc.* **1987**, *109*, 5035.

44. Muehldorf, A. V.; Van Engen, D.; Warner, J. C.; Hamilton, A. D. *J. Am. Chem. Soc.* **1988**, *110*, 6561.

45. Chang, S. A.; Hamilton, D. *J. Am. Chem. Soc.* **1988**, *110*, 1318.

46. Tecilla, P.; Chang, S.-K.; Hamilton, A. D. *J. Am. Chem. Soc.* **1990**, *112*, 9586.

47. Busch, D. H. *Coord. Chem. Rev.* **1990**, *100*, 119.

48. Martell, A. E.; Smith, R. M. *Critical Stability Constants;* Plenum: New York, 1974, 1975, 1977, 1976, 1982; Vols. 1-5.

49. Margerum, D. W.; Cayley, G. R.; Weatherburn, D. C.; Pagenkopf, G. K. *Coordination Chemistry;* ACS Monograph 74; Martell, A. E., Ed.; Amer. Chem. Soc.:Washington, 1978,p1f.

50. Cabbiness, D. K.; Margerum, D. W. *J. Am. Chem. Soc.* **1970**, *92*, 2151.

51. Hinz, F. P.; Margerum, D. W. *Inorg. Chem.* **1974**, *13*, 2941.

52. Jones, T. E.; Zinner, L. L.; Diaddario, L. L.; Rorabacher, D. B.; Ochrymowycz, L. A. *J. Am. Chem. Soc.* **1975**, *97*, 7163.

53.   Busch, D. H. *Proceedings of the First Hanford Separation Science Workshop*, **1993**, II.9-II.14.
54.   Boschmann, E.; Weinstock, L. M.; Carmack, M. *Inorg. Chem.* **1974**, *13*, 1297.
55.   Mason, S. F.; Peacock, R. D. *J. Chem. Soc., Dalton Trans.* **1973**, *226*.
56.   Childers, R. F.; Wentworth, R. A. D. *Inorg. Chem.* **1969**, 8, 2218.
57.   Melson, G. A.; Wilkins, R. G. *J. Chem. Soc.* **1963**, 2662.
58.   Taylor, L. T.; Busch, D. H. *J. Am. Chem. Soc.* **1967**, *89*, 5372.
59.   Busch, D. H. *Chem. Eng. News* June 29, **1970**, p 9.
60.   Busch, D. H.; Farmery, K.; Goedken,V.; Katovic, V.; Melnyk, A. C.; Sperati, C. R.; Tokel, N. *Adv. Chem. Ser.* **1971**, *100*, 44.
61.   Cram, D. J.; deGrandpre, M. P.; Knobler, C. B.; Trueblood, K. N. *J. Am. Chem. Soc.* **1984**, *106*, 3286.
62.   Chang, C. K. *J. Am. Chem. Soc.* **1977**, *99*, 2819.
63.   Meade, T. J.; Busch, D. H. *Prog. Inorg. Chem.* **1985**, *33*, 59.
64.   Kobayashi, N.; Akiba, U.; Takator, K.; Ueno, A.; Oas, T. *Heterocycles* **1982**, *19*, 2011.
65.   Collman, J. P.; Zhang, X.; Hembre, R. T.; Brauman, J. I. *J. Am. Chem. Soc.* **1990**, *112*, 5356.
66.   Meade, T. J.; Alcock, N. W.; Busch, D. H. *Inorg. Chem.* **1990**, *29*, 3766.
67.   Meade, T. J.; Takeuchi, K. J.; Busch, D. H. *J. Am. Chem. Soc.* **1987**, *109*, 725.
68.   Meade, T. J.; Kwik, W. -L.; Herron, N.; Alcock, N. W.; Busch, D. H. *J. Am. Chem. Soc.* **1986**, *108*, 1954.
69.   Busch, D. H.; Stephenson, N. A. *J. Incl. Phen. Mol. Recog. Chem.* **1989**, 7, 137.
70.   Lindsey, J. S.; Schreiman, I. C.; Hsu, H. C.; Kearney, P. C.; Marguerattaz, A. M. *J. Org. Chem.* **1987**, *52*, 827.

RECEIVED December 6, 1993

Chapter 13

Coordination Chemistry of Pigments and Dyes of Historical Interest

Mary Virginia Orna[1], Adrienne W. Kozlowski[2], Andrea Baskinger[1], and Tara Adams[1]

[1]Department of Chemistry, College of New Rochelle, New Rochelle, NY 10805
[2]Department of Chemistry, Central Connecticut State University, New Britain, CT 06050

Many pigments and dyes of historical interest exhibit the capability of forming Werner-type coordination complexes. The relationships between the structures and color characteristics of these colorants are reviewed in this paper, with particular attention to the work of Paul Pfeiffer, a student and long-time assistant of Alfred Werner's who did extensive work in the characterization and application of Werner's coordination theory to alizarin-related dyes.

The production and use of pigments and dyes in ancient, medieval, and modern times until the middle of the nineteenth century gradually evolved from the craftsman's art into a science. In the dyeing of textiles it was recognized even in prehistoric times that plant ashes and other materials such as lime and clay were important auxiliaries that conferred fastness and often color variation. Gradually dyers came to realize that these natural materials contained metals such as aluminum, tin, and iron and that it was the metals' presence that determined the desirable properties of the dyebath. Pigments, on the other hand, were essentially insoluble colorants applied to a substrate with the use of a medium such as water, egg yolk, or oil. Many colorants are used as either a dye or a pigment, the simplified generalized rule of thumb being that a pigment is water-insoluble and applied via a medium whereas a dye is water-soluble and penetrates or adheres directly to its substrate when applied from an aqueous medium (1).

Natural Colorants Before Perkin

Yellow Colorants. The most important yellow dye in ancient and medieval times was weld, a flavone (Figure 1a) derivative extracted from the seeds, stems, and leaves of *Reseda luteola L.*, commonly known as dyer's rocket. This colorant is resistant to atmospheric oxidation, rendering it quite lightfast and hence extremely popular and useful. In combination with the blue dye woad it was used to produce the Lincoln green made famous by Robin Hood and his merry men. Unlike weld, quercitron, a flavonol derivative (Figure 1b), is much more susceptible to degradation by light and was not as important (2). Safflower yellow, derived from carthamin, was often used as a surrogate for Spanish saffron, a polyene extracted from the stigmas of *Crocus sativus* (3).

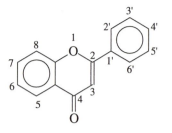

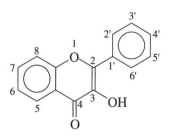

Figure 1a. Flavone (2-Phenyl-4H-1-   Figure 1b. Flavonol (3-Hydroxy-
benzopyran-4-one)   flavone)

Table I lists some important examples of each type of yellow dye discussed above. The structures of both luteolin and quercetin, the principal coloring matters of weld and quercitron, respectively, suggest that 4,5-type chelates could form with metal ions impregnated in the fibers to be dyed. Undoubtedly, the formation of these chelates lent stability to the colors of the dyed fibers, but their importance as chelates seems to be as limited at the importance of their respective dyes.

Blue and Purple Colorants. The only natural blue dyes in antiquity were indigo and woad, both containing the identical principal coloring matter, indigo, or indigotin (Figure 2), probably the oldest coloring matter known. The chemical identity of this colorant was, of course, unknown until the advent of modern chemistry, and both materials were thought to be distinctly different from one another. Although woad has always been associated with its vegetable origin, it was once thought, at least in England, that indigo was of mineral origin. The indigo-bearing plant, *Indigofera tinctoria*, was formerly grown all over the world, but the synthetic product has replaced the vegetable product since 1900. Woad is obtained from *Isatis tinctoria*, a herbaceous biennial indigenous to southern Europe (*4*), and contains as little as 1/30 the amount of indigo in its coloring matter. Although indigo does not form coordination compounds, it is mentioned here because of its historical importance. It is to this day one of the few naturally occurring dyes in wide use.

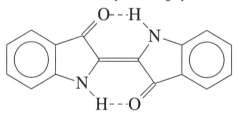

Figure 2. Indigo [2-(1,3-Dihydro-3-oxo-2H-indol-2-ylidene)-
1,2-dihydro-3H-indol-3-one], blue colorant used as a vat dye

Prussian blue is a notable coordination compound in that it is the first modern pigment to have a known history and established date of preparation (*4*). The most

Table I. Some Important Yellow Dyes

Dye	Chemical Compound	Source	Structure
Weld	Luteolin	Flavone dye extracted from the seeds, stems, and leaves of *Reseda luteola L.* (Dyer's Rocket)	
Quercitron	Quercetin	Flavonol dye from the bark of the North American oak, *Quercus tinctoria nigra*	
Safflower	Carthamin	Chalcone dye from dried petals of *Carthamus tinctorius* (Dyer's Thistle)	

commonly used formula is $Fe_4[Fe(CN)_6]_3$. The intense blue color arises from an intervalence charge transfer band involving $Fe^{2+} + Fe^{3+} \rightarrow Fe^{3+} + Fe^{2+}$.

The purple of antiquity was obtained from various genera of Mediterranean shellfish. The most important of these were *Phyllonotus, Thais, Dicathais, and Bolinus* (Linnaean terminology), but known commonly as *Murex* and *Purpura*. The one dye that is common to all of these muricids is 6,6′-dibromoindigo, a deep purple indigotin derivative which differs from the parent compound by the presence of two bromine atoms. The dye is extracted as a yellowish secretion from the hypobranchial gland of the shellfish (12,000 molluscs yield approximately 1.5 g of the dye), which contains the dye precursor, indoxyl sulfate, usually as the potassium salt (Figure 3) with a variety of substituents at the 2 and 6 positions, depending upon the species of mollusc (*5*). Under the influence of light and air, the precursor changes from yellow-

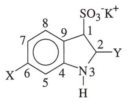

Figure 3. Potassium indoxyl sulfate, dye precursor of Tyrian Purple
(X = H, Br; Y = H, SCH $_3$, SO$_2$CH$_3$)

ish, to green, to blue, and finally to purple within a few minutes. The colored matter is fairly insoluble in water and must be reduced to the water-soluble leuco form in the presence of fibers and then reoxidized to form the colored dye on the fiber. The whole process is known as vat dyeing (*6*). Debromination of the purple dye yields the parent compound, indigotin. Varying proportions of the dibromoindigo, monobromoindigo, and nonbrominated indigo yield varying shades of purple to blue.

The historic interest in this dye is evidenced by the fact that Pliny the Elder (A.D. 23–79) devotes six chapters of his celebrated *Natural History* to its production, including treatises on the nature and kinds of shellfish, the history of its appearance in Roman outerwear, the high cost of the dye, and the methods used to obtain and process it (*7*). Furthermore, volumes, and indeed, doctoral dissertations have been written on its role in the production of the Biblical blue, or tekhelet, but the nature of this mysterious blue remains doubtful to this day (*8*).

Black Colorants. Satisfactory synthetic black dyes are scarce even to this day. Most modern black shades are obtained by mixing two or more dyes together. Thus logwood, the only important black dye of the pre-Perkin era, still enjoys usage in the dyeing of silks and leather. Logwood, derived from the heartwood of the Central American tree, *Hematoxylon campechiancum L.*, was imported to Europe in the early sixteenth century. It was a relatively unimportant red dye until the French chemist Michel Eugène Chevreul (1786–1889) discovered that it combined with metallic salts to give colored lakes. The principal coloring constituent is hematein (Figure 4), which, when combined with chromium, yields the black shades for which logwood is renowned. The structure of the hematein-chromium chelate complex is yet to be elucidated, but it is thought that it has a macromolecular structure in which the chromium ions link hematein molecules together by chelation (Figure 5).

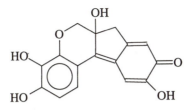

Figure 4. Hematein [6a,7-Dihydro-3,4,6a,10-tetrahydroxybenz
[b]indeno[1,2-d]pyran-9(6H)-one]

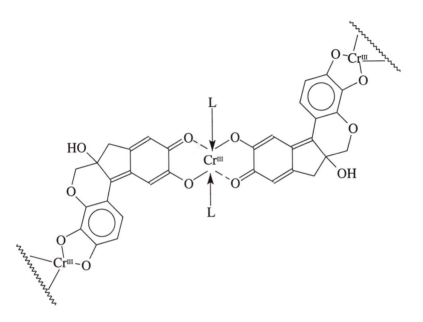

Figure 5. Logwood, a black chroman-type dye (L = Ligand)

Red Colorants. The most important class of pre-Perkin colorants, from the coordination chemistry point of view, is the red colorants. All of the major red colorants, whether of animal or vegetable origin, are derivatives of anthraquinone (See Table II). The principal red coloring matter of madder is alizarin, or 1,2-dihydroxyanthraquinone. Natural madder contains a considerable amount of another colorant, purpurin, or 1,2,4-trihydroxyanthraquinone, which accounts for the various shades of red to purplish-red that can be obtained when different sources of madder are used. In almost every source that describes dyeing with anthraquinone derivatives, the use of a metallic salt is mentioned. The principal metals used were aluminum, iron, tin, and in later years, chromium. The resulting color on the fiber depended in large part upon the metal used since the dyes chelated with the metal ions. The colors of different chelate compounds with alizarin and cochineal are shown in Table III.

Table II. Anthraquinone-Based Red Colorants

Dye	Structure
Anthraquinone	
Madder or Alizarin. Roots of the *Rubia tinctorum* plant. Roots were known as "alizari," hence alizarin	
Cochineal (Carminic Acid). Female insect, *Coccus cacti*, which lives on Prickly Pear cactus, found in Mexico. 200,000 insects yield 1 kg. of dye	
Kermes (Kermesic Acid). Female scale insects, *Coccus ilicis*, which infect the Kermes oak	

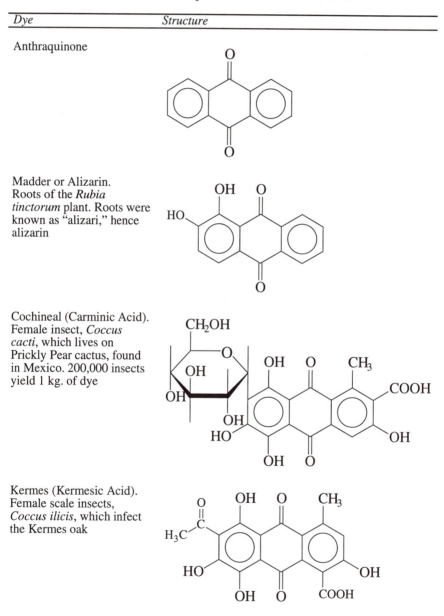

Cochineal is a red colorant derived from female scale-insects; its chief coloring matter is carminic acid, the structure of which is shown in Table II.

Table III. Colors of Chelate Compounds of Alizarin and Cochineal

Mordant	Alizarin	Cochineal
None	Brownish-Yellow	Scarlet Red (aq. soln.)
Aluminum	Red	Crimson
Tin	Pink	Scarlet
Chromium	Puce Brown	Purple
Iron	Brown	Purple
Copper	Yellow Brown	

SOURCE: Adapted from ref. 9.

Figure 6 illustrates the formation of 2:1 neutral complexes of divalent metal ions and 1-hydroxyanthraquinones. Figure 7 illustrates the possibility of the formation of polymeric complexes with 1,4- and 1,5-dihydroxyanthraquinones with divalent metal ions.

An interesting variation on the chelate formation of alizarin derivatives is the production of Turkey red, a brilliant red dye known from ancient times. Recipes for its production insist not only on aluminum, but on lime, in order to achieve the red-purplish color and the high fastness properties for which it is famed. Its probable structure is shown in Figure 8 (*10*). Some modern mordant dyestuffs derived from alizarin are shown in Table IV.

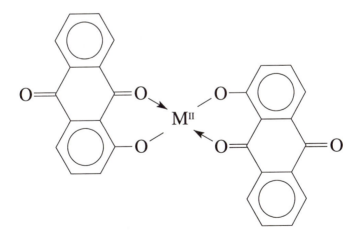

Figure 6. 1-Hydroxyanthraquinones form neutral 2:1 complexes with divalent metals such as Cu & Zn

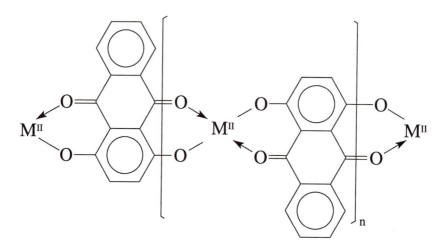

Figure 7. 1,4- and 1,5-dihydroxyanthraquinones form
polymeric complexes with divalent metals

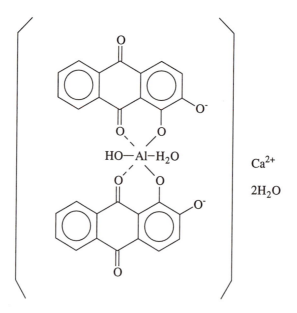

Figure 8. Probable Structure of Turkey Red 1:2 Complex of Aluminum with Alizarin.

Table IV. Mordant Dyestuffs Derived from Alizarin

Dye	Structure
Alizarin Orange A	
Brilliant Alizarin Bordeaux R	
Alizarin Red PS	
Borate ester of 1,2-dihydroxy-anthraquinone. Treatment with acetic anhydride yields 1-hydroxy-2-acetoxyanthraquinone, the 1-hydroxy group being protected against electrophilic attack because of coordination to boron	

Application of Werner's Coordination Theory to Alizarin-Related Dyes

Paul Pfeiffer (1875–1951), Alfred Werner's student and long-time assistant, published in *Liebig's Annalen* in 1913 (*11*) an extensive paper describing the application of Werner's coordination theory to alizarin-related dyes. Through a detailed series of experiments involving various substituted quinones, he showed that alizarin coordinates to tin through *both* the carbonyl oxygen and the adjacent 1-hydroxyl group, as illustrated in Figure 9. He particularly pointed out how this reaction is an example of Heinrich Ley's inner complexes.

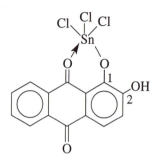   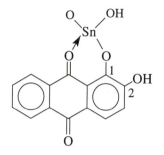

Figure 9. Metal complex formation with hydroxyanthraquinones

Figure 10. Compound used by Pfeiffer in dyeing experiments

Pfeiffer's Experimental Observations. Pfeiffer knew that $SnCl_4$ reacts smoothly on heating with a variety of 1-hydroxyanthraquinones. For example, alizarin (Table II) reacted to give a greenish-brown solution from which a violet-black powder crystallized. His analysis showed that only three chlorides remained per tin atom, as represented by his structure shown in Figure 9. The dark powder dissolved in aqueous ammonia (NH_3) to give a dark violet solution; in alcohol it gave an orange solution which yielded orange needles upon addition of water.

The chlorides in the original complex (Figure 9) could be replaced by oxide and hydroxide by dissolving the powder in pyridine and adding a small amount of water. Figure 10 illustrates the structure that Pfeiffer assigned to the orange-red needles obtained from this reaction. This product dissolved in ammonia to give a deep orange-red solution which was stable on heating. Pfeiffer determined that silk and wool were dyed red by this solution, but that cotton was not.

Pfeiffer's observation that alizarin dye lakes (soluble dyes precipitated on insoluble substrates) contained alkaline earth metals (notably calcium from lime) as well as coordinated tin or other mordant metals such as chromium, aluminum, or iron led to his investigations of the role of the 2-hydroxyl groups in anthraquinone dyes. He collected a large body of data showing that strong bases neutralize 2-hydroxyl groups in preference to 1-hydroxyl groups. Furthermore, he noted that the color of the dye is deepened when a 2-hydroxyl group is present in its ionized form. When the 2-hydroxyl group is absent, the tin chloride derivatives of 1-hydroxyanthraquinone are red, while alizarin derivatives which possess an ionized 2-hydroxyl group are violet-black.

This study of Pfeiffer's neatly fit mordant dyes into Werner's coordination theory by differentiating between inner complex formation with tin and ionic salt formation with strong bases. Pfeiffer did not pursue any further research on dyes.

Metal Ions as Mordants

The effect of metal ions as mordants for anthraquinone dyes has been noted for many years. Alizarin was usually dyed on wool mordanted with aluminum to give red shades, or mordanted with chromium which produced browns and violet-browns. A typical dyeing recipe involved the slow heating of wool, alum and potassium acid tartrate, followed by addition of alizarin and calcium acetate.

Chrome Mordants. Chrome mordants have been used extensively, giving a brighter color than other transition metals. The usual recipe involved heating the wool with potassium dichromate ($K_2Cr_2O_7$) in a dilute sulfuric acid solution before adding the dye. In the dyeing process amino acid residues are coordinated to Cr(III) in the mordanted wool, but the Cr(III) can be removed with oxalate. Hartley (*12*) has proposed that the mechanism for this chrome process involves a series of two-electron reductions, from Cr(VI) to Cr(IV) and from Cr(IV) to Cr(II). Cr(II) is labile and coordinates to the wool. It is subsequently oxidized to Cr(III) in air. It has long been known by dyers that the chromium uptake by wool is much faster from Cr(VI) solutions that from Cr(III) solutions. The reduction of Cr(VI) is believed to occur by reaction with disulfide linkages in the wool, although sometimes reducing agents such as formic or lactic acids or potassium hydrogen tartrate can also play a role.

Spectroscopic Evidence. In the case of M = Cr(III) infrared spectra suggest that the Cr(III) is coordinated to the protein carboxyl groups (*13*); in addition, Hartley's data show that sulfur is not coordinated. The esterification of the carboxylic acid group reduces chromium uptake, but blocking amines by dinitrophenylation does not inhibit this process (*14*).

While the effects of mordanting on the color properties of alizarin and cochineal dyes have been widely noted, spectra of the mordanted dyes and the shifts induced by various metals do not appear to have received much attention.

Uncoordinated alizarin shows a broad, intense ultraviolet envelope with a low energy absorption at 430 nm. This has been assigned to an $n \rightarrow \pi^*$ transition. Labhart (*15*) studied the effects of various substituents on the spectrum of anthraquinone and noted that groups in the 1-position, which are capable of hydrogen bonding to the carbonyl group, cause the absorption to move to longer wavelengths -- what dyers call the bathochromic shift.

This shift is the same effect observed with metallized derivatives of alizarin. In either case electron delocalization is enhanced, and the absorption wavelength maximum moves farther into the visible region. Ionizing a second hydroxyl group in the adjacent 2-position further enhances electron donation into the ring system and increases the intensity of the absorption as well as its bathochromic shift. These observations help to explain the empirically discovered benefits of using alkaline earth salts of the mordanted dye.

Use of alizarin dyes continues today with substituted molecules such as those given in Table IV.

Conclusion

Although the chemistry of ancient dyes and mordanting processes was developed empirically, today we understand much of it in terms of modern coordination theory. However, many aspects of color shifts still await investigation.

Literature Cited

1. Price, R. In *Comprehensive Coordination Chemistry: The Synthesis, Reactions, Properties & Applications of Coordination Compounds*; Wilkinson, G., Ed.; Pergamon Press: New York, NY, 1987; Vol. 6; pp 35-94.
2. Gregory, P. F.; Gordon, P. *Organic Chemistry in Colour*; Springer-Verlag: New York, NY, 1983.
3. *The Merck Index, 11th Ed.*; Budavari, S., Ed.; Merck & Co., Inc.: Rahway, NJ, 1989.
4. Gettens, R. J.; Stout, G. L. *Painting Materials: A Short Encyclopaedia*; Dover Publications: New York, NY, 1966.
5. Elsner, O. In *Dyes in History and Archaeology, No. 10*; Rogers, P. W., Ed.; Textile Research Associates: York, UK, 1992; pp 11-16.
6. Koren (Kornblum), Z. C. In *Colors from Nature: Natural Colors in Ancient Times*; Sorek, C.; Ayalon, E., Eds.; Eretz Israel Museum: Tel Aviv, Israel, 1993; pp 15*-31.*
7. Meldola, R. *J. Soc. Dyers Colour.* **1910**, *26*, pp. 103-111.
8. *The Royal Purple and the Biblical Blue*; Spanier, E., Ed.; Keter Publishing House: Jerusalem, Israel, 1987.
9. Peters, R. H. *Textile Chemistry, Vol. III: The Physical Chemistry of Dyeing*; Elsevier Scientific Publishing Co.: New York, NY, 1975; p 649.
10. Kiel, E. G.; Heertjes, P. M. *Rec. Trav. Chim.* **1965**, *84*, 89 ff.
11. Pfeiffer, P. *Justus Liebig's Annalen der Chemie* **1913**, *398*, 137 ff.
12. Hartley, F. *J. Soc. Dyers Colour.* **1969**, *85*, 66 ff; **1970**, *86*, 209 ff; *Aust. J. Chem.* **1969**, *22*, 229 ff.
13. Hartley, F. *Aust. J. Chem.* **1968**, *21*, 2723 ff.
14. Meekel, L. *Textilveredlung* **1967**, *2*, 715 ff.
15. Labhart, H. *Helv. Chim. Acta* **1957**, *40*, 1410 ff.

RECEIVED October 18, 1993

Chapter 14

The Importance of Non-Bonds

Michael Laing

Department of Chemistry, University of Natal, Durban 4001, South Africa

The term "coordination complex" almost invariably brings to mind a transition metal atom to which are bonded six atoms in an octahedron. We now automatically concentrate our thinking on these metal-to-ligand bonds and tend to assume that they are all-important in the behavior of the compound. In fact, the stoichiometry, coordination geometry, and reactivity of many complexes is dictated not by the metal and the atoms to which it is bonded, but by the bulk of parts of the ligands *remote* from the bonding center: the Non-Bonds between the ligands.

Alfred Werner was born French in Mulhouse, Alsace; he grew up and received his schooling in German as an outcome of the Franco-Prussian War; and he was a citizen of Switzerland when he became famous. Although he achieved greatness as an inorganic coordination chemist, he began his career as an organic chemist. Thus it seems appropriate to begin this discussion of Non-Bonds with the important contributions of the great Russian crystallographer, A.I. Kitaigorodskii, to an understanding of the principles governing the structure of crystals of organic compounds.

Close Packing

In 1961 the English translation of Kitaigorodskii's classic book was published (*1*). In it he emphasized that molecules have a well-defined volume and shape and that the atoms have a "hard" contact surface which prevents two non-bonded atoms from approaching closer than the sum of their contact or van der Waals radii. (The well-known Leybold and CPK models incorporate this principle.) He concluded that it is the thickness and shape of the molecules that dominate their mode of packing in the solid state. They pack together in the crystal by a "bump-in-hollow" process, the projections of one molecule fitting into the hollows of adjacent molecules, invariably

0097–6156/94/0565–0177$08.00/0

with each molecule in contact with six others in a layer, to give material of maximum density with a minimum energy (2) (Figures 1 and 2). It is important to understand that there is NO essential physical difference between the non-bonded repulsions between atoms within a molecule and these intermolecular repulsions between molecules in the solid state (3). Thus, for example, while two methyl groups within a coordination compound may repel each other, they will still pack together by the bump-in-hollow process, with their C-H bonds acting like gears or knuckles. This process considerably reduces the effective van der Waals volume of the group from the typical value of about 2 Å for the radius of a randomly or freely rotating methyl group. The implications of Kitaigorodskii's ideas are enormously wide-ranging because they apply not only to organic molecules and crystals but to every coordination compound.

Steric Approach Control

In 1957 Dauben described effectively the same concept but from an entirely different point of view — that of the organic chemist looking at the reduction of a carbonyl group to an alcohol (4). He found that the conformation of the OH group produced by reduction of the C=O group by the BH_4^- ion was determined by the pathway of LEAST hindrance to C(3) (Figure 3). If the approach from above, **A**, was open, then the H atom went onto C(3) from above and the subsequently formed OH group was *axial* down, the energetically LESS favoured conformation. The easier the approach from above, the more likely it was to get the thermodynamically less stable *axial* hydroxyl group below. If bulky groups were present on C(2), C(4), or C(10) above, then the attack came from below and the OH group was formed on the same side as the bulky groups. He termed this phenomenon "Steric Approach Control" — control by kinetics, where the first-formed compound was the thermodynamically less favored isomer.

Rate of Reaction

Basolo gives a beautiful example of the effect of the physical blocking of the reaction pathway and thus the reduction in reactivity at the metal center (5). In the series of compounds *trans*-[PtCl(PEt$_3$)$_2$L] where L is a phenyl ring, *ortho* substituted, the rate of nucleophilic substitution of Cl by pyridine systematically decreases as the bulk of the *ortho* substituent increases:

L	k_{obs}, sec^{-1}
phenyl	1.2×10^{-4}
o-tolyl	1.7×10^{-5}
mesityl	3.2×10^{-6}

The cause is easily pictured: the "top" of the metal atom is screened from the incoming nucleophile by the bulk of the methyl group on the phenyl ring, which is remote, chemically inert, and in no way bonded to the metal atom. The methyl groups simply block the pathway for the incoming ligand (Figure 4).

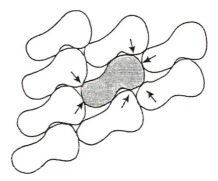

Figure 1.
Packing of randomly shaped molecules by a simple bump-in-hollow process.

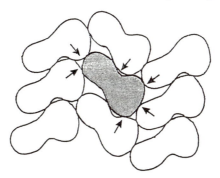

Figure 2.
Randomly shaped molecules close-packed by "glides" to yield a structure with a rectangular cell. Note how each molecule touches six others.

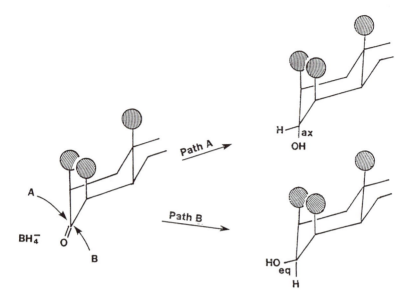

Figure 3.
If the axial groups at C(2), C(4), and C(10) are small, the BH_4 will approach along pathway **A**. If the groups are bulky, the attack is along path **B**.

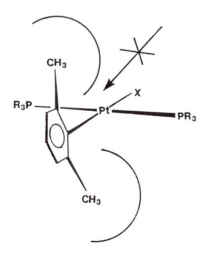

Figure 4.
A schematic diagram of the compound *trans*-[PtCl(PEt$_3$)$_2$L], L = mesityl, showing how the bulk of the two methyl groups blocks access to the platinum atom.

Cone Angle for Phosphine Ligands

As a result of his studies of the rates of substitution of metal carbonyls by various phosphines and of the equilibrium constants for decomposition of the substituted complexes, Tolman concluded that in many cases it was not electronic bonding effects that control the reactions but that it was rather the bulk of the phosphine ligand that dictates the lability of the compounds (*6*). He developed the concept of Cone Angle for a ligand, defined in terms of an idealized $M-PX_3$ system, where the M-P distance was 2.28 Å, and the PX_3 ligand was constructed of CPK models. The Cone Angle Θ was the apex angle of the ideal cone, which had its apex at the metal atom and just touched the van der Waals surface of the peripheral atoms of the ligand, usually hydrogens (*7*). Typical values of Θ, the Cone Angle, are: PF_3 104°; PMe_3 118°; PPh_3 145°; $P(t\text{-But})_3$ 182° (Figure 5). Subsequently, modifications have been suggested for estimating the Cone Angle of asymmetric ligands of the class PXYZ (*8 -11*). Also, the effective Cone Angle as observed and measured from crystal structures often is considerably smaller than the ideal. This is a result of the bump-in-hollow gearwheel effect as well as of compression of the Metal-P-R angles. Nevertheless, the principle is correct, and the concept is of great value, as is made clear in a recent review by McAuliffe (*12*). The importance of Cone Angle in metal-catalyzed reactions in organic synthesis has also been elegantly described (*13*).

cis- or *trans*-[PtCl$_2$(PR$_3$)$_2$]?

Forty years ago Chatt and Wilkins studied the *cis-trans* equilibria for these square planar complexes of platinum (*14*). It was evident that bond energies favored the *cis* isomer. This is caused by the π back-bonding from the metal d-orbital being concentrated on only one phosphorus atom. The isomer which has the greater Pt-P double-bond character will have the larger overall bond strength. This strengthening of the Pt-P bond will be favored by the arrangement Cl-Pt-PR$_3$. Thus the *cis* square planar isomers will be favored if bonding effects ALONE were in control (In the *trans* case two phosphorus atoms would of necessity be competing for the favors of the same d-orbital). The effect is also clearly seen in the *facial* and *meridional* isomers of [IrCl$_3$(PMe$_2$Ph)$_3$] (*15*). In the *fac* isomer the mean bond lengths are: Ir-Cl 2.46 Å; Ir-P 2.29 Å. In the *mer* isomer, the mean length for the *trans* pair of Ir-P bonds is 2.37 Å, while that for the *trans* pair of Ir-Cl bonds is 2.36 Å. On the other hand, the lengths of the unique bonds of the *trans* Cl-Ir-P moiety are: Ir-Cl 2.43 Å; Ir-P 2.28 Å (values very close to those found in the *facial* isomer). The Ir-P bond is considerably shorter and stronger, indicating that M-P bond strengths favor the *facial* isomer for [MX$_3$(PR$_3$)$_3$] as well as the *cis* isomer for [MX$_2$(PR$_3$)$_2$].

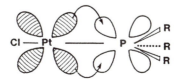

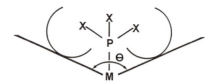

Figure 5.
The Cone Angle Θ, as defined for a typical phosphine ligand PX_3.

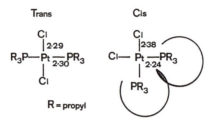

R = propyl

Figure 6.
The *cis* and *trans* isomers of [PtCl$_2$(PPr$_3$)$_2$] showing Pt-P and Pt-Cl bond lengths and how the van der Waals envelopes of the *cis* phosphines interfere.

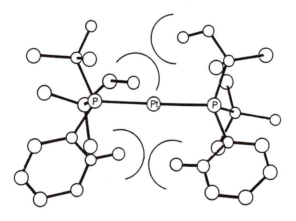

Figure 7.
The molecule [Pt{PPh(*t*-But)$_2$}$_2$], showing how the van der Waals surfaces of the hydrogen atoms block access to the platinum atom.

However, if the alkyl groups on the phosphines are large, the van der Waals repulsions between the groups will cause the square planar compound to isomerize to yield the *trans* isomer, which does not suffer from inter-ligand compression strain. A good example is [PtCl$_2$(PPr$_3$)$_2$] which can be prepared as the pure *cis* isomer, but on equilibration in benzene for half an hour, it yields an equilibrium mixture that contains 97% of the *trans* isomer (5). Non-bonds are clearly beating bonds. The larger the Cone Angle of the phosphine ligand, the more the *trans* isomer is favored (Figure 6).

Phosphine Complexes of Platinum and Palladium

Compounds like [Pt(PPh$_3$)$_4$] and [Ni(PMe$_3$)$_4$] are easily made (*16*). However, in solution [Pt(PPh$_3$)$_4$] dissociates, forming the three-coordinate complex [Pt(PPh$_3$)$_3$]. Determination of the crystal structure of [Pt(PPh$_3$)$_4$] shows quite certainly that the metal atom is four- coordinate with P-Pt-P angles close to 109^0, *i.e.*, it is tetrahedral (*17*). This appears to contradict Tolman's value of 145^0 for the Cone Angle of triphenylphosphine. In fact, it does not, because in this tetrakis case the phenyl rings are locked together like gears. Nevertheless, there are severe compression strains, which cause dissociation of the complex in solution. By the same token, the inter-ligand repulsions in the compound [Pt(P(cyclohexyl)$_3$)$_3$] are surprisingly small, a result of the "intermeshing" of the cyclohexyl groups reducing the effective Cone Angle far below Tolman's estimate of about 170^0 (*18*).

If the phosphine ligand is large enough, only two ligands can be fitted onto the platinum atom. The compound [Pt{PPh(*t*-But)$_2$}$_2$] is an example (*19*). The P-Pt-P backbone is linear, and the metal atom is shielded from the outside world by the organic moieties, thus rendering the platinum atom inert to attack. The Cone Angle of PPh(*t*-But)$_2$ is 170° (Figure 7).

Addition of O$_2$ to [Platinum(0)(phosphine)$_2$]

It is possible to prepare a variety of diphosphine complexes of platinum(0) and palladium(0) (*19*). Some of these compounds are sensitive to oxygen, forming compounds of the class [Pt(PR$_3$)$_2$O$_2$], while others are quite inert to oxygen (*11*). The explanation is simply given by consideration of the Cone Angles of the phosphine ligands involved. Where the ligand has a sufficiently large Cone Angle, *e.g.*, PPh(*t*-But)$_2$, Θ = 170^0, the compound is inert because there is neither a pathway for the oxygen molecule to approach the metal atom nor is there place in the coordination shell about the metal atom if the O$_2$ should come within bonding distance. When the Cone Angle is smaller, the compound is easily formed, *e.g.*, [Pt(PPh$_3$)$_2$] immediately reacts with oxygen to yield [Pt(PPh$_3$)$_2$O$_2$] (*20*).

The compound [Pt{PPh(*t*-But)$_2$}$_2$O$_2$] can be made by an indirect route (*21*) and the crystal structure shows beautifully how the anisotropy of the ligand allows the compound to exist. The phenyl rings of the two phosphine ligands lie face-to-face, just touching, in a typical Kitaigorodskii close-packing mode, thus reducing the effective Cone Angles of the ligands to give a P-Pt-P angle of 113°. It is impossible to fit the two ligands onto the metal in any other orientation (Figure 8).

Reactive [RuH(PMe$_2$Ph)$_5$]$^+$

A number of cationic hydride complexes of the type [RuHL$_5$]$^+$, L = P(OR)$_3$, were prepared in 1974. They were remarkably inert in solution. Subsequently, the compound [RuH(PMe$_2$Ph)$_5$]$^+$, with a far bulkier phosphine ligand, was found to be highly reactive in solution and a precursor to a large range of new Ru(II) complexes formed by the substitution of one or more of the phosphine ligands (*22*). The crystal structure of the PF$_6^-$ salt clearly showed the severe compression strains of the PMe$_2$Ph ligands, the "interlocking" of the methyl and phenyl groups of adjacent ligands to effectively reduce the Cone Angle, and (in some senses more interesting) the way in which the H atom on the ruthenium atom was shielded from the outside world by the van der Waals bulk of the phenyl and methyl groups of four of the PMe_2Ph ligands. What should be a labile H atom is rendered inert by non-bonds (Figure 9).

"Umbrella" Effect in the Purple Isomer of $[Ru(S_2CH)(PMe_2Ph)_4]^+$

Reaction of $[RuH(PMe_2Ph)_5]PF_6$ with CS_2 yields an orange dithioformato complex $[Ru(S_2CH)(PMe_2Ph)_4]^+$ (23), which rearranges in boiling methanol to a purple isomer (24). This isomer had three PMe_2Ph ligands arranged facially on the ruthenium atom, while the P atom of the fourth was bonded to the carbon of the CS_2 group. The effective Cone Angles of the PMe_2Ph groups are larger in the purple isomer with a coordination number of 5 than in the orange isomer with a coordination number of 6, indicating that relief of the non-bonded repulsions accompanied the transfer of the unique PMe_2Ph group. This rearrangement is reversible if the purple isomer is heated with $P(OR)_3$ ligands of smaller Cone Angle, giving an orange dithioformato compound with the two smaller ligands as the equatorial *cis* pair, *trans* to the two S atoms of the S_2CH moiety. Reducing the steric compressions in the equatorial plane thus allows the Ru-P bonds to become favored (Figure 10).

This example shows how delicate is the balance between bonds and non-bonds; how fine the line is between the electronic covalent bonding forces and the repulsions due to the bulk of inert groups.

More interesting, however, is the inertness of the vacant coordination site in the purple isomer. The reason is simple: the Ru atom is protected by a phenyl ring "umbrella," which completely blocks access to the Ru atom. The Ru···C distances exceed the sum of their van der Waals radii, but the gap is nevertheless small enough to exclude approach of either solvent or reactant molecules (Figure 11).

16-Electrons or 18-Electrons?

Reduction of $[\{RhCl(CO)(PhO)_2PN(Et)P(OPh)_2\}_2]$ with Zn/Hg in methanol under CO gas yields the unexpected: NOT the symmetrical structure with four CO and two 18-electron rhodium atoms but $[Rh_2(CO)_3\{(PhO)_2PN(Et)P(OPh)_2\}_2]$, a compound with only three CO groups, all terminal (25). The crystal structure shows that one Rh atom is 5-coordinate and obeys the EAN rule, while the other is square planar with a 16-electron configuration. The back of this Rh atom is completely shielded by a phenyl ring which blocks all access, thus preventing reaction with the CO molecules in solution. The 16-electron configuration is thus protected by a distant phenyl ring "umbrella", remote from the rhodium atom (Figure 12).

Complexes of $CoCl_2$ with 4-Substituted Pyridines — Tetrahedral or Tetragonal?

Ethanolic solutions of compounds $[CoCl_2(4-Rpy)_2]$ are deep blue, but for some substituents, R, violet-pink crystals are formed, while for other substituents the crystals are deep blue (suggesting that only in these cases is the species in the crystal identical

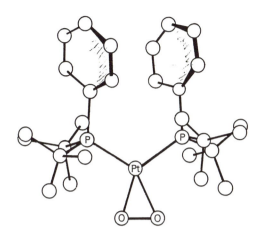

Figure 8.
The molecule [Pt{PPh(*t*-But)$_2$}$_2$O$_2$]. The two phenyl rings lie face to face, just touching.

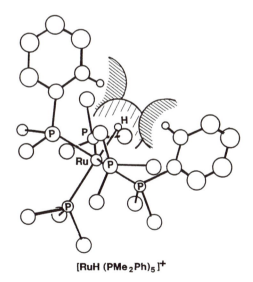

[RuH (PMe$_2$Ph)$_5$]$^+$

Figure 9.
The cation [RuH(PMe$_2$Ph)$_5$]$^+$, showing how the van der Waals surfaces of the hydrogen atoms on the two phenyl rings protect the H atom that is bonded to the ruthenium.

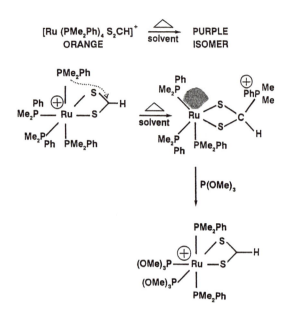

Figure 10.
A schematic pathway showing how reaction of the "purple" isomer with phosphite ligands of small Cone Angle will cause reversal of the reaction and thus yield the "orange" isomer.

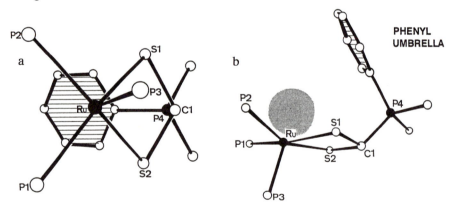

Figure 11.
(a) Top view and (b) Side view of the "purple" isomer of $[Ru(S_2CH)(PMe_2Ph)_4]^+$, showing how the phenyl ring bonded to P4 protects the back of the ruthenium atom.

with that in solution) (*26*). Violet crystals are formed where R = H, CN, vinyl, or phenyl. Determination of the crystal structures gives an explanation (*27, 28*).

In the crystals of the violet compounds the molecules are stacked to give a polymeric ladder of -CoCl$_2$CoCl$_2$- atoms. Each Co atom is bonded to four Cl atoms in a square plane, with the pyridine ring perpendicular to this plane, giving the cobalt a coordination number of 6 with a tetragonal geometry. The Co $\cdot \cdot \cdot$ Co repeat distance is about 3.7 Å, which can accommodate the pyridine ring whose thickness is about 3.4 Å. Thus this polymeric stacking will occur for any of these [CoCl$_2$(4-Rpy)$_2$] complexes as long as the van der Waals diameter of the group R is 3.7 Å or less. If R is larger, then it becomes impossible to stack the groups R while maintaining the infinite CoCl$_2$ chain ladder. Whatever bonding advantages the chain of CoCl$_2$ atoms may have in the tetragonally coordinated violet structures, it is insufficient to overcome the non-bonded repulsions between the groups R, once their diameter exceeds the critical value of about 4 Å. As a result, the compound simply crystallizes as individual blue tetrahedral molecules, with the cobalt atom having the coordination number of four, which is what occurs for [CoCl$_2$(4Me-py)$_2$] (*27*).

Interestingly, these compounds are isostructural with the zinc analogues in which the zinc atom is always tetrahedrally coordinated in both solution and in the crystal, independent of the bulk of the group R (Figure 13).

Acetylacetonate Derivatives of Nickel(II)

The bonding and stereochemistry of d^8 nickel(II) is susceptible to the bulk of groups remote from the metal atom. The anhydrous acetylacetonate is trimeric, [Ni$_3$(ACAC)$_6$], green, and paramagnetic, with each nickel atom bonded to six oxygen atoms (*29*) (Figure 14). If the methyl groups of the ACAC ligands are replaced by tertiary butyl groups, the anhydrous compound is red, diamagnetic, and monomeric, [Ni(*t*-ButACAC)$_2$] (*30*) (Figure 15).

It is the bulk of the tertiary butyl groups that prevents the formation of the complicated arrangement of ligands that is required for the "trimeric" structure (Figure 16). Of course, the mode of packing of the individual trimeric molecules in the crystal is exactly as predicted by Kitaigorodskii: Bump-in-Hollow.

Neocuproin : 2,9-Dimethyl-1,10-phenanthroline and Copper

A final example of how bulk can make a ligand specific, not only to the metal but also to its oxidation state, is the case of neocuproin and copper(I). This ligand cannot coordinate with Cu^{2+} to give a square planar complex [CuL$_2$]$^{2+}$ because of the collisions between the methyl groups of the two ligands. However, reduction of d^9 Cu^{2+} to d^{10} Cu$^+$ immediately and quantitatively yields a colored product [CuL$_2$]$^+$, in which the four Cu-N bonds are arranged in a tetrahedron with the two ligands lying in mutually perpendicular planes. In this arrangement the methyl groups no longer collide and the complex is stable (*31*), so much so that it can be used for the quantitative determination of copper in solution (Figure 17).

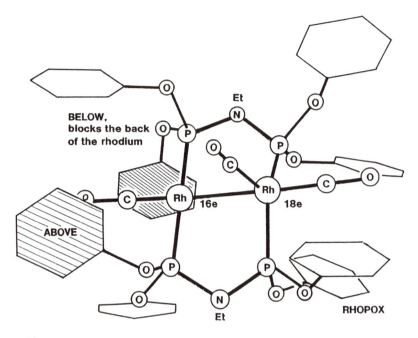

Figure 12.
The structure of $[Rh_2(CO)_3\{(PhO)_2PN(Et)P(OPh)_2\}_2]$, showing how the square planar
Rh atom is protected by the phenyl ring, which blocks the approach of any potentially
reactive species.

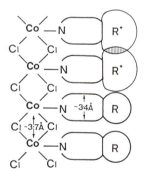

Figure 13.
A schematic diagram of the crystal structure of a typical violet $[CoCl_2(4-Rpy)_2]$
compound. When the van der Waals diameter of the group R* is about 4 Å, they will
have to interpenetrate if the molecules pack in this chain-like mode. This is impossible.

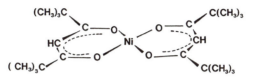

Figure 14.
The trimeric complex [Ni$_3$(ACAC)$_6$], with the planar ACAC groups represented by the curved lines and cross-hatching.

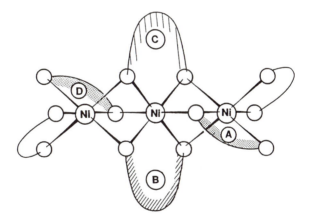

Figure 15.
The square planar molecule [Ni(*t*-ButACAC)$_2$].

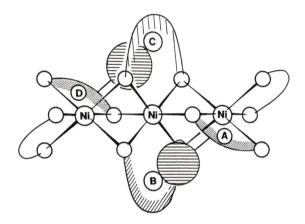

Figure 16.
The hypothetical trimeric isomer of [Ni(*t*-ButACAC)$_2$] cannot form because of collision of the tertiary butyl group on ring A with ring B and the tertiary butyl group on ring D with ring C.

Figure 17.
The complex [Cu(neocuproin)$_2$]$^+$, showing the tetrahedral coordination about the Cu(I) ion and the manner in which the methyl groups are accommodated.

Conclusion

The examples presented above are not "recent" but were chosen deliberately with the old Russian adage in mind: "The best of the new can often be found in the long-forgotten past, which in turn must inspire the future." From these simple examples can be learned many valuable lessons, all of which are encompassed by one simple concept: it is the bulk of the groups NOT directly bonded to the metal atom that can cause decomposition to relieve compression, can speed up reactions, and can cause inertness by blocking the reactant's pathway. Thus the non-bonded repulsions between those parts of ligands remote from the metal center can have a critical influence on the behavior of the metal atom and can control the structure and reactivity at the metal center of that coordination compound. This is the importance of Non-Bonds.

Literature Cited

1.   Kitaigorodskii, A. I. *Organic Chemical Crystallography*; Consultants Bureau: New York, 1961.
2.   Laing, M. *S. Afr. J. Science* **1975**, *71*, 171-175.
3.   Kitaigorodsky, A. I. *Chem. Soc. Rev.* **1978**, 133-163.
4.   Dauben, W. G.; Fonken, G. J.; Noyce, D. S. *J. Am. Chem. Soc.* **1956**, *78*, 2579-2582.
5.   Basolo, F.; Pearson, R. G. *Mechanisms of Inorganic Reactions*, 2nd ed.; John Wiley: New York, **1967**; pp 371, 387, 424.
6.   Tolman, C. A. *J. Am. Chem. Soc.* **1970**, *92*, 2956-2965.
7.   Tolman, C. A. *Chem. Rev.* **1977**, *77(3)*, 313-348.
8.   Ferguson, G.; Roberts, P. J.; Alyea, E. C.; Khan, M. *Inorg. Chem.* **1978**, *17*, 2965-2967.
9.   Immirzi, A.; Musco, A. *Inorg. Chim. Acta* **1977**, *25*, L41-L42.
10.   Porzio, M.; Musco, A.; Immirzi, A. *Inorg. Chem.* **1980**, *19*, 2537-2540.
11.   Yamamoto, Y.; Aoki, K.; Yamazaki, H. *Inorg. Chem.* **1979**, *18*, 1683-1687.
12.   McAuliffe, C. A. In *Comprehensive Coordination Chemistry*; Wilkinson, G. Ed.; Pergamon: Oxford, 1987; Vol. 2, pp 1015-1029.
13.   Bartik, T.; Heimbach, P.; Schenkluhn, H. *Kontakte* (Merck): Darmstadt **1983**(1), 16-27.
14.   Chatt, J.; Wilkins, R. G. *J. Chem. Soc.* **1952**, 273, 4300; **1956**, 525.
15.   Robertson, G. B.; Tucker, P. A. *Acta Cryst.* **1981**, *B37*, 814-821.
16.   Malatesta, L.; Ugo, R.; Cenini, S. In *Werner Centennial;* Kauffman, G. B., Ed.; American Chemical Society Advances in Chemistry Series No. 62: Washington, 1967; p 318.
17.   Andrianov, V. G.; Akhrem, I. S.; Chistovalova, N. M.; Struchkov, Yu. T. *Russ. J. Struct. Chem.* **1976**, *17*, 111-116.
18.   Immirzi, A.; Musco, A.; Mann, B. E. *Inorg. Chim. Acta* **1977**, *21*, L37-L38.

19. Otsuka, S.; Yoshida, T.; Matsumoto, M.; Nakatsu, K.
     J. Am. Chem. Soc. **1976**, *98*, 5850-5857.
20. Kashiwagi, T.; Yasuoka, N.; Kasai, N.; Kakudo, M.; Takahashi, T.;
     Hagihara, N. *Chem. Commun.* **1969**, 743.
21. Yoshida, T.; Tatsumi, K.; Matsumoto, M.; Nakatsu, K.;
     Nakamura, A.; Fueno, T.; Otsuka, S. *Nouveau J. de Chim.* **1979**,
     3, 761-774.
22. Ashworth, T. V.; Nolte, M. J.; Singleton, E.; Laing, M.
     J. Chem. Soc. (Dalton) **1977**, 1816-1822.
23. Laing, M. *Acta Cryst.* **1978**, *B34*, 2100-2104.
24. Ashworth, T. V.; Singleton, E.; Laing, M. *Chem. Commun.* **1976**,
     875-876.
25. Haines, R. J.; Meintjies, E.; Laing, M.; Sommerville, P.
     J. Organometallic Chem. **1981**, *216*, C19-C21.
26. Graddon, D. P.; Heng, K. B.; Watton, E. C. *Aust. J. Chem.*
     1968, *21*, 121-135.
27. Laing, M.; Carr, G. *Acta Cryst.* **1975**, *B31*, 2683-2684.
28. Laing, M.; Carr, G. *J. Chem. Soc. A.* **1971**, 1141-1144.
29. Bullen, J. G.; Mason, R.; Pauling, P. *Inorg. Chem.* **1965**,
     4, 456.
30. Cook, C. D.; Cheng, P.-T.; Nyburg, S. C. *J. Am. Chem. Soc.*
     1969, *91*, 2123.
31. Cheng, K. L.; Ueno, K.; Imamura, T. *Handbook of Organic
     Analytical Reagents*; CRC Press: New York, 1982; pp 331-333.

RECEIVED March 15, 1994

Chapter 15

Effective Atomic Number and Valence-Shell Electron-Pair Repulsion

60 Years Later

Michael Laing

Department of Chemistry, University of Natal, Durban 4001, South Africa

It is now 60 years since Sidgwick showed that the stoichiometry of many coordination compounds could be deduced from his Effective Atomic Number formalism and almost 40 years since Gillespie showed that the geometric structure of compounds of the main group elements could be correctly predicted from his Valence Shell Electron Pair Repulsion approach. These two seemingly independent approaches to bonding and structure can be combined. One observes that a simple form of the VSEPR rules will also correctly give the geometry of a large number of transition metal complexes if these compounds obey the EAN formalism.

The Beginnings

In 1934 Nevil Vincent Sidgwick published a paper in which he established the well-known Effective Atomic Number rule for the bonding and stoichiometry of transition metal carbonyl compounds (*1*). For the metals of even atomic number Z, he observed that if each carbonyl group bonded to the metal atom is considered to "donate" two electrons to the metal atom (in its oxidation state of zero), then the sum of the atomic number, Z, of the metal plus two times the number of CO groups attached to it equals the atomic number of the next heavier inert gas. This worked beautifully for the 3d metals, *i.e.*, Cr in $Cr(CO)_6$, Fe in $Fe(CO)_5$, and Ni in $Ni(CO)_4$. All had an Effective Atomic Number of 36, Z of krypton. This idea was quickly extended to deal with transition metals with odd atomic numbers and to groups that donated one or three electrons, *e.g.*, Mn in $Mn_2(CO)_{10}$ (to give a Mn—Mn covalent bond) and Co in $Co(CO)_3(NO)$ and Fe in $Fe(CO)_2(NO)_2$ in which the NO molecule contributed 3 electrons.

Six years later Sidgwick published a famous paper based on his Bakerian Lecture (*2*). In this truly remarkable paper he described how the shape of the molecule of a simple molecular compound of a nonmetallic element is determined by the number of bonds, hence pairs of electrons, surrounding the central atom.

0097–6156/94/0565–0193$08.00/0

Molecules with four pairs of electrons are tetrahedral, $e.g.$, CCl_4, molecules with five pairs are trigonal bipyramidal, $e.g.$, PF_5, and molecules with six pairs are octahedral, $e.g.$, SF_6. This new idea contradicted the widely accepted Octet Rule of Langmuir (3) and Lewis (4); the molecules did not have octets, nor did they have closed electron shells of 18 electrons as did the atoms of the heavier inert gases and the metal atoms in the carbonyl compounds that obey the EAN rule.

VSEPR is Born

Thus was laid the foundations for the famous *Quarterly Review* published in 1957 by the two Rons, Gillespie and Nyholm (5), a paper that has had a greater influence on the teaching of structural chemistry at undergraduate level than any other publication during the past 40 years. Not only did it summarize the advances over the 20 years since Sidgwick and Powell's original ideas, but it also led to the development of what was to become the VSEPR method for deducing and rationalizing the structures of molecules. Moreover, the new ideas now encompassed molecules with mixed double and single bonds, $e.g.$, SO_2Cl_2, as well as giving a rationale for molecules with one or more lone pairs of electrons, $e.g.$, SF_4, ICl_3, and IF_5, and an internally self-consistent explanation of the shapes of the AB_4 molecules of the main group elements, $e.g.$, CF_4, SF_4, and XeF_4 (6 - 10).

By a strange coincidence, it was exactly 30 years ago that the ACS held its 144th National Meeting in Los Angeles — 31 March to 5 April, 1963 (11). On the Tuesday morning Ron Gillespie gave a presentation of his newly developed "VSEPR Theory of Directed Valency" (Abstract No 15, page 7E, Division of Chemical Education). He was severely attacked by R. E. Rundle (Abstract No 17, page 8E), who declared that the VSEPR approach was too naive and that the only approach to molecular structure was by the "more exact" Molecular Orbital method. After some discussion, Ron Gillespie then said, "Xenon hexafluoride has just been prepared (by Malm $et\ al.$; Abstract No 26, page 10K). You predict the shape of the molecule, and I will predict the shape. Then when the structure is determined, we will see who is correct" (12). We all know the result. History has proved that Ron Gillespie was correct. Bartell's infrared study subsequently showed that XeF_6 is NOT a perfect octahedron in the gas phase, as Rundle had predicted from the MO method, but has a distorted structure as was predicted by the VSEPR formalism (13). This correct prediction was described by Ron Gillespie in his contribution to the Werner Centennial Symposium at the 152nd American Chemical Society National Meeting held in New York, September 12 - 16, 1966 (14).

Why the VSEPR approach should be so successful has been muchly discussed; whether the electron pairs are truly similar in energy and whether they repel by either simple electrostatic forces or by the Pauli Exclusion principle (15). In his comparison of the VSEPR "points-on-a-sphere" formalism with results of Molecular Orbital computations of potential energy surfaces, Bartell concluded that "the VSEPR model somehow captures the essence of molecular behavior"(16).

The Application of VSEPR to Coordination Compounds

That such a simple model should yield such consistently correct predictions is remarkable. The question is: Is it applicable to compounds of transition metals? The answer is "Yes," if one limits the discussion to only those molecules that obey the EAN rule. One observes an interesting, surprising — and beautiful — phenomenon. These molecules have geometric structures that appear to depend solely on the number of ligands that are bonded to the metal atom, *i.e.*, the number of pairs of electrons in directed atomic orbitals: they obey a form of the VSEPR model.

Carbonyl Compounds

Consider first the simple carbonyl compounds of the 3d transition metals. $Ni(CO)_4$ is tetrahedral, $Fe(CO)_5$ is trigonal bipyramidal, and $Cr(CO)_6$ is octahedral. Their geometric structures appear to depend solely on the number of ligands that are bonded to the metal atom in accordance with the simplest VSEPR approach. The electrons NOT in the covalent metal—C bonds apparently play no role in determining the geometric structure; they behave as if they were not there. These electrons are certainly not "lone pairs" in the VSEPR sense of having spatial influence; they seem to be "spherical" about the metal atom — structurally "inert" pairs. Yet there are 10 of these electrons in $Ni(CO)_4$, 8 in $Fe(CO)_5$, and 6 in $Cr(CO)_6$.

Four Ligands

As long as the different species obey the EAN formalism, their geometries can be correctly predicted by the VSEPR model. Moreover, isoelectronic species are isostructural, independent of the charge carried. For example, $Ni(CO)_4$, $[Co(CO)_4]^-$, $[Fe(CO)_4]^{2-}$, $Fe(NO)_2(CO)_2$, $MnCO(NO)_3$, $Co(CO)_3NO$, and $[Pd(PPh_3)_4]$ are all tetrahedral, all have 4 covalently bonded ligands and, it appears, have 10 d-electrons uninvolved in determining the shape.

However, it is not only this class of compound that "fits." All compounds of the type $[ZnX_2(4\text{-Rpyridine})_2]$ (X = halogen; R = alkyl or aryl group), are tetrahedral. Similarly, both of the complex cations $[Cu(I)(2,2\text{-biquinoline})_2]^+$ and $[Cu(I)(2,9 \text{ dimethyl-1},10\text{-phenanthroline})_2]^+$ are tetrahedral. Their commonality is ten 3d electrons, seemingly not involved in the bonding and certainly not affecting the geometry.

Five Ligands

Isoelectronic and isostructural compounds with 5 ligands include $Fe(CO)_5$, $Mn(CO)_4NO$, $[Mn(CO)_5]^-$, $[Ru(CO)_3(PPh_3)_2]$, $Fe(PF_3)_5$, $[CoCl_2NO\{P(CH_3)_3\}_2]$, and the complexes formed between Co(I) and nitriles, *e.g.*, $[Co(NCMe)_5]^+$. We now have in these compounds 8 d-electrons acting as disinterested spectators, while the five bonding pairs obey the VSEPR model and yield a trigonal bipyramidal geometry.

Six Ligands

The classic case of octahedral geometry includes all possible types of complex, with the metal atoms in a variety of oxidation states : $Cr(CO)_6$, $[V(CO)_6]^-$, $Mo(PF_3)_6$, $[Co(NO_2)_3(NH_3)_3]$, $[Co(CN)_6]^{3-}$, $[Re(CO)_5Br]$, $[RhCl_3(PPh_3)_3]$, $[PtCl_6]^{2-}$, $[Mn_2(CO)_{10}]$, and the isoelectronic $[Co_2(CN)_{10}]^{6-}$, $[IrCl(O_2)CO(PMe_3)_2]$, and a host of others. Now it is only 6 d-electrons that are uninvolved.

Nine Orbitals, Nine Ligands

Given that in the transition metals there are 9 valence orbitals available for bonding, one s, three p, and five d (implying a maximum of 18 electrons which then obey the EAN formalism), one can naively expect a maximum of 9 ligands covalently bonded to the metal atom. There is a classic case of such a complex whose geometry exactly fits what the VSEPR theory predicts: $[ReH_9]^{2-}$ — a tricapped trigonal prism. In this case there are no disinterested electrons; all pairs making up the EAN rule are in the Re—H bonds (or so it seems).

Eight Ligands

In the field of large coordination numbers different geometries can be energetically very similar (17). This applies especially to coordination number 8. The atoms bonded to the metal can be arranged at the corners of a cube, as a square antiprism, or as a dodecahedron. The compound $[Mo(CN)_8]^{4-}$ exemplifies this class and is dodecahedral, which the VSEPR would predict as very likely, but we are now stretching the theory to its very limit. This $[ML_8]$ structure appears to have 2 electrons not involved in dictating the geometry.

Seven Ligands

We have seen the cases of 0, 2, 6, 8, and 10 "disinterested" electrons; we now must consider the case of 4. This is the coordination complex with 7 ligands bonded to the metal, and, interestingly, this is the geometry where VSEPR founders, even for simple molecules like IF_7 and XeF_6 (18,19). What is the most favored geometry: "capped octahedral" (1, 3, 3); "capped trigonal prismatic" (1, 4, 2); "pentagonal bipyramidal" (1, 5, 1)? IF_7 is (1, 5, 1); $[NbF_7]^{2-}$ and $[TaF_7]^{2-}$ are (1, 4, 2); $[ZrF_7]^{3-}$ and $[HfF_7]^{3-}$ are (1, 5, 1); and $[NbOF_6]^{3-}$ is (1, 3, 3). There is clearly little to choose between the energies of the three idealized geometries. Moreover, none of these coordination complexes obey the EAN rule. There are, however, several examples that do satisfy the EAN formalism: $[Mo(CN)_7]^{5-}$ and $[Re(CN)_7]^{4-}$; and $[OsH_4(PEt_2Ph)_3]$ and $[IrH_5(PEt_2Ph)_2]$. All of these complexes are pentagonal bipyramidal — the 4 unbonded electrons do not disturb the geometry, which is correctly predicted by the VSEPR approach.

Limitations

Of course, the attainment of large coordination numbers is often limited by the bulk of the ligand, defined by its Cone Angle Θ (*20*); and if this is large enough, it alone will totally dominate the geometry of the complex. Two simple examples are: trigonal planar [Pd(PPh$_3$)$_3$], ligand Cone Angle Θ = 145°, and linear [Pt{P(*t*-But)$_3$}$_2$], ligand Cone Angle Θ = 182°. These complexes do not obey the EAN rule, but they have the simple geometry deduced from the VSEPR approach: two pairs, collinear; three pairs, trigonal planar.

It is well at this stage to recall the basic premise of the VSEPR approach: the electron pairs are similar in energy and repel by either simple electrostatic forces or by the Pauli exclusion principle (*15*). Within the d-block transition metals, this implies that the 4s, 4p, and 3d electrons should be of similar energy if the model is to work. We know that this is not true, and certainly the energy differences will be greater for metals like Sc, Ti, Zr, Zn, and Hg (*21*). Worse and different problems exist for the elements Rh, Ir, Pd, Pt, and Au, which exhibit the robust square planar 16-electron structure and which participate in oxidative addition and reductive elimination reactions. Similarly, the geometry of the common low-spin square planar d^8 compounds of Ni(II) (like Ni(DMG)$_2$ and [Ni(CN)$_4$]$^{2-}$), which do not obey the EAN rule, cannot be deduced from the VSEPR approach.

One might ask whether there are coordination complexes of the heavy base metals that simultaneously obey both the EAN and VSEPR rules. This question is further complicated by the existence of the Inert Pair Effect (*22, 23*) — another concept coined by Sidgwick (*24*).   Tetraethyllead cannot be correctly called a coordination compound although it does fulfill the EAN formalism. There is one nice example that does satisfy the requirements: HgI$_4$$^{2-}$, the tetrahedral colorless complex formed by dissolving the brilliant orange, insoluble HgI$_2$ in KI solution. Tetrahedral [Zn(NH$_3$)$_4$]$^{2+}$ and [Zn(CN)$_4$]$^{2-}$ also fall into this category.

Is There a Straightforward Theoretical Explanation?

There is no facile explanation for the success of the EAN formalism (*25*) nor for the apparently self-consistent geometric behavior of the various coordination complexes described above. The relationship between total electron count, number of bonded ligands, number of non-bonding electrons and molecular geometry is complex and has been discussed at length (*26-29*). To tease a full understanding out of all the data, self-consistent as well as contradictory, is to hope for too much.

One could of course say:

"This is *NOT* the VSEPR theory at all. It is solely a steric or electrostatic scheme that happens to work on this carefully chosen group of coordination compounds." However, where there are no sterically active lone pairs, surely this is precisely what the simplest version of the VSEPR formalism is. What is remarkable is that so simple a model, originally designed solely for covalent bonded compounds of nonmetals, should succeed for so many coordination complexes of transition metals!   The only requirement for the success of this particular form of the VSEPR approach seems to be that the compounds obey the EAN rule.

We inorganic coordination chemists will be forever grateful for this combination of the visionary ideas of Nevil Vincent Sidgwick and the painstaking development of these ideas by Ron Gillespie that has given to us this remarkable pair of valuable tools — EAN and VSEPR.

Literature Cited

1.   Sidgwick, N. V.; Bailey, R. W. *Proc. Roy. Soc.* **1934**, *144*, 521-537.
2.   Sidgwick, N. V.; Powell, H. M. *Proc. Roy. Soc. A.* **1940**, *176*, 153-180.
3.   Langmuir, I. *J. Am. Chem. Soc.* **1919**, *41*, 868.
4.   Lewis, G. N. *J. Am. Chem. Soc.* **1916**, *38*, 762.
5.   Gillespie, R. J.; Nyholm, R. S. *Quart. Revs. Chem. Soc.* **1957**, *11*, 339-380.
6.   Gillespie, R. J. *J. Chem. Educ.* **1963**, *40*, 295-301.
7.   Gillespie, R. J. *J. Chem. Educ.* **1970**, *47*, 18-23.
8.   Gillespie, R. J. *Molecular Geometry*; Van Nostrand: London, 1972.
9.   Gillespie, R. J.; Hargittai, I. *The VSEPR Model of Molecular Geometry*; Allyn and Bacon: Boston, 1991.
10. Gillespie, R. J. *Chem. Soc. Revs.* **1992**, 59-69.
11. Collected Abstracts, 144th National ACS Meeting, Los Angeles, 31 March to 5 April, 1963; American Chemical Society: Washington, 1963.
12. Laing, M. I was present at the presentation by Ron Gillespie and remember well the discussion and his prediction that XeF_6 "will NOT have a perfect octahedral geometry."
13. Gavin, R. M.; Bartell, L. S. *J. Chem. Phys.* **1968**, *48*, 2466.
14. Gillespie, R. J. In *Werner Centennial;* Advances in Chemistry Series No. 62; Kauffman, G. B., Symposium Chairman; American Chemical Society: Washington, 1967; pp 221-228.
15. Bader, F. W.; Gillespie, R. J.; MacDougall, P. J. *J. Am. Chem. Soc.* **1988**, *110*, 7329-7336.
16. Bartell, L. S. *J. Am. Chem. Soc.* **1984**, *106*, 7700-7703.
17. Kepert, D. L. In *Comprehensive Coordination Chemistry*; Wilkinson, G., Ed.; Pergamon: Oxford, 1987; Vol 1, pp 31-107.
18. Boggess, R. K.; Wiegele, W. D. *J. Chem. Educ.* **1978**, *55*, 156-158.
19. Burbank, R. D.; Jones, G. R. *J. Am. Chem. Soc.* **1974**, *96*, 43-48.
20. Tolman, C. A. *Chem. Rev.* **1977**, *77*, 313-348.
21. Mitchell, P. R.; Parish, R. V. *J. Chem. Educ.* **1969**, *46*, 811-814.
22. Pyykkö, P.; Descaux, J.-P. *Acc. Chem. Res.* **1979**, *12*, 276-281.
23. Pyykkö, P. *Chem. Rev.* **1988**, *88*, 563-594.
24. Sidgwick, N. V. *The Chemical Elements and Their Compounds*; Clarendon Press: Oxford, 1950; pp 287, 481, 617, 795.
25. Chu, S. Y.; Hoffmann, R. *J. Phys. Chem.* **1982**, *86*, 1289-1297.
26. Burdett, J. K. *Molecular Shapes*; John Wiley: New York, 1980; pp 170-238. N.B. page 171: "... the VSEPR approach does not work for transition metal species." This statement is most certainly true for the high-spin compounds of the 3d-series for which the crystal field theory is usually invoked and also for the many unusual or transient species like $Fe(CO)_4$ and $Cr(CO)_5$ which do not obey the EAN formalism.
27. Albright, T. A.; Burdett, J. K.; Whangbo, M. H. *Orbital Interactions in Chemistry*; John Wiley: New York, 1985; pp 277-338.
28. Mingos, D. M. P.; Hawes, J. C. *Structure and Bonding* **1985**, *63*, 1-63.
29. Mingos, D. M. P.; Zhenyang, L. *Structure and Bonding* **1990**, *72*, 73-111.

RECEIVED March 15, 1994

Chapter 16

Brief History of the Thermodynamics of Complex Equilibria in Solution

Mihály T. Beck

Department of Physical Chemistry, Lajos Kossuth University, Debrecen 10, 4010 Hungary

Although a number of overall stability constants were determined as early as the 1890s, the basic principles of the determination of the equilibrium constants of stepwise complex formation were recognized only in 1914, independently by Jaques and Niels Bjerrum. The experimental and computational problems were solved in the 1940s mainly through the investigations of Jannik Bjerrum and Ido Leden. The application of computers helped enormously to solve the computational problems. The introduction of microcalorimetry to the study of stepwise complex equilibria gave a new impetus to the progress. The history of this most important field clearly demonstrates the interplay between theoretical and experimental approaches in the development of science.

The Beginnings

As is well known, early studies of coordination compounds primarily involved thermodynamically stable and kinetically inert complexes. These studies, mainly by Werner, Blomstrand, and Jørgensen, made possible the formulation of the coordination theory by Werner and led to the discovery of the spatial structure of complexes. The resolution of optical isomers, obviously the most spectacular evidence for the octahedral configuration of certain complexes, could be achieved only by the study of inert complexes. Consequently, the study of complex equilibria at this time was of secondary importance.

Although recognition of the principle of electrolytic dissociation and knowledge of the law of mass action are obviously prerequisites for the treatment of complex equilibria, one may find some results among the earlier studies indicating that in the solutions of certain metal salts the formation of some sort of "complex" must be considered. As far as we know, this term was first used by Hittorf in connection with the discovery of certain anomalies in the transference numbers observed in the study of

0097–6156/94/0565–0199$08.00/0
© 1994 American Chemical Society

cadmium salts (1). He found that in the case of cadmium iodide the
transference numbers defined by him depend on the concentration of
the salt and may be even negative, which means that under certain
conditions negatively charged complex ions, moving towards the anode
during electrolysis, are formed from positively charged metal ions
by overcompensation of the charges by the anions bonded to these
cations. As early as 1814 Porrett observed that in the electrolysis
of "prussides", potassium and iron moved in opposite directions, the
iron going toward the anode.

Ostwald (2) combined the Arrhenius theory of electrolytic
dissociation with the law of mass action and calculated the
dissociation constants of various weak acids from the results of
conductivity measurements. The existence of complex ions could be
deduced from distribution experiments (3) and solubility behavior
(4) as well as from rate studies, and several equilibrium constants
were determined.

Although the formation of a series of complexes was obvious to
chemists, at the beginning only the determination of the composition
of the predominating species and their overall equilibrium constants
appeared accessible. Bodländer (5) developed a general
potentiometric method for the determination of these
characteristics. The essence of this method is that the free metal
ion concentration is calculated from the EMF of an appropriately
constructed concentration cell. If the mass action law is applied to
the dissociation of $M_q L_r$

$$K = \frac{[M]^q [L]^r}{[M_q L_r]} \tag{1}$$

For two solutions with different concentrations, therefore

$$\frac{[M]^q_1}{[M]^q_2} = \frac{[M_q L_r]_1}{[M_q L_r]_2} \frac{[L]^r_2}{[L]^r_1} \tag{2}$$

If experiments are made at two different total metal ion
concentrations and at the same total ligand concentrations while the
ligand is in high excess

$$[L]_1 = T_{L_1} \quad \text{and} \quad [L]_2 = T_{L_2} \tag{3}$$

If the concentration of L is kept constant:

$$\frac{[M]^q_1}{[M]^q_2} = \frac{[M_q L_r]_1}{[M_q L_r]_2} \tag{4}$$

The EMF of the concentration cell is then

$$E = \frac{RT}{nF} \ln \frac{[M]_1}{[M]_2} = \frac{RT}{nF} \ln \left(\frac{[M_q L_r]_1}{[M_q L_r]_2} \right)^{1/q} \tag{5}$$

Since $q[M_q L_r] = T_M$, *i.e.*, the excess of ligand means that practically the entire amount of the central ion is present in complex form, and the value of q can be calculated, while the value of r/q may be determined through variation of ligand concentration at constant T_M.

As a matter of course, no information is provided by this approach on the composition and the equilibrium constant of the series of complexes present in the solution under different conditions.

It was first pointed out by Jaques (6), in the appendix to his book that appeared in 1914, that dissociation constants of the various complexes can in principle be calculated from electrode potential measurements. The total concentration of the metal ion is the sum of the concentrations of each species. Assuming the formation of N mononuclear complexes only, this sum is

$$T_M = [M] + [ML] + \ldots + [ML_N] \tag{6}$$

Considering the dissociation constants of each of the complexes and changing the concentration of the ligand while keeping the concentration of the total metal ion concentration constant, a set of equations is obtained:

$$T_M = [M]_1 \left[1 + \frac{[L]_1}{K_1} + \frac{[L]_1^2}{K_1 K_2} + \ldots + \frac{[L]_1^N}{K_1 K_2 \ldots K_N} \right]$$

$$T_M = [M]_2 \left[1 + \frac{[L]_2}{K_1} + \frac{[L]_2^2}{K_1 K_2} + \ldots + \frac{[L]_2^N}{K_1 K_2 \ldots K_N} \right] \tag{7}$$

$$T_M = [M]_N \left[1 + \frac{[L]_N}{K_1} + \frac{[L]_N^2}{K_1 K_2} + \ldots + \frac{[L]_N^N}{K_1 K_2 \ldots K_N} \right]$$

Jaques was aware of the difficulties of the procedure, which are primarily due to the fact that the free ligand concentration is not known; the measurements must therefore be carried out at high ligand excess so that the quantity of ligand bound in the complexes is negligibly small. Surprisingly, however, the principle was not applied before the forties. Jaques' name did not appear after 1914. His book was published immediately before the outbreak of the World War I, which may explain why it did not result in as much interest as it deserved. The stepwise formation of chloro complexes

of aluminum was observed by Heyrovský (7) while following the
hydrolysis of aluminum chloride potentiometrically.

The fact that an entire series of complexes must indeed be
considered was first demonstrated with absolute certainity by Niels
Bjerrum in 1915. He investigated the thiocyanato complexes of
chromium(III) (8). (This paper of 70 pages was originally published
in Danish with a summary in French. An English version of the
summary was published in a collection of selected papers by Niels
Bjerrum (9).) Like the chromium(III) complexes in general, the
thiocyanato complexes of chromium(III) are formed and dissociate in
slow reactions, i.e., they are inert - in Niels Bjerrum's
terminology, robust complexes. This condition makes possible the
separation of the constituents of the solution and the determination
of their concentrations without changing the equilibrium
concentrations. Bjerrum prepared ten solutions of different
concentrations and kept them for a few days at 50 °C until
equilibrum was attained. Analytical determinations were then made at
room temperature where dissociation of the complexes is very slow
(in fact years would be necessary to attain equilibrium). Bjerrum
reported the distribution diagram of the complexes. As Sillén wrote
many years later: "To him it was quite obvious, that once one knows
what the law of mass action is and knows the formation constants for
the various complexes, it is a matter of simple algebra to calculate
the concentration of each complex in any solution, and it would be
a waste of paper, and an insult to the reader, to give any detail of
the calculations" (10).

The Period of Dormancy

After World War I, however, investigations of the properties
of electrolyte solutions were directed primarily at the
clarification of long-range interactions, i.e., the description of
the properties of electrolytes using the activity concept.
Interestingly, Niels Bjerrum took the pioneering steps in this
direction too (11). (This paper was published in Danish. An English
translation of the full paper can be found in the Selected Papers
volume (9).) These investigations soon led to imposing results,
but the success of the Debye-Hückel theory quite overshadowed (12)
the study of complex equilibria. It was believed that the latter
need be taken into account only exceptionally, and deviations
from the laws of solution could be interpreted not through the
chemical equilibria described by the law of mass action, but
entirely in terms of long-range electrostatic interaction between
the ions. Although McBain and Van Rysselberghe (13) pointed out in
1928 that "It must remain a primary aim to determine the actual
molecular species present and their real concentrations", relatively
little quantitative work was reported between the two wars. It was
recognized by mainly British and Scandinavian chemists that the
either-or view is not correct and that both complex formation and
long-range interactions have to be considered. These investigations
were particulary successful concerning ion-pair formation. For these
studies Davies' book (14) provides an excellent review.

The most important publications dealing with multicomponent
systems were those of Jannik Bjerrum on copper(II)-ammine
complexes (15-17) and of Møller on iron(III)- thiocyanato complexes
(18).

Dramatic Progress

The discovery of the glass electrode substantially facilitated the measurement of pH, which gave Jannik Bjerrum the idea of studying complex equilibria pH-metrically (19). The significance of Bjerrum's work is illustrated by the fact that a second edition of his thesis appeared in 1957 and that it was published in Russian and in Japanese. It is extremely rare, indeed, that a Ph.D. thesis is published several times.

In the same year as Bjerrum's thesis was published, Leden (20) described a method based on the measurement of electrode potentials for the determination of the stability constants of the halogeno complexes of cadmium(II) based on the measurement of electrode potentials. This method essentially solved the problem raised by Jaques. While Bjerrum's book was published in English (taking into account that Denmark was under German occupation, this was a brave deed), Leden's paper was published in German and his thesis two years later in Swedish (21).

The more general character of Bjerrum's approach, on the one hand, and the publication language, on the other hand, resulted in Bjerrum's work making a greater impact on the development of the study of complex equilibria than that of Leden. After publication of these theses, development took place at a rate virtually unparalleled in the history of inorganic chemistry. This is at once obvious if the small number of quantitative equilibrium data determined prior to 1941 is compared with the first collections of stability constants published in 1957 and 1958 (22, 23). A few years later the data referring to the complexes of organic and inorganic ligands was published together (24) and the number of data available nearly doubled. It is worth mentioning that the first term applied to complex equilibria was *dissociation* or *instability* constant, following the practice of acid-base equilibria. However, Scandinavian authors always used the reciprocal value and called it *stability constant*, a term which has been generally accepted. For a time k and K were used to designate the stability constant. Leden was probably the first to suggest b as the the symbol for the stability constant and β for the stability product. Although β became generally accepted, K has remained as the symbol for the stability constant.

In this renaissance in coordination chemistry the discovery of new elements, the preparation of an exceedingly high number of ligands of new types, and, perhaps to the greatest extent, the new and very high demands of analytical chemistry, as well as the emergence of bioinorganic chemistry have played crucial roles.

It was most important that in 1945 Schwarzenbach (25) at the Eidgenössische Technische Hochschule in Zürich and Calvin (26) at the University of California, Berkeley began equilibrium studies of chelate complexes. Calvin dealt with these studies very briefly, but in Schwarzenbach's school in Zürich and in Martell's school at Clark University in Worcester, Massachussetts a great number of Ph.D. students and postdoctoral fellows made systematic investigations. Equilibria involving multidentate ligands, both in aqueous and nonaqueous media, have since then been studied in many laboratories. Martell and Calvin's monograph (27), in which a long chapter dealt with complex equilibria, and Schläfer's (28) and Rossotti and

Rossotti's books (29), devoted entirely to the determination of stability constants, provided a great impetus to thermodynamic studies of metal complexes.

In the rapid development of the study of complex equilibria the introduction of computers into the numerical evaluation of the experimental results also played a decisive role. Hugus was probably the first to use electronic computers for the evaluation of stability constants. In 1961 he gave a lecture on this approach at the 6th ICCC in Detroit, Michigan (30). Initially, Sillén and his school were the most active group in this field (31). It is worth remembering the title of the series of their papers High Speed Computers as a Supplement to Graphical Methods, which clearly indicates that they regarded graphical analysis of the experimental data to be of fundamental importance.

Sillén made two other important contributions to the thermodynamics of complex equilibria. First, he and his collaborators meticulously studied the stabilities of polynuclear hydroxo complexes (32). Second, they were among the first in applying and developing microcalorimetry in the study of complex equilibria (33). Although in Yatsismirkii's book (34), published in 1956, many numerical data can be found; most of them are calculated ΔG, ΔH, and ΔS values. The introduction of thermistors made possible experimental calorimetric studies of complex equilibria. The most important pioneers in this field were Jordan, Hume, Izatt, and Christensen in the USA, Sartori and Paoletti in Italy, and Sillén, Schlyter, and Johansson in Sweden.

The increasing number of stability constants available have made possible the finding of correlations. The fact that chelate complexes are always more stable than the complexes of the corresponding monodentate ligands was discovered independently by Martell and Calvin (26) and by Schwarzenbach (35). The term chelate effect was coined by Schwarzenbach. Among the many correlations found, the Irving-Williams order of the complexes of transition metal ions (36) is rather general and was theoretically rationalized.

The determination of stability constants offered an apparently easy and not very expensive research field, so it was unavoidable that in the gold rush some rather careless studies have been published, and collections of stability constants were heavily contaminated by unreliable data. Therefore the Equilibrium Commission of the International Union of Pure and Applied Chemistry began to publish a series of critically evaluated stability constants (37). A great number of families of metal complexes were treated in this continuing series. (To date 12 reviews were published in either separate volumes or in Pure and Applied Chemistry.) Critically selected stability constants are offered in the volumes prepared by Martell and Smith (38).

A number of monographs were published in the last quarter-century dealing with the principles and practical aspects of the determination of different types of complex equilibria (39-44).

Finally, it may be mentioned that the chemistry of simultaneous equilibria has developed in three independent ways. Almost in parallel with development of stepwise complex equilibria, a method was elaborated for the treatment of the equilibria of multicomponent gas reactions (*45,46*). Essentially this problem is mathematically equivalent to the description of simultaneous complex equilibria in solution. Both from mathematical and chemical points of view, studies of successive complex equilibria closely related to the investigation of complex equilibria of biopolymers (*47*). This three-pronged advance had the obvious consequence that numerous important and mutually valuable results were attained independently of one another. This is yet a further example of the general feature that progress in science does not follow the most economical route.

Literature Cited

1. Hittorf, J.H. *Pogg. Ann.* **1853,** *89,* 177.
2. Ostwald, W. *Z. phys. Chem.* **1888,** *2* , 36.
3. Dawson, H.M.; McCrae, J. *J. Chem. Soc.* **1900, 77,** 1239.
4. Bodländer, G. *Z. phys. Chem.* **1892,** *9,* 770.
5. Bodländer, G., *Festschrift zum 7o. Geburtsag von R. Dedekind;* Vieweg: Braunschweig, Germany, 1901; pp 153-165.
6. Jaques, A. *Complex Ions in Aqueous Solutions;* Longmans: London, England, 1914.
7. Heyrovský, J. *J. Chem. Soc.* **1920,** *97, 27.*
8. Bjerrum, N. *Kgl. Danske Videnskab. Selskab. Naturvidenskab. Math. Afdel.* **1915,** *(7)* 12, 147.
9. Bjerrum, N. *Selected Papers;* Einar Munksgaard: Copenhagen, Denmark, **1949.**
10. Sillén, L.G. In *Essays in Coordination Chemistry;* Schneider, W.; Anderegg, G.; Gut, R., Eds.; Birkhäuser Verlag: Basel, Switzerland, **1949,** pp 24-38.
11. Bjerrum, N. *Fys. Tidskr.* **1916,** *16,* 59.
12. Sillén, L.G. *J. Inorg. Nucl. Chem.* **1958,** *8,* 176.
13. McBain, J.W.; Van Rysselberghe, P.J. *J. Am. Chem. Soc.* **1928,** *50,* 3009.
14. Davies, C.W. *Ion Association;* Butterworths: London, England, **1962.**
15. Bjerrum, J. *Kgl. Danske Videnskab. Selskab. Mat.-fys. Medd.* **1931,** *11, No. 5.*
16. Bjerrum, J. *Kgl. Danske Videnskab. Selskab. Mat.-fys. Medd.* **1932,** *11, No. 1o.*
17. Bjerrum, J. *Kgl. Danske Videnskab. Selskab. Mat.-fys. Medd.* **1934,** *12, No. 15.*
18. Møller, M. *Studies on Aqueous Solutions of the Iron Thiocyanates;* Dana Bogtrykkeri: Copenhagen, Denmark, 1937.
19. Bjerrum, J. *Metal Ammine Formation in Aqueous Solution;* Haase: Copenhagen, Denmark, **1941.**
20. Leden, I. *Z. phys. Chem.* **1941,** *A 188,* 160.
21. Leden, I. *Potentiometrisk Undersökning av Några Kadmiumsalters Komplexitet;* Gleerupska Univ.- Bokhandeln.: Lund, Sweden, **1943.**
22. *Stability Constants of Metal-ion Complexes, Part I. Organic Ligands;* Bjerrum, J.; Schwarzenbach, G; Sillén, L.G., Eds.; The Chemical Society: London, England, **1957.**
23. *Stability Constants of Metal-ion Complexes, Part II. Inorganic Ligands;* Bjerrum, J.; Schwarzenbach, G; Sillén, L.G., Eds.; The Chemical Society: London, England, **1958.**

24. *Stability Constants of Metal-ion Complexes;* Martell, A.E.; Sillén, L.G.; Eds.; The Chemical Society: London, England, **1964**.
25. Schwarzenbach, G. *Helv. Chim. Acta* **1945**, *28*, 828.
26. Calvin, M.; Wilson, K.W. *J. Am. Chem. Soc.* **1945**, *67*, 2003.
27. Martell, A.E.; Calvin, M.; *Chemistry of the Metal Chelate Compounds;* Prentice-Hall: New York, NY, USA, **1952**.
28. Schläfer, H.L.; *Komplexbildung in Lösung,* Springer: Berlin, Germany, **1961**.
29. Rossotti, F.J.C.; Rossotti, H. *The Determination of Stability Constants;* McGraw Hill: New York, NY, USA, **1961**.
30. Hugus, Jr., Z.Z. In *Advances in the Chemistry of the Coordination Compounds:* Kirschner, S., Ed.; Macmillan: New York, NY, USA, **1961**; pp 379-390.
31. Sillén, L.G. *Acta Chem. Scand.* **1962**, *16*, 159.
32. Sillén, L.G. *Nature* **1947**, *16o*, 715.
33. Sillén, L.G.; Schlyter, K. *Acta Chem. Scand.* **1959**, *13*, 355.
34. Jazimirski (Yatsimirskii), K.B. *Thermochemie von Komplexverbindungen;* Akademie Verlag: Berlin, Germany, **1956**.
35. Schwarzenbach, G. *Helv. Chim. Acta* **1952**, *35*, 2344.
36. Irving, H.; Williams, R.J.P. *Nature* **1948**, *162*, 746.
37. Beck, M.T.; *Pure Appl. Chem.* **1977**, *59*, 127.
38. Martell, A.E.; Smith, R.M. *Critical Stability Contants, Vol. 1: Amino Acids;* Plenum Press: New York, NY, USA, **1974**. (To date five more volumes were published.)
39. Butler, J.N. *Ionic Equilibrium;* Addison-Wesley: Reading, MA, USA, **1964**.
40. Nancollas, G.H. *Interactions in Electrolyte Solutions;* Elsevier: Amsterdam, London, New York, **1966**.
41. Beck, M.T. *Chemistry of Complex Equilibria;* Van Nostrand Reinhold: London, England, **1970**.
42. Martell, A.E.; Motekaitis, R.J. *Determination and Use of Stability Constants;* VCH Publishers: New York, NY, USA, **1988**; Second Edition, 1992.
43. Burgess, J. *Ions in Solution: Basic principles of Chemical Interactions;* Ellis Horwood: Chichester, England, **1990**.
44. Beck, M.T.; Nagypál, I. *Chemistry of Complex Equilibria;* Ellis Horwood: Chichester, England, **1990**.
45. Klein, M. In *Practical Treatment of Coupled Gas Equilibria;* Jost, W., Ed.; *Physical Chemistry;* Academic Press: New York, NY, USA, **1971**; Vol. 1, p 489.
46. Smith, W.R. *Ind. Eng. Chem. Fundamentals* **1980**, *19*, 1.
47. Poland, D. *Cooperative Equilibria in Physical Biochemistry;* Clarendon Press: Oxford, England, **1978**.

RECEIVED October 14, 1993

Chapter 17

Stabilization of Unstable d-Metal Oxidation States by Complex Formation

K. B. Yatsimirskii

Institute of Physical Chemistry, Academy of Sciences of Ukraine, prospect Nauki 31, Kiev, 252039 Ukraine

Alfred Werner studied the "strengthening of the primary valence force by the saturation of the secondary valence force," but he failed to explain fully this phenomenon. To elucidate the stabilization of unstable oxidation states by complex formation two aspects should be taken into account: the thermodynamic stability of complexes and their kinetic redox lability. Such factors as ligand field stabilization energy for M^n and M^{n-1} and the geometry of the donor atoms' spatial orientation also affects the stability of a given oxidation state. Macrocyclic ligands are especially suitable for the stabilization of unstable metal oxidation states, both from the thermodynamic and kinetic viewpoints.

The stabilization of unstable d-metal oxidation states by complex formation has been studied for many years as one of the important problems of coordination chemistry. Alfred Werner paid attention to this, writing, "as a very peculiar phenomenon of the strengthening of primary valence by means of secondary valence forces, saturation has been often observed. The essence of this phenomenon has not been clear until now" (*1*). He then gave some examples of stabilization by formation of oxide and chloride complexes in the cases of Fe(VI), Mn(III), and Pb(IV). He pointed out that very unstable CoX_3 salts can be stabilized by the coordination of ammonia molecules. Similarly, silver(II) compounds may be isolated only as the tetrakis(pyridine) adduct $[Ag(py)_4]S_2O_8$ (*1*).

Today inorganic chemists do not use such terms as "primary" (*Haupt*) and "secondary" (*Neben*) valence but instead speak of "oxidation state" and "coordination number." One modern popular definition for the term "oxidation state" is: "The oxidation state of a metal in a complex is

0097–6156/94/0565–0207$08.00/0

simply the charge that the metal would have on the ionic model.... Once we have the oxidation state, we can immediately obtain the corresponding d configuration. This is simply the number of d electrons which would be present in the free metal ion that corresponds to the oxidation state we have assigned" (2). The d^n configuration may be confirmed by experimental methods such as ESR, UV-VIS spectroscopy, etc.

The primary characteristic of d-transition metals is their ability to assume several oxidation states with different stabilities. Of special interest is the stabilization of unstable oxidation states of transition metals, which is of great significance in explaining the essence of "strengthening of the main valence by means of saturation of secondary valence forces" (coordination number) (1).

The coordination number is equal to the number of ligand atoms which are directly bonded to the central metal atom. In some cases (mainly in organometallic compounds) some difficulties exist in determining the coordination number, e.g., in ferrocene.

It is clear that Werner's "strengthening of the main valence force" means the stabilization of unstable oxidation states by complex formation. This problem has at least three interconnected aspects: (1) structural (electronic and geometric), (2) thermodynamic, and (3) kinetic. The compatibility in donor-acceptor interaction is one of the most important aspects of the problem. Ligands can be classified according to their donor-acceptor ability into at least three groups: (i) molecules and ions possessing only one lone electron pair: NH_3, NR_3, CH_3^-, etc., $-\sigma_A$ group (single sigma donor group); (ii) molecules and ions incorporating donor atoms with two or more electron pairs. These ligands can act as σ- and π-donors. Many ligands fall into this group: F^-, OH^-, OH_2, $RCOO^-$, etc.; and (iii) molecules and ions with lone electron pairs and vacant π^*- or d-orbitals suitable for accepting electrons from metal atoms and π-back bond formation. These ligands can be denoted as the $\sigma_A\pi_A\pi_A$ group. There are many ligands that belong to this group. CO, CN^-, and NCR are examples of ligands with vacant π-orbitals, while SR_2, Cl^-, Br^-, I^-, etc. are ligands with vacant d-orbitals.

A similar classification can be proposed for d-metal ions, and thus many cases of electron structure compatibility in complexes can be explained. For instance, Co^{3+} possesses a low-spin state and in an octahedral field has the $(t_{2g})^6$ electronic configuration. It is clear that this ion can act only as a σ-acceptor as π-orbitals are filled. This causes the very strong complex formation of Co^{3+} with σ_A-ligands (log K for $[Co(NH_3)_6]$ is 34.36). The neighboring ions (Mn^{2+} and Fe^{3+}) have vacancies on the t_{2g} orbitals and thus can interact with $\sigma_A\pi_A$ ligands (OH^-, H_2O, etc.) which compete with ammonia ligands. The compatibility can be observed also in the case of phenanthroline and α, α'-dipyridyl complexes. These ligands form more stable $Fe^{2+}(d^6)$ complexes than the analogous complexes with neighboring ions (Mn^{2+}, Co^{2+}, and even with Fe^{3+}, d^5-configuration).

The ligand field stabilization energy (LFSE) plays an important role in the stabilization of some complexes with a given electronic structure of metal ions, *i.e.*, with a given oxidation state. Most coordination compounds of the d-metals have a coordination number of 6 and an octahedral arrangement of ligand donor atoms in the coordination sphere. The LFSE has maximum values for ions with d^3 (V^{2+}, Cr^{3+}), low-spin d^6 (Mn^+, Fe^{2+}, Co^{3+}, *etc.*) and d^8 (Co^+, Ni^{2+}, Cu^{3+}) electron configurations. For a different geometry of coordination polyhedron LFSE possesses a maximum for different electronic structures. Therefore the LFSE can be considered as a measure of metal-ligand compatibility.

Ligand geometry plays an important role in the stabilization of unstable metal oxidation states whenever the possibility of forming one, two, or three metal-chelate rings exists. This circumstance plays an especially governing role in macrocyclic complexes where metal ions are encapsulated in the cavities of macrocyclic ligands. Each cavity has a definite dimension, and thus there must be a "correspondence" between the dimension of the macrocyclic hole and the metal ionic radius. The ionic diameter and the size of the macrocyclic ligand cavity must match each other. Here both thermodynamic and kinetic stabilization must be observed. The correspondence of electronic and geometric structures is realized by the definite structure of the coordination sphere and the corresponding electronic structure.

An examination of metal ions which are stable in aqueous solution will elucidate the importance of compatibility factors. The oxidation state 1+ can be observed only for ions with the stable nd^{10} electronic shell (second ionization potential is high): Cu^+, Ag^+, and Au^+. Other M^+ ions with d-electrons reduce water to free hydrogen. The observed aqua ions with 2+ and 3+ oxidation states are:

n	d^1	d^2	d^3	d^4	d^5	d^6	d^7	d^8	d^9	d^{10}
3	–	–	V^{2+}	Cr^{2+}	Mn^{2+}	Fe^{2+}	Co^{2+}	Ni^{2+}	Cu^{2+}	Zn^{2+}
4	–	–	–	Mo^{2+}	Tc^{2+}?	Ru^{2+}	Rh^{2+}	Pd^{2+}	Ag^{2+}	Cd^{2+}
5	–	–	–	–	–	–	–	Pt^{2+}	Au^{2+}	Hg^{2+}
3	Ti^{3+}	V^{3+}	Cr^{3+}	Mn^{3+}	Fe^{3+}	Co^{3+}	–	–	–	Ga^{3+}
4	–	–	Mo^{3+}	Tc^{3+}?	Ru^{3+}	Rh^{3+}	–	Ag^{3+}	–	In^{3+}
5	–	–	–	–	Os^{3+}	Ir^{3+}	–	(Au^{3+})	–	Tl^{3+}

The ions at the beginning of the d-series are strong reducing agents (*e.g.*, V^{2+}, Cr^{2+}, Ti^{3+}, and V^{3+}), whereas the last metal ions in these series are oxidants. We assume that Tc^{2+} and Tc^{3+} may exist in aqueous solution, but they have not yet been observed. Some of the 4d- and 5d-ions can form metal-metal bonds and thus may be identified in dimeric and cluster forms (Mo_2^{4+}, Rh_2^{4+}, and possibly Au_2^{4+}).

The d-metal complexes in unstable oxidation states may possess high or low redox potential values. These values depend on the ionization potentials (I) and on the difference between the solvation free energy for n+ and (n-1)+ charged ions (ΔG_n and ΔG_{n-1}, respectively):

$$E_{n/n-1} = I_n - (\Delta G_n - \Delta G_{n-1}) - \text{const} \qquad (1)$$

where const is a value dependent on the reference electrode and in the case of the normal hydrogen electrode is equal to 4.55 eV (439 kJ) (3).

The $E_{n/n-1}$ values can be regulated by complex formation and for the $M^nL/M^{n-1}L$ pair can be expressed by the following equation:

$$E^C_{n/n-1} = E_{n/n-1} + (RT/F)(\ln K_{M^{n-1}L} - \ln K_{M^nL}) \qquad (2),$$

where $E^C_{n/n-1}$ is the redox potential for the pair $M^nL^n/M^{n-1}L$, which depends on the difference in the logarithms of the stability constants or on the ratio $K_{M^{n-1}L}/K_{M^nL} = \eta$, which can be more or less than 1. In most cases this ratio (η) exceeds 1, but for the ligands with soft donor atoms (thioether group, phosphine, etc.) $\eta < 1$. The ratio η depends on the nature of the metal and ligand in the complex. For instance, the rather "hard" ligand NH_3 forms coordination compounds that are more stable with M^{n+} than with $M^{(n-1)+}$, but there are exceptions (e.g., log $K_{Cu(NH_3)_2^{2+}} >$ log $K_{Cu(NH_3)_2^+}$). For the hard bases (F^-, OH^-, $RCOO^-$, etc.) the ratio $\eta > 1$ in all cases. The carbon monoxide molecule (CO, with the dominating resonance form $C\equiv O^+$) and the cyanide ion (CN^- with the strong donor ability of the highest occupied σ-orbitals and lowest unoccupied π^*-orbitals) form $\sigma\ L \rightarrow M$ and $\pi\ M \rightarrow L$ bonds. Such compounds as $K_5[Mn(CN)_6]$ and $K_5[Re(CN)_6]$ illustrate the stabilization of Mn^+ and Re^+ owing to the synergetic effect (the electronic structures of the central atoms are $3d^6$ and $5d^6$, respectively).

The stabilization of the 1+ oxidation state may be realized by the encapsulation of metal ions into the cavity of macrocyclic ligands containing azomethine or thioether groups. Only one Fe^+ macrocyclic complex has been described in the literature (5). Since 1972 no reports concerning the stabilization of Fe^+ by macrocyclic complex formation have appeared in the literature.

The macrocyclic complexes $[M^I([9]aneS_3)_2]^+$ (where $[9]aneS_3$ = 1,4,7-trithiacyclononane) have been obtained for Co^+ ($3d^8$) and Rh^+ ($4d^8$). These ions have electronic structures which best fit the O_h symmetry of the arrangement of S-donor atoms.

Much more information is available for Co^+ tetraazamacrocyclic complexes. The cobalt(I) corrin systems (compounds modeling vitamin B_{12}) have been most thoroughly studied.

Macrocyclic complexes $[M^I([9]aneS_3)_2]^+$ with d^7 and d^9 electronic configurations are not yet known, although many Ni(I) macrocyclic complexes are more stable than the corresponding Co(I) complexes (*4*).

Many macrocyclic complexes of $Ni^+(3d^9)$ and $Pd^+(4d^9)$ are known. Polyazamacrocyclic ligands with methylated N-donor atoms form rather stable complexes with these metals.

There are many examples of the stabilization of unstable 3+ oxidation states by complex formation with macrocyclic ligands. Ni^{3+} and Cu^{3+} are stabilized by complex formation with tetraazamacrocyclic ligands. The most universal ligand for the stabilization of unstable oxidation states is $[9]aneS_3$ (L), which forms coordination compounds of general type ML_2^{n+} with many unstable ions, e.g., Ag^{2+}, Pd^{3+}, and Pt^{3+}. The stabilization of Hg^{3+} in the tetraazamacrocyclic complex $[Hg(Cyclam)]^{3+}$ was established in 1976 (*6*).

Thus the stabilization by complex formation of such ions that do not exist in aqueous solution (Mn^+, Fe^+, Co^+, Ni^+, Rh^+, Pd^+ and Ni^{3+}, Cu^{3+}, Pd^{3+}, Pt^{3+}, Hg^{3+}) has been well established. However, there are still no data about the stabilization by complex formation of metals in the oxidation state 1+: Os^I, Ir^I, and Pt^I.

In all cases the stabilization of unstable metal oxidation states depends on the ratio of stability constants $K_{M^{n-1}L}/K_{M^nL}$, which regulates the redox potential in accordance with equation 2 (thermodynamic stabilization).

Finally, the kinetic aspect of this problem exists. The essence of kinetic stabilization is the screening of the central atom from attack by different active molecules (H_2O, O_2, *etc.*). For this purpose bulky groups in ligands are desirable.

For instance, norbornyl anions can stabilize cobalt in the 5+ oxidation state ($Co(Nor)_4^+$), *i.e.*, a $e^4t_2^0$ electronic structure (in the field of T_d symmetry) (*7*). Nickel(I) complexes are stable when alkylated nitrogen donor atoms are incorporated into the macro ring.

Redox-unstable complexes may be stabilized by the coordination of additional ligands. For instance, many unstable nickel(I) and cobalt(I) tetraazamacrocyclic complexes can be stabilized by additional coordination, forming mixed ligand complexes containing macrocyclic ligands and such π-acceptors as CO and PR_3.

Thus the essence of the "strengthening of primary valence by means of saturation of secondary valence forces" mentioned by Werner or the "stabilization of unstable oxidation states by complex formation" can be explained by three different reasons: (i) the electronic and geometric structure compatibility (correspondence) for the central ions and ligands; (ii) the thermodynamic characteristics — redox-potential, which depends on the ratio of the stability constants of the complexes with the same composition but different oxidation states; and (iii) the kinetic characteristics (inertness of the complex), which depend on the screening of central metal atoms by different bulky ligands.

Literature Cited

1. Werner, A. *Neuere Anschauungen auf dem Gebiete der anorganischen Chemie*, 5th ed.; F. Vieweg und Sohn: Braunschweig, 1923.
2. Crabtree, R. H. *Organometallic Chemistry of the Transition Metals*; Wiley: New York, 1988; p 30.
3. Yatsimirskiĭ, K. B. *Theor. Exper. Chem.* **1986**, *25*, 280 (in Russian).
4. Yatsimirskiĭ, K. B. *Russ. J. Inorg. Chem.* **1991**, *30*, 2010 (in Russian).
5. Rillema, D. P.; Endicott, J. F.; Papaconstantinou, E. *Inorg. Chem.* **1971**, *10*, 1739.
6. Deming, R. L.; Allred, A. L.; Dahl, A. R.; Herlinger, A. W.; Kestner, M. O. *J. Am. Chem. Soc.* **1976**, *98*, 4132.
7. Byrne, E. K.; Theopold, K. H. *J. Am. Chem. Soc.* **1987**, *109*, 1282.

RECEIVED April 6, 1994

Chapter 18

Oxidation States and d^q Configurations in Inorganic Chemistry

A Historical and Up-to-Date Account

Jesper Bendix[1], Michael Brorson[2], and Claus E. Schäffer[1]

[1]Department of Chemistry, H.C. Ørsted Institute, University of Copenhagen, Universitetsparken 5, DK–2100 Copenhagen Ø, Denmark
[2]Chemistry Department A, Building 207, The Technical University of Denmark, DK–2800 Lyngby, Denmark

Oxidation numbers can be traced back to Alfred Werner. Oxidation states -- and the oxidation numbers labeling them -- are sometimes based upon a combination of conventions and electron bookkeeping. However, to focus upon this formal aspect of oxidation states is to deprive inorganic chemistry of one of its major means of classification of its subject material. For most chemical systems in which oxidation states are of interest, there is general agreement about which oxidation numbers should be used. In transition metal chemistry the one-to-one relationship between d^q configuration and oxidation state for completely ionic systems survives the effects of configuration interaction and covalency; what remains is the useful concept of a classifying d^q configuration. This concept can be made the basis for a mathematical model, the parametrical d^q model, that parametrically embodies a quantitative ligand-field description as well as Slater-Condon-Shortley theory. This model allows new quantitative comparisons of all the classical empirical parameters.

When Alfred Werner wrote his famous paper whose centennial we commemorate at this symposium, chemistry had long outgrown the stage of being a mere collection of experimental facts. The main reason why Werner is considered the father of coordination chemistry is the fact that he successfully extended the idea of the tetrahedral carbon atom to the rest of chemistry. Chemistry as a whole became three-dimensional.

In the present paper we focus on another of Werner's accomplishments that derives directly from his idea of the central atom: his development of the *coordination theory* and of coordination nomenclature together with his essential foundation of the concept of *oxidation state*.

Through ligand-field theoretical results and new knowledge about local stereochemistry around central atoms, these accomplishments have led to the use of *atomic electron configurations* as one of the most important classificatory tools in inorganic chemistry. This is the key point of this paper together with its further illustration of how qualitative results, based upon atomic electron configurations, can be quantified.

0097–6156/94/0565–0213$08.00/0

A Brief Status Report from around 1893

A hundred years ago results from physics and physical chemistry had already influenced the conceptual status of inorganic chemistry. In the present context, it may be noted, in particular, how the experimental study of electrolysis processes had led to the concepts of cations, anions, and electrochemical equivalents. An important conclusion from these studies was, for example, that the monovalency of silver and the divalency of copper in their normal salts were more than just stoichiometric attributes. This conclusion, based upon *integers*, gives rise to the most important class of statements in chemistry, which we would like to call: *qualitative in a strong sense.* We shall see further examples of this kind of statement below in connection with oxidation states, atomic electron configurations, and ground state specifications.

Chemistry in aqueous solution was always important for the development of the concepts of inorganic chemistry. Thus the concept of oxidation, which, of course, was originally associated with addition of oxygen, was soon used also for abstraction of hydrogen because it was difficult to differentiate between these two events in an aqueous medium. Furthermore, the concepts of acidic and basic anhydrides were early used to define important, nonredox links.

To add some documentation to this account, which is not a historical investigation in a strict sense, we refer to one of the most prominent chemists of the time, Wilhelm Ostwald. In his inorganic chemistry textbook (1900) (*1a*) Ostwald states (*1b*) that there is a consensus to include also under reduction and oxidation the uptake and loss of hydrogen, respectively. Moreover, he says that ferric ion is formed from ferrous ion through the reaction with any "oxidation agent" (*1c*) and, on the same page: the opposite of oxidation is called reduction. Furthermore, he states: with our meaning, reduction also means a decrease of the positive or increase of the negative ionic charge. Seen through the eyes of a chemist in 1993, this latter statement is as close to the concept of oxidation number as it could be, especially when one considers that ferric and ferrous ions were written as $Fe^{\cdots}$ and $Fe^{\cdot\cdot}$ even though this was before the advent of the electron. Nevertheless, a little further on in Ostwald's monograph where the subject is redox equations (*1d*), one finds this table for manganese:

manganous series	$Mn(OH)_2$	divalent
manganic series	$Mn(OH)_3$	trivalent
manganese peroxide, $MnO_2 + 2\ H_2O$	$Mn(OH)_4$	tetravalent
manganate series, $H_2MnO_4 + 2\ H_2O$	$Mn(OH)_6$	hexavalent
permanganate series, $HMnO_4 + 3\ H_2O$	$Mn(OH)_7$	heptavalent

together with the following table for sulfur

sulfuric acid	=	SO_4H_2
sulfurous acid, $H_2SO_3 + H_2O$	=	SO_4H_4
sulfur, $S + 4\ H_2O$	=	SO_4H_8
hydrogen sulfide, $H_2S + 4\ H_2O$	=	SO_4H_{10}

Ostwald pointed out that oxidation of hydrogen sulfide to sulfuric acid requires $10 - 2 = 8$ oxidation units while permanganate furnishes $7 - 2 = 5$ when the manganous ion is formed. It is clear that Ostwald was aiming directly at redox equations here and that he had realized that only *differences* between integers were relevant for this purpose.

The difference in substance of the concept of oxidation number for a main-group and a transition-group element has survived the passage of a hundred years; most chemists today agree that this concept is somewhat more formal for sulfur than it is for manganese.

Werner's Contributions to Nomenclature: Introduction of Oxidation States

During Werner's time compositional nomenclature for binary compounds had already been agreed upon in much the same way as today, and names such as manganese dichloride and manganese monooxide were in common use to express information about stoichiometric compositions only. For complex compounds, this type of nomenclature had simply been extended as in the notational example: 3 KCN,Fe(CN)$_3$. Werner realized that this notation could be modified to become a nomenclature including *structural* information, when this was available, and potassium hexacyanoferriate is his ingenious proposal (2). This is an example of what is today referred to as additive nomenclature or coordination nomenclature, as opposed to the substitutional nomenclature of organic chemistry.

However, Werner also realized that nomenclature embodied a perspective far beyond the mere naming and structural characterization of each particular substance. It also could be used to support the thinking process during attempts at rationalizing the variegated facts of chemistry. In this context Werner was before his time when he suggested that the suffix -*ic* always ought to refer to trivalent metal ions. The following scheme is reprinted from his book: *Novel Conceptual Contributions to the Field of Inorganic Chemistry (2)*.

MeX	*a*-compounds	(monovalent)
MeX$_2$	*o*-compounds	(divalent)
MeX$_3$	*i*-compounds	(trivalent)
MeX$_4$	*e*-compounds	(tetravalent)
MeX$_5$	*an*-compounds	(pentavalent)
MeX$_6$	*on*-compounds	(hexavalent)
MeX$_7$	*in*-compounds	(heptavalent)
MeX$_8$	*en*-compounds	(octavalent)

With this nomenclature (2,3) MoF$_6$ becomes molybdon fluoride and K$_2$PtCl$_6$ becomes potassium hexachloroplateate. Werner's suggestion is, of course, equivalent to that used today under the name of Stock nomenclature (4) or oxidation-number nomenclature.

There exists an alternative to the Stock nomenclature called the Ewens-Bassett nomenclature (5) or the charge-number nomenclature. This is somewhat less convention loaded. For example, K$_2$PtCl$_6$ has the charge number name potassium hexachloroplatinate(2-), which only involves the assumption that potassium, which behaves cationically in solution, is potassium(1+). This is not a very audacious assumption for a chemist.

The *Grand Sum Rule* for Oxidation Numbers

If one has the restricted aim of avoiding inconsistencies, particularly in connection with finding the number of moles of electrons involved in a redox process, *i.e.*, the number of redox equivalents involved in the associated *reaction scheme* or *redox equation*, it is only necessary to have one condition fulfilled: the *grand sum rule* or the axiom for oxidation numbers.

> *Grand sum rule*: The sum of the oxidation numbers within a given chemical entity must equal the net charge on this entity.

For example, if the process is the oxidation of K$_2$CS$_3$ to K$^+$, CO$_2$, and HSO$_4^-$ in aqueous acidic solution, it is permissible to analyze the situation of the reactant K$_2$CS$_3$ by attributing the oxidation numbers IV and VI to the atoms, carbon and sulfur, respectively, with the consequence that potassium must be assigned the oxidation number -XI. Oxidizing K_2CS_3 then amounts to oxidizing two moles of K from -XI to I, that is, 24 redox equivalents (moles of electrons); this is the same result as that obtained by the conventional way of viewing K_2CS_3 as three S of oxidation number -II to be oxidized to VI.

The above axiom may also be viewed as the most liberal definition imaginable for oxidation numbers, and one might call a set of oxidation numbers, chosen for a particular purpose and restricted only by the condition of the grand sum rule, a set of *ad hoc* oxidation numbers. *Ad hoc* oxidation numbers may well be fractional, but this is of no interest in this paper's context.

Oxidation Numbers in Inorganic Chemistry

Inorganic chemistry is at the same time a collection of facts and the science of rationalizing these facts. For this purpose it uses a body of more or less well-defined concepts. Oxidation state is one of the most important ones among these. Although there is a large number of inorganic compounds to which the concept does not usefully apply (6), the following picture can still be used to emphasize its general importance.

To a large extent inorganic chemistry can be classified by a three-dimensional skeleton whose dimensions are characterized by three numbers: primarily the group and period numbers of the Periodic Table and secondarily the oxidation number that adds an extra dimension to it.

Even though the grand sum rule with the K_2CS_3 example in the previous section was found chemically useful for a limited purpose, it was also found to be too mathematical and too unrestricted to provide results of any general interest for chemistry.

However, if this sum rule is applied to a restricted system consisting of a monoatomic entity, it leads to a conclusion that is unique, almost trivial, and yet of general chemical interest. For example, Fe^{3+} is Fe^{III}, and after that comes the chemistry, almost invited by the tradition. Fe^{3+} is often used synonymously with $[Fe(OH_2)_6]^{3+}$ even though the chemical difference is stoichiometrically and energetically enormous. The situation is that Fe^{3+} may be used with two entirely different meanings -- as a symbol for the atomic ion and for the aqua ion. What is important in the present context is the fact that going from monoatomic Fe^{3+} to $[Fe(OH_2)_6]^{3+}$ is not conceived as a redox process. The example illustrates an amazingly general principle in chemistry that allows the chemistry of an individual element to be subclassified into the chemistries of this element's individual oxidation states. Think of all the reactions that one can perform with Fe^{3+} without even considering that a redox process should have taken place; think in particular about coordination of a large variety of ligands, neutral or charged. The situation is not nearly so clear with S^{6+} as the example, but even here it is agreed that there exists a chemistry of S^{VI}. This is how oxidation numbers add a third dimension to the Periodic Table.

It is characteristic of the science of chemistry that the wonderful perspective just outlined has its limits within sight. In the following sections we shall touch upon some of the less obvious limits from the point of view of nomenclature, spectroscopy, stereochemistry, and model theory. Our examination will lead to a reappraisal of the concept of oxidation states when applied to transition metals.

Oxidation Numbers Used in IUPAC Nomenclature

In all unambiguous cases of oxidation numbers, the IUPAC has not felt that definitions of the oxidation numbers used in nomenclature are necessary (7). However, its recommendations for the cases in fuzzy areas may serve as good examples of the "softness" of chemistry. IUPAC focuses essentially upon two such areas: coordinated H and coordinated NO.

Regarding H, the oxidation number for nomenclature shall be I and -I when H is connected with nonmetals and metals, respectively, except in hydride complexes when it shall be -I. While these recommendations are useful for their purpose, they have some consequences that are conceptually curious. For example, the following two reactions and others of a similar kind become redox reactions:

$$HCo(CO)_4 = H^+ + Co(CO)_4^- \tag{1}$$
$$B_2H_6 + 2H^- = 2BH_4^- \tag{2}$$

The recommendations also touch upon organometallic chemistry by suggesting the generalization from the ligand, CH_3^-, methanide, which is a consequence of the H^I convention for the nonmetal C to other R^- ligands of the alkyl and aryl type.

Regarding NO, the IUPAC says that it shall be regarded as a neutral ligand. This implies that the isoelectronic ions $[Fe(CN)_5(NO)]^{2-}$ and $[Fe(CN)_5(CO)]^{3-}$ contain Fe^{III} and Fe^{II}, respectively, along with many other analogous examples. We shall return to a discussion of these in the section on oxidation states and classifying configurations.

Oxidation Numbers and Oxidation States

From a purely linguistic point of view, oxidation state is an attribute, while oxidation number is a denotation for or designation of an attribute. Therefore the third dimension in the skeleton of chemistry, mentioned above, has oxidation states as its body and oxidation numbers as the indices of the third coordinate. These oxidation numbers need not necessarily be the same as those recommended by the IUPAC for nomenclature. Rules of nomenclature are sometimes too rigid and sometimes phase-shifted with respect to time relative to current use in chemistry, but their aim is always to avoid ambiguity and inconsistency. Regarding oxidation numbers, unanimity has been achieved in the vast majority of cases.

A book has been written with the title of the present section (8). In this book and in a number of papers (6,9,10), including one from the present symposium, Christian Klixbüll Jørgensen has pointed out and illustrated how chemical conventions about oxidation states, particularly for transition metal complexes, have been given substance by developments since the 1950s. Spectroscopic, magnetic, and stereochemical properties of metal complexes have been found, independently of covalency/ionicity, to be connected with preponderant electron configurations -- as Jørgensen calls the configurations used for classification -- and thereby with oxidation states. A strong link has thus been established from intuitive oxidation states, through those formal states that derive from one-sided sharing of the electron-pairs of the coordinative bonds to the description of ground states (and low-lying excited states) in terms of classifying electron configurations. This description, which includes the connection between oxidation state and classifying configuration, merits further discussion and will be the subject of the next section.

Oxidation Numbers and Classifying Configurations

The conception of the Periodic Table as consisting of 18 groups has been one of the most controversial IUPAC initiatives in recent times (7). It is difficult for a chemist, and particularly for a main group chemist, to have to place boron, carbon, and nitrogen in groups 13, 14, and 15, respectively. It may help, however, to realize that the numbering arises from focusing upon the first long period, the one containing the

3d transition elements. Recent analyses regarding chemical entities containing transition metals show that these (with certain predictable or expectable exceptions (6)) can be associated with a classifying d^q or f^q configuration. This result can, once an nd^q (n = 3, 4, or 5) configuration has been established, be expressed by the pure electron-count or bookkeeping equation

$$q = g - z \qquad (3)$$

where g is the new IUPAC group number and z the oxidation number characterizing the spectroscopically (8) or stereochemically (9) defined oxidation state. It should be noticed, though, that establishing nd^q as the characterizing configuration is not trivial. For the most common, typical, formal oxidation states of the transition metals, atomic spectroscopy, ligand-field spectra and magnetism go hand in hand in supporting nd^q. However, for low oxidation states, $z < 2$, a direct comparison with atomic spectroscopy would have led one to expect configurations of the type $d^{q-a}s^a$ ($a = 2$, or, sometimes, $a = 1$), but here the assumption of a characterizing d^q configuration of the chemical systems is supported by diamagnetism and by a very consistent connection between stereochemistry and a simple bonding model. We shall now illustrate equation 3 with some chemical examples.

Copper has $g = 11$, trivalent copper has $z = 3$, and for the number of d electrons we have accordingly $q = 8$. Cu^{III} exists as other d^8 systems with two stereochemistries, square planar as typical of Pd^{II}, Pt^{II}, and Au^{III} and octahedral as typical of Ni^{II}. In square planar $[Cu^{III}\{IO_5(OH)\}_2]^{5-}$ the ground state may, in accordance with analogous systems, be labeled as $(d_{yz})^2(d_{zx})^2(d_{xy})^2(d_{z^2})^2$, $^1A_{1g}(D_{4h})$, while in octahedral $[CuF_6]^{3-}$, the label is $(d_{yz})^2(d_{zx})^2(d_{xy})^2(d_{z^2})(d_{x^2-y^2})$, $^3A_{2g}(O_h)$. These labels, which, of course, agree with the diamagnetism and paramagnetism, respectively, consist of a preponderant electron configuration and a symmetry type for the ground state of the system. The double labels are examples of property characterizations that are qualitative in a strong sense.

Another example is the isoelectronic and isosteric series

$$Mn(CO)_4^{3-}, Fe(CO)_4^{2-}, Co(CO)_4^-, Ni(CO)_4 \qquad (4)$$

where both g and z increase by one unit across the series, thus leaving $q = 10$ for all four cases according to equation 3. In these back-bonding systems the 18-electron rule and the ($q = 10$)-description are almost synonymous; the coordination number is four, the coordination geometry is tetrahedral, and the ground state double label is $(e)^4(t_2)^6$, $^1A_1(T_d)$. However, this example is less clearcut because in these cases one neither has a ligand-field spectrum nor some especially characteristic magnetic properties to support a choice of q and thereby an oxidation state.

For the ($q = 8$)- and the ($q = 10$)-examples given here, the associated z values agree with the IUPAC oxidation numbers for nomenclature. The same is true for the example $HCo(CO)_4$, which by having $g = 9$ and $z = 1$, becomes a d^8 system as $Fe(CO)_5$, but here the property justification is even more limited (a distorted and a nondistorted trigonal bipyramid, respectively) than in the pure carbonyl examples. The conclusion is to some extent based upon the feeling that typical bonds of metals to hydrogen are hydridic in character for electronegativity reasons. In any case, the reaction of equation 1 remains an acid dissociation reaction, which is also a redox disproportionation reaction.

In the constellation metal-NO, nomenclature cannot distinguish offhand between linear and bent ligation at nitrogen, for the reason alone that this structural knowledge is not always available at the time when a name is required for a given compound. The IUPAC recommendation (7) that electroneutral NO shall always be considered to make up the ligand in the definition of the oxidation numbers used for nomenclature has the chemical basis that the molecule NO is viewed as the ligand in much the same way as is the molecule NH_3. However, a purely chemical basis does not take into account that NO is special by having an odd number of electrons.

We want to comment on stereochemically-based oxidation states (9) in the metal-NO systems where nitrogen is linearly ligating. These systems are, for example, the members of the isoelectronic series

$$Cr(NO)_4, \ Mn(CO)(NO)_3, \ Fe(CO)_2(NO)_2, \ Co(CO)_3(NO), \ Ni(CO)_4 \qquad (5)$$

which is also isoelectronic with series 4. Here the mainstream view is that these systems obey the 18-electron rule, and we do not object to that. However, this view either requires NO to be a three-electron donor ligand or NO^+ to be the ligand in a usual two-electron donor sense. These requirements are identical for pure, electronic bookkeeping, but they are different as far as oxidation numbers are concerned. We think that the three-electron donor view is *ad hoc*, lacks esthetic appeal, and -- perhaps worse -- is destructive of the beauty of the 18-electron rule by making it unnecessarily formal. In our opinion this rule should be viewed as a sum rule in which the number of electrons in orbitals, classified as central ion orbitals that are π-bonding with respect to the individual metal to ligand subsystems, is added to the number of electrons in orbitals, classified as ligand orbitals that are σ-bonding with respect to these same subsystems. With this molecular-orbital view of the 18-electron rule it is, in a chemical stabilization context, understandable that it works the better the more π-back-bonding its d electrons and the more σ-donating its ligand electrons. For example, the members of the two series 4 and 5 have four ligands, each providing two bonding σ-electrons, and, in addition, all of these systems have a central atomic d^{10} system which, since both e and t_2 orbitals in tetrahedral coordination have suitable symmetries to engage in π-bonding, provides 10 π-bonding electrons. This view, which entails the synonymous ideas of isoconfigurational systems and systems whose central atomic parts are isoelectronic, embodies the idea of the two-electron donor ligand NO^+. However, it does not involve any preconceived notions about the charge distribution in the molecules concerned (see the next section of this paper). With NO^+ ligands the isoelectronic species $[Fe(CN)_5(NO)]^{2-}$ and $[Fe(CN)_5(CO)]^{3-}$, which, as we saw above, have IUPAC-nomenclature oxidation numbers of III and II, respectively, both obtain values of g, z, and q equal to 8, 2, and 6, respectively, and may thus be called centrally isoelectronic. Their diamagnetism supports this d^6 classification, but their energy-low electron transfer spectra preclude the observation of the spin-allowed ligand-field transitions, which otherwise are the typical additional indicators for low-spin d^6 systems.

The Rise and Partial Fall of the Concept of the Orbital: The Classifying l^q Configuration and the Oxidation State

Most of the developments described above occurred before the advent of the electron in chemistry. Then came the golden years of wave mechanics with one-electron wave functions (orbitals), the Pauli principle, the building-up principle (*Aufbauprinzip*) and, even before the end of the 1920s, the idea that chemistry had now become a question of computation was proposed. Not many years later the best conceivable method of describing atomic and molecular systems in terms of fixed orbitals with one or two electrons in each was invented and named the Hartree-Fock description. The appearance of electronic computers made the method practical for heavy systems, such as atomic 3d transition-metal ions, as early as about 1960. The results for d^q spectra were enthusiastically received, mainly because it was wonderful to see that such calculations could actually be done. On second thought and on further development of computational chemistry, the orbital or one-electron picture of chemistry became blurred. However, the many-electron consequences of the orbital picture remained crystal clear, firmly anchored on symmetry, and supported by experimental facts.

We shall illustrate the current status of the orbital picture by first considering, as examples, atomic d^q systems, whose d^q spectra have been completely observed as far as J levels are concerned (11). The focus will be on the d^2 systems, which, compared to other d^q systems, are the simplest ones to discuss. For d^2, nine levels are expected on the basis of symmetry, and nine are observed. Moreover, the increasing energy order of the levels is invariably $^3F_2$, $^3F_3$, $^3F_4$, $^1D_2$, $^3P_0$, $^3P_1$, $^3P_2$, $^1G_4$, $^1S_0$ except for some minor exceptions that occur for systems with large spin-orbit coupling splittings. Furthermore, the energies of these nine levels can in every case be parametrized, almost within the experimental accuracy, by using a model in which five energy parameters represent the energies of the five multiplet terms ^3F, ^1D, ^3P, ^1G, and ^1S, and one energy parameter, ζ_{nd}, represents the *one-electron* spin-orbit coupling splitting. This type of model, which is called a PMTζ model where PMT stands for parametrical multiplet term (11), applies quantitatively well to all atomic d^q systems.

The situation then is the following. The qualitative predictions of the one-electron picture are perfect; the quantitative parametric fitting of energy level data with a model that accounts for *all* the d^q configuration fine-structure by means of a *single* empirical one-electron parameter is almost perfect. This would seem to provide strong support for the validity of the one-electron picture. However, logic does not work that way.

The orbital picture can be criticized as a physical model at a level as fundamental as the virial theorem. This theorem, which is valid for any stationary state, requires the state's total energy to be equal to its kinetic energy, apart from the sign. Since in an atomic model containing an invariant electronic core plus a number of equivalent l electrons (l = p, d, or f), the kinetic energy must have a fixed magnitude, then, by the virial theorem, the same must be true for the total energy. Thus, in this model, an energy separation of the l^q levels is not allowed. The physics around this literal model has also been analyzed at the much less sophisticated, but on the other hand still quantitative, Hartree-Fock level with the most atrocious results (17,18).

The reason why the strict d^q concept breaks down while its consequences qualitatively remain valid has to do with the symmetry contents of d^q. It is a symmetry property that the configuration, d^q, gives the energy levels that it does and that the symmetries of these levels survive the blurring mixing of d^q with other configurations. What still remains to be understood is the fact that a quantitative parametrization of the d^q energy levels, based upon the q-electron-function concept, as mentioned above, is *extremely* successful. It is, however, exactly this fact that justifies the concept of the classifying d^q configuration.

Until now, our examples have been atomic. Although it is symmetry arguments of much the same kind that apply to molecular systems as well, it is in the context of oxidation states illustrative to include an example involving the charge transfer that takes place through bond formation, that is, the interaction known as covalency. For this purpose we choose the typical low-spin d^8 system $[Pd^{II}Cl_4]^{2-}$. Here the concepts of the linear combination of atomic orbitals molecular orbital model are useful for our discussion but not in any way necessary for the qualitative conclusion based upon it. Using the LCAO-MO picture of $[Pd^{II}Cl_4]^{2-}$, the orbital classifiable as $d_{x^2-y^2}$ is strongly σ-antibonding. It has the highest energy among the d orbitals and since $[Pd^{II}Cl_4]^{2-}$ is a *low-spin* d^8 system no electrons occupy this orbital. The emptiness combined with its partial ligand orbital character implies that the ligands have been partially deprived of their electronic density by the bond formation, and, concertedly, the genuine $d_{x^2-y^2}$ orbital has been partially populated with ligand electrons.

However, this covalency-type charge transfer has no influence upon the integral number of electrons that enters the building-up principle and thus has no influence upon the determination of the oxidation state. Therefore the charge separation in $[FeF_6]^{3-}$ containing Fe^{III} may well be larger than that in $[FeO_4]^{2-}$ containing Fe^{VI}. The important point is that the spectroscopically-defined oxidation states for Fe are *independent* of the degree of ionicity and are based upon the experimental facts that the complexes are the high-spin systems with ground terms $^6A_{1g}(O_h)$ and $^3A_2(T_d)$, and accordingly have the classifying configurations d^5 and d^2, respectively. Likewise, the charge separation in $Fe(CO)_2(NO)_2$ may be quite small because of back-bonding even though the stereochemical oxidation state of iron here is -II. It thus seems as if the largest charge separations are associated with *intermediate* oxidation states, and the examples make it probable that covalency depends on several parameters that almost necessarily are mutually dependent. This is unfortunate for a concept in the hard natural sciences but still rather typical of chemistry, and this is why we call our subject the "soft" science of chemistry.

So far the discussion has focused on the qualitative aspects of the concept of a classifying d^q configuration and thereby of spectroscopically well-defined oxidation states. The final sections of this paper will exemplify some quantitative aspects.

Sum Square Splitting and Root Mean Square Splitting as Measures of Configurational Splitting

There exists a complete mathematical l^q Hamiltonian model (*11*) that is in essence symmetry-based and parametrical. The domain of functions on which the parametrical Hamiltonian operator acts is the collection of all the states of the l^q configuration. This operator is traceless, which means that the sum of its diagonal matrix elements is equal to zero, independently of the choice of function basis. Accordingly, the sum of its eigenvalues is also equal to zero. The Hamiltonian is thus said to be barycentered, which means that it has been designed to account for the splitting of the configuration but not for its average energy, that is, its energy relative to an external zero point of energy.

Once an l^q function basis has been chosen, the Hamiltonian operator has a parametrical Hamiltonian matrix, an energy matrix, set up in this basis. If the basis is chosen so that the matrix is real, the sum of the squares of all the matrix elements is independent of the choice of function basis and is therefore a property of the operator itself. This property is called the operator's norm square. Because the norm square can be calculated in any basis, it can also be calculated in the eigenbasis, which means that it is equal to the sum of the squares of the eigenenergies, measured relative to their barycenter. This is why such a norm square is often referred to by us as the *sum square splitting*. If the norm square is divided by the number of states of the configuration, the square root of the resulting quantity is the *root mean square splitting* of the configuration (*11*).

If each empirical parameter, P, is associated with a coefficient operator, $\hat{Q}_P$, the Hamiltonian, $\hat{H}$, can be written as

$$\hat{H} = \sum_{all\ P} \hat{Q}_P\ P \tag{6}$$

and if the coefficient operators are chosen to be mutually orthogonal, the Hamiltonian's norm square obtains the simple generalized Pythagorean form

$$\langle \hat{H} | \hat{H} \rangle = \sum_{all\ P} \langle \hat{Q}_P | \hat{Q}_P \rangle\ P^2 \tag{7}$$

which means that the norm square consists of a sum of terms, where each one belongs to an individual empirical parameter.

It is useful to think about two Hamiltonian operators, the complete mathematical one, $\hat{H}_{math}$, and the model one, $\hat{H}_{model}$. $\hat{H}_{math}$ contains a *complete* set of empirical parameters, while $\hat{H}_{model}$ contains a subset of these parameters. It is therefore possible to write equation 8:

$$\chi_w^2 = \langle \hat{H}_{math} | \hat{H}_{math} \rangle - \langle \hat{H}_{model} | \hat{H}_{model} \rangle \tag{8}$$

In the case where the set of observations is complete from the experimental material, one can immediately obtain the value for the norm square of the empirical Hamiltonian; it is equal to the sum of the squares of all the barycentered energies, each weighted by the degeneracy number of the energy level concerned. The result of the calculation is equal to $\langle H_{math} | H_{math} \rangle$, which represents an upper limit to $\langle H_{model} | H_{model} \rangle$. Thus in this case there exists a kind of variational theorem. The practical problem is that of determining those values for the empirical parameters contained in H_{model}, which minimize the difference between $\langle H_{math} | H_{math} \rangle$ and $\langle H_{model} | H_{model} \rangle$, χ_w^2 of equation 8, called the weighted variance. The final result can be viewed as a decomposition of $\langle H_{model} | H_{model} \rangle$ into terms associated with the individual empirical parameters according to an expansion of the form of equation 7.

Examples of Contributions to Configurational Splittings

Our first example concerns an analysis of the spectrum of gaseous Ni^{2+}. The experimental value for the *sum square splitting*, $\langle H_{math} | H_{math} \rangle$, of the characterizing d^8 configuration is equal to 55.158 μm^{-2}, corresponding to a *root mean square splitting* of 1.1071 $\mu m^{-1} = 11071$ cm^{-1}. In the PMTζ model $\langle H_{math} | H_{math} \rangle$ is resolved into the sum of four terms where the first one represents the spin and seniority splittings, parametrized by D and $D \perp$; the second one the orbital-symmetry splitting, parametrized by B and $B \perp$; the third one the spin-orbit splitting, parametrized by ζ_{3d}; and the fourth one, the residual term, χ_w^2. These terms are equal to 36.8585, 17.7650, 0.5347, and 0.000085 μm^{-2}, respectively. This means that only 0.000085/55.158 $\approx$ 10^{-4}% of the experimental value for the sum square splitting is left unaccounted for in terms of χ_w^2 and that the spin-orbit coupling accounts for only 0.97% of this sum square splitting (*11*).

It is possible to make a contraction of the PMTζ model so that it degenerates into the conventional Slater-Condon-Shortley model (*11*), with our notation the SCSζ model. In doing that, the first and the second term mentioned above become parametrized by the conventional parameters, D, the spin-pairing energy parameter, and, B, the Racah parameter (*11*). In the above analysis the D term accounts for 36.8099 and the B term for 17.7029 μm^{-2} so that including the spin-orbit coupling term, it is seen that the SCSζ part of the PMTζ model accounts parametrically for 99.8% of the sum square splitting of the d^8 configuration of Ni^{2+}. This is in spite of the fact that the physical basis of the SCSζ model holds the virial theorem in contempt.

The second example, the approximately octahedral chromium(III) complex, *trans*-$[CrBr_2(NH_3)_4]^+$, is chosen from the ligand-field area. Here the cubic field, described by the traditional parameter Δ, and the two symmetry-independent tetragonal fields, described by the parameters $\Delta(e)$ and $\Delta(t_2)$, will be quantitatively compared. For this complex the values (*12*) for the three parameters, Δ, $\Delta(e)$, and $\Delta(t_2)$ are 1.819, 0.445, and -0.100 μm^{-1}, respectively. The definition of the parameters is given in Figure 1, from which the norm squares of the corresponding coefficient operators can also be calculated (*cf.* equation 7). For the operator associated with $\Delta(e)$ the norm square is, for example, $(\frac{1}{2})^2 + (-\frac{1}{2})^2 = \frac{1}{2}$ so that the contribution of this parameter to the sum square splitting of the d orbital level is $\frac{1}{2}(0.445)^2 = 0.099$ μm^{-2}. For Δ and $\Delta(t_2)$ these contributions are similarly found to be 3.97 and 0.007 μm^{-2}, respectively, so that the sum square splitting of the total ligand field is 4.07 μm^{-2}, of which the cubic term makes up 97.5%. This result provided (*13*) for the first time a piece of quantitative justification for the common colloquial reference to this kind of tetragonal complex as an *octahedral* complex. Comparison of the $\Delta(e)$ contribution, 0.099 μm^{-2}, which in the angular-overlap-model interpretation is due to the difference between Br^- and NH_3 with respect to their σ-antibonding effects upon the d orbitals, with the $\Delta(t_2)$ contribution, 0.007 μm^{-2}, which accounts for the π-antibonding difference, furthermore gives a quantitative measure of the dominating influence of the σ-antibonding difference upon the tetragonality.

As the third and final example, d^q configuration sum square splittings, represented in Figure 2 as areas of circles, are analyzed for a series of high-symmetry transition metal systems. The effective one-electron operator contributions due to the ligand field are represented by the area of the shaded sectors, the effective two-electron operator contributions by that of the unshaded sectors. These latter contributions are usually referred to as interelectronic repulsion contributions, but because of the breakdown of this interpretation by the virial theorem, we here refer only to their formal two-electron operator character. A comparison of the data for $[Cr(OH_2)_6]^{3+}$ and $[Co(OH_2)_6]^{3+}$ shows that the ligand field of the water molecules gives rise to the same 56% of the sum square splitting in both of these two trivalent metal complexes of the d^3 and the d^6 configurations, respectively. The ligand field of the cyanide ions, on the other hand, gives a relatively higher contribution in the d^6 system $[Co(CN)_6]^{3-}$ than in the d^3 system $[Cr(CN)_6]^{3-}$. We have interpreted this (*14*) as caused by the d^6 configuration being particularly susceptible to π-backbonding (18 electron rule).

For the d^7 configuration of Co^{II} there are two issues that we find noteworthy. These relate to the possibility of comparing individual contributions to the total sum square splitting within one system, here the $[CoI_4]^{2-}$ system, and to the possibility of comparing the total sum square splittings of different systems, here gaseous Co^{2+} and the complex $[CoI_4]^{2-}$. Regarding the first issue, we note that the ligand field in $[CoI_4]^{2-}$ is responsible for only 5% of the total sum square splitting, a fact, which is qualitatively in agreement with the expectations for a tetrahedral complex, and in particular for one containing ligands that are placed in the lowest end of the spectrochemical series. Regarding the second issue, which is of special relevance to this paper, it is seen that the absolute magnitude of the total sum square splitting *decreases* as one goes from the gaseous ion to the tetraiodo complex in spite of the fact that the ligand field in the latter case gives its *additional* contribution. This is due to what is known as *nephelauxetism*, that is, the quite general phenomenon that the two-electron operator contributions to configurational splittings are smaller in complexes than in the corresponding gaseous ions (*15,16*). Qualitatively the same is observed if comparison is made between two d^7 gaseous ions, the first being of lower charge than the second. The first gaseous ion, which necessarily belongs to an element that has a lower atomic number, is thus able to mimic our centrally isoelectronic complex, that is, the d^7 complex, as far as the two-electron operator terms are concerned. Traditionally speaking, the lower charge gives rise to a lower interelectronic repulsion due to an expansion of the electronic cloud (cloud expansion = nephelauxetism). The comparison with a (centrally) isoelectronic, gaseous ion, which we have just made, is the atomic spectroscopical and therefore nonchemical, yet empirical background for interpreting nephelauxetism as an indicator of covalency, which is charge transfer without change of oxidation state.

Conclusion

We have been concerned with the concept of the classifying d^q configuration in atomic and in molecular transition-metal systems. We have discussed the oxidation states which are intimately connected with such d^q configurations and which sustain large variations in nephelauxetism. These nephelauxetic variations are the variations in the

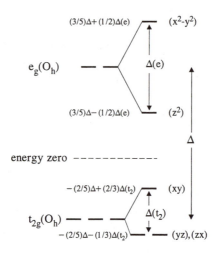

Figure 1. Definition of ligand-field symmetry parameters for a complex of tetragonal symmetry. Δ represents the cubic field as usual; $\Delta(e)$ and $\Delta(t_2)$ represent the tetragonal field that requires two independent variables.

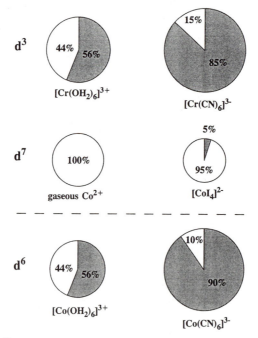

Figure 2. Pie diagrams representing configuration *sum square splitting* and its decomposition into one-electron and two-electron opertor parts. Within a given d^q configuration (and its complementary configuration d^{10-q}) the total areas of the pies may be mutually compared as representing configuration *sum square splittings*. The unshaded sectors represent those fractions of the sum square splittings which is due to two-electron operator contributions (traditionally conceived as interelectronic repulsion), the shaded sectors those due to the one-electron operator contributions (ligand field).

configurational splitting due to the effective two-electron operator contributions. They mimic charge transfer. This kind of charge transfer is extremely close to what is called covalency, and the remarkable thing is that a considerable amount of charge transfer can take place without a change in the oxidation state of the central atomic system, that is, without the whole system undergoing an intramolecular redox reaction. Charge transfer without electron transfer may be the rationalization of the "black holes" of chemistry that consist of unexpectedly high oxidation states of central ions bonded to reducing ligands, *e.g.*, $[Cu^{III}\{S_2CN(CH_2CH_3)_2\}_2]^+$.

By using orthogonal operators acting on the classifying d^q configurations, quantification of configurational splittings allows comparisons to be made involving quantities that until now were almost incomparable. It is expected that orthogonal operators will find new useful applications in connection with other techniques used to investigate and rationalize inorganic chemistry. Moreover, it is hoped that this account will help authors of textbooks to realize that *oxidation numbers* are not purely formal quantities but in many cases labels that refer to *oxidation states* with quite definite and chemically important properties.

Acknowledgment

One of the present authors (C.E.S.) would like to express his gratitude to George B. Kauffman for giving him the opportunity to participate in this symposium and meet old friends from a time when fundamental research could still be funded.

Literature Cited

1.   (a) Ostwald, W. *Grundlinien der anorganischen Chemie*; Wilhelm Engelmann: Leipzig, 1900; (b) p 142; (c) p 575; (d) p 602.
2.   Werner, A. *Neuere Anschauungen auf dem Gebiete der anorganischen Chemie*; Friedrich Vieweg und Sohn: Braunschweig, 1905; p 13 and p 73.
3.   Groth, P. *Einleitung in die chemische Kristallographie*; Engelmann: Leipzig, 1904.
4.   Stock, A. *Z. Angew. Chem.* **1919**, *32*, 373-374; Kauffman, G.B.; Jørgensen, C.K. *J. Chem. Educ.* **1985**, *62*, 243-244.
5.   Ewens, R.V.G.; Bassett, H. *Chem. Ind.* **1949**, *68*, 131-139; Kauffman, G.B.; Jørgensen, C.K. *J. Chem. Educ.* **1985**, *62*, 474-476.
6.   Jørgensen, C.K. *Z. Anorg. Allg. Chem.* **1986**, *540/41*, 91-105.
7.   International Union of Pure and Applied Chemistry, *Nomenclature of Inorganic Chemistry: Recommendations 1990 ("Red Book")*; Leigh, G.J., Ed.; Blackwell Scientific Publications: Oxford, 1990.
8.   Jørgensen, C.K. *Oxidation Numbers and Oxidation States*; Springer-Verlag: Berlin, 1969.
9.   Jørgensen, C.K. *Chem. Phys. Lett.* **1969**, *3*, 380-382.
10. Jørgensen, C.K. *Chimia* **1974**, *28*, 605-608.
11. Bendix, J.; Brorson, M.; Schäffer, C.E. *Inorg. Chem.* **1993**, *32*, 2838-2849.
12. Glerup, J.; Mønsted, O.; Schäffer, C.E. *Inorg. Chem.* **1976**, *15*, 1399-1407.
13. Brorson, M.; Damhus, T.; Schäffer, C.E. *Comments Inorg. Chem.* **1983**, *3*, 1-34.
14. Josephsen, J.; Schäffer, C.E. *Acta Chem. Scand., Ser. A* **1977**, *A31*, 813-824.
15. Schäffer, C.E.; Jørgensen, C.K. *J. Inorg. Nucl. Chem.* **1958**, *8*, 143-149.
16. Jørgensen, C.K. *Prog. Inorg. Chem.* **1962**, *4*, 73-124.
17. Vanquickenborne, L.G.; Pierloot, K.; Görller-Walrand, C. *Inorg. Chim. Acta* **1986**, *120*, 209-213.
18. Jørgensen, C.K. *J. Less-Common Met.* **1989**, *148*, 147-150.

RECEIVED May 2, 1994

Chapter 19

Coordination Based on Known Free Ligands, Moderate Dissociation Rate, Weaker Electron Affinity of Central Atom Than Ionization Energy of Ligand, and Quantum Paradoxes

Christian K. Jørgensen

Section of Chemistry, University of Geneva, 30 Quai Ansermet, CH 1211 Geneva 4, Switzerland

Most minerals were described by Berzelius as adducts of binary oxides (a concept close to double salts). Werner pointed out that most doublecyanides,many complexes of ammonia and/or of amines,and indirectly even some aqua ions have very slow rates of dissociation contrary to those salt hydrates formed as by freezing of a melt. The era 1916-1984 of electron-pair bonds ignored the inconsistency of the "covalent" or "electrovalent" dichotomy; crystallography confirmed most (but not all) of Werner's intuitions; and lip-service was rendered to quantum mechanics as the Chemical Theory of Everything (and the rest). Today we are in a cataclysmic turmoil,separating the factual observables from pedagogic-model chemistry nearly as much as astrophysics from astrology. Arguments related to Werner complexes are presented.

Coordination compounds have definite,rather shortfrequently long-lived links,each between (usually one) central atom and one atom of some monatomic (halides; chalcogenides; argon or krypton atom) or with unidentate ligands containing several atomic nuclei (water; NH_3; CN^-; CO; NCS^-; CH_3; pyridine; etc.) or alternatively,links to two,three, four,.. atoms of an (at least potentially) multidentate ligand (such as carbonate; acetate; oxalate; biological or synthetic amino acids and aminopolycarboxylate anions; dithiocarbamates; dithiophosphates; WS_4^{2-}; nitrite; nitrite; etc.).This is not a definition in the sense of a mathematical,or even of an encyclopedic definition; the concept of "coordination" has been applied,with enthusiasm or with some reluctance,to huge groups of novel compounds (to a large extent, organometallic species with at least one direct M-C link) and below, the special problem of linkage of M to H^+; H^-; and H_2; to NO^+, NO,and NO^-; to carbonium vs. carbanion ligands; and suspiciously long links, are shortly discussed. Even carbon forms the aqua ion $H_2OCH_3^+$ (i.e. protonated methanol),the ammonia complex $H_3NCH_3^+$ (isoelectronic with ethane and diprotonated hydrazine) and is surrounded by four As atoms in a tetrahedral $C(das)_2^{4+}$ bis-complex of the diarsine ligand studied

0097–6156/94/0565–0226$08.00/0

by F.G.Mann and by R.S.Nyholm,and by four mercury atoms in $C(HgI)_4$ containing only 0.9 weight percent carbon.Therefore,some extremists argue that perhaps all inorganic compounds are coordinated,with minor exceptions like diatomics X_2, cyclic systems (S_8; Te_4^{2+}; As_4 known from skudderudite $CoAs_3$),and clusters in three dimensions (P_4 and C_{60}) containing one element.As I interpret <u>Neuere Anschauungen auf dem Gebiete der anorganische Chemie</u>,Werner was reluctant to call binary compounds coordinative,and as seen in the next three sections,his perception of ternary and quaternary compounds is diffusely mixed with his novel concept of coordination. Today polyatomic ions are fully recognized to be isoelectronic with neutral molecules (e.g.,octahedral AlF_6^{3-},SiF_6^{2-},PF_6^{-},SF_6,ClF_6^{+}; tetragonal-pyramidal TeF_5^{-},IF_5,XeF_5^{+}; tetrahedral WO_4^{2-},ReO_4^{-},and OsO_4) readily accepted as "complexes".

Berzelius,Idealized Composition of Minerals,and Stoichiometry

The century between Lavoisier's execution in 1794 and Werner's appoint-ment in 1893 as extraordinary professor at the University of Zürich represented a growing bifurcation (and estrangement) between organic and inorganic (French: <u>chimie minérale</u>) (becoming less pronounced after 1900,in part because both Sophus Mads Jørgensen,Marcel Delépine,William J.Pope,and Werner himself started as organic chemists). In one way,it is a paradox that stoichiometry did not become the major concern for gaseous species.Lavoisier's paradigms (<u>1</u>) of a certain number (31 in his book of 1789; 76 in 1896,the year that radioactivity was discovered) of indestructible,intransmutable elements with strictly additive mass would attract attention to the (at least) five nitrogen oxides,the multitudinous volatile sulfur compounds,among other subjects.Actually, the "Big Science" large-scale demonstration in Paris of the decomposi-tion and synthesis of water only provided an approximate weight ratio H:O not far from 1/7.Dalton is not only more widely known for his studies of color blindness,he also had a schoolboy hope that the World cannot be nasty and complicated.Thus he assumed the <u>simplest</u> formulae OH and NH (in absence of other evidence) for water and ammonia.By a stroke of bad luck,one of Dalton's few quite careful measurements of gas density (as a function of temperature) involved brown NO_2 in rapid equilibrium with N_2O_4.This became later a prime argument against the hypothesis of Avogadro,originally almost unknown and entirely marginal. Then,in 1860,a major controversy broke out,leading to the arrière-garde defense of equivalent weights [at least having the good taste to be twice as large for Cu(I), Sn(II),and Cr(III) as for Cu(II), Sn(IV),and Cr(VI),but having the ratio 5:3 for nitrogen,phosphorus,arsenic,and antimony].For our purpose,it is relevant that $ClH \bullet NH_3$ and $Cl_2 \bullet PCl_3$ were described as "molecular compounds" [as we still call quinhydrone] because of anti-Avogadro behavior on heating.The main benefit of the "law" of Avogadro (the number of moles n was incorporated into PV = nRT of Boyle and Mariotte) was the possibility for establishing the Periodic Table from atomic weights (as would be unsuccessful,using equivalent weights) in spite of the latter quantities being closer [in a naively-positivistic perspective] to operational science.

It is evident that chemists get their materials exclusively from minerals,the ocean,the atmosphere,coal,crude oil,plants,and from animals (and organic chemists tend to emphasize the four latter sources, not conceptually,but the much easier separation,even of rare compounds). Elements were frequently discovered by attempts to analyze a rare or novel mineral.Thus tellurium was discovered 1782 in sylvanite $AuAgTe_4$.

At least 22 elements were discovered in oxide and silicate minerals: uranium (1789), zirconium (1789), yttrium and fourteen ($\underline{2}$) lanthanides Ln (1794 to 1901),beryllium [glucinium](1798), niobium [columbium] (1801), tantalum (1802), thorium (1828),and finally hafnium (in 1923). As still seen in mineralogical museums,the quantitative analysis (assuming additive masses) of oxides are given as,$\underline{e.g.}$,beryl being 3 $BeO \cdot Al_2O_3 \cdot 6 \ SiO_2 = Be_3Al_2Si_6O_{18}$ or (in view of the inevitable fact that any experimental quantity has an error limit,it being in the second or the thirteenth decimal digit),the "rare earth" mineral gadolinite x $Y_2O_3 \cdot (1-x) \ Ln_2O_3 \cdot 2 \ BeO \cdot FeO \cdot 2 \ SiO_2$ [many differing Ln; reminding us about the analysis of a portion of ham and eggs as $C_{50}H_{122}Cl_3N_7Na_5O_{19}PS_3$,to be compared with the stoichiometry known from the crystal structure $C_{63}H_{88}CoN_{14}O_{14}P$ of vitamin B 12].The cubic alum $K_2O \cdot Al_2O_3 \cdot 4 \ SO_3 \cdot 24 \ H_2O$ and the mineral epsomite $MgO \cdot SO_3 \cdot 7 \ H_2O$ are known from crystal structures to be $K[Al(OH_2)_6](SO_4)_2 \cdot 6 H_2O$ and $[Mg(OH_2)_6]SO_4 \cdot H_2O$.Especially since more than 85 percent of several thousand recognized minerals ($\underline{3}$),frequently named after scientists, are mixed oxides (of which a large majority is silicates) not to speak about their,far higher,total weight proportion in the outer crust,the "museum style" oxide sum deriving from Jöns-Jakob Berzelius is very appropriate.About 10 percent of (frequently much rarer minerals) are mixed sulfides of which the textbook example is argyrodite Ag_8GeS_6 (the only known primary mineral of germanium).When boiled with strong hydrochloric acid,6 percent of the weight leaves as evaporated $GeCl_4$.

Werner was less enthusiastic about Berzelius' motivation for writing anhydrous calcium sulfate $CaO \cdot SO_3$ and hydrated gypsum as $CaO \cdot SO_3 \cdot 2 \ H_2O$.Berzelius insisted that chemical bonding of typical inorganic compounds is connected with an (overall neutral) charge distribution,later ramified to three or seven kinds of numerical electronegativity ($\underline{4}$-$\underline{7}$)and in extreme cases approximated as the Madelung potential keeping fully charged ions Ca^{2+}, S^{6+}, and O^{2-} together ($\underline{8}$). Berzelius had been very impressed by Davy (1806 and 1807) electrolyzing molten,anhydrous salts and preparing highly reactive,metallic Na, K, Ca, Sr,and Ba.The inorganic slogan that "chemical bonding is essentially electric" deepened the trenches around organic ideas,where bonds exist between equivalent carbon atoms in ethane and in diamond,like bonds in H_2, N_2, and P_4 needed for saving the hypothesis of Avogadro. This paradigm ($\underline{9}$-$\underline{11}$) lived from 1916 to 1984 as G.N.Lewis' concept that each chemical bond is a pair of electrons,with some rhetoric effort in aromatic compounds and organic colorants,sometimes superposing many Lewis structures.

The pioneer work on X-ray diffraction by W.L.Bragg in 1913 demonstrated that NaCl has no preferred relation between pairs;each Na is surrounded by a regular octahedron of six Cl,and $\underline{vice \ versa}$. Almost all solid inorganic materials do not contain molecules,but extended lattices like hexagonal corundum Al_2O_3 and ruby $Al_{2-x}Cr_xO_3$, blue sapphire $Al_{2-2x}Fe^{II}_xTi^{IV}_xO_3$,and black ilmenite $FeTiO_3$.Many salts related to Werner's ideas,such as the cubic types K_2PtCl_6 and $[Ni(NH_3)_6]I_2$ are similar to CaF_2 and the (almost insoluble) salt $[Co^{III}(NH_3)_6](Rh^{III}Cl_6)$ is related to CsCl.

Did Werner Change His Idea About "Hauptvalenz" and "Nebenvalenz" ?

Crystallographic studies of ternary compounds became available only a
a year after Werner's death in 1919.Starting with regular (or almost
regular) tetrahedral disposition around the central carbon or nitrogen
atom in $C(RR'R''R''')$ and $N(RR'R''R''')^+$ having two optically active,
long-lived enantiomers when all four R groups are different (however,
N-H bonds can be very labile),Werner proposed that many metallic
elements form tetrahedral complexes. Among the colored species,cobalt
shows numerous Co(II) examples (although their absorption spectra (12)
determined by the instantaneous nuclear distribution may suggest quite
pronounced deviations from the point-group T_d) and the organometallic
MR_4 (M= silicon,germanium,and tin) and cations MR_4^+ (M= phosphorus,
arsenic,and antimony) are quite similar to C and to N^+.
      Some of the most persistent octahedral complexes (including various
enantiomers) are formed by cobalt(III),where the Blomstrand chain-
formulae $(MNH_3NH_3NH_3..$ analogous to $CH_3CH_2CH_2..)$ preferred by Sophus M.
Jørgensen (but who repudiated them in 1907) are reinterpreted as
octahedral $X_nCo(NH_3)_{6-n}^{(3-n)+}$ [many of the slightly unconventional
formulae used in this paper are motivated by illustrating direct M-X
contacts,avoiding to appear as broken bonds (as far as can be avoided
in a formula written on a straight line)].
      Soon analogous rhodium(III),iridium(III),and platinum(IV) complexes
(with absorption bands at higher photon energies,and hence differing
colors,if not almost white) were characterized.Chromium(III) seems very
similar to cobalt(III) but can show problems of photochemical reactions
or formation of $Cr(OH)_3$ in presence of basic ligands.Werner suggested
octahedral coordination to be quite common.The isomorphous salts of
SiF_6^{2-}, TiF_6^{2-},$ONbF_5^{2-}$,and $O_2WF_4^{2-}$ studied around 1860 to 1868 by
Marignac in Geneva provided such examples,and soon after Moissan had
prepared F_2 by electrolysis of K^+FHF^- in 1886,gaseous SF_6,SeF_6,MoF_6,
RuF_6,RhF_6,WF_6,OsF_6 (originally believed to be OsF_8),IrF_6,PtF_6,and
UF_6 were characterized,evaporating very readily.
      There is no doubt that Werner overestimated the predominance of
the three other regular "Pythagorean" polyhedra (and that even today,
chemists contemplate the surfaces of any polyhedron,rather than the
number of apices giving the coordination number $\underline{N}$).The reviewer hardly
is aware of monomeric,truly O_h chromophores MX_8 (highly stabilized by
the collective Madelung potential (2,13) in CaF_2 type CeO_2,ThO_2,...
through CfO_2).Icosahedral MX_{12} is rampant in boron chemistry (14) and
close to the idealized symmetry in salts of $La(O_2NO)_6^{3-}$,$Ce(O_2NO)_6^{3-}$,
$Nd(O_2NO)_6^{3-}$,and $Th(O_2NO)_6^{2-}$ involving bidentate nitrate,like the green
volatile $Co(O_2NO)_3$ isoelectronic with $Co(O_2CO)_3^{3-}$. $\underline{N}$ = 16 is known in
organo-uranium complexes $ClU(C_5H_5)_3$ and $U(C_8H_8)_2$.Cuboctahedral N = 12
is known both in cubic close-packed metals and the strontium site of
$SrTiO_3$ and,to a good approximation,the K site in K_2PtCl_6.

      Werner had the conceptual problem that $PtCl_4$ is a very elusive
compound (and $IrCl_4$ has never been isolated free of H_3O^+ and/or of OH^-
bridges),but anions $PtCl_6^{2-}$ were used for detecting K^+ and,since 1861,
Rb and Cs.In his first writings on K_2PtCl_6,the Hauptvalenz (primary
valency) of Pt is 4,giving with 1+1 of the two K^+ a total sum of +6,
rationalizing the presence of six Cl^- anions,each with -1.

It is true,of course,that macroscopic chunks of matter are quite extremely close to electroneutrality.the Nebenvalenz (secondary or "auxiliary" valency) of a given element in a definite primary valency is the (preferred) number $\underline{N}$ of immediate-neighbor ligating atoms X. For X atoms distributed on a sphere with radius R,a repulsion between X atoms depending on their mutual distance r,and proportional to $r^{-0.7}$, $r^{-1},..,r^{-4}$ strongly favors octahedral $\underline{N}$ = 6.For $\underline{N}$ = 5,the trigonal (of point-group D_{3h}) and tetragonal (Cheops) pyramids(C_{4v}) are not widely different in energy. Square coordination (N = 4) is known from palladium(II), platinum(II), gold(III),and the main-group bromine(III), iodine(III),and xenon(IV) and should be a few percent less stable ($\underline{7}$) than a regular tetrahedron. On the other hand,$\underline{N}$ = 7,9,10,and 11 meets very weak,and rather unpredictable discrimination between the (rather low-symmetry) coordination alternatives,in agreement with the quite indeterminate $\underline{N}$ and point-group of moderately large cations,such as calcium(II),strontium(II),yttrium(III),all the lanthanides(III),the large thorium(IV),and uranium(IV), For $\underline{N}$ = 8 Archimedean antiprisms always are more stable than a cube,but with a certain choice of x as parameter,lower (non-tetragonal) alternatives can be even more stable, as probably verified for molybdenum(IV) in $Mo(CN)_8^{4-}$.For $\underline{N}$ = 12 the high-symmetry icosahedral MX_{12} is marginally (0.2 percent) more stable than cuboctahedral,also called cubic tetrakaidecahedral ($\underline{7},\underline{15}$).

Putting on the eyeglasses of an organic chemist,electrovalent aspects would seem less primary than the covalent aspects in a salt like $N(CH_3)_4^+ ClO_4^-$.A large polyatomic cation forms salts nearly as stable with $B(C_6H_5)_4^-$,"tosylate" $CH_3C_6H_4SO_3^-$,or $(C_5H_5N)IrCl_5^-$ as with Br^- or Cl^-. In Werner's earlier publications the charge of N^+ would "shine through" the methyl shielding and attract the minus charge (wherever present in perchlorate). On the other hand,the need of four immediate neighbors to both nitrogen and chlorine(VII) would seem dominant,almost as in carbon chemistry,making the Nebenvalenz the most important.Werner slowly stuck his neck in,arguing that all the six Pt-Cl bonds in $PtCl_6^{2-}$ are,in any case,equivalent (i.e., no observations can disclose a varying proportion of Hauptvalenz and of Nebenvalenz).It is worth noting that Werner applied a spherical model for the "intensity" of the Nebenvalenz;this is,in no sense,a precursor of the long procession of models related to the hypothesis of angular s-,p-,and d-functions becoming hybridized (with miraculously similar radial functions),cf.($\underline{15},\underline{16}$).

On one point,Werner may have been less prepared to analyze a highly controversial contemporary opinion than this reviewer is.It is clear that Svante Arrhenius' suggestion that nearly all salts in water (among the exceptions are BrHgBr, NCHgCN,and $(H_2O)_3RhCl_3$) are dissociated into cations and anions is assumed in Werner and Miolati's work on electric conductance in water of $(O_2N)_nCo(NH_3)_{6-n}^{(3-n)+}$ in 1894. Nevertheless,a clear acceptance of "The hypothesis of essentially full ionic dissociation in water has been widely confirmed" would have assisted in understanding Nebenvalenz (in a way emerging as the triumphant "valency" after Arrhenius had removed the role of previous Hauptvalenz in water and a few similar solvents).The grumbling by chemists (including S.M.Jørgensen) was kept active by those few physical chemists who had such a respect for Kohlrausch conductances of single ions that the (somewhat decreased) conductivity(per mole) of almost saturated $Na^+ Cl^-$ was interpreted as degrees of dissociation around 0.6 rather   than such solutions differing from pure water.

We are not counting here the number of geometrical isomers.It was more innovative that Werner pointed out Salzisomere (saline isomers, having the same stoichiometric ratios) that can be far more numerous than the monomeric isomers.For example,the neutral,quadratic Pt(II) complex of neutral L ligands (e.g.,unidentate NH_3) and X^- anions(e.g. Cl^-) occurs as cis- and trans-L_2PtX_2 (whereas tetrahedral H_2CX_2 shows a unique isomer).However,any salt formed from cations such as L_3PtX^+ or PtL_4^{+2} and from anions $LPtX_3^-$ and/or PtX_4^{2-} provides identical stoichiometry 1Pt: 2L: 2X,like the two neutral isomers.Main-group crystalline compounds such as $PCl_4^+PCl_6^-$ are also,in a sense,such Salzisomere. Applying high pressure to I_2 crys tals might produce species such as $I_3^+I_3^-$,but a more likely result is metallic behavior, as one may predict also for (hypothetical) Li^+Li^- with LiH structure. Some high-pressure CsI and Xe phases are discussed (15)and related to the electric conductivity of supercritical hot mercury or caesium vapor,becoming metallic at a characteristic density.

At the dawn of crystallography,it may be mentioned that early ideas of a force vitale in carbon atoms is not needed to produce enantiomers; in 1914 Werner separated the two enantiomers of the brown cation $Co[(OH)_2Co(NH_3)_4]_3^{6+}$ as predicted for three bidentate "ligands",each containing one cobalt(III): cis-$(OH)_2Co(NH_3)_4^+$

Double Salts, Aqua Ions, and Ligands Bonded by Hydrogen

The fashion of writing (frequently rather "idealized") formulae for minerals as sums of (not very large) integers times the formulae for oxides (accepting with the elderly Berzelius that the atomic weight ratio O : H as close to 16 rather than 8) is easily extended to double salts prepared in the laboratory. Hydrated potassium ferrocyanide and tomato-red potassium ferricyanide can be written as $4KCN \cdot Fe(CN)_2 \cdot 3H_2O$ and $3KCN \cdot Fe(CN)_3$,but such formulae do not disclose how slowly ferricyanide loses CN^- nor how much slower the dissociation of ferrocyanide is. The formula is like that for a double salt such as alum or the schoenite (Tutton salt) type $(NH_4)_2O \cdot MO \cdot 2SO_3 \cdot 6H_2O$(as known from Mohr's salt,an air-stable iron(II) compound) with a crystal structure $(NH_4)_2[Fe(OH_2)_6](SO_4)_2$ without implying that ammonium oxide exists.

Nobody has really thought that H_3PO_4 consists of $3H_2O \cdot P_2O_5$(the neutral molecules are the "complex" $OP(OH)_3$ and the "cluster-shaped" P_4O_{10}).This comment is somewhat unfair;the oxides in Berzelius minerals are accounting tokens.However,there is little doubt that both S.M.Jørgensen and Werner became interested in what we call the ion $Co(NH_3)_6^{3+}$ because it does not form NH_4^+ with strong acids, as do all the blue $(H_2O)_{6-n}Ni(NH_3)_n^{2+}$ complexes,the colorless zinc(II)- and cadmium(II)- ammonia complexes studied thoroughly by Jannik Bjerrum (17).This is also true for the dark blue copper(II) known since medieval times [today considered (15,18) as $Cu(NH_3)_4(OH_2)_2^{2+}$ and $Cu(NH_3)_5^{2+}$ with distinct visible spectra,but anhydrous $Cu(NH_3)_4^{2+}$ is pink like the red copper(II) tetrakis(succinimidate) anion at pH 14.

Before 1780 it was known that pink cobalt(II) in aqueous ammmonia forms a dark product now known to be a peroxo-bridged dimeric complex of Co(III): $(H_3N)_5CoOOCo(NH_3)_5^{4+}$.It does not lose all NH_3 on acidification;modest yields of brick-red $(H_2O)Co(NH_3)_5^{3+}$ and orange $Co(NH_3)_6^{3+}$ could be recovered.The former complex was the first aqua ion to be recognized as an instantaneously reversible indicator [or Brønsted acid] with pK = 5.7,deprotonated to plum-colored hydroxo $HOCo(NH_3)_5^{2+}$. Unidentate nitrate occurs in $(O_2NO)_3Co(NH_3)_3$ crystals,dissolving as $(H_2O)_3Co(NH_3)_3^{3+}$ in aqueous solution.

Werner proved that some complexes react very slowly,rhodium(III) approaching,and iridium(III) sometimes exceeding,the robustness of typical organic compounds. Jannik Bjerrum (17) found active charcoal to be an excellent catalyst,establishing within hours many cobalt(III) equilibria,and Delépine found that the presence of ethanol allows both $RhCl_6^{3-}$ and $RhBr_6^{3-}$ to react with pyridine to form trans-$RhX_2(py)_4$ millions of times more rapidly than in the absence of alcohols.

Werner believed that H_2O and NH_3 are not very different as ligands. Today even the 7 + 3 = 10 geometrical isomers of octahedral chromium (III) with n ammonia and (6-n) aqua ligands,have all been characterized. **However, it is less true that the maximum N_{max} always is the same for H_2O and NH_3. The magnesium(II), aluminum(III), and zinc(II) aqua ions** in solution are all octahedral,but Zn(II) is tetrahedral $Zn(NH_3)_4^{2+}$ in strong aqueous ammonia (17)much like $Hg(NH_3)_4^{2+}$,but $Cd(NH_3)_6^{2+}$ is also known in crystalline salts. Silver(I) aqua ions are seen from spectra (19) to be $Ag(OH_2)_4^+$,but $H_3NAgNH_3^+$ is linear,much like the very strong ammonia complex of mercury(II).

The major surprise during the last 20 years,as far as dissolved aqua ions are concerned,is their wide range of exchange time (20) with solvent water,varying from below nanoseconds to above (the extrapolated values of) centuries (much like the ^{18}O isotope exchange of the two oxo ligands in the linear,nonprotonable uranyl ion OUO^{2+} (21)in the dark). H_3O^+,like all other aqua ions,is involved in an intricate network of hydrogen bonding in solution. On an instantaneous picture,it is close to $O(-H\cdot\cdot,OH_2)_3^+$,but it exchanges its proton every two picoseconds. Since chemically pure water is 55 molar H_2O and 10^{-7} $\underline{M}$ in both H_3O^+ and OH^-,a given water molecule,on the average,goes through the avatar of being protonated once every millisecond (22).

The Eigen temperature-jump technique gives precise reaction rates, but it is often difficult to know which reaction is measured. As far as the neighborhood of a cation of a metallic element is concerned,the most informative technique is a detailed study of neutron diffraction **by D_2O and H_2O solutions containing ions, as reviewed by Enderby (23). Lithium(I) has an average life-time 10^{-11} s (at 25°C, like all the following values) and is probably $Li(OH_2)_6^+$. Octahedral aqua ions of Mg(II).** Al(III),and Ga(III) show life-times of $2\cdot10^{-6}$s; 6 s; and $5\cdot10^{-4}$ s, respectively.As already shown by Niels Bjerrum in 1909,the kinetics of $Cr(OH_2)_6^{3+}$ with anions in solution is strongly speeded up by increased pH above 0.5 because of the rapid exchange of OH-bridged and monomeric OH^- complexes.The Cr-O exchange at pH 0.5 attains (the ^{18}O,but not the ^2D value) of 40 hours. $Mn(OH_2)_6^{2+}$ and $Ni(OH_2)_6^{2+}$ have life-times of $2\cdot10^{-8}$ s and $4\cdot10^{-4}$ s.Despite evidence for stronger covalent bonding, copper(II) aqua ions exchange far more rapidly (10^{-10} s compatible with 4 short Cu-O,one long Cu-O,and one verylong Cu-O bonds readily exchanging their roles).

Quadratic complexes of platinum(II) are normally very slow,and Elding ($\underline{24}$)prepared $Pt(OH_2)_4^{2+}$ and various $XPt(OH_2)_3^+$,but reaction rates (especially photochemistry) can increase with traces of Pt(IV) or Pt(III). From this point of view,palladium(II) has some Cu(II)-like tendencies. Yellow $Pd(OH_2)_4^{2+}$ has ($\underline{25}$)almost the Cu_{aq}^{2+} spectrum at twice as high photon energy and quite intense bands,like the mixed $(H_2O)_{4-n}Pd(NH_3)_n^{2+}$ and the ethylenediamine complexes $(H_2O)_2Pd(en)^{2+}$ and ($\underline{18},\underline{25},\underline{26}$)the colorless $Pd(en)_2^{2+}$.Like typical Cr(III)-ammonia complexes,the mixed aqua-ammonia Pd(II) complexes lose NH_3 in weakly acidic solution by a unimolecular dissociation. The rhodium(III) kinetics of $(H_2O)_{6-n}RhCl_n^{(3-n)+}$ has been studied ($\underline{27}$) at 65°C;orange **hexaaqua ions are very likely to exchange ^{18}O during several years at 25°C.** If all parasitic reactions can be avoided, pale yellow ($\underline{28}$) $Ir(OH_2)_6^{3+}$ may take centuries. The slow aqua ions are quite weak Brønsted acids with pK about 5 (aluminum at 4),but in solution,their slow formation of polymeric hydroxo-bridged complexes may occur at _much_ lower pH,as found by Niels Bjerrum for "hidden basic" Cr(III) and by Lene Rasmussen ($\underline{25}$) for palladium(II) at pH 0.3 to 1.

Reasons have been given ($\underline{7}$) for assuming only the oxidation state H(I) for hydrogen bonded to highly electronegative elements F,Cl,Br,I,O,S,Se (perhaps Te),and N. It is advantageous (but not absolutely compelling) to assume hydride(-I) bonded to all metallic, and to all non-specified elements,including carbon.Tc(VII) and Re(VII) form TcH_9^{2-} and ReH_9^{2-}. Hydride bridges occur in the Cr(0) complex $(OC)_5CrHCr(CO)_5^-$ analogous to iodide-bridged $(OC)_5CrICr(CO)_5^-$ .The boronate ion BH_4^- (isoelectronic with CH_4) easily forms diborane with two hydride bridges $H_2BH_2BH_2$ and functions as a bidentate ligand in $\underline{N}$ = 6 $Al(H_2BH_2)_3$ and as a tridentate ligand in $Zr(H_3BH)_4$ having $\underline{N}$ = 12 like $U(BH_4)_4$ syncrystallized in hafnium(IV) boranate,but the undiluted U(IV) compound has $\underline{N}$ = 14. For ligands containing several boron atoms, see Greenwood and Earnshaw ($\underline{14}$).

The alkali metal hydrides crystallize in the NaCl structure (like AF) for all A = Li,Na,K,Rb,Cs showing $\underline{N}$ = 6 for both A^+ and H(-I).The cubic perovskites $BaLiH_3$ and $EuLiH_3$ have $\underline{N}$ = 12 (cuboctahedral) M(II) and $\underline{N}$ for octahedral lithium(I),and also $\underline{N}$ = 6 for hydride,four M-H distances on two Cartesian axes being 1.4142 times longer than the two Li-H distances. A somewhat comparable compound is yellowish-red Be_2C (nonmetallic,CaF_2 type) having the conventional $\underline{N}$ = 4 for Be(II) but cubal $\underline{N}$ = 8 for carbon(-IV).Crystallographers use the word "cubic" for "having all three Cartesian axes physically equivalent" [even for the lowest such point-group T_h exemplified by pyrite FeS_2 showing equally pronounced striation in all three directions].Thus, we need a word for being like a cube,"cubal",having itself the O_h point-group like a regular octahedron,or cuboctahedral MX_{12}.

Some organometallic molecules have two hydride ligands at very short mutual distance. Evidence is growing ($\underline{29},\underline{30}$)that they may show isomers due to further shortening,forming a H_2 molecule back-bonded in much the same way as the C_2H_4 molecule in Zeise's platinum(II) complex $Cl_3PtC_2H_4^-$.It may be argued ($\underline{7}$) that analogous C_2F_4 ligands are better described as $C_2F_4^-$ bound to Pt(IV) in such a way that the F_4 (or the H_4) plane is further away from the Pt than are the two carbon atoms. Olefins are rapidly exchanged with Zeise's anion.

What is a Genuine Ligand ? Reversibility and Isolated Existence

For good reasons,handbooks like "Gmelin" distinguish between the formation (Bildung) from elusive or expensive starting materials (or with ludicrously low yields) and the one or few ways of preparation (Darstellung),being practically feasible and economical (and in recent years ecologically soft).This distinction applies also to extraction of rarer elements (e.g.,as metals,oxides,or chlorides) from secondary and (if any) primary minerals. Time-evolution is accelerating and highly modified by use of by-products,disposal of waste and undesired volatiles,and demand,compared to the primordial availability.

Obviously,a molecule or polyatomic ion containing 10 or 20 nuclei can be divided (on paper) many different ways. The attention given to metallic elements (increasing from 7 to at least 80 during the last 1000 years) twisted the description to ligands being considered as bonded to central atoms [but bromine(III),iodine(III),platinum(II), and gold(III) are not strikingly different in quadratic BrF_4^-,ICl_4^-, XeF_4,$PtCl_4^{2-}$ and $AuCl_4^-$].It was discussed above why exchange of H_2O ligands can be so rapid in aqua ions that a structure is blurred out by fluxional behavior (the term used by organometallic chemists for, e.g.,rapid fluctuation of Hg(II) being bonded,at a given instant,to 2 out of 10 carbon atoms in $Hg(C_5H_5)_2$ not being a "sandwich" like Mg(II),Fe(II),Fe(III),Co(III),Ni(II),Ni(III),Ru(II),Rh(III),Os(II), **and Ir(III) cyclopentadienides). Actually, large ions with low positive charge such as K(I), Rb(I), Cs(I), Ba(II), and possibly Tl(I) have an** indefinite N of neighbor oxygen nuclei on instantaneous pictures obtained (20,23) by neutron diffraction. As already pointed out by N. Bjerrum in 1909,$Cr(OH_2)_6^{3+}$ is a full-fledged Werner complex (its H_2O exchanging ^{18}O nuclei with a half-life close to 10^5 seconds).Evidence (18,25)for fluctionality [though less than in copper(II)] is available from spectra and formation constants for palladium complexes,and perhaps also for gold(III) and posttransitional tellurium(IV), thallium(I),lead(II),and bismuth(III) in solids and solution.

In addition to water,many ligands (such as ammonia,amines,amino acids,A^+ salts of CN^-,ambidentate NCS^-,the halides X^-,dithiocarbamates $R_2NCS_2^-$) are today readily purchased or prepared from chemicals which were uncommon in 1900.This option has a strong impact on which part of a species studied are considered as ligands. However,one can,in a few minutes,produce a long series of paradoxes. Is it suitable that F_3CCH_3 is a complex of the carbanion CF_3^- with CH_3^+? Is $Te(CH_3)_3^+$ a complex of tellurium(IV) or Te(-II) ? On the whole (7),there is a trend to see carbanion ligands on metallic elements,including Sb(III),W(VI), Pt(IV),and Au(III),but carbonium ligands to those elements having H(+I) hydrides,perhaps extended to phosphorus and arsenic.We should not be bashful that it is,to some extent,an autocratic decision.

One must realize that it would be very difficult to recover the free ligand from Co(III),Rh(III),Ir(III),Pt(IV),and Pt(II)-ammonia or cyanide complexes; only the smooth continuity to less irreversibly bonded Cr(III),Pd(II),and Hg(II) complexes makes this taxonomy almost palatable.Organisms can produce iron complexes like cytochrome c, hemoglobin,cobalt(III) in vitamin B 12,and plants can produce Mg(II) complexes like chlorophyll.For some years,it was considered difficult to remove the central atom from the planar,heterocyclic systems with strong absorption bands in the visible (like phthalocyanine),and much progress was made on <u>template synthesis</u> (<u>in vitro</u>) such as the class of alicyclic tetramines "cyclams" and related multidentate,sterically strained polyamines (<u>31</u>),first prepared by condensing acetone at 30°C with orange,anhydrous $Ni(en)_2^{2+}$ in the form of the (shock-sensitive) perchlorate.Although nickel(II) usually can be removed as NiS with aqueous Na^+HS^-,it is no longer considered necessary for a ligand to be accepted that it can be synthesized free of a central atom.

A much more difficult problem is that of doubtful ligands taking <u>one</u> electron at the time.Related to the great popularity of the Lewis paradigm (<u>7</u>,<u>9-11</u>) between 1916 and 1984,well-behaved ligands had to choose modifications removing or adding <u>two</u> electrons at a time,as seen above with CR_3^+ and CR_3^- or H(I) and H(-I).However,two series of O_2; O_2^- (superoxo); and O_2^{2-} (peroxo) and NO^+[isoelectronic with CO and CN^-]; NO; and NO^- are plausible ligands,to recall oxidative dimerization(<u>32</u>) of two RS^- to RSSR.The colloquial name "innocent ligands"(<u>7</u>) has been introduced for ligands not preventing their central atom in having a well-defined oxidation state z [of which 317 noncatenated,nonmetallic cases were known for the first 104 elements (<u>33</u>)in 1986,but 18 others are added,mainly because 14 new,negative z have been observed (<u>34</u>,<u>35</u>) for d-group CO complexes,reaching down to Ti(-II),V(-III),Cr(-IV), Zr(-II),Nb(-III),Mo(-IV),Rh(-III),Hf(-II),Ta(-III),W(-IV),and Ir(-III).◆ The next most efficient ligand for negative z is PF_3(<u>36</u>).Au(-I) known from Cs^+Au^- is a monatomic ion like Na^- obtained by disproportionation of metallic sodium (<u>37</u>) in unreactive solvents by a multidentate poly-ether ("cryptate") forming very strong complexes with Na(I).Dilute solutions of Na in liquid ammonia are blue and contain solvated **single** electrons,but strong solutions also contain Na^-.

Beginning in 1962,noble gas [Ng] compounds in positive z as Kr(II). Xe(II),Xe(IV),Xe(VI),Xe(VIII),and Rn(II) [if not Rn(VI)] have been extensively reported (<u>38</u>)and also catenated green Xe_2^+ (in mass spectra, all Ng_2^+,NgH and many other species are detected). These Ng compounds have been treated by refined quantum chemistry (<u>39</u>,<u>40</u>) and by more qualitative arguments (<u>41</u>).The brightly colored $ArCr(CO)_5$,$KrCr(CO)_5$,and $XeCr(CO)_5$ in cool matrices have Ng(0) ligands (<u>42</u>,<u>43</u>)as have $KrMo(CO)_5$ and $KrMn(CO)_5^+$. Both gaseous and uncoordinated $Cr(CO)_5$ are blue;yellow $ArCr(NN)_5$and $Ar_2Cr(NN)_4$ and gaseous $CoAr_6^+$ have been discussed (<u>43</u>).

It would seem that the theme of noninnocent ligands (<u>7</u>)is best connected with "a given compound containing a definite ligand" <u>via</u> "the central atom has (much) lower electron affinity than the ligand Highest Occupied Orbital ionization energy".This statement is plausible for electron affinity and ionization energy <u>in situ</u> but runs contrary to the observed trend of most electron transfer spectra (<u>2</u>,<u>7</u>,<u>21</u>,<u>44)</u> to start above 1.5 eV,if not (in the violet) 3 eV,and also in apparent conflict with photoelectron (ESCA) results showing the ionization energy of the loosest bound ligand orbital frequently (<u>2</u>,<u>16</u>)**to be** several eV <u>lower</u> than the ionization energy of a <u>partly filled</u> 3d or 4f shell, This abolishes our feeling that "covalent bonding is the strongest when the binding orbitals have similar energy" [whatever that means (<u>45</u>)] but does not surprise an atomic spectroscopist accustomed to nonadditive orbital energies in differently occupied configurations.

What has Quantum Chemistry Done for Complexes-- Total Atomic Energies
Modified in the Fourth Digit; Correlation Energies and Multidimensional
Potential Surfaces; and Coffee Table Textbooks

Most transition-group (or more precisely,Z above 20) chemists have
relations to quantum chemistry of two textures: the Melanesian Cargo
Cult that all the results are dumped on our tables the year after the
university gets its 9 gigadollar computer and also the professional
pride that the heats of atomization H_a must be large (since they are
tabulated with 3-4 digits).
    One should start the debriefing by noting that compounds do not,
strictly speaking,consist of atoms (any more than the huge majority
of inorganic solids consist of neutral molecules). Seen from the
standpoint (45) of nonrelativistic Schrödinger (1926) equations [S.e.],
materials studied by chemists consist of (practically point-shaped)
nuclei characterized by integers $Z_1,Z_2,...,Z_N$ (giving the nuclear
charges in protonic units) accompanied by K^* electrons (large systems
have K^* very close to the sum of Z values,but this is not a stringent
rule for a few nuclei). All electrons are super-identical;they carry
no social security numbers,and when two electrons have met,they cannot
remember who is who.
    Historically speaking,atomic spectra describe differences between
discrete energy levels (45,46) of one Z nucleus and K electrons with
a total ionic charge z = (Z - K) either being positive and below Z,
zero (neutral atoms),or (for most,but not all,elements M) z = -1
(exceptions are He,Be,N,Ne,Mg,Ar,Mn,Kr,..,Hg,.. refusing to accept
an electron). No gaseous M^{2-} is stable (not even O^{2-} or S^{2-}) and
detaches instantaneously an electron to form M^-.The time-independent
S.e. shows an eigen-value E enormously negative for large K relative
to the ionization energy I_1 of neutral atoms known to vary between
3.894 eV for caesium (Z = 55) and 24.587 eV for helium (1 eV = 96487
J/mol = 23061 cal/mol).For not fully understood reasons (47,48) the
total binding energy of Z electrons (-E in S.e. is the sum over all
consecutive values from I_1 to I_Z) is between 3 and 2 percent higher
than the Gaspar approximation $E_G = Z^{2.4}(13.6$ eV) for atoms from Z =
6 through 47 (silver). If the Rydberg constant is replaced by a free
parameter,14.0 eV ameliorates the agreement. Relativistic effects(49)
make the actual ground state energy 1 % more negative at germanium
(Z = 32) and 10 % for curium (Z = 96). For comparison,m_0c^2 of an
electron is 511000 eV. Descending a hierarchy of energies of interest
to the chemist,a plausible expression (45,48) for the closed-shell
effect in noble gas atoms,Z = 10,18,36,54,and 86, decreases from 42.6
to 17.6 eV,approximately (100 eV) times $Z^{-0.4}$.
    The heat of atomization (per atom) named H_a is 7.5 eV for both
diamond and graphite and varies for metallic elements from 0.64 eV
(mercury) to 8.8 eV (tungsten). Only 14 diatomic molecules (CO,N_2,ten
other monoxides MO,CN and BF) have H_a above 4 eV.A few organic carbon-
rich molecules, gaseous BF_3, OCO, SiF_4,and UF_6 (but not SF_6) have
H_a around 5 eV. Inorganic solids (50) involving one Z value above 37
can show very enhanced H_a values for oxides and fluorides of Sr,Zr,Ba,
La,Ce,Lu,Hf,Ta,Th (the highest value is 8.0 eV for crystalline ThO_2),
and of U,possibly because of correlation effects defined by the

positive quantity $-E_{corr}$ being the energy difference between the ground state E in the Hartree-Fock approximation ($\underline{45},\underline{48}$)and the (slightly more negative) E of S.e. with full configuration intermixing,This correlation energy vanishes for K = 1 and turns out to be roughly $-E_{corr} = Z^{1.2}$(0.7 eV) for typical neutral atoms. $-E_{corr}$ tends to be 1 to 5 eV higher for gaseous molecules (compared to the constituent atoms) and so much higher for F_2 than for two fluorine atoms that F_2 would dissociate spontaneously in the Hartree-Fock model.

It is legitimate that chemical consumers prefer one-electron energies in orbital models rather than total many-electron energies. (Atomic spectroscopists had the same ($\underline{7},\underline{46}$)preference). Photoelectron spectra of solids ($\underline{2},\underline{51}$) and of gaseous molecules ($\underline{2},\underline{52-54}$) provided much clearer ideas of one-electron energies (distinctly not additive for differently occupied atomic ($\underline{16}$)or molecular orbitals).Extended Hückel treatment,going from binary molecules benzene,anthracene,etc. to heterocyclic pyridine,pyrazine,etc.can be used also in "ligand" field theory as a pragmatic parametrization of inorganic facts but is miles away from S.e. by neglecting configuration interaction in the general sense of abandoning the Hartree-Fock approach.

A referee kindly drew attention to the evaluation of densities of valence electrons and (at most one) partly filled shell in a parametrized pseudo-potential representing the almost rigid inner shells (modifying their ionization energies,as seen in photoelectron spectra,but hardly their radial distributions). This concept goes back to central fields ($\underline{46}$) derived since 1928 from the Rydberg (1895) formula ($\underline{45}$) for one external electron added to a set of spherical shells closer to the nucleus. Conceptually,the Hartree treatment (of which the Hartree-Fock treatment is a refinement,abandoning spherical symmetry of the influence of the other electrons on a given electron) should have been superior;unfortunately,it is not in systems with some 20 electrons or more,them accommodating one or more nuclei.

Coming finally to genuine chemistry,we may accept the Born-Oppenheimer approximation ($\underline{45}$) and replace the potential curve (for each many-electron state) of a diatomic species,in a system with N nuclei having (3N-6) mutually independent internuclear distances. Hence each potential surface occurs in a (3N-5)-dimensional space($\underline{55}$). Both MX_6 and any other surface for 7 nuclei subsist in 16 dimensions, and 9 nuclei (e.g.,the isomers CH_3CH_2OH and CH_3OCH_3) in 22 dimensions. Among textbook publishers,there is a terrible craving for cheaper imitations,e.g.,the two-dimensional "reaction path" and luminescence with one "breathing vibration". This reviewer is also worried about results valid only for the exact solution of S.e. such as the Hellmann-Feynman theorem or the rush ($\underline{56},\underline{57}$) to the Density Functional Theory.Here a solution can only be recognized when the S.e. has been solved.In any case,it may be a remote consolation that many aspects ($\underline{58}$) of stellar spectra,detailed astrophysics,the origin of the Z elements,and the behavior of "elementary" particles are in an even more pronounced upheaval. The Lewis paradigm (1916) was the last chemical paradigm,not referring to quantum mechanics,but several chemists like these tawdry surrogates,even with quantum glazing.

Acknowledgments

I am grateful to the late Ms. Lene Rasmussen and our late Professor
Jannik Bjerrum for opening the doors in 1950 to actual inductive
complex chemistry; to his successor Claus E.Schäffer and to
Professor Renata Reisfeld for fruitful adaptation of my early
aspirations to be an atomic spectroscopist; and to Professor Virginia
**Trimble for accepting my even earlier interest in astronomy with a
perspective perhaps more vertiginous on quantum paradoxes than 10^{22}
km distances; and finally, to Professor George B. Kauffman for inviting
me to participate in the Denver symposium.**

Literature Cited

1. Jørgensen,C.K.; Kauffman,G.B. Struct. Bonding 1990, 73, 227.
2. Jørgensen,C.K. In Handbook on the Physics and Chemistry of Rare
   Earths; Gschneidner,K.A.; Eyring,L.,Eds; North-Holland:
   Amsterdam,1988,Vol. 11,pp 197-292.
3. Dana,E.S. Textbook of Mineralogy,4th ed;John Wiley:New York,1963.
4. Pauling,L. J.Am.Chem.Soc. 1932, 54, 3570.
5. Mulliken,R.S. J.Chem.Phys. 1934, 2, 782.
6. Sanderson,R.T. J.Chem.Educ. 1988, 65, 112 , 227.
7. Jørgensen,C.K. Oxidation Numbers and Oxidation States;
   Springer: Berlin and New York,1969.
8. Rabinowitch,E.; Thilo,E. Periodisches System, Geschichte und
   Theorie; Enke: Stuttgart,1930.
9. Lewis,G.N. J.Am.Chem.Soc. 1916, 38, 762.
10. Jørgensen,C.K. Chimia 1974, 28, 605.
11. Jørgensen,C.K. Chimia 1984, 38, 75.
12. Reisfeld,R.; Chernyak,V.; Eyal,M.; Jørgensen,C.K.
    Chem.Phys.Lett. 1989, 164, 307.
13. Jørgensen,C.K. Radiochim. Acta 1983, 32, 1.
14. Greenwood,N.N.; Earnshaw,A. Chemistry of the Elements;
    Pergamon: Oxford,1990.
15. Jørgensen,C.K. Top.Current Chem. 1989, 150, 1.
16. Jørgensen,C.K. Chimia 1974, 28, 6.
17. Bjerrum,J. Metal Ammine Formation in Aqueous Solution,2nd ed.;
    P.Haase: Copenhagen,1957.
18. Jørgensen,C.K. Top.Current Chem. 1975, 56, 1.
19. Texter,J.; Hastreiter,J.J.; Hall,J.L. J.Phys.Chem. 1983, 87, 4690.
20. Hunt,J.P.; Friedman,H.L. Progr.Inorg.Chem. 1983, 30, 359.
21. Jørgensen,C.K; Reisfeld,R. Struct.Bonding 1982, 50, 121.
22. Giguère,P.A. J.Chem.Educ. 1979, 56, 571.
23. Neilson,G.W.; Enderby,J.E. Adv.Inorg.Chem. 1989, 34, 195.
24. Hellquist,B.; Bengtsson,L.A.; Holmberg,B.; Persson,I.;
    Elding,L.I. Acta Chem.Scand. 1991, 45, 449.
25. Rasmussen,L.; Jørgensen,C.K. Acta Chem.Scand. 1968, 22, 2313.
26. Jørgensen,C.K.; Parthasarathy,V. Acta Chem.Scand 1978, A 32, 957.
27. Pavelich,M.J.; Harris,G.M. Inorg.Chem. 1973, 12, 423.
28. Gajhede,M.; Simonsen,K.; Skov,L.K. Acta Chem.Scand. 1993, 47, 271.
29. Kubas,G.J. Acc.Chem.Res. 1988, 21, 120.
30. Jessop,P.G.; Morris,R.H. Coord.Chem.Rev. 1992, 121, 155.
31. Comba,P. Coord.Chem.Rev. 1993, 123, 1.

32. Jørgensen,C.K.; Kauffman,G.B. Chimia 1987, 41, 150.
33. Jørgensen,C.K. Z.Anorg.Allg.Chem. 1986, 540, 91.
34. Beck,W. Angew. Chem. Int. Ed. Engl. 1991, 30, 168.
35. Ellis,J.E. Adv. Organomet. Chem. 1990, 31, 1.
36. Nixon,J.F. Adv.Inorg. Chem. 1985, 29, 41.
37. Dye,J.L. Progr, Inorg, Chem. 1984, 32, 327.
38. Selig,H.; Holloway,J.H. Top.Current Chem. 1984, 124, 33.
39. Frenking,G.; Koch,W.; Deakyne,C.A.; Liebman,J.F.; Bartlett,N.
    J.Am.Chem.Soc. 1989, 111, 31.
40. Frenking,G.;Cremer.D. Struct.Bonding 1990, 73, 17.
41. Jørgensen,C.K.; Frenking,G. Struct. Bonding 1990, 73, 1.
42. Perutz,R.N.; Turner,J.J. J.Am.Chem.Soc. 1975, 97, 4791.
43. Jørgensen,C.K. Chem. Phys. Lett. 1988,153, 185.
44. Jørgensen,C.K. Progr. Inorg. Chem. 1970, 12, 101.
45. Jørgensen,C.K. Comments Inorg.Chem. 1991, 12, 139.
46. Condon,E.U.; Shortley,G.H. Theory of Atomic Spectra,2nd ed.;
    Cambridge University Press: Cambridge,1953.
47. Gaspar,R. Acta Universitatis Debreceniensis de Ludovica Kossuth
    Nominatae Series Physica et Chimica 1971,pp 7-57.
48. Jørgensen,C.K. Chimia 1988, 42, 21.
49. Pyykkö,P. Chem.Rev. 1988, 88, 563.
50. Jørgensen,C.K. Chimia 1992, 46, 444.
51. Jørgensen,C.K. Struct.Bonding 1976, 30, 141.
52. Turner,D.W.; Baker,C.; Baker,A.D.; Brundle,C.R. Molecular
    Photoelectron Spectroscopy; Wiley-Interscience: London,1970.
53. Bock,H,; Solouki,B. Angew. Chem. Int. Ed. Engl. 1981, 20, 427.
54. Green,J.C. Struct. Bonding 1981, 43, 37.
55. Jørgensen,C.K. In OSA Proceedings on Tunable Solid State Lasers;
    Shand,M.L.; Jenssen,H.P.,Eds.; Optical Society of America:
    Washington,DC, 1989, Vol. 5,pp 252-257.
56. Weber,J.; Huber,H.; Weber,H.P. Chimia 1993, 47, 57.
57. Chattaraj,P.K.; Parr,R.G. Struct. Bonding 1993, 80, 11.
58. Jørgensen,C.K. Comments on Astrophysics 1993 , 17, 49.

RECEIVED October 14, 1993

Chapter 20

The Chelate, Macrocyclic, and Cryptate Effects

Arthur E. Martell[1] and Robert D. Hancock[2]

[1]Department of Chemistry, Texas A&M University,
College Station, TX 77843–3255
[1]Department of Chemistry, The University of Witwatersrand, Wits 2050,
Johannesburg, South Africa

The factors which contribute to the chelate, macrocyclic, and cryptate effects are described. These include the dilution effect, translational entropy, intrinsic basicities of donor atoms, coulombic attractions and repulsions of charged ions and groups, and covalent character of the coordinate bonds. Reduction in hydration/solvation energies and the related coulombic repulsions of the donor atoms of multidentate ligands are factors which result from preorganization of the ligands for complexation of metal ions. A not insignificant part of the chelate and macrocyclic effects is due to increase in the basicities of the donor atoms that occurs on ring formation as well as to reduction of the steric repulsions of alkyl groups. The effects of ring size on stabilities are described. These factors are illustrated with stabilities of chelates, macrocyclic complexes, and cryptates taken from work of the authors and from the literature.

It is a pleasure and an honor to be able to able to express our ideas concerning coordination chemistry in solution at this coordination chemistry centennial symposium. The outstanding work of John C. Bailar, Jr. and Jannik Bjerrum, together with the contributions of David P. Mellor, Francis P. Dwyer, and Ronald S. Nyholm, established the foundations of coordination chemistry and were major factors in the growth in this interesting field of chemistry. To these may be added the names of Lars Gunnar Sillén and Gerold Schwarzenbach, who made outstanding contributions to the formation and stabilities of metal complexes in solution. Their contributions served to establish the field during a period of time which we like to look upon as the golden age of coordination chemistry. Today there is a resurgence of interest in coordination compounds in solution because of their applications to biological systems, environmental problems, and many other fields. The introduction of macrocycles and cryptates has further increased the number and variety of organic ligands that can be designed for metal complex

0097–6156/94/0565–0240$08.00/0
© 1994 American Chemical Society

formation. Also, there is the stimulating effect of high speed computers, which make possible calculations of complex systems to a degree that was not attainable at the time of Schwarzenbach and Sillén. Therefore we look forward to further expansion of this important field of inorganic chemistry.

The factors that influence the stabilities of metal complexes in solution involve both enthalpic and entropic effects. The principal enthalpic effect is the energy of formation of the coordinate bond between the metal ion and the ligand and involves the covalent tendencies of both the metal ion and the ligand for bond formation. In addition to this the energy of charge neutralization is very important, especially for the combination of positive metal ions with ligands having negative charge. In opposition to these effects are the enthalpies of desolvation of both the metal ion and of the ligand because formation of the complex involves displacement of a number of molecules of the solvent. Entropy effects, which occur on combination of metal ions and ligands to form complexes, involve differences in the freedom of motion of the metal ion, of the ligand, and of the complex that is formed. The freedom of motion of the metal ion is, for the most part, translational entropy, which may or may not change on complex formation. Also, metal ions have negative entropy because of the solvated molecules around the metal ions, which are more or less frozen in position due to solvation effects. The ligands usually have considerable freedom of motion, which is greatly decreased on complex formation. The ligand assumes a somewhat different conformation in the complex than it has in solution in the free state. Ligands may be preorganized for complex formation if the donor groups are in a position favorable for complex formation. When the donor groups are in the position approximately required by the metal ion for forming the coordination complexes, both the heats of complex formation and the entropies of complex formation are more favorable. The mutual repulsions that keep monodentate donors apart in solution are partially overcome in building the donor atoms into a single multidentate molecule, and the heat of complex formation is therefore more favorable. Also, the entropy of the ligand is made considerably more negative by a built-in rigidity, which results in the donor groups being more or less frozen into a position favorable for complex formation.

The purpose of this paper is to describe the principal factors that influence the stabilities of the complexes formed in solution. This involves heats and entropies of complex formation and the factors which favor a more negative enthalpy and a more positive entropy. Also, a number of misconceptions that have been adopted in the recent literature are described and replaced by a more reasonable explanation of the chelating tendencies involved.

The Classical Chelate Effect

The first explanation of the chelate effect was given by Schwarzenbach (*1*), who compared the stabilities of complexes formed by monodentate ligands with the stabilities of complexes formed by multidentate ligands, which produce one or more chelate rings. In this model Schwarzenbach assumed that the enthalpy of complex formation is the same for two monodentate ligands that combine with the metal ion as for a didentate ligand in which two donor groups of the same kind are joined together in an organic molecule. Therefore the enthalpy of replacement of two monodentate donor groups by a didentate ligand containing similar donor groups is zero. The model also assumed that the internal entropies of the metal ion, ligand, and the complexes formed are the same so that the differences in entropies of rotation and vibration in each of the molecules involved can be neglected. Therefore when a didentate donor replaces two monodentate donors in this model, the only change is that there is one more molecule on the right hand side of the equation than on the left. Therefore the replacement of two monodentate donors in a metal complex by a didentate ligand containing the same donor atoms results in the formation of one solute molecule, and the entropy increase of the reaction is the translational entropy of that additional molecule of solute.

Although Schwarzenbach showed that the effect is an entropy effect resulting in higher stability of the metal chelate, it remained for Adamson (2) to show that the increase in entropy is the entropy of translation for the additional mole of solute at its standard state of one molal. This results in an increase of 7.9 entropy units (cal/deg/mole) per additional mole of solute in this system; for n chelate rings the increase in entropy would be 7.9 n entropy units. At 25.0 °C this results in about 2 log K units per chelate ring formed. As will be shown below, however, it is not always true that there is an increase of one solute molecule per chelate ring formed. In any case, it should be remembered that the increase in entropy is due to the increase in the number of solute molecules and not to the number of chelate rings.

It should be emphasized that the increase in entropy of 7.9 entropy units per chelate ring is an approximation and is true only for the model used by Schwarzenbach. This model involved several assumptions including the assumption that all the internal rotations and vibrations contributing to the entropy of the complex and of the free ligand remain the same. This, of course, is not true when real systems are involved. Also, the model assumes that the coordinate bonds formed between the donor groups and the metal are not affected when the donor groups are joined together to form a chelating ligand. This, of course, can never be achieved in real systems. Therefore the increase in entropy associated with chelate ring formation must be considered an approximation. However, as indicated by several examples below, it is an approximation that seems valid in many cases.

The Dilution Effect

The so-called dilution (3) effect illustrated by Table I indicates that chelate compounds are more stable than analogous complexes in dilute solution. If the Schwarzenbach model is used, a chelate may be considered to be more stable than the corresponding complex by about two log units per chelate ring formed. Thus the metal chelates and complexes in their standard state may be assigned the overall stability constants (β values) shown in Table I. If one assumes a value of β of 10^{12} for the complex MA_6, then the value of β for a didentate donor which has three chelate rings would be 10^{18}, and with a hexadentate donor which forms five chelate rings β would be 10^{22}. The formulations of the β values used and the units of β show that the metal complex is very sensitive to its concentration in solution, while the chelates are less sensitive to that factor. With the assumptions that have been made it is possible to determine the degree of dissociation in dilute solution. In very dilute solution (10^{-6} M) the complex MA_6 seems to be

100% dissociated, while the metal chelate containing five chelate rings is only slightly dissociated. There can be no question that a metal chelate is more stable in very dilute solution. Thus for a complex to retain its integrity and remain undissociated in very dilute solution, it is necessary that the metal ion be bound by a number of donor sites, giving rise to a number of chelate rings.

Table I. Comparison of Dissociation of Metal Complexes and Metal Chelates in Dilute Solution

Equilibrium Quotient	β	Total Concentration of Complex and Metal Ion					
		1.0 M		1.0 x 10^{-3} M		1.0 x 10^{-6} M	
		pM^a	% Diss.	pM^a	% Diss.	pM^a	% Diss.
$\dfrac{[MA_6]}{[M][A]^6}$	10^{12} M^{-6}	16.67	2.1x10^{-15}	3.23	57.9	6.00	100
$\dfrac{[MB_3]}{[M][B]^3}$	10^{18} M^{-3}	19.43	3.7 x 10^{-18}	13.43	3.7x10^{-9}	7.49	3.2
$\dfrac{[ML]}{[M][L]}$	10^{22} M^{-1}	22.00	1x10^{-20}	22.00	1x10^{-17}	22.00	1x10^{-14}

^{a} for 100% excess ligand.

Thermodynamics of Metal Chelates and Macrocyclic Complexes

The thermodynamics of formation of metal chelates of homologs of EDTA (ethylenediaminetetraacetic acid) with various metal ions is indicated in Table II (*4*), which shows that the major factor determining the stabilities of these

complexes is the increase in entropy that occurs on complex formation. For the formation of five chelate rings the Schwarzenbach model would indicate an increase in entropy of about forty units provided that the octahedral complex contains five chelate rings in each case. The increases in entropy that are shown in Table II are about the expected value or somewhat higher. The further increase in entropy can be explained on the basis of the release of solvated water molecules from the ligand donor groups as well as from the metal ion. The increase in entropy occurs when the solvent molecules, which have negative entropies due to their being held by the ligand donor atoms or by the metal ion, increase their entropy to the zero values of the solvent when they are released.

The decrease in enthalpy that occurs on complex formation is seen in Table II also to be important, especially for the transition metal ions and for zinc(II). This is due mainly to covalent bonding between the metal ions and the nitrogen atoms of the diamine bridge plus a contribution from the coloumbic attraction between the positive metal ions and the negative donor groups of the ligand. The decrease in enthalpy is seen to be a contributor to the stability of the complex in all cases except the magnesium(II) complexes. The magnesium(II) ion apparently is solvated by the small water molecules to a greater extent than the combined effects contributing to the enthalpy of the complexes formed. Thus it is seen that for Mg(II) and two of the examples shown for La(III), formation of diaminotetraacetate complexes in solution is driven by the large entropy increase that takes place on complex formation.

Table II. Thermodynamics of Formation of Metal Chelates of EDTA Homologs[a]

		Metal Ion					
n^b	Quantity[c]	Mg^{2+}	Ca^{2+}	La^{3+}	Ni^{2+}	Cu^{2+}	Zn^{2+}
2	log K	8.8	10.6	15.5	18.4	18.8	16.5
	ΔH^o	3.3	-6.1	-2.9	-7.4	-8.3	-4.6
	ΔS^o	52	28	61	59	58	60
3	log K	6.3	7.3	11.3	18.1	18.8	15.2
	ΔH^o	9.1	-1.7	3.8	-6.7	-7.7	-2.3
	ΔS^o	59	27	64	60	60	62
4	log K	6.3	5.6	9.2	17.3	17.2	15.0
	ΔH^o	8.5	-0.9	1.9	-7.0	-6.5	-3.5
	ΔS^o	57	23	48	56	57	57
5	log K	5.2	4.6	9.0	13.8	16.1	12.6
	ΔH^o	-	-	-	-6.7	-10.9	-2.7
	ΔS^o	-	-	-	41	37	49

[a] Data from ref. 4 at $\mu = 0.10\ M$ and $T = 25.0\ ^oC$. [b] n is the number of carbon atoms between the iminodiacetate moieties of the ligands. [c] Units: ΔH, kcal/mole; ΔS, cal/deg/mole.

The thermodynamics of replacement of ammonia by ethylenediamine in Ni(II) and Cu(II) complexes is presented in Table III (4). The formation of an en complex would represent the formation of one chelate ring, two en complexes would represent the formation of two chelate rings, and three en complexes would represent three chelate rings. It is seen from Table III that the increases in entropy associated with these reactions are not greatly different from the increase, 7.9 n, that would be predicted by the classical chelate effect. Therefore

Table III. Thermodynamic Contributions to the Chelate Effect in Complexes of Ethylenediamine with Cu(II) and Ni(II)

	ΔG^a	ΔH^a	ΔS^a	7.9 n
Chelate Effect				
Ni(en)	-3.4	-1.2	7	7.9
Ni(en)$_2$	-8.0	-2.7	18	15.8
Ni(en)$_3$	-12.70	-4	29	23.7
Cu(en)	-4.0	-2.0	6	7.9
Cu(en)$_2$	-9.8	-3.5	21	15.8

[a] Data from ref. 4. Units: ΔG, ΔH, kcal/mole; ΔS, cal/deg/mole.

entropy is an important factor in determining the increase in stability of the ethylenediamine complexes over those of the corresponding ammonia complexes. However, the data also show that there is a considerable contribution from the enthalpy of reaction and that the formation of coordinate covalent bonds accounts for considerable additional stability of the complex. Apparently, the coordinate bonds formed by ethylenediamine are stronger than those formed by ammonia. In Table IV the stabilities of the sodium(I), potassium(I), and barium(II) complexes of 18-crown-6 are compared with the corresponding stabilities of the complexes formed by the open-chain analog, pentaglyme (4,5). Here it is seen that nearly all of the macrocyclic effect, which is the increase in stability due to the formation of the macrocyclic ring, is an enthalpy effect, and the entropy is close to zero and generally quite small. Therefore the coordinate bonds in the 18-crown-6 complexes are stronger than those in pentaglyme.

Table IV.   Thermodynamics of Complex Formation of Na^+, K^+, and Ba^{2+} with 18-Crown-6 and Its Open-Chain Analog (pentaglyme) in 100% Methanol

		Na^+	K^+	Ba^{2+}
log K	18-crown-6	4.36	6.06	7.04
	pentaglyme	1.44	2.1	2.3
	log K (MAC)	2.92	3.96	4.74
ΔH^a	18-crown-6	-8.4	-13.4	-10.4
	pentaglyme	-4.0	-8.7	-5.6
	ΔH (MAC)	-4.4	-4.7	-4.8
ΔS^a	18-crown-6	-8	-17	-3
	pentaglyme	-7	-20	-8
	ΔS (MAC)	-1	3	5

a Units: ΔH, kcal/mole; ΔS, cal/deg/mole.

Similarly, in the formation of macrocyclic complexes the macrocyclic effect of the four nitrogen macrocycles such as cyclam (Table V) (4,5) is also seen to be mostly enthalpy, and the entropy increases are close to zero and therefore insignificant. It should be noted that in the formation of the macrocyclic rings indicated in Tables IV and V there is no increase in the number of solute molecules formed, even though there is an increase in one chelate ring in the macrocycle over the open-chain analog. The main difference between the macrocyclic ligand and the open-chain ligand is the additional alkylene bridge between two of the donor atoms to form the ring. The classical chelate effect is due entirely to the formation of additional solute molecules, and when complex formation is not accompanied by formation of solute, there is no entropy increase. Thus the classical chelate effect is not a factor in the increased stability of the macrocycle. Therefore the use of the classical chelate effect and of the concept of an increase of 7.9 entropy units per chelate ring formed should be used with caution because it is really the formation of an additional mole of solute per chelate ring formed that is important. Formation of the macrocyclic complexes illustrated in Tables IV and V indicates that the coordinate bonds formed by the macrocyclic ligand are more stable than those of the open-chain analogs. Therefore the macrocyclic ligands form stronger coordinate bonds and are more basic than their open-chain analogs.

Table V. Thermodynamics Contributions to the Macrocyclic Effect in Complexes of Tetraazamacrocycles[a]

		Cu(II)	Ni(II)	Zn(II)
log K	cyclam	27.2	22.2	15.5
	2,3,2-tet	23.2	16.1	12.7
	log K (MAC)	4.0	6.1	2.8
ΔH^b	cyclam	-32.4	-24.1	-14.8
	2,3,2-tet	-26.5	-17.9	-11.6
	ΔH (MAC)	-5.9	-6.2	-3.2
ΔS^b	cyclam	12	21	21
	2,3,2-tet	16	15	19
	ΔS (MAC)	-4	6	2

[a] Since stability constants and heats were variously reported at μ = 0.1, 0.5, and 1.0, all differences were matched to similar conditions. [b] Units: ΔH, kcal/mole; ΔS, cal/deg/mole.

Increased Basicity of the Ligand Due to Alkylation

The increased basicity in the gaseous state caused by alkylation is illustrated in Figure 1 (5), which shows that methanol is more basic than water and that dimethyl ether is still more basic. The same applies to methylation of ammonia. An increase in basicity is seen as the hydrogens on ammonia are replaced by methyl groups, and the most basic of all is trimethylamine.

The increase in basicity due to alkylation is not clearly observed in aqueous solution because of solvation effects. In aqueous solution the basicity of trimethylamine seems to be about the same as that of ammonia. In the case of ammonia the positive ion formed by protonation, the ammonium ion, is stabilized by hydrogen bonding. For the trimethylammonium ion this kind of hydrogen bonding is repressed by the methyl groups, which cannot hydrogen bond to the solvent. Therefore the stabilization of the trimethylammonium ion by the solvent is much less than that of the ammonium ion, and the pK of protonated trimethylamine is reduced accordingly in water solution. However, trimethylamine should intrinsically form much stronger coordinate bonds with metal ions than does ammonia.

Although trimethylamine should form more stable metal complexes than ammonia, steric repulsion between methyl groups becomes important when there is more than one trimethylamine coordinated to a metal ion. Comparatively, ethylenediamine does not show this steric repulsion although the alkylation of the nitrogens by the ethylene bridge is nearly as great as would be the case for two methylamines. Thus the formation of an ethylene bridge is sterically efficient, and the repulsions between the methyl groups of coordinated methylamines is greatly reduced when methyl groups are replaced by the alkylene bridge in ethylenediamine. Therefore ethylenediamine is a more basic donor than two ammonias would be because of alkylation, and this is the main reason for the greater stability of the macrocyclic ring over the open-chain analog. The same result cannot be achieved by methylating the open-chain analog to produce more basic terminal oxygens or nitrogens because of the steric repulsions between the methyl groups.

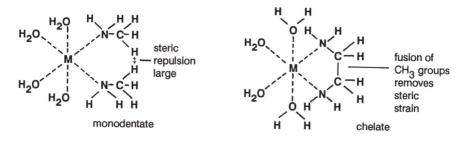

monodentate   chelate

The effect of increased alkylation on the stabilities of complexes can be seen readily from the following series of copper complexes containing four ammonias, two ethylenediamines, one 2,3,2-tet, and finally cyclam as ligands (6). All complexes are analogous in that four nitrogens are coordinated to the Cu(II) ion, but the effect of increased alkylation on the heat of reaction is seen to be considerable and the macrocyclic effect, which in this case involves 4.7 kcal per mole in going from 2,3,2-tet to cyclam, is seen to be largely due to increased alkylation.

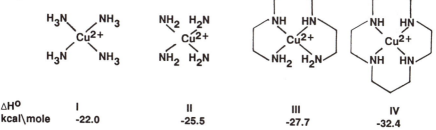

ΔH° kcal\mole	I -22.0	II -25.5	III -27.7	IV -32.4

A similar effect could not be achieved by increased alkylation of the donor nitrogens. For example there is a dramatic drop in the formation constant when cyclam is methylated to give tetramethylcyclam as indicated in Table VI (5). This is due to extensive steric repulsion between the methyl groups, which also serve to force the complex formed by cyclam with Cu(II) into a less stable *trans*-1 conformation (the steric repulsions between the carbon groups in the more stable *trans*-III conformation would be even greater), discussed in the next section.

Table VI. N-Alkylation Leads to a Drop in Log K when Cyclam is Methylated to Give TMC^a

	cyclam	TMC	Δ Log K	Ionic Radius (Å)
log K [Cu(II)]	27.2	18.3	8.9	0.65
log K [Ni(II)]	22.2	8.6	13.6	0.69
log K [Co(II)]	12.7	7.6	5.1	0.72
log K [Zn(II)]	15.5	10.4	5.1	0.74
log K [Cd(II)]	11.7	9.0	2.7	0.97
log K [Pb(II)]	11.3	(~7.5)	(~3.8)	1.21

^{a} $\mu = 0.10\,M$, T = 25.0 °C.

The Size-Fit Concept

It should be pointed out that the most stable conformation for a macrocyclic metal complex is not always one in which the metal ion is in the center of the cavity, *i.e.* is in the plane of the four nitrogens of a four nitrogen macrocycle such as cyclen (a twelve-membered four nitrogen macrocycle [12]aneN4). For example, a small metal ion may fit into the cavity of this macrocycle as indicated by the following *trans*-III conformation. With a larger metal ion the same macrocycle can assume a different conformation, which will allow it to complex the metal ion quite strongly, as indicated by the *trans*-I conformation. Thus a metal ion may rise above the plane of the donor atoms provided by the macrocycle to form quite stable complexes.

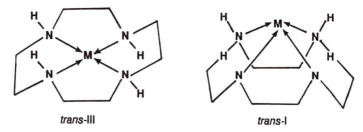

trans-III   *trans*-I

Therefore the notion that macrocyclic complexes favor a metal ion which produces the best fit in the cavity provided is subject to considerable doubt because there are alternative explanations for the maximum stability of a metal complex in a series such as is shown in Figure 2 (5). For small metal ions the stabilities would be expected to be fairly low because of the difficulty of the macrocycle in coordinating all the donor atoms to the metal ion because of the strain involved. Also a small metal ion may not take advantage of all the donor atoms provided by a macrocycle. For larger metal ions one would expect a decrease in stability due to the fact that the coordinate bonds themselves are longer and have lower stability. Between these two effects one would expect a

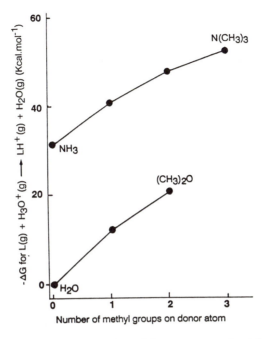

Figure 1. The effect of methylation of donor atom on basicity in the gas phase.

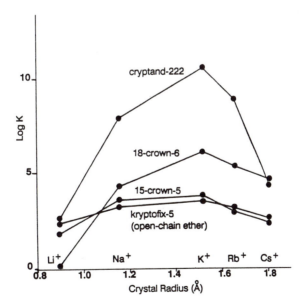

Figure 2. Formation constants, log K, in methanol of a variety of ligands, complexing with the alkali metal ions, as a function of crystal radius of the metal ion.

maximum in stability to be achieved. It is not necessary that the metal ions fit into the macrocyclic cavity. In the case of cryptates the maxima are much greater, and one can visualize a higher stability for the metal ion which best fits into the cavity formed by the three-dimensional ligand. In the case of cryptates, therefore, we have a combination of the effects of metal ion size on strength of coordinate bonding and steric strain in the ligand and the effect produced by whether or not the metal fits into the cavity provided by the ligand.

Preorganization

A familiar example involving partial preorganization of the ligand can be found in the comparison of the tetranegative anions of EDTA and *trans*-cyclohexanediaminetetraacetate (CDTA). In free EDTA solution the carboxylate ions are as far away from each other as they can get, and they have the flexibility to accomplish this. Also, the nitrogen atoms, although they involve lesser repulsive forces, are no doubt in the *trans* position. In the CDTA anion the nitrogens and the carboxylates again try to be as far away as possible, but they are forced together to a considerable extent by the ligand itself. Therefore they are much closer to the position that they would occupy in complexing and surrounding the metal ion, and the repulsive forces of the polar groups of the ligand are partially overcome in the free ligand. The result of the partial preorganization of CDTA is that the stability constants of the complexes formed are 10^2 to 10^3 greater than the corresponding complexes of EDTA.

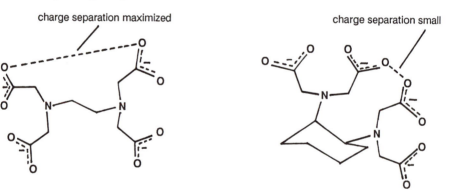

charge separation maximized   charge separation small

EDTA anion   CDTA anion

Although macrocyclic and macrobicyclic ligands are generally considered to be more highly preorganized than their open-chain analogs, they can frequently be in conformations that are not amenable to metal complex formation, and the free ligand must change its conformation in order to form stable complexes. For example, the macrocyclic ligand 18-crown-6 (*7-9*), and the macrobicyclic ligand cryptand-222 (*5,10*) in the free state are not in the conformations in which they exist in the corresponding metal complexes but are folded in on themselves to minimize the repulsions between polar groups and to maximize hydrophobic bonding. These ligands must open up to form the complexes as indicated below:

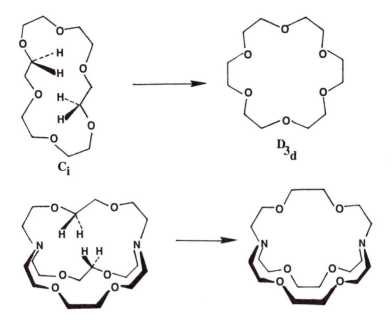

An outstanding example of a preorganized macrocyclic ligand is the four-nitrogen macrocycle shown below (11,12). The stability of its copper complex is $10^{21.5}$, almost ten orders of magnitude higher than that of the analogous open-chain compound, also shown. The large increase in stability is due to the forced preorganization of the ligand resulting from the macrocyclic ring. The open-chain ligand does not have the conformation shown in the figure. The piperazine moiety in the free ligand is in the more stable chair form and it has to be forced into the much less stable boat form in order to form the metal complex. Thus the repulsive forces between the nitrogen atoms of the open-chain analog that must be overcome in bringing the donor atoms into position around the metal ion are much greater than those encountered in the macrocyclic ligand. The adverse effect of these repulsive forces have already been overcome in the synthesis of the macrocyclic ligand.

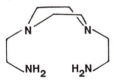

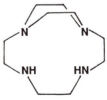

The Size of the Chelate Ring

The conformations of highest stability of ethylenediamine and trimethylenediamine complexes of metal ions are shown below. The ethylenediamine ligand forms a five-membered chelate ring, while that of trimethylenediamine forms a six-membered chelate ring Through the use of molecular mechanics, the "bite size" of the five-membered chelate ring, *i.e.*, the distance between the donor atoms, is greater than that of the six-membered chelate ring and the angle which is subtended between the donor groups and the metal ion in the five-membered chelate ring is shorter with a greater metal ion donor distance than in the six-membered chelate ring (6). The cyclohexane model used for the six-membered ring shows a larger angle between the donor groups and the metal ion but a smaller coordinate bond distance. Therefore it would seem that for small metal ions the six-membered chelate ring would form the more stable complexes, while for larger metal ions the five-membered ring would tend to be more stable. As the size of the metal ion increases further, one would expect a drop in stability as a general phenomenon. However, for small metal ions, going from a five- to a six-membered ring (in other words, increasing the ring size) would result in an increase in the stability.

The stabilities of the four-nitrogen metal chelates shown in Figures 3 and 4 illustrate the trend in stabilities of the complexes that occur when the size of the chelate ring is increased and the metal ion size is also increased (4). In Figure 3 it is seen that the stabilities of the complexes of Cu(II), Ni(II), and even Zn(II) and Cd(II) increase when an additional carbon atom is inserted in the ligand. In other words, if one goes from triethylenetetramine to 2,3,2-tet there is an increase in stability for the smaller metal ions. The general trend of stability, however, is seen to be one of decreasing stability as the size of the metal ion increases.

Also, Figure 4 shows that as the size of one of the rings of the four nitrogen macrocycle is increased from two to three to four to five the general trend is a decrease in stability (4,6). However, in going from a five-membered ring in the case of an ethylene bridge to a six-membered ring for a propylene bridge the stabilities of complexes of both low-spin and high-spin nickel(II) and of copper(II) are seen to increase.

Conclusions

The concepts described in this paper provide an understanding of the general principles governing the stabilities of metal complexes, chelates, macrocyclic complexes, and cryptate complexes.

1.   Completely preorganized ligands are rare; they maximize stability through enthalpic and entropic effects. Macrocyclic ligands are sometimes poorly preorganized.

2.   Preorganization is observed for chelating ligands as well as macrocycles and cryptands.

3.   Higher stabilities of macrocyclic and cryptand complexes are due in large part to alkylation of donor atoms.

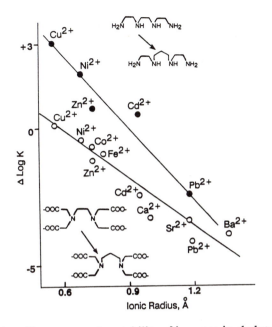

Figure 3. The effect on complex stability of increase in chelate ring size from 5-membered to 6-membered in open-chain ligands, as a function of metal ion size. $\Delta \log K = \log K_{ML}$ (6-membered) - $\log K_{ML}$ (5-membered).

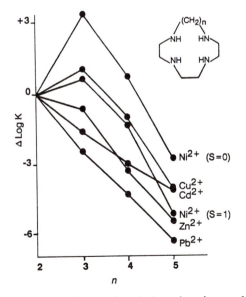

Figure 4. The effect of increasing chelate ring size on the stability constants of metal complexes of tetraaza macrocyclic ligands.

4.   Alkyl groups bridging donor atoms in macrocyclic and cryptand ligands are sterically efficient.

5.   The size of the macrocyclic "cavity" does not prevent the ligand from strongly binding larger metal ions.

6.   Small metal ions prefer 6-membered chelate rings and large metal ions prefer 5-membered chelate rings. This applies to macrocyclic and cryptate complexes as well as open-chain chelating ligands.

7.   The "size-match" selectivity claimed for macrocyclic ligands is due in large part to decreasing affinity with increasing size of the metal ion and increased steric strain in complexing small metal ions. This effect is observed in open-chain chelating agents as well as in their cyclic analogs.

Acknowledgments

This research was supported by a grant, No. A-259, from The Robert A. Welch Foundation.

Literature Cited

1.   Schwarzenbach, G. *Helv. Chim. Acta* **1952**, *35*, 2344.
2.   Adamson, A. W. *J. Am. Chem. Soc.* **1954**, *76*, 1578.
3.   Martell, A. E. *Rev. Trav. Chim.* **1956**, *6*, 75; *Adv. Chem. Ser.* **1967**, *62*, 272.
4.   Smith, R. M; Martell, A. E. *Critical Stability Constants*, Plenum: New York, 1974, 1975, 1976, 1977, 1982, 1989; Vols. 1-6.
5.   Hancock, R. D.; Martell, A. E. *Comment Inorg. Chem.* **1988**, *6*, 237.
6.   Hancock, R. D.; Martell, A. E. *Chem. Rev.* **1989**, *89*, 1875.
7.   Wipff, G.; Weiner, P.; Kollman, P. *J. Am. Chem. Soc.* **1982**, *104*, 3249.
8.   Ranghino, G.; Romano, S.; Lehn, J. M.; Wipff, G. *J. Am. Chem. Soc.* **1985**, *107*, 7873.
9.   Doxsee, K. M.; Wierman, H. R.; Weakley, J. T. R. *J. Am. Chem. Soc.* **1992**, *114*, 5165.
10. Cram, D. J.; Kaneda, R.; Helgeson, R. C.; Brown, S. B.; Knobler, C. B.; Maverick, E.; Trueblood, K. N. *J. Am. Chem. Soc.* **1985**, *107*, 3645.
11. Hancock, R. D.; Dobson, S. M.; Evers, A.; Wade, P. W.; Ngwenya, M. P.; Boeyens, J. C. A.; Wainwright, K. P. *J. Am. Chem. Soc.* **1988**, *110*, 2788.
12. Hancock R. D.; Evers, A.; Ngwenya, M. P.; Wade, P. W. *J. Chem. Soc. Chem. Commun.* **1987**, 1129.
13. Hancock, R. D.; Ngwenya, M. P. *J. Chem. Soc. Dalton Trans.* **1987**, 2911.

RECEIVED February 8, 1994

ISOMERISM

Chapter 21

Linkage Isomerism of Thiocyanate Bonded to Cobalt(III)

From Alfred Werner to the Present Day

D. A. Buckingham

Department of Chemistry, University of Otago, P.O. Box 56,
Dunedin, New Zealand

The attempted preparation of the $[CoCl(SCN)(en)_2]^+$ and $[CoCl(NCS)(en)_2]^+$ isomers by Werner and Bräunlich is reviewed, as is their reformulation as $cis/trans$ isomers of $[CoCl(NCS)(en)_2]^+$. The aqueous solution isomerization of the $[CoSCN(NH_3)_5]^{2+}$ ion (spontaneous, OH^- catalyzed, Hg^{2+} and Ag^+ catalyzed) is reviewed with emphasis being placed on competition by other species in solution ($N^{14}CS$, N_3^-, NO_3^-, H_2O). A detailed account is given of the **solid state** isomerization of $trans$-$[CoSCN(NH_3)_4(^{15}NH_3)](N^{14}CS)_2$ where the first products of heating at 60°C are $trans$-$CoNCS^{2+}$ (40%, intramolecular), cis-$CoN^{14}CS^{2+}$ (30%, lattice), and cis-$CoS^{14}CN^{2+}$ (30%, lattice). A crystal structure of the reactant requires two different Co(III) cations to be involved and suggests the coordination of one of four different lattice NCS^- ions. A new nomenclature for substitution reactions at a metal center is proposed.

The ambidentate nature of thiocyanate coordination was first recognized by Alfred Werner in a paper published together with Bräunlich in 1899 (1). Today, the alternative CoNCS and CoSCN bonding modes are well known, as are bridging CoNCSCo and CoNCSM systems (M = Cr(III), or another metal). In this talk I will briefly describe Werner's contributions and then concentrate on our own work since this, I believe, leads to a more intimate understanding of substitution mechanism at a metal center. This comes about by using the CoSCN → CoNCS rearrangement as a mechanistic probe since the rearrangement process is not exclusive under most experimental conditions and neighboring groups which are not formally attached to the metal compete for it from within the encounter sphere about the octahedral ion. This allows these and other reactions to be monitored against the backdrop of the rearrangement.

But first, some bond lengths and angles (Table I). These values differ little from complex to complex, with the isothiocyanate structure (1) being essentially

0097–6156/94/0565–0256$08.00/0

linear with a Co-N bond length very similar to that found for Co(III)-amines; i.e., it is reasonably short. The thiocyanate structure (2), on the other hand, is decidedly "bent" at the S atom (as would be expected from simple valence bond theory), and the Co-S bond is quite long, being similar to that found for coordinated Cl⁻ ($\approx$ 230 pm). Thus the two bonding modes are quite different, with S coordination resembling true "anionic" bonding and N coordination resembling an inert amine system (although still ionic). In a geometric sense the more weakly bound, bent, thiocyanate structure (2) may be considered part-way toward the more strongly bound, linear, isothiocyanate structure (1). Crystal structures of the Co(III) cations in [CoSCN(NH$_3$)$_5$]Cl$_2$.H$_2$O and [CoNCS(NH$_3$)$_5$]SO$_4$ are shown in Figures 1 (A and B).

Table I. Bond Lengths and Angles of Thiocyanates

System	Bond Length (pm)	Angle (°) at N/S
CoNCS	192 - 195	179 (2)
CoSCN	227 - 231	105 (2)

CoNCS bonding is normally the thermodynamically favored form demonstrating the "hardness" of Co(III) (note that protonation occurs as HNCS), but occasionally S coordination occurs. Thus both [CoNCS(CN)$_5$]$^{3-}$ and [CoSCN(CN)$_5$]$^{3-}$ are known in the solid state (as their n-Bu$_4$N$^+$ (2) and K$^+$ (3) salts, respectively), and an equilibrium situation is found with various dimethylglyoximato systems in solution (4-6) (3 and 4), equation (1). Both systems exhibit the "softer" characteristics of Co(III) coordination. However, all saturated amine ligand systems adopt CoNCS coordination, and CoSCN complexes are only produced via "capture" processes whereby a poor ligand (e.g. NO$_3$⁻, ClO$_4$⁻, CF$_3$SO$_3$⁻) vacates a coordinatation site, which can then be captured by the S (or N) end of SCN⁻ or by the solvent (equation (2)).

The CoSCN complex (7) is usually stable under neutral conditions and at ambient temperature, and separation from (8) is possible by fractional crystallization or by chromatographic methods. The two isomers can easily be distinguished spectrally, with (7) having extensive charge transfer absorptions in the near UV (< 400 nm), whereas (8) does not. Also IR spectroscopy can be used to distinguish these isomers. The most recent general account of thiocyanate coordination is that of Burmeister (7).

Werner, and NCS⁻/SCN⁻ Isomerism

In 1899 Werner and Bräunlich reported (1) the synthesis of two isomers of the [CoCl(thiocyanate)(en)$_2$]$^+$ ion, a **red** (11) and a **purple** (14) form (equation (3)). They assigned these to S (thiocyanato) and N (isothiocyanato) coordination, respectively, largely on the basis that the red form completely loses thiocyanate

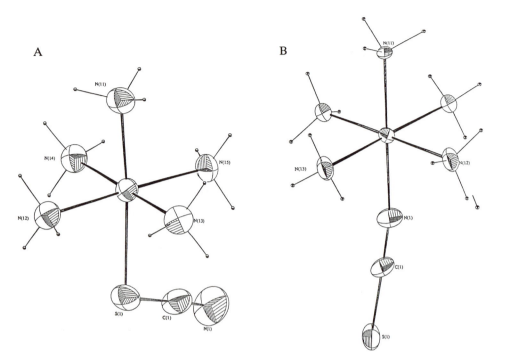

Figure 1. Perspective views of the cations in **A** [CoSCN(NH$_3$)$_5$](NCS)$_2$ and **B** [CoNCS(NH$_3$)$_5$]SO$_4$.1.5H$_2$O showing "bent" (107.0°) thiocyanate and close to linear (170°) isothiocyanate coordination. Thermal ellipsoids are drawn at the 50% probability level.

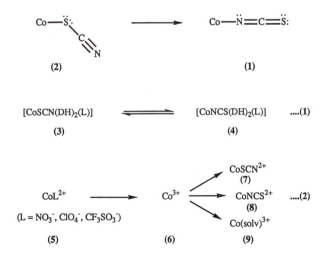

under oxidizing conditions (to give (12) and then (13)), whereas the purple form is converted to coordinated ammonia (15). In a subsequent paper (8) Werner revised this assignment to one of *cis/trans* isomerism because both (11) and (14) could be converted into **the same** isothiocyanato ammine complex, $[CoNCS(NH_3)(en)_2]^{2+}$ (16), using liquid ammonia (equation (4)), and he correctly predicted that (11) → (14) isomerism is very unlikely to occur under the low temperature conditions. Other reactions described in Werner's paper (8) (equation (5)) supported the *cis/trans* assignment (equation (5)), and the purple *trans*-CoNCS isomer (14) was shown to be more stable towards loss of coordinated NCS⁻ than the red *cis*-CoNCS isomer (11).

Werner attempted oxidation experiments on a number of other $CoNCS^{2+/1+}$ systems using H_2O_2, Cl_2, and fuming HNO_3, and he showed (1,8-10) that the degree of conversion to $CoNH_3$ depends both on the complex and the oxidizing agent. Some lose NCS⁻ completely (as do all Cr(III)NCS systems that I am aware, (11)), while others smoothly formed coordinated ammonia.

In summary, it is quite remarkable that at this very early time in the history of coordination chemistry Werner recognized the possibility of CoNCS/CoSCN linkage isomerism and spent much time and effort investigating it.

Preparation of $[CoSCN(NH_3)_5]^{2+}$

The first authentic CoSCN complex, $[CoSCN(NH_3)_5]Cl_2 \cdot H_2O$ (20), was prepared and characterized in 1969 (12). It was prepared (equation (6)) by base hydrolysing $[Co(ONO_2)(NH_3)_5]^{2+}$ (19) in a supersaturated NaSCN solution (~ 10 M), neutralizing the solution after about 10 half-lives for base hydrolysis, and isolating (20) and (21) on Dowex 50W X 2 cation exchange resin ((20) : (21) = 4:1); the isomers were then separated by fractional crystallization. It was also found possible to separate them by using a long ion-exchange column. Subsequently, for analytical purposes, HPLC has been shown to be very effective (13).

Reactions of $CoSCN^{2+}$ Complexes

I do not have time in this talk to describe in detail the results of all our studies in this area. I will only have time to describe our most recent investigation, the thermal isomerization of crystalline *trans*-$[CoSCN(NH_3)_4(^{15}NH_3)](N^{14}CS)_2$. However, before I do this I will briefly summarize what we have learned about reactions in aqueous solution.

The Spontaneous Reaction of $[CoSCN(NH_3)_5]^{2+}$ (14). This reaction is very slow at 25°C in water but eventually produces 9.1% $[Co(NH_3)_5OH_2]^{3+}$ (equation (7)). In the presence of $N^{14}CS^-/S^{14}CN^-$ small amounts of both $CoN^{14}CS^{2+}$ and $CoS^{14}CN^{2+}$ are directly formed from the reactant, i.e., not via $CoOH_2^{3+}$, and these amounts are similar to those found using reactant complexes such as $CoOClO_3^{2+}$ and $CoOSO_2CF_3^{2+}$ which also hydrolyze spontaneously, but much more rapidly. The interference is that competition for $N^{14}CS^-/S^{14}CS^-$ is insensitive to the leaving group and to its subsequent chemistry; in the present case the leaving group largely re-

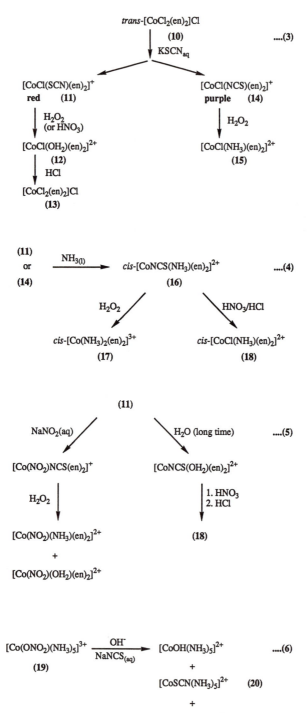

trans-[CoCl₂(en)₂]Cl
(10)   ....(3)

KSCN$_{aq}$

[CoCl(SCN)(en)₂]⁺   [CoCl(NCS)(en)₂]⁺
red   **(11)**   **purple**   **(14)**

H₂O₂
(or HNO₃)   H₂O₂

[CoCl(OH₂)(en)₂]²⁺   [CoCl(NH₃)(en)₂]²⁺
(12)   **(15)**

HCl

[CoCl₂(en)₂]Cl
(13)

(11)
or $\xrightarrow{\text{NH}_{3(l)}}$ *cis*-[CoNCS(NH₃)(en)₂]²⁺   ....(4)
(14)   **(16)**

H₂O₂   HNO₃/HCl

cis-[Co(NH₃)₂(en)₂]³⁺   *cis*-[CoCl(NH₃)(en)₂]²⁺
(17)   **(18)**

(11)

NaNO₂(aq)   H₂O (long time)   ....(5)

[Co(NO₂)NCS(en)₂]⁺   [CoNCS(OH₂)(en)₂]²⁺

H₂O₂   1. HNO₃
                                    2. HCl

[Co(NO₂)(NH₃)(en)₂]²⁺   **(18)**

+

[Co(NO₂)(OH₂)(en)₂]²⁺

[Co(ONO₂)(NH₃)₅]³⁺ $\xrightarrow[\text{NaNCS}_{(aq)}]{\text{OH}^-}$ [CoOH(NH₃)₅]²⁺   ....(6)
(19)

+

[CoSCN(NH₃)₅]²⁺   **(20)**

+

[CoNCS(NH₃)₅]²⁺   **(21)**

enters as $CoNCS^{2+}$. Also, the results show that the isomerizing thiocyanate ligand is preferred over that in the $CoSCN^{2+}.S^{14}CN^-$ ion pair. Thus if bond dissociation precedes entry of $S^{14}CN^-/N^{14}CS^-$ (or solvent), the two thiocyanates never become equivalent. This suggests that the immediately released SCN^- is in an (vibrationally) excited state and is preferred because of this. It may well remain π-bonded to the metal as has been suggested previously (*12*).

Alkaline Hydrolysis of *trans*-$[CoSCN(NH_3)_4(^{15}NH_3)]^{2+}$ (*15*). This reaction (without the stereochemical label) had been investigated previously and its kinetic aspects reported (*12*). However, more careful HPLC investigations of the products have now shown that the amount of $CoNCS^{2+}$ formed (equation (**8**)) depends on the electrolyte conditions and temperature. More $CoNCS^{2+}$ is formed in the **absence** of electrolytes and at a **lower** temperature. The stereochemistry of the two products is also affected. In the presence of N_3^-, increasing amounts of CoN_3^{2+} are formed, but this product competes for the solvent ($CoOH^{2+}$) rather than for $CoNCS^{2+}$. The formation of $CoNCS^{2+}$ is truly intramolecular, and its stereochemistry remains unchanged. However large amounts of inversion, i.e., *cis* product, occur with $CoOH^{2+}$ and CoN_3^{2+}. Such results suggest that N_3^- competes largely for H_2O entry on the "backface" of the complex while re-entry of NCS^- competes with H_2O on the "frontface." Once again, if an intermediate is formed following cleavage of the Co-SCN bond, its lifetime is extremely short; it does not allow for translational diffusion or rotation of nearest neighbors.

The Hg^{2+}-Catalyzed Reaction (*16,17*). This study was carried out using the t-$[CoSCN(NH_3)(tren)]^{2+}$ and t-$[CoX(NH_3)(tren)]^{2+}$ (X = Cl, Br) complexes. The former gives the product distribution of equation (**9**) in the absence of anions and the distribution given by equation (**10**) for the NO_3^--dependent reaction. Other anions (ClO_4^-, $CF_3SO_3^-$) also compete for the isomierization pathway as well as for solvent entry. For t-$[CoX(NH_3)(tren)]^{2+}$ (X = Cl, Br) very similar amounts of $CoONO_2^{2+}$ are formed by the NO_3^--dependent pathway (55-57%) as for the $CoSCN^{2+}$ complex with the remainder in this case being $CoOH_2^{3+}$. Such results are interpreted as H_2O occupying a position in the second coordination sphere otherwise occupied by the N terminus of the rotating thiocyanate ligand, replacing it as a product once the Co-X bond is broken.

The Ag^+-Catalyzed Reaction (*16*). This reaction was not studied in detail, but a two-term rate law $k_{obs} = k_1[Ag^+] + k_2[Ag^+]^2$ was found, with the product distributions in 1 M $NaClO_4$ being given by equation (**11**) and (**12**). The first-order term was interpreted as attachment of Ag^+ to the S atom, and the second-order pathway to attachment of Ag^+ to both S and N, thereby preventing recoordination as $CoNCS^{2+}$. The NO_3^- anion was found to compete for both paths.

Solid State Isomerization of *trans*-$[CoSCN(NH_3)_4(^{15}NH_3)](N^{14}CS)_2$

In the first publication reporting a Co(III)-thiocyanate complex (*12*) it was found that heating $[CoSCN(NH_3)_5]Cl_2.H_2O$ under the infrared lamp or in the absence of

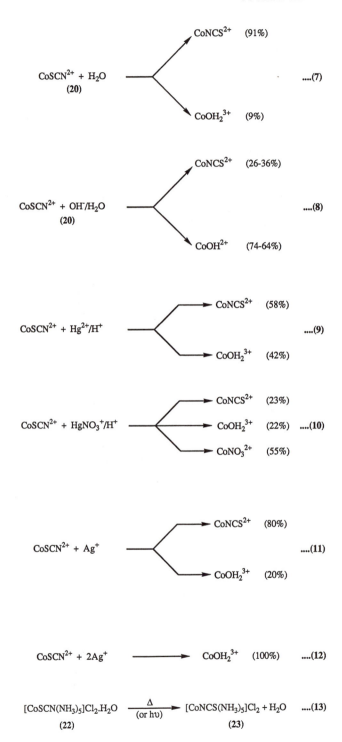

$CoSCN^{2+} + H_2O$
(20)

$\longrightarrow$ $CoNCS^{2+}$ (91%)

$CoOH_2^{3+}$ (9%)

....(7)

$CoSCN^{2+} + OH^-/H_2O$
(20)

$\longrightarrow$ $CoNCS^{2+}$ (26-36%)

$CoOH^{2+}$ (74-64%)

....(8)

$CoSCN^{2+} + Hg^{2+}/H^+$

$\longrightarrow$ $CoNCS^{2+}$ (58%)

$CoOH_2^{3+}$ (42%)

....(9)

$CoSCN^{2+} + HgNO_3^+/H^+$

$\longrightarrow$ $CoNCS^{2+}$ (23%)

$CoOH_2^{3+}$ (22%) ....(10)

$CoNO_3^{2+}$ (55%)

$CoSCN^{2+} + Ag^+$

$\longrightarrow$ $CoNCS^{2+}$ (80%)

$CoOH_2^{3+}$ (20%)

....(11)

$CoSCN^{2+} + 2Ag^+$ $\longrightarrow$ $CoOH_2^{3+}$ (100%) ....(12)

$[CoSCN(NH_3)_5]Cl_2 \cdot H_2O$ $\xrightarrow[\text{(or h}\upsilon\text{)}]{\Delta}$ $[CoNCS(NH_3)_5]Cl_2 + H_2O$ ....(13)
(22)   **(23)**

light at 80°C caused the purple-violet crystals to change to an orange color. Ion exchange chromatography confirmed that isomerization to $[CoNCS(NH_3)_5]Cl_2$ had taken place. The reaction went to completion, and no Cl^- ion was incorporated from the lattice. Also, no $[Co(NH_3)_5OH_2]^{3+}$ was formed (equation (13)). Such aspects have been subsequently verified using the sensitive HPLC technique, and a similar reaction has been shown to occur with crystalline $[CoSCN(NH_3)_5]Br_2$, which does not contain lattice H_2O (*18*). It is not known whether the reverse isomerization CoNCS $\rightarrow$ CoSCN can be photochemically promoted (as can the nitro to nitrito isomerization, $CoNO_2 \rightarrow CoONO$), but X-rays have been shown to facilitate slowly the CoSCN $\rightarrow$ CoNCS change at liquid nitrogen temepratures so the reaction appears to be more than thermally motivated.

Following this early study Snow and Boomsma (*19*) determined the crystal structure of $[CoSCN(NH_3)_5]Cl_2.H_2O$ and also showed that the solid state product crystal has a structure similar to that of $[CoNCS(NH_3)_5]Cl_2$ crystallized from aqueous solution, <u>viz.</u>, it is disordered with the five NH_3 ligands and NCS^- being indistinguishable within the lattice. They suggested two possible mechanisms to account for this disorder : (i) a "head to tail" intramolecular rearrangement about each Co center followed by random reorientation of each $CoNCS(NH_3)_5^{2+}$ unit within the lattice (considerable translational and rotational movement would be necessary for this to occur); or (ii) initial dissociation to give $Co(NH_3)_5^{3+} + SCN^-$, followed by random re-introduction of NCS^- from the same, or from a nearby, dissociation. The second mechanism was preferred since less movement within the crystal would be required, but ionic Cl^- could not be included, and each $Co(NH_3)_5^{3+}$ unit would need to exist for some time since the process is not autocatalytic. However, in a subsequent paper (*20*) a somewhat different result was found on heating the ionic thiocyanate salt, $[CoSCN(NH_3)_5](N^{14}CS)_2$. In this case some mixing of coordinated and ionic thiocyanate occured with at least half the product containing lattice $N^{14}CS$. This supported a dissociative process for a substantial part of the reaction but with re-entry of anion being less specific.

We have looked again (*18*) at the reaction of the thiocyanate salt using crystalline (as opposed to freeze-dried) reactant and a *trans* $^{15}NH_3$ label to follow stereochemical change as well as ionic $N^{14}CS^-$ to follow incorporation of thiocyanate from the lattice. The stereochemical results of heating crystals of *trans*-$[CoSCN(NH_3)_4(^{15}NH_3)](N^{14}CS)_2$ at 60°C are given in Figure 2 as a function of % reaction. (Unfortunately the reactant contained a 12% *cis* impurity, but this was allowed for.) These results were obtained using ^1H-NMR by analyzing the separated (HPLC) $CoNCS^{2+}$ product and $CoSCN^{2+}$ reactant at different times. Examples of ^1H-spectra are given in Figures 3 and 4 where the *cis* and *trans* $^{15}NH_3$-proton doublets are shown. While the $CoNCS^{2+}$ product shows a slight increase in the *trans* $\rightarrow$ *cis* rearrangement as the reaction progresses, **there is a significant *trans* $\rightarrow$ *cis* change in the remaining CoSCN^{2+}**. These stereochemical changes are accompanied by the incorporation of ionic $N^{14}CS^-$ into the product, $(CoN^{14}CS^{2+})$, and $S^{14}CN^-$ into the reactant, $(CoS^{14}CN^{2+})$, and the stereochemical change in the final $CoNCS^{2+}$ product agrees well with the total amount of $N^{14}CS^-$ incorporated from the lattice (~ 60%).

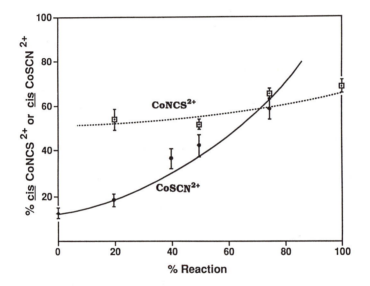

Figure 2. Observed and calculated percentages of *cis*-CoSCN²⁺ (——) and *cis*–CoNCS²⁺ (--------) as a function of % reaction on heating *trans*–[CoSCN(NH₃)₄(¹⁵NH₃)](NCS)₂ (containing a 12% *cis* impurity) at *ca.* 60°C. Experimental uncertainties are given as error bars on the experimental points.

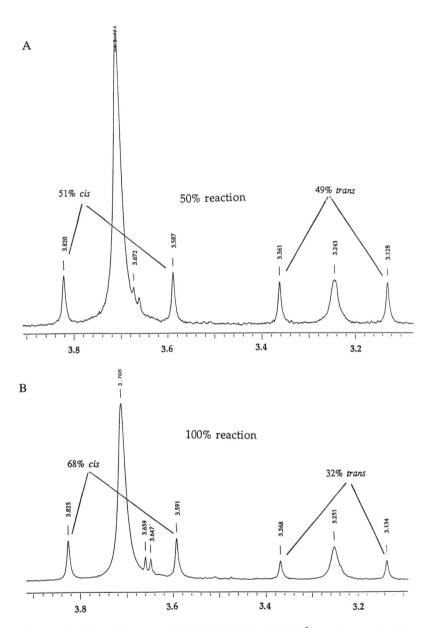

Figure 3. ^{1}H NMR spectra (300 MHz) of the $CoNCS^{2+}$ product on heating *trans*-[$CoSCN(NH_3)_4(^{15}NH_3)$](NCS)$_2$ at *ca.* 60°C after (**A**) 50% reaction and (**B**) after 100% reaction. The *trans*- and *cis*-$^{15}NH_3$ doublets are identified and are centered on the broader $^{14}NH_3$ resonance.

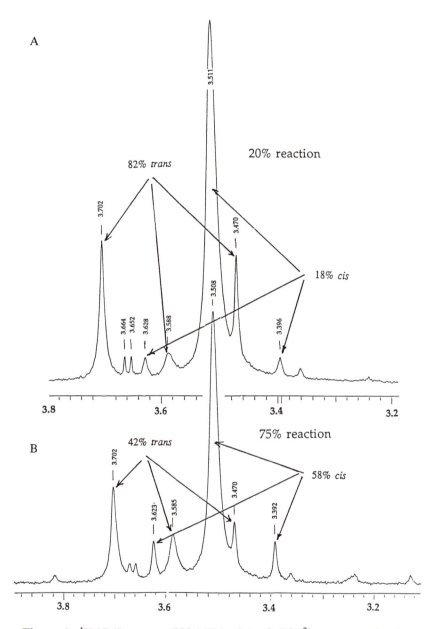

Figure 4. ^1H NMR spectra (300 MHz) of the $CoSCN^{2+}$ reactant on heating *trans*-$[CoSCN(NH_3)_4(^{15}NH_3)(NCS)_2$ at *ca.* 60°C after (**A**) 20% reaction and (**B**) after 75% reaction. The *trans*-$^{15}NH_3$ doublet now appears downfield of the *cis*-$^{15}NH_3$ doublet with both doublets being centered on broader $^{14}NH_3$ singlets. A small amount of a $CoNCS^{2+}$ impurity is present in both spectra.

Such results mean that some 60% of the product derives from "backside" attack of lattice $N^{14}CS^-$ and $S^{14}CN^-$; the results cannot be accommodated by a complete, or even a partial, scrambling process. However, they can be accommodated by a combination of three pathways, equation (14), with the immediate products comprising 40% $CoNCS^{2+}$ without stereochemical change or incorporation of $N^{14}CS^-$ from the lattice, and about equal amounts of *cis*-$CoS^{14}CN^{2+}$ and *cis*-$CoN^{14}CS^{2+}$ containing ionic thiocyanate and involving stereochemical change. The intermediate *cis*-$CoS^{14}CN^{2+}$ product is, of course, chemically indistinguishable from the *trans*-$CoSCN^{2+}$ reactant and goes on to produce more *cis*-$CoN^{14}CS^{2+}$.

A crystal structure of the $[CoSCN(NH_3)_5](NCS)_2$ reactant gives a good explanation of these results. It crystallizes in the monoclinic space group $P2_1/n$, with **two** separate Co(III) cations in the asymmetric unit. There are therefore four dissimilar ionic thiocyanates. One of these stands out as an excellent candidate for subsequent coordination to each of the Co(III) centers. Figures 5A and 5B show the arrangement of the ionic thiocyanates in the lattice with those "backface", *i.e.*, remote from the coordinated SCN, being shaded since they are the important ones. S(4)C(4)N(4) on the backface to both units is unique. S(4) is the closest sulfur atom to Co(1) (433.4 pm), and N(4) is the closest donor atom of any kind to Co(2) (376.7 pm) (Table II). However, probably more importantly S(4) and N(4) subtend angles of 94.8° and 167.7°, respectively, towards the two Co(III) centers, which means that the S and N lone pairs point almost exactly at the metal. Also, these atoms are located almost exactly down three-fold axes comprising three of the ammine ligands (Figures 6A and 6B). With a $(t_{2g})^6(e_g)^0$ configuration such axes represent positive holes towards the metal and hence ideal locations for electron pairs. It therefore appears that the $S(4)C(4)N(4)^-$ anion is beautifully poised on low energy paths to the two Co(III) cations. The isomerization yields can then be apportioned as in equation (15).

The energetics of S and N coordination from that lattice are therefore similar (they did not need to be), and remote "backside" entry occurs in competition with head-to-tail "frontside" N-entry of previously S-bound thiocyanate. Such competitive three-way entry is also found in solution (*vide supra*).

A reanalysis of the crystal structure of $[CoSCN(NH_3)_5]Cl_2.H_2O$ suggests that the failure of lattice Cl⁻ to enter the coordination sphere on heating this material (see beginning of this section) is not due to more distant Cl⁻ ions or to a poorer positioning of Cl⁻ with respect to the three-fold axes of the octahedron, but to a more extensive network of H-bonds to the ammine ligands. Thus the electron lone pairs of the Cl⁻ ion are less available to the Co(III) center, and head-to-tail intramolecular isomerization of coordinated SCN is favored. Such nonincorporation of lattice Cl⁻ is accompanied by retention of crystal packing and nonfracturing, whereas the $[CoSCN(NH_3)_5](NCS)_2$ crystal fractures in the final stages of heating.

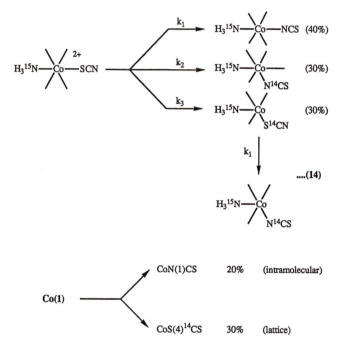

....(14)

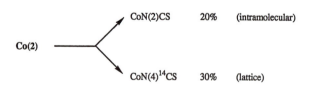

....(15)

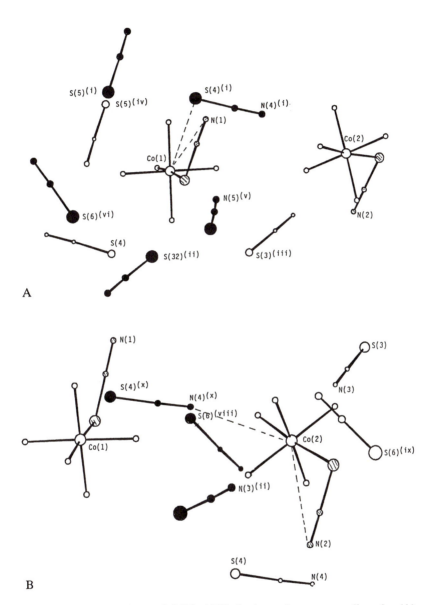

Figure 5. Perspectives of SCN⁻, NCS⁻ lattice anions surrounding the (**A**) Co(**1**) and (**B**) Co(**2**) cations in crystalline [CoSCN(NH₃)₅](NCS)₂. "Backface" anions (with respect to the Co-N bond) are shaded, whereas "frontface" anions are left open. The dashed lines show suggested pathways for entry of the unique S(4)C(4)N(4)⁻ lattice anion and for intramolecular entry of coordinated NCS.

Table II.   Non-Bonded Distances (< 5 Å) of Ionic SCN⁻, NCS⁻
and Coordinated SCN from Co(1,2)

Interaction	Translation	Distance (Å)	Orientation (perpendicular distance from CoN$_4$ mean plane) (Å)
Co(1) Ionic			
Co(1)---N(5)	(v)	4.103(5)	Backface (2.700(6))
Co(1)---S(4)	(i)	4.334(2)	Backface (2.856(1))
Co(1)---S(4)	(0)	4.476(2)	Frontface (1.482(1))
Co(1)---S(6)	(vi)	4.528(2)	Backface (3.023(2))
Co(1)---S(5)	(iv)	4.563(1)	Frontface (1.521(2))
Co(1)---S(5)	(i)	4.593(2)	Backface (3.002(2))
Co(1)---S(3)	(iii)	4.616(1)	Frontface (1.582(1))
Co(1)---S(3)	(ii)	4.630(2)	Backface (2.848(1))
Coordinated			
Co(1)---N(2)	(i)	4.126(7)	Frontface (0.695(1)); intermolecular
Co(1)---N(1)	(0)	4.138(6)	Frontface (3.377(5)); intramolecular
Co(2) Ionic			
Co(2)---N(4)	(x)	3.767(5)	Backface (2.038(6))
Co(2)---N(3)	(ii)	3.807(6)	Backface (2.521(6))
Co(2)---N(6)	(ix)	4.148(7)	Frontface (1.876(7))
Co(2)---N(6)	(viii)	4.227(7)	Backface (2.755(7))
Co(2)---S(3)	(0)	4.456(2)	Frontface (1.493(1))
Co(2)---N(3)	(0)	4.813(2)	Frontface (2.997(6))
Co(2)---S(4)	(0)	4.849(2)	Frontface (0.474(1))
Co(2)---N(4)	(0)	4.952(5)	Frontface (1.691(6))
Coordinated			
Co(2)---N(1)	(vii)	3.819(5)	Backface (2.226(5)) intermolecular
Co(2)-N(2)	(0)	4.089(5)	Frontface (3.269(5)); intramolecular

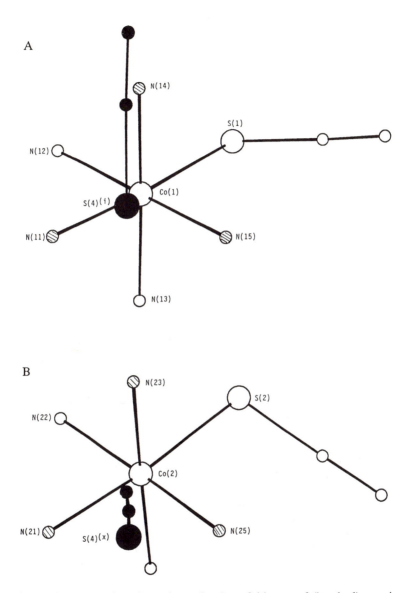

Figure 6. Perspective views down the three-fold axes of (hatched) ammine ligands for (**A**), "backface" entry of S(4)C(4)N(4)⁻ to Co(**1**), and (**B**) "backface" entry of N(4)C(4)S(4)⁻ to Co(**2**), showing the orientations of this unique lattice anion.

Conclusions

It is obvious that in the solid state lattice NCS⁻ and SCN enter the octahedral coordination sphere "backside" to the isomerizing thiocyanate ligand, i.e., with inversion. This process is not exclusive and competes for "frontside" intramolecular isomerization. A similar situation occurs in alkaline solution where N_3^- and H_2O enter (in part), with inversion, and in competition with the isomerization process. However, for the Hg^{2+}- and Ag^+-catalyzed reactions NO_3^- (and other anions) as well as H_2O compete "frontside", with retention of configuration. The spontaneous reaction also occurs with retention (at least for H_2O entry).

We conclude that substitution of Y^- (in CoY^{n+}) by X^- or H_2O is controlled by the immediate availability of these groups, i.e., that preassociation of entering groups **is essential**. It also appears necessary for electron lone-pairs to be correctly oriented towards the electrofugic Co^{3+} center, and that preassociation in the sense of distance availability is not sufficient. This certainly seems to hold for the solid state reaction. We view substitution as largley being controlled by entropic considerations; it is little influenced by the nucleophilic (or basic) properties of the entering group or by the eventual stability of the immediate substitution product.

It then becomes uncertain as to whether the substitution process should be considered stepwise, since there may be no intermediate energy state of sufficient lifetime to be considered as a separate chemical entity. Bond dissociation (**D**, see below) appears to be the major energetic feature, but anions (Y^-) have more influence on this than was previously recognized (the concept of constant ionic strength has been a distraction for investigations with cationic complexes). Reaction rates are also affected by the associated anion as well as by the solvent. The solvent, and anions derived from it, end up as coordinated products, so that their influence on the rate of reaction is also paralleled by their eventual coordination. The question then is : does their entry into the primary coordination sphere **coincide** with their affects on the loss of leaving group? This question has not yet been answered to our satisfaction.

Our preference in the meantime is for bond breaking (**D**) to precede bond formation (**A**) for octahedral complexes, at least for those complexes with a closed $(t_{2g})^6(e_g)^0$ electron configuration. The full t_{2g} shell is likely to prevent additional electron donation to the metal until one of these orbitals is freed (by dissociation). The lowest energy path to substitution lies down the three-fold axes of the octahedron between opposing electron pairs. Bonding may be initiated by electron rehybridization such that the new octahedral ligand field is reoriented along the previous three-fold axes. Such electronic redistributions will be instantaneous on the translational and rotational time scales (Born Oppenheimer), and the making of a new bond will be on the vibrational time scale ($\sim 10^{-13}$ s). The lifetime of the dissociated five-coordinate intermediate is not likely to be longer than this. Thus we view substitution as a $D^{\ddagger*}A$ process (see below) if it is accepted that a five-coordinate intermediate is allowable between the breaking of the bond to the leaving group, electron reorientation, and the making of the new bond to the entering group.

In my view a more exact naming system for substitution reactions at a metal center is now needed. The chemistry described in this talk requires a more detailed picture than the old I_a, I_d classification of Langford and Gray (*21*), which has served us well as long as we did not inquire too much into the time sequences of bond-making and bond-breaking within the inner coordination sphere. Now the sequences of bond-breaking and bond-making and the time delay between them needs to be more carefully detailed. We need some way of describing these alternatives.

The recently suggested names for organic reactions recommended by the IUPAC provides an excellent starting point (*22,23*). I have adopted this in the following short-hand abbreviations for the substitution mentioned in this talk. I will, in a separate account, describe these designations in more detail.

	Name	Description
Solid State Reaction	*intra*-$D_N{}^{\ddagger}A_N$	intramolecular isomerization
	$D_N{}^{\ddagger}A_{N(ip)}$	substitution with inversion
Spontaneous Reactions (in solution)	*intra*-$D_N{}^{\ddagger*}A_N$	intramolecular isomerization
	$D_N{}^{\ddagger*}A_S$	substitution by associated solvent
	$C_{ip}+D_N{}^{\ddagger*}A_N$	substitution by capture of a preassociated nucleophile
Induced (e.g. Hg^{2+}) Reactions (in solution)	A_e+*intra*-$D_N{}^{\ddagger*}A_N+D_e$	association of electrophile, followed by intramolecular isomerization, followed by dissociation of electrophile
	$A_e+D_N{}^{\ddagger*}A_S+D_e$	association of electrophile, followed by capture of solvent, followed by dissociation of electrophile
	A_e+*cyclo*-$D_N{}^{\ddagger*}A_N+D_e$	association of electrophile, followed by capture of a nucleophile attached to the electrophile, followed by dissociation of electrophile

Literature Cited

1.   Werner, A.; Bräunlich, F. *Anorg. Chem.* **1899**, *22*, 127 and 154.
2.   Gutterman, D.F.; Gray, H.B. *J. Am. Chem. Soc.* **1969**, *91*, 3105; **1971**, *93*, 3364.
3.   Stotz, I.; Wilmarth, W.K.; Haim, A. *Inorg. Chem.* **1968**, *7*, 1250.
4.   Norbury, A.H.; Shaw, P.E.;Sinha, A.I.P. *J. Chem. Soc.* **1970**, 1080.
5.   Epps, L.A.; Marzilli, L.G. *Chem. Commun.* **1972**, 109.

6. Dodd, D.; Johnson, M.D. *J. Chem. Soc. Dalton Trans.* **1973**, 1218.
7. Burmeister, J.L. *Coord. Chem. Rev.* **1990**, *105*, 77.
8. Werner, A. *Ann. Chem.* **1912**, *386*, 22, 41, and 192.
9. Werner, A. *Ber.* **1901**, *34*, 1733; **1907**, *40*, 774.
10. Werner, A. *Ann. Chem.* **1912**, *386*, 39.
11. Pfeiffer, P. *Ber.* **1904**, *37*, 4257.
12. Buckingham, D.A.; Creaser, I.I.; Sargeson, A.M. *Inorg. Chem.* **1970**, *9*, 655.
13. Buckingham, D.A. *J. Chromatogr.* **1984**, *313*, 93.
14. Buckingham, D.A.; Clark, C.R.; Liddell, G.F. *Inorg. Chem.* **1992**, *31*, 2909.
15. Brasch, N.E.; Buckingham, D.A.; Clark, C.R.; Finnie, K.S. *Inorg. Chem.* **1989**, *28*, 4567.
16. Buckingham, D.A.; Clark, C.R.; Gaudin, M.J. *Inorg. Chem.* **1988**, *27*, 293.
17. Buckingham, D.A.; Clark, C.R.; Webley, W.S. *Inorg. Chem.* **1991**, *30*, 466.
18. Buckingham, D.A.; Clark, C.R.; Liddell, G.F.; Simpson, J. *Aust. J. Chem.* **1993**, *46*, 503.
19. Snow, M.R.; Boomsma, R.F. *Acta Cryst.* **1972**, *B28*, 1908.
20. Snow, M.R.; Thomas, R.J. *Aust. J. Chem.* **1974**, *27*, 1391.
21. Langford, C.H.; Gray, H.B. *Ligand Substitution Processes*; Benjamin : New York, 1965; *Chem. & Eng. News* **April 1, 1968**, *46*, 68.
22. Commission on Physical Organic Chemistry, IUPAC, *Pure and Appl. Chem.* **1989**, *61*, 23; **1989**, *61*, 57.
23. Guthrie, R.D.; Jencks, W.P. *Accts. Chem. Res.* **1989**, *22*, 343.

RECEIVED December 27, 1993

Chapter 22

Optical Activity in Coordination Chemistry

Bodie Douglas

Department of Chemistry, University of Pittsburgh,
Pittsburgh, PA 15260–3995

The demonstration of the optical activity of octahedral complexes was important in confirming Alfred Werner's intuitive ideas about coordination chemistry. Early work involved the resolution of complexes characterized by optical rotations. Modern instruments for optical rotatory dispersion were developed first, but circular dichroism (CD) spectra proved to be more useful. CD has been a powerful tool for detailed studies of the stereochemistry of octahedral complexes. Contributions to rotational strength of chelate ring conformational, configurational, and vicinal contributions are additive. Chiral metal complexes are now used in enantioselective synthesis of chiral pharmaceuticals.

Alfred Werner's ideas *(1)* were firmly based on his vision of the stereochemistry of metal coordination compounds. He was very successful in substantiating his ideas by the isolation of the number of isomers expected for octahedral complexes. This approach did not eliminate all other possibilities.

Geometrical and Optical Isomers

Two isomers of $[CoXY(en)_2]^{n+}$ complexes (where en = $H_2NC_2H_4NH_2$) were well known. Werner believed them to be *cis* and *trans* isomers, and he thought that there should be corresponding isomers *(2)* of $[CoXY(NH_3)_4]^{n+}$. Sophus Mads Jørgensen had explanations for the two isomers of the chloride of ethylenediamine complexes as shown below. Jørgensen expected only one form of $[CoCl_2(NH_3)_4]^{2+}$, and only the green (*trans*) isomer was known. In 1907 *(2)* Werner isolated the violet isomer *cis* - $[CoCl_2(NH_3)_4]^+$. Jørgensen abandoned his criticism of Werner's proposals.

In 1897 Werner recognized that a compound such as $[Co(C_2O_4)(en)_2]^+$ should be resolvable into optical isomers, and he discussed this in a publication in 1899 *(3)*. He made many attempts to resolve cobalt complexes. His American student Victor L. King carried out some 2000 fractional crystallizations without resolution. Finally, the resolution of *cis* -$[CoCl(en)_2(NH_3)]^{2+}$ and *cis* -$[CoBr(en)_2(NH_3)]^{2+}$ was successful in

0097–6156/94/0565–0275$08.00/0

Praseo
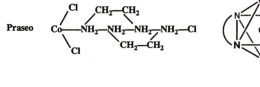

Violeo
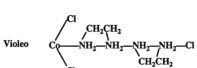

Jørgensen

Werner

1911 using bromocamphorsulfonic acid as the resolving agent without many recrystallizations *(4)*. Resolution of $[Co(en)_3]^{3+}$ was achieved in 1912 *(5)*. Werner must have been confident that his visions were proven valid. However, there were still skeptics. The optical activity of organic compounds was well established, and some felt that the optical activity could be caused by the organic ligands. It was not suggested as to how ethylenediamine, a symmetrical ligand, could give optical activity. Of course, nothing was known of the chiral conformations of ethylenediamine chelate rings.

Jørgensen prepared *(6)* a compound called anhydrobasic tetraammine diaquodiammine cobalt chloride, which he believed to be a dimer involving an oxo bridge. Werner found that OH^- could not be bonded to Co in the usual way since it did not react with dilute mineral acid to give an aqua complex. Decomposition of the complex by reaction with mineral acid gave three moles of *cis*-$[Co(H_2O)_2(NH_3)_4]X_3$, 1 mole of CoX_2, and $\frac{1}{2}$ mole of X_2. He concluded *(5)* that the complex cation was tetranuclear:

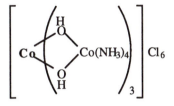

The wide acceptance of his ideas was shown by his receipt of the Nobel Prize in chemistry in 1913, the first Swiss chemist to be so honored. The optical resolution of $[Co\{(OH)_2Co(NH_3)_4\}_3]^{6+}$ in 1914 must have satisfied all skeptics. This was the first complex ion without carbon to be resolved *(7)*. Examples of optically active complex ions without carbon are still rare. Werner reported the optical rotation from 478 to 675 mμ as the first qualitative (samples racemized) optical rotatory dispersion (ORD) curve for a well characterized coordination compound.

John C. Bailar, Jr., the father of American coordination chemistry, who has been eulogized at this symposium, was trained as an organic chemist. He became interested in the possibilities for isomerism in inorganic compounds, leading to his distinguished career in coordination chemistry. Bailar *(8,9)* observed that opposite configurations of $[Co(CO_3)(en)_2]^+$ were formed from optically active *cis*-$[CoCl_2(en)_2]^+$ under different experimental conditions. He suggested that a Walden-type inversion can occur in coordination compounds as well as in organic compounds. He characterized compounds by optical rotations and related the configurations of reacting complexes and products using optical rotations or ORD curves. Frank Dwyer said that he had doubts about the reality of a Walden-type inversion. He followed the kinetics of substitution reactions and was pleased to find *(10)* that inversion does occur through a trans displacement process involving Ag^+ and OH^-.

Bailar realized that racemization reactions could occur without the dissociation of ligands. He proposed a mechanism involving the twisting of one triangular octahedral face of Δ-$[M(AA)_3]$ by $120°$ about the C_3 axis to form Λ-$[M(AA)_3]$. This is known as the Bailar or trigonal twist. A rhomboid twist is known as the Ray–Dutt twist. In some cases racemization results from ligand dissociation.

Optical Rotatory Dispersion and Circular Dichroism

Werner used optical rotations to characterize complexes and to check for complete resolution of octahedral complexes. Sometimes more than one wavelength was needed because the rotation might be zero at a particular wavelength. The ORD curve for $[Co\{(OH)_2Co(NH_3)_4\}_3]^{6+}$ reported (7) in 1914, shows a range of molecular rotations from $+47,000°$ (547.5 mμ) to $-22,000°$ (643.5 mμ) with $0°$ at 617.5 and 495 mμ. ORD curves were used in later work on complexes to relate configurations and to establish whether the configuration was retained or inverted during substitution reactions.

Aimé Cotton (11,12) was the first to study ORD in detail within the region of absorption. He observed that in the region of an absorption band right and left circularly polarized light is absorbed to different extents. In this region the optical rotation changes abruptly giving a characteristic shaped curve, with a peak and a trough of opposite sign. This is known as an "anomalous" optical rotatory dispersion curve, the ORD curve. The behavior is not anomalous, but distinctly different from the "normal" dispersion curve observed outside an absorption region. These normal curves, described by the Drude equation, are really tails of the anomalous dispersion curves. The different interactions of right and left circularly polarized light is known as the Cotton effect. The difference in absorption of the components ($\Delta \varepsilon = \varepsilon_l - \varepsilon_r$) is circular dichroism (CD), and the difference in indices of refraction ($n_l - n_r$) causes the rotation of the plane of linearly polarized light.

Cotton constructed an apparatus to measure optical rotations and CD over a range of wavelengths. He established Cotton effects for potassium chromium tartrate and potassium copper tartrate solutions. These were the first ORD and CD studies of coordination compounds, but the complexes were not isolated. For a long period there was little activity in ORD or CD studies of complexes. Kuhn (13) was interested in metal complexes in connection with his ideas about the theory of optical activity (14). J. P. Mathieu (15,16) presented combined studies of ORD, CD, and absorption spectra of many complexes of Co, Cr, Pt, Rh, and Ir during the 1930s. After his work there was no activity in CD studies of complexes for many years.

The development of good ORD instruments encouraged numerous studies. We began ORD studies of cobalt complexes in the late 1950s, but we had difficulties in resolving the spectral components for individual transitions. The complex, overlapping curves complicated the process. The rotation observed for a sugar or for tartaric acid in the visible region is a composite from tails (normal dispersions) of allowed transitions in the ultraviolet region. A CD peak drops to zero not far from the maximum so there is overlap only of peaks close in energy. The CD spectra give much clearer information than ORD curves, greatly simplifying the resolution of components. It was obvious that we should study CD instead of ORD. No CD instruments were available, and the construction of Cotton's apparatus was not simple. Fortunately, Grosjean and Legrand (17) developed the dichrographe in 1960, and it was soon available commercially. In a few years CD studies largely displaced ORD studies.

Chiral Contributions of Ligands

Transition metal complexes of optically active ligands display ORD and CD curves in the region of the metal $d \rightarrow d$ absorption region. Substitution of Co^{2+} for Zn^{2+} in carboxypeptidase provides a "spectroscopic probe". CD studies of the Co^{2+} $d \rightarrow d$ transitions provide useful information about the environment at the metal ion site.

Shimura (18) reported two Cotton effects in the region of the lower energy $d \rightarrow d$ absorption band for $[Co(L\text{-leuc})(NH_3)_4]^{2+}$. The contribution of L-leucine was referred to as a "vicinal effect". C. T. Liu resolved $[Co(glycine)(en)_2]^{2+}$ and $[Co(L\text{-}palan)(en)_2]^{2+}$. The glycine complex and one diastereoisomer of the L-phenylalanine complex $\{(-)_{546}\text{-}[Co(L\text{-}palan)(en)_2]^{2+}\}$ showed very similar CD curves throughout the visible region. Both displayed one CD peak in the lower energy absorption band. The other diastereoisomer $\{(+)_{546}\text{-}[Co(L\text{-}palan)(en)_2]^{2+}\}$ showed two CD peaks of opposite signs (Figure 1). Different intensities and shapes were expected for the two diastereoisomers but not different numbers of CD peaks. The resolutions were repeated with the same results. The CD spectrum of unresolved $[Co(L\text{-}palan)(en)_2]^{2+}$ showed three CD peaks $(-,+,-)$ in the lower energy absorption band region. The CD spectrum of the complex ion $[CoL\text{-}palan(NH_3)_4]^{2+}$ was very similar to that of unresolved $[CoL\text{-}palan(en)_2]^{2+}$. This, taken as the "vicinal effect" of L-palan, could be subtracted from the CD curves for the two diastereoisomers of the L-phenylalanine complex to give essentially mirror-image CD curves, corresponding to those of the enantiomers of $[Co(gly)(en)_2]^{2+}$. The contributions for CD curves are additive for the "vicinal effect" of the optically active ligand and the configurational contribution (Δ or Λ) for the complex *(19)*.

A complex such as 1,2-propanediaminetetraacetic acid (pdta) is stereospecific. The Co^{III} complex of S-pdta is $\Lambda\text{-}(-)_{546}\text{-}[Co(S\text{-}pdta)]^-$. Most optically active ligands yield a preferred diastereoisomer but not exclusively. Dwyer and Sargeson *(20)* isolated all of the mixed en-(–)pn complexes with Co^{III} [(–)-pn is (–)-1,2-propanediamine]. The contributions to the optical activity from the vicinal effect of pn and from the overall configurational (Δ or Λ) effect of the complex was found to be additive *(21)*.

A thorough analysis of the conformations of chelate rings was given in a classic paper by Corey and Bailar *(22)*. The energy differences for various combinations of ligand conformations predicted from their treatment have been consistent with experimental results.

Spectroscopy

Absorption bands in the ligand field ($d \rightarrow d$) region are very broad because of vibronic broadening. For octahedral ($\mathbf{O}_h$) complexes of Co^{3+} (d^6) and Cr^{3+} (d^3) there are two allowed transitions to triply degenerate (T_{1g} or T_{2g}) excited states. Lowering the symmetry gives splitting of the energy levels. For effective $\mathbf{D}_3$ or $\mathbf{D}_{4h}$ symmetry there are two components for each of these transitions. For still lower symmetry, such as $\mathbf{C}_{2v}$, $\mathbf{C}_2$, or $\mathbf{C}_1$, there are three components for each transition; the degeneracy is completely removed. The broad absorption bands obscure splittings in most cases. Usually the effect observed for lower symmetry is the appearance of a shoulder on the first (lower energy) absorption band with no effect for the second (higher energy) band. In extreme cases such as *trans*-$[CoF_2(en)_2]^+$ two absorption peaks appear in the first band region.

Interpretation of ORD and CD Spectra. The curve shapes for ORD and the significant contributions of transitions differing greatly in energy make the separation of components difficult. CD peaks are not broadened like absorption bands, and they overlap only with adjacent peaks. The peaks can be positive or negative, and this can help in the spectral resolution of components.

For $[Co(en)_3]^{3+}$ the symmetry is D_3, but the absorption spectrum shows no more splitting than for $[Co(NH_3)_6]^{3+}$ (O_h). The absorption spectrum of $[Co(en)_3]^{3+}$ is consistent with effective O_h symmetry. In fact, in the absence of any splitting, the spectrum provides no basis for using lower symmetry. Mathieu (16) displayed absorption, ORD, and CD in the same figure for $[Co(en)_3]^{3+}$. In the region of the lower energy absorption band there is one well-defined ORD curve and a slight irregularity that can be noted if one knows where to expect a second ORD component. The CD spectrum displays two well-resolved peaks with opposite signs in the first band region. One CD peak appears in the second band region, as expected, since one of the two transitions expected is forbidden. CD spectra generally reveal more splitting than is observed in the absorption or ORD spectra.

Three parameters, $\overline{\nu}_{max}$ (energy), $\Delta\varepsilon_{max}$ (intensity), and half-width, describe a CD peak. The corresponding ORD curve can be calculated from these parameters. Unless the splittings are great enough to produce components that are well enough defined, curve analysis is uncertain. For three components there are nine variable parameters. The remark has been made that with five variables one can draw an elephant.

There are few examples of Co^{3+} complexes that reveal three components in the first absorption band region for complexes with symmetry lower than D_3 or D_{4h}. The splitting of the second band is usually smaller, and these peaks commonly have low intensities.

Splitting Patterns for CD Spectra. The $[Co(L\text{-}aa)(en)_2]^{2+}$ complexes containing optically active amino acids (L–aa) display one CD peak for one diastereoisomer or two CD peaks for the other diastereoisomer in the first band region (21). $[Co(L\text{-}aa)(NH_3)_4]^{2+}$ has three CD peaks with alternating signs (–,+,–) and this pattern is seen for the unresolved $[Co(L\text{-}aa)(en)_2]^{2+}$ (see Figure 1). This should indicate that the degeneracy of $[A_{1g} \rightarrow T_{1g} (O_h)]$ is removed completely by the C_1 symmetry. The dominance of some peaks can cause two or one peak to appear.

For $[Co(edta)]^-$ there are two CD peaks in the first absorption band region, but three weak CD peaks in the second absorption band region (Figure 2). The complex $[Co(malonate)_2(en)]^-$ was prepared (23) as a ligand field model of $[Co(edta)]^-$. They both have cis -$Co(N)_2(O)_4$ arrangements with C_2 symmetry. There are three prominent CD peaks of alternating signs in the first and second absorption band regions for $[Co(malonate)_2(en)]^-$. Probably this actual splitting pattern is obscured by one of the components being canceled or covered for $[Co(edta)]^-$.

Further evidence of the presence of three components in the first absorption band region is revealed by an extensive series of Co^{3+} complexes of edta–type ligands (24). The usual pattern is two CD peaks of opposite sign in the first absorption band region. Complexes of edta–type ligands with four acetate arms have one CD peak at lower energy than that of the absorption maximum, and the other CD peak has higher energy than that of ε_{max}. For complexes of ligands with acetate and propionate arms both CD peaks occur on the lower energy side of the absorption maximum, or for two of this group the higher energy CD peak is at about the same energy as that of ε_{max}:

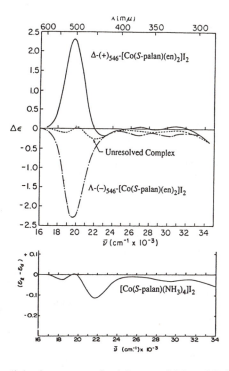

Figure 1 Circular dichroism spectra for $(+)_{546}$- and $(-)_{546}$-[CoS-palan(en)$_2$]I$_2$ and [CoS-palan(NH$_3$)$_4$]I$_2$.

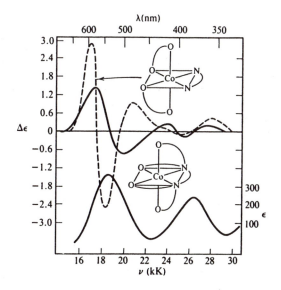

Figure 2. Absorption and circular dichroism spectra for $(-)_{546}$K[Co(edta)] · 2H$_2$O and the circular dichroism spectrum for $(-)_{546}$K[Co(mal)$_2$(en)] · 2H$_2$O. (Reproduced with permission from reference 34. Copyright 1994 Wiley.)

Four acetate arms		Acetate and propionate arms	
First abs. max.	 |	 |	 |
			or
CD peaks	 |		 |		 |	
	Energy →	Energy →	Energy →

These patterns for the two groups should be different combinations of three components. For *trans* (O_6)-[Co(1,3-pddadp)]⁻ *(25)* and [Co(*S,S*)-edds]⁻ *(26)* (1,3-pddadp is 1,3-propanediamine-*N,N* '-diacetate-*N,N* '-di-3-propionate ion; (*S,S*)-edds is *S,S*-ethylenediamine-*N,N*'-disuccinate ion) there are three well resolved CD peaks in the region of the first absorption band.

Chiral Metal Complexes in Enantioselective Synthesis *(27,28)*

Substitution reactions of resolved metal complexes can occur with retention or with inversion. Commonly, some activity is lost because of racemization. Since the late 1960s there have been developments in metal complexes containing chiral ligands which react catalytically in highly enantioselective organic reactions. Small amounts of the chiral complexes can produce much larger amounts of products of high enantiomeric purity. This is referred to as chemical multiplication of chirality.

Diethylzinc in the presence of (−)-3-*exo* -(dimethylamino)isoborneol

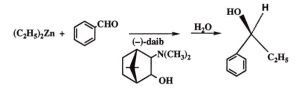

[(−)-daib] reacts with benzaldehyde in toluene to produce (*S*)-1-phenyl-1-propanol in 97% yield with 98% enatiomeric purity *(29,30)*. The reaction can occur with only partially resolved daib (15% enrichment of the (−) isomer) to produce the product with 98% enantiomeric purity *(31)*. This process involving a partially resolved catalyst, called amplification of chirality, results from much greater turnover efficiency of the chiral catalytic system as compared with the *meso* isomer formed from the racemate.

Wilkinson's catalyst, [RhCl{PC$_6$H$_5$)$_3$}$_3$], modified by substitution of chiral phosphines for triphenylphosphine, can bring about enantioselective hydrogenation. The hydrogenation of the olefin shown below is reduced in the presence of [Rh(dipamp)]⁺ in the production of *L*-DOPA. *L*-DOPA is used in the treatment of

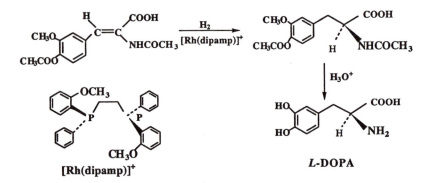

Parkinson's disease. A similar hydrogenation reaction with the catalyst $[Rh(pnnp)]^+$ produces L-phenylalanine used in the production of aspartame $[(S,S)-$ aspartylphenylalanine methyl ester], the artificial sweetener.

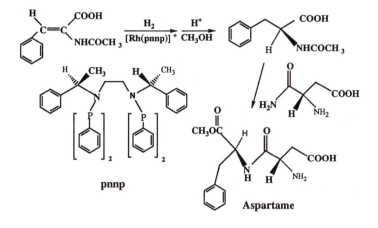

pnnp

Aspartame

There are many other enantioselective reactions catalyzed by chiral metal complexes. These include complexes of various chiral phosphines, tartrate ion, and others as ligands. Chiral cyclopentadienyl metal complexes *(32)* also are used. Chiral ligands are not limited to those containing an asymmetric carbon. $[Ru(binap)]^{2+}$ is a catalyst for enantioselective hydrogenation. In some cases such as $[Rh(dipamp)]^+$ and $[Rh(pnnp)]^+$ well defined complexes are used. In many cases the reaction is carried out in the presence of a chiral ligand and a metal compound. The reaction involving $Zn(C_2H_5)_2$ above is one. Other examples include those catalyzed by Ti^{IV} in the presence of tartrate ion and OsO_4 in the presence of chiral ligands. All of the Group VIII (Groups 8, 9, and 10) metals, Cu, Zn, and other transition metals have been used for enantioselective syntheses. Enzymes are usually specific for a particular reaction and a particular substrate. Some of the chiral metal catalysts, such as $[Ru(binap)]^{2+}$ as a hydrogenation catalyst, are quite general.

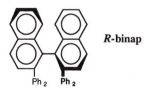

R-binap

Ph$_2$ Ph$_2$

Synthesis of Chiral Pharmaceuticals. Biologically active compounds are usually chiral. Synthetic drugs produced in the past were usually racemic products. Commonly, only one enantiomer is active. Production of the active enantiomer permits the use of lower dosage with possible reduction of side effects. Applications of chiral metal complexes in enantioselective syntheses are increasing rapidly in the pharmaceutical industry (33).

Literature Cited

1. Werner, A. *Z. anorg. Chem.* **1893**, *3*, 267. For a discussion and annotated English translation see Kauffman, G. B. *Classics in Coordination Chemistry, Part 1: The Selected Papers of Alfred Werner*, Dover: New York, NY, 1968; pp 5-8.

2. *Werner, A. Ber.* **1907**, *40*, 4817. For a discussion and annotated English translation see Kauffman, G. B. *op. cit.;* pp 141-154.

3. Werner, A. *Z. anorg. Chem.* **1899**, *21*, 145.

4. Werner, A. *Ber.* **1911**, *44*, 1887. For a discussion and annotated English translation see Kauffman, G. B. *op. cit.*; pp 155-173.

5. Werner, A.; Berl, E.; Zinggeler, E.; Jantsch, G. *Ber.* **1907**, *40*, 2103.

6. Jørgensen, S. M. *Z. anorg. Chem.* **1898**, *16*, 184.

7. Werner, A. *Ber.* **1914**, *47*, 3087. For a discussion and annotated English translation see Kauffman, G. B. *op. cit.*; pp 175-184.

8. Bailar, Jr, J. C.; Auten, R. W. *J. Am. Chem. Soc.* **1934**, *57*, 774.

9. Bailar, Jr, J. C.; McReynolds, J. P. *J. Am. Chem. Soc.* **1939**, *61*, 3199.

10. Dwyer, F.; Sargeson, A. M.; Reid, I. K. *J. Am. Chem. Soc.* **1963**, *85*, 1215.

11. Cotton, A. *Compt. Rend.* **1895**, *120*, 989, 1044.

12. Cotton, A. *Ann. Chim. Phys.* **1896**, *8*, 347.

13. Kuhn W.; Bein, K. *Z. Physik. Chem.* (B) **1934**, *24*, 335.

14. Kuhn W.; Bein, K. *Z. Anorg. Allgem. Chem.* **1934**, *216*, 321.

15. Mathieu, J. P. *Bull. Soc. Chim. France* **1936**, (5) *3*, 476; **1937**, *4*, 687; **1939**, *6*, 873.

16. Mathieu, J. P. *J. Chim. Phys.* **1936**, *33*, 78.

17. Grosjean M.; Legrand, M. *Compt. Rend.* **1960**, *251*, 2150.

18. Shimura, Y. *Bull. Chem. Soc. Japan* **1958**, *31*, 315.

19. Liu C. T.; Douglas, B. E. *Inorg. Chem.* **1964**, *3*, 1356.

20. Dwyer F. P.; Sargeson, A. M. *J. Am. Chem. Soc.* **1959**, *81*, 5272; Sargeson, A. M. In *Chelating Agents and Metal Chelates;* Dwyer, F. P.; Mellor, D. P., Eds.; Academic Press: New York, NY 1964; p 200.

21. Douglas, B. E. *Inorg. Chem.* **1965**, *4*, 1561.

22. Corey E. J.; Bailar, Jr., J. C. *J. Am. Chem. Soc.* **1959**, *81*, 2620.

23. Douglas, B. E.; Haines, R. A.; Brushmiller, J. G. *Inorg. Chem.* **1963**, *2*, 1194.

24. Douglas, B. E.; Radanovic, D. J. *Coord. Chem. Rev.* **1993**,*128* , 139.

25. Radanovic, D. J.;Trifunovic, S. R.; Cvijovic, M. S.; Maricondi, C.; Douglas, B. E. *Inorg. Chim. Acta* **1992**, *196*, 161.

26. Jordan, W. T.; Legg, J. I. *Inorg. Chem.* **1974**, *13*, 2271.

27. Noyori, R. *Science* **1990**, *248*, 1194; **1992**, *258*, 584.

28. Scott, J. W., *Topics in Stereochem.* **1989**, *19*, 209.

29. Noyori, R.; Kitamura, M. *Angew. Chem. Int. Ed. Engl.* **1991**, *30*, 49.

30. Kitamura, M.; Suga, S.; Kawai, K.; Noyori, R. *J. Am. Chem. Soc.* **1986**, *108*, 6071.

31. Kitamura, M.; Okada, S.; Suga, S.; Noyori, R. *J. Am. Chem. Soc.* **1989**, *111*, 4028.

32. Halterman, R. L. *Chem. Rev.* **1992**, *92*, 965f.

33. Stinson, S. C., "Chiral Drugs", *Chem. Engr. News* **1992**, September 28, 46.

34. Douglas, B.; McDaniel, D. H.; Alexander, J. J. *Concepts and Models of Inorganic Chemistry*; 3rd ed.; Wiley: New York, NY, 1994; p 467.

RECEIVED December 14, 1993

Chapter 23

Inorganic Optical Activity

Department of Chemistry, University of Wales, Cardiff, CF1 3TB Wales,
United Kingdom

This account plaits three major inorganic threads: stereochemistry and optical
activity; qualitative analysis; and aqueous coordination chemistry.
Knowledge of the phenomena of optical activity has grown largely through
inorganic systems, from Biot (quartz), through mineralogy (Haidinger –
circular dichroism) and crystal properties (*e.g.*, the spontaneous resolution of
sodium chlorate), to the Cotton effect and the resolutions of Werner's cationic
hexol and Mann's anion. The stereochemistries of oligosulfides of the
platinum metals are described. $[PtS_x]^{2-}$ (*e.g.*, x = 15) can be resolved through
diastereo- isomeric salts with $(+)-[Ru(bipy)_3]^{2+}$. When preparations of
oligosulfides of platinum are controlled carefully, the sole product is
optically active $(NH_4)_2[PtS_{17}]\cdot 2H_2O$. This is a unique spontaneous
appearance of optical activity in an inorganic molecule.

This paper traces the development of optical activity in inorganic compounds to the
point where Alfred Werner was able to use optical resolution to such striking effect in
proving the octahedral stereochemistry of several metal ions by resolving chelated
compounds, in particular the fully inorganic cation now called "Werner's hexol". This
achievement came to fruition *(1)* in the years just prior to World War I.

The present account brings together the themes which led to that first resolution of
a purely inorganic compound, defined as containing no carbon of any kind. These
themes are: first, optical activity and its relation to stereochemistry; second, an aspect
of qualitative analysis related to the complexation of metal ions, and, finally, chelation
in aqueous coordination chemistry. The paper concludes by tracing the development
of inorganic optically active compounds since Werner's time.

Optical Activity

The notion of handedness in physical science arose from John Frederick William
Herschel's descriptions *(2)* of quartz. (Parenthetically, it is worth noting here his
discovery of the complexation of silver ions by thiosulfate and his comment that the
resulting solutions tasted sweet. The eutropic $[Au(S_2O_3)_2]^{3-}$ is also bioactive in
"myochrysin", the chrysotherapeutic agent for rheumatoid arthritis.) The now well
developed relationship between shape and chiroptical spectroscopy has roots in the

0097–6156/94/0565–0286$08.00/0
© 1994 American Chemical Society

almost contemporaneous findings of handedness in minerals and Jean-Baptiste Biot's discovery *(3)* of the rotation of linearly polarized light by liquids.

During the hundred-odd years between Biot's first measurements of optical rotation and Werner's resolutions of chiral coordination compounds, this relationship between chiral shapes and optical properties waxed and waned. Chemical systems containing colored transition metal ions lent themselves rather readily to studies of this kind made by eye because magnetic-dipole transitions are common: the optical densities are low and dissymmetries often high.

Many of the major developments in optical activity have come from inorganic systems. For example, the great mineralogist Wilhelm Karl Haidinger observed *(4)* circular dichroism in amethystine quartz. This was contemporaneous (in the 1840s) with the intensive studies of the complexation of copper(II) ions by natural acids. Charles-Louis Barreswils and Hermann Fehling focused on (+)-tartaric acid: Jean-Baptiste Joseph Dieudonné Boussingault, the father of food chemistry, on "Leimzucker" (glycine). These cupric aminoacidates have had a leading role in the development of chemistry (chelation—the "Inner Komplexe" of Heinrich Ley and Giuseppe Bruni; the determination of relative optical configuration by Paul Pfeiffer; the "biuret" reaction; isolation of protein hydrolysates; copper bands for rheumatism). It is remarkable that Boussingault's simple compound *(5)* of 1841, $[Cu(NO_3)(H_2O)(glycinate)]$, was not made again *(6)* for 150 years.

In one sense, it was likewise odd that the chiroptical properties of Fehling's solution were not studied for 50 years, until the time of Aimé Cotton, Professor of Physics at the Sorbonne. By then, in the 1890s, measurements of chiroptical properties as a function of wavelength had become rare, the interest of a few physicists. (It is worth remembering that, though the phenomena of the Cotton effect *(7)* of polarized light in the region of an absorption band were known, i.e., the anomalous rotatory dispersion, the unequal absorption of left- and of right-circularly polarized light, and the ellipticity of the emergent initially linearly polarized light, it was not until 1915 that Bruhat defined circular dichroism using a differential extinction coefficient, $\Delta\varepsilon$, ε_L-ε_R). This change may have been due to the invention of the Bunsen burner, which gave a ready source of relatively monochromatic light, in the 1860s. Coupled with the colorless nature of the many natural products beginning to be studied in the developing field of organic chemistry, this gave the extraordinary dominance of $[\alpha]_D$ as the measured chiroptical property. The relationship between the physics and the chemistry of shape, so strong in the French tradition, became weak. In a related way, although Kelvin coined the word "chirality" in the 1880s, it was not widely taken up by stereochemists generally for the best part of a century. In the context of inorganic optical activity, this dominance of $[\alpha]_D$ was ignored by a few, including Werner himself, and, a little later, F.M. Jaeger *(8)* and (independently) I. Lifschitz, both of Groningen, who routinely measured Cotton effects by optical rotatory dispersion (ORD).

The present-day relationship between molecular shape and effects of polarized light was more fully anticipated in the work, on the one hand, of Stotherd Mitchell *(9)* in Glasgow on production of optical activity by photo-reactions of racemic solutions (including Fehling's) using circularly polarized light and, on the other, by direct measurement in the 1930s of the circular dichroism of Werner complexes by Mathieu *(10)*. Karl Freudenberg's great work *(11)* of 1933 was perhaps the last to bring together all aspects of stereochemistry.

One other aspect of inorganic morphology concerns the handedness of crystals. Indeed, Werner employed this in work on the chirality of crystals of $K_3[Rh(C_2O_4)_3]$·

$4\frac{1}{2}H_2O$ and of $[M(C_2O_4)(en)_2]Br$ (M = Cr or Co). The organic chemist F.R. Japp (of the Japp-Klingemann reaction) first systematically drew together, in his Presidential Address *(12)* to the Chemistry Section of the British Association for the Advancement of Science, the possible "Origins of Asymmetry". Work on the statistics of the crystallization of enantiomorphous substances was quite common. Eakle had noted *(13)* in 1896 that aqueous sodium periodate crystallized from sodium nitrate solutions with left-handed crystals in large excess, and there were similar findings by Gregoire Wyrouboff *(14)* on silicotungstates. A popular subject for such experiments was sodium chloride, $NaClO_3$. In early experiments, of 938 sealed tubes, 433 gave only right-handed crystals, 411 only left-handed, and 94 gave both. The general experience *(15)* is that, where only a few larger crystals form, there is usually a dominance of one hand, but where many smaller crystals form, the distribution is more even.

Qualitative Analysis

As part of his magisterial survey of mineral compositions Berzelius devised many analytical procedures which have stood the test of time. Yellow ammonium sulfide was used to part the various metal sulfides precipitated from acid aqueous solutions by sulfane. Some dissolved. This property was known to many if not all nineteenth-century chemists. The German K.A. Hofmann discovered several fascinating coordination compounds, including platinum blue, "Hofmann's compound", $K_3[Fe(CN)_5(OH_2)]$, and the famous clathrate compound of benzene, $Ni(CN)_2NH_3 \cdot C_6H_6$. He undertook a systematic study *(16)* of the complex compounds which could be isolated from solutions of metal sulfides in yellow ammonium sulfide. These included simple salts of anions like $[CuS_4]^{3-}$, $[AuS_3]^-$, $[PdS_{11}]^{2-}$, $[PtS_{10}]^{2-}$, and $[PtS_{15}]^{2-}$, and $[IrS_{15}]^{3-}$. The rhodium eutrope of the last has been made only recently *(17)*; the structure of its ammonium salt is known *(18)*. Despite this and the great current interest in these oligosulfides, Hofmann's iridium compound has not been studied since his day.

Werner's formulations of these thioanions are interesting; he mentioned *(19)* the salt, made *(20)* in 1879 by Sophus Mads Jorgensen, $[CrCl(NH_3)_5](S_5)$, in which the doubly charged pentasulfide ion is simply a counter-ion. However, Werner accepted Drechsel's then plausible suggestion *(21)* that the pentasulfide ion (actually linear) is structurally related to sulfate and thiosulfate:

$$K_2 \begin{bmatrix} S & & S \\ & S & \\ S & & S \end{bmatrix} \quad ; \quad K_2 \begin{bmatrix} S & & O \\ & S & \\ O & & O \end{bmatrix} \quad ; \quad K_2 \begin{bmatrix} O & & O \\ & S & \\ O & & O \end{bmatrix}$$

Werner's formulations of Hofmann's compounds were consequently written *(19)* as:

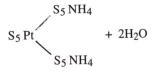

$$S_5 Pt \overset{\displaystyle S_5\,NH_4}{\underset{\displaystyle S_5\,NH_4}{\diagdown}} \quad + \; 2H_2O$$

At almost the same time the correct formulation of Werner's hexol as a trischelate led him to its optical resolution. The earliest suggestion that Hofmann's pentakaidekasulfidoplatinate ion should be formulated as the ammonium salt of the trischelated

$$(NH_4)_2[Pt(S_5)_3]\cdot 2H_2O$$

seems to be that *(22)* of Nevil Vincent Sidgwick.

Chelation

Werner's initial and monumental article *(23)*, whose centenary we celebrate here, "Beitrag zur Konstitution anorganischen Verbindungen", elucidated the nature of the metal-ammines, with ammonia as an uncharged ligand. Within a short time the stereochemical utility of linking the N-donors into the didentate 1,2-diaminoethane metal-ammines, (en), again uncharged, was utilized. Very rapidly the concept of chelation of charged ligands emerged from the work of Bruni *(24)* and of Ley *(25)* on the seemingly anomalous properties of bisglycinatocopper(II).

Werner himself later spoke *(26)* of his own work on acetylacetonates of platinum as having anticipated this concept of "Innerkomplexe", in which the charged (main valency) and uncharged (subsidiary valency) ligands are part of one and the same molecule. Certainly the stereochemical consequences of chelation at a metal ion, whatever the nature of the ligand, whether uncharged en, singly-charged α-amino-acidate, or doubly-charged oxalate, were widely appreciated by 1911.

This appreciation led to a rapid increase in the range of resolved chelated ions, like $[Fe(bipyridyl)_3]^{2+}$ or $[Co(C_2O_4)_3]^{3-}$. However, the "hexol" remained the only resolved inorganic molecule until Frederick G. Mann *(27)* in 1933 achieved the resolution of the anion cis-$[Rh(HNSO_2NH)_2(OH_2)_2]^-$ using the cation of an optically active amine (α-phenylethylamine) as the resolving agent.

Optically Active Platinum Polysulfides

Sidgwick's suggestion *(22)* that Hofmann's sulfur-rich salt contained the $[Pt(S_5)_3]^{2-}$ ion was borne out by a crystallographic study, using X-ray diffraction *(28)*.

The simplest compositions are $[Pt(S_5)_2]^{2-}$ and $[Pt(S_5)_3]^{2-}$. Several salts of both ions are known, prepared in aqueous solution as in equations 1 and 2, following the route *(16)* of Hofmann and Höchtlen:

$$[PtCl_4]^{2-} + 2(NH_4)_2S_5 \rightarrow [Pt(S_5)_2]^{2-} + 4NH_4Cl \qquad (1)$$

$$[PtCl_6]^{2-} + 3(NH_4)_2S_5 \rightarrow [Pt(S_5)_3]^{2-} + 6NH_4Cl \qquad (2)$$

There are several complications endemic in this system. First, as was common experience from qualitative analysis with arsenic or tin, reagents of either oxidation state, with an excess of yellow ammonium sulfide, give the same higher oxidation state:

$$[Pt(S_5)_2]^{2-} + 2S_5^{2-} \rightarrow [Pt(S_5)_3]^{2-} + S_2^{2-} + S_3^{2-} \qquad (3)$$

Indeed, this tris pentasulfidoplatinate(IV) product of equation (3) is remarkably easy to obtain, being the chief product *(29)* of reactions designed to yield $[PtS_x]^{2-}$, where x > 15. Likewise, the ostensibly very stable compound $[PtCl_2(bipy)]$ reacts under mild conditions with aqueous yellow ammonium sulfide as in equation (4) to give *(30)* the oxidized homoleptic $[Pt(S_5)_3]^{2-}$:

$$[PtCl_2(bipy)] + 4S_5{}^{2-} \rightarrow [Pt(S_5)_3]^{2-} + bipy + 2Cl^- + S_2{}^{2-} + S_3{}^{2-} \qquad (4)$$

Further, quite apart from this redox activity, the oligosulfides ($S_x{}^{2-}$ and $HS_x{}^-$) of yellow ammonium sulfide are very labile with changing pH. The chain length varies as pH changes, presumably because the values of pK_a for the dissociations of equation 5 vary markedly depending on the value of x:

$$HS_xH \rightleftharpoons H^+ + HS_x{}^- \rightleftharpoons H^+ + S_x{}^{2-} \qquad (5)$$

Yellow ammonium sulfide at pH 9 contains more $S_6{}^{2-}$ than it does at pH 10. Partly in consequence, the complex anions with platinum also vary. Whereas $(NH_4)_2[PtS_{15}]\cdot 2H_2O$ is the product from preparations at pH 10.5, the product from ostensibly similar solutions at pH 9.5 is $(NH_4)_2[PtS_{17}]\cdot 2H_2O$.

An aged solution of this ammonium heptakaidekasulfidoplatinate(IV) dihydrate undergoes disproportionation as in equation (6):

$$3[PtS_{17}]^{2-} \rightarrow 2[PtS_{18}]^{2-} + [PtS_{15}]^{2-} \qquad (6)$$

The new complex anion $[PtS_{18}]^{2-}$ gave *(31)* a crystalline salt with the tetraphenyl-phosphonium ion. The anions were $[Pt(S_6)_3]^{2-}$, with three seven-membered chelate rings; the angles within the rings, S-Pt-S, were much larger than 90°. These systems containing platinum ions complexed by polysulfides are thus chemically rather more complicated than expected.

There are also physical complications arising from the details of crystal packing. The salt $(NH_4)_2[Pt(S_5)_3]\cdot 2H_2O$ is dimorphic *(32)*, there being a red-orange form and a quite distinct brown-maroon form. The unit-cell of the second form *(33)* is about twice as large as that of the red-orange crystals. In both, the complex anion has three-fold symmetry, each chelate ring PtS_5 having a chair-conformation. The same structural features are found *(18)* in the crystal of the new compound containing rhodium(III), $(NH_4)_3[Rh(S_5)_3]\cdot 3H_2O$. As in so many other salts the replacement of ammonium by potassium gives equivalent dimorphism in $K_2[Pt(S_5)_3]\cdot 2H_2O$. The dimorphic potassium salts are isomorphous with the ammonium pair.

In the several solid compounds studied by X-ray diffraction only one conforma-tion of the trischelated anion $[M(S_5)_3]^{n-}$ (M = Pt, n = 2; M = Rh, n = 3) has been seen; this has three chair-form rings MS_5, related by a three-fold axis through the metal ion. In solution, however, these six-membered rings are flexible so other conformations exist where the three-fold axis is lost. Indeed, the [195]Pt chemical shift observed for solutions of $(NH_4)_2[Pt(S_5)_3]\cdot 2H_2O$ at room temperatures broadens *(34)* on cooling and then splits into two sharp signals (with intensity ratio 30:1), indicating the existence of a second conformer.

Optical Resolutions. In the same manner as for other kinetically inert complex ions with the d^6 electronic configuration, $[Pt(S_5)_3]^{2-}$ can be resolved into its enantiomers. Pasteur's method of forming diastereomeric salts with a chiral counter-ion was applied. The resolving agent, (+)-tris-2,2'-bipyridylruthenium(II), was chosen for two reasons. It has a double charge {to balance that of $[PtS_{15}]^{2-}$}, and it can be made by a dissymmetric synthesis. Indeed, an aqueous solution of $(NH_4)_2[Pt(S_5)_3]\cdot 2H_2O$ on treatment with this resolving agent gave *(35)* one insoluble diastereomer. This separation seems ideal; one diastereomeric salt is precipitated fully and the other not at all, giving optically pure $[Pt(S_5)_3]^{2-}$ in one operation. The ostensibly similar 1,10-phenanthroline complex cation, (+) $[Rh(phen)_3]^{2+}$, also gave separation, albeit less clean, when used as a resolving agent for $[PtS_{15}]^{2-}$. The optically resolved $[PtS_{15}]^{2-}$ ion shows no racemization in the dark.

This new inorganic optically active compound has several novel features. Werner's prototype, his "hexol", contained cobalt, a first-row element, and was a cation. F.G. Mann's rhodate, while an anion, contained a second row element. (+) - $[Pt(S_5)_3]^{2-}$ contains a third-row element. Further, this thioplatinate(IV) shares with the helicene hydrocarbon series the distinction of manifesting molecular optical activity while containing atoms of only two elements. Of course, enantiomorphism is well known in crystalline binary assemblies (as in cinnabar, HgS, or quartz, SiO_2) and even in elements (like tellurium).

Unique Spontaneous Inorganic Optical Activity. The known values of x in isolated salts of $[PtS_x]^{2-}$ are 15, 17, and 18. The first and last are homoleptic $[Pt(S_5)_3]^{2-}$ and $[Pt(S_6)_3]^{2-}$, respectively. The X-ray photoelectron spectra of the heptakaidekasulfidoplatinate(IV), which can be made optically active, suggests *(36)* that it is to be formulated $[Pt(S_6)_2(S_5)]^{2-}$ rather than $[Pt(S_5)_2(S_7)]^{2-}$. Resolution has been achieved for x = 15, but not yet for x = 18. For x = 17, a remarkable finding is that the product $(NH_4)_2[PtS_{17}]\cdot 2H_2O$ is optically active. Yellow ammonium sulfide is made by dissolving elemental sulfur in strong aqueous ammonia using hydrogen sulfide. Sodium hexachloroplatinate(IV) is dissolved in this solution, and hydrochloric acid added until the pH is 9.2. The product of crystallization is optically active $(NH_4)_2[PtS_{17}]\cdot 2H_2O$.

This spontaneous occurrence of molecular optical activity in a purely inorganic compound is unique. Presumably, either enantiomer might be obtained in any single experiment. In the author's laboratory, the first few experiments indeed yielded crystals of one or the other enantiomer, apparently at random. However a single enantiomer, (+), has subsequently been dominant. The phenomenon is presumably related to the handed seeding found for sodium chlorate *(15)* and, by Werner 80 years ago, for cationic complexes of 1,2-diaminoethane (en) *(37)*. Elucidating this new inorganic optical activity more fully will no doubt extend our appreciation of Werner's extraordinary stereochemical insights.

Literature Cited

1. Werner, A. *Neuere Anschauungen auf dem Gebiete der anorganischen Chemie,* 3rd edition; F. Vieweg und Sohn: Braunschweig, 1913.
2. Herschel, J. F. W. *Trans. Cambridge Phil. Soc.* **1822**, *1*, 43.
3. Biot, J-B. *Nouv. Bull. Soc. Philom.* **1815**, *4*, 26.
4. Haidinger, W. K. *Ann. Phys.* **1847**, *70*, 531.
5. Boussingault, J. B. J. D. *Ann. Chim. Phys.* **1841**, 257.
6. Davies, H. O.; Gillard, R. D.; Hursthouse, M. B.; Mazin, M. A.; Williams, P. A., *J. Chem. Soc., Chem. Comm.* **1992**, 226.
7. Cotton, A. *Ann. Chim. Phys.* **1896**, *8*, 347.

8. Jaeger, F. M. *Optical Activity and High Temperature Measurements*; the George Fisher Baker Lectures at Cornell University; McGraw-Hill: New York, NY, 1930.
9. Mitchell, S. *The Cotton Effect;* G. Bell and Sons: London, 1933.
10. Mathieu, J-P. *Bull. Soc. Chim. France*, **1936**, *3*, 476.
11. Freudenberg, K. *Stereochemie*; Franz Deuticke: Leipzig und Wien, 1933.
12. Japp, F. R. *Report of the 68th Meeting, Brit. Ass. Adv. Sci., Bristol* **1898**; John Murray: London, 1899; p 813.
13. Eakle, A. S. *Z. Krystallogr. Mineral.* **1896**, *26*, 562.
14. Wyrouboff, G. *Zentralblatt.* **1898**, *98*, II, 90.
15. Gillard, R. D.; Jesus, J. P. *J. Chem. Soc. Dalton* **1979**, 1779.
16. Hofmann, K. A.; Höchtlen, F. *Chem. Ber.* **1903**, *36*, 3090.
17. Krause, R. A. *Inorg. Nucl. Chem. Lett.* **1971**, *7*, 973.
18. Cartwright, P.; Gillard, R. D.; Sillanpaa, E. R. J.; Valkonen, J. *Polyhedron* **1987**, *6*, 1775.
19. Werner, A. ref. 1; p 151.
20. Jørgensen, S. M. *J. prakt. Chem.* **1879**, *20*, 136.
21. Drechsel, E. *Z. anorg. Chem.* **1900**, *25*, 407.
22. Sidgwick, N. V. *The Chemical Elements and Their Compounds*; Oxford University Press: London, 1950; Vol. 2; p 1622.
23. Werner, A. *Z. anorg. Chem.* **1893**, *3*, 267; for a discussion and annotated English translation see Kauffman, G. B. *Classics in Coordination Chemistry, Part 1: The Selected Papers of Alfred Werner;* Dover: New York, NY, 1968; pp 5-88.
24. Bruni, G.; Fornara, C. *Atti R. Acad. dei Lincei Roma* **1904**, *13*, II, 26.
25 Ley, H. *Z. Elektrochem.* **1904**, *10*, 954; for a discussion and annotated English translation see Kauffman, G. B. *Classics in Coordination Chemistry, Part 3: Twentieth-Century Papers (1904-1935);* Dover: New York, NY, 1978; pp 5-19.
26. Werner, A. ref. 1; p 237.
27. Mann, F.G. *J. Chem. Soc.* **1933**, 412.
28. Jones, P. E.; Katz, L. *Acta Crystallogr.* **1969**, *B25*, 745.
29. Cartwright, P.; Gillard, R. D.; Sillanpaa, E. R. J.; Valkonen, J. *Polyhedron* **1991**, *10*, 2501.
30. Collins, J.; Gillard, R. D.; Hursthouse, M. B. *Polyhedron* **1993**, *12*, 255.
31. Sillanpaa, R.; Cartwright, P. S.; Gillard, R. D.; Valkonen, J. *Polyhedron* **1988**, *7*, 1801.
32. Krause, R. A.; Wickenden-Kozlowski, A.; Cronin, J. L. *Inorg. Synth.* **1982**, *21*, 12.
33. Evans, E. H. M.; Richards, J. P. G.; Gillard, R. D.; Wimmer, F. L. *Nouveau J. Chimie* **1986**, *19*, 783.
34. Gillard, R. D.; Riddell, F. G.; Wimmer, F. L. *J. Chem. Soc., Chem. Comm.* **1982**, 332.
35. Gillard, R. D.; Wimmer, F. L. *J. Chem. Soc., Chem. Comm.* **1978**, 936.
36. Buckley, A. N.; Wouterlood, H. J.; Cartwright, P. S.; Gillard, R. D. *Inorg. Chim. Acta.* **1988**, *143*, 77.
37. Werner, A. *Ber.* **1911**, *44*, 1887; **1914**, *47*, 3087; for discussions and annotated English translations see Kauffman ref. 23; pp 155-173 and 175-184, respectively.

RECEIVED December 27, 1993

Chapter 24

Chirality in Coordination Compounds

Alex von Zelewsky, Pascal Hayoz[1], Xiao Hua, and Paul Haag

Institute of Inorganic Chemistry, University of Fribourg, Pérolles,
CH–1700 Fribourg, Switzerland

Alfred Werner conjectured as early as 1899 that octahedrally coordinated metal complexes should occur in nonidentical mirror image isomers. For such objects, Lord Kelvin, in 1893, had coined the adjective "chiral", a term never used by Werner. It can be proved by examination of the original sample of $[Co(NO_2)_2(en)_2]Br$, prepared by Edith Humphrey, a Ph.D. student of Werner's, that crystals of optically pure samples were obtained in Werner's laboratory as early as 1899 or 1900. However, Werner did not publish the first successful resolution of an octahedral metal complex until 1911. Presently, interest in chirality in coordination compounds is booming, mainly because of the importance of coordination compounds in enantioselective homogeneous catalysis. Other interesting applications are enantioselective interactions of chiral coordination species with bio-molecules, and the stereoselective synthesis of multicenter systems.

Although the concept of *chirality* was introduced by Lord Kelvin (*1*) in the same year that Alfred Werner published his landmark paper in coordination chemistry, *i.e.,* 1893 (*2*), Werner never used this term. Instead, he used the German word *Spiegelbildisomerie*, or he spoke about "asymmetrical metal atoms". The conjecture that *Spiegelbildisomerie* should occur in coordination compounds with octahedral coordination appears for the first time in Werner's publications in 1899 (*3*). Werner's later work of shows that he had a very clear understanding of what chirality means and how it would manifest itself experimentally. His investigations of the tetranuclear *hexol* complexes of cobalt are particularly elegant and enlightened (*4*). It is therefore surprising that Werner overlooked the occurrence of a case of spontaneous separation of a chiral complex into its enantiomers upon crystallization. This was investigated from a historical point of view in a series of publications "Overlooked Opportunities in Stereochemistry" by Bernal and Kauffman (*5-6*). As they point out, a British woman, Edith Humphrey, had already prepared in 1899, or at the latest, in 1900 the compound $[Co(NO_2)_2(en)_2]Br$, which separates spontaneously into the enantiomeric forms upon crystallization. But this went unnoticed by Alfred Werner.

[1]Current address: Department of Chemistry, Stanford University, Stanford, CA 94305

0097–6156/94/0565–0293$08.00/0

The Historic Sample of $[Co(NO_2)_2(en)_2]Br$

The Royal Chemical Society invited learned societies from around the world to its 150th anniversary celebration in London in spring 1991. There was a special ceremony for the donation of the gifts to the president of the society, Sir Rex Richards. The present author represented Switzerland at this occasion. We had obtained from the Institute of Organic Chemistry of the University of Zürich (Prof. H.-J. Hansen), where the Werner collection is located, a sample of the cobalt complex prepared by Werner's British co-worker. The measurement of the CD-spectrum of a solution obtained from one crystal of the original sample shows clearly its optical activity (Figure 1). After 90 years on the shelves of the University of Zürich, the compound had not changed in any way.

The nicely crystalline compound can now be seen, displayed in a teakwood-box with a transparent insert, in the exhibition room of the Royal Chemical Society, Burlington House, London.

Today's Challenges in Chirality of Coordination Compounds

Although chirality has been a subject of continuous interest in coordination chemistry since Werner's time, it has not reached the mature state of development that it has in organic chemistry. Stereochemistry of species with coordination number > 4 is inherently much more involved than the organic stereochemistry predominantly determined by the tetrahedron. It is therefore not surprising that even today some basic problems connected with the chirality of coordination compounds are not yet completely solved.

Control of the Axial C_2 and C_3 Chirality in OC-6; Towards Chiral Building Blocks in Coordination Chemistry.

Since the first successful resolution of the $[CoBr(NH_3)(en)_2]^{2+}$ ion by Victor L.King in Werner's laboratory, numerous racemic complexes of the type $M(L^{\wedge}L)_2X_2$ (C_2-axial chirality) and $M(L^{\wedge}L)_3$ have been separated into enantiomers. Although some progress has been made in the understanding of packing properties of diastereomeric salts of such chiral complex ions with chiral counter ions (7), the success of the resolution by this method is still largely a matter of trial and error. Yet chiral complexes of this type can be of great importance for several purposes.

It has been shown, *e.g.*, that the interaction of metal complexes with DNA is enantiomerically discriminative (8). The controlled synthesis of one or the other desired optical isomer of two enantiomeric forms of one complex can therefore be of central interest for biochemical purposes.

The design and subsequent synthesis of polynuclear metal complexes as supramolecular species for molecular devices (9, 10) has recently evoked considerable interest. Figure 2 shows an example of a trinuclear species recently investigated for intramolecular energy- and electron-transfer processes.

One of the major problems in the characterization of such species is the occurrence of a great number of isomers. Nature has chosen to built up biopolymers as, *e.g.*, proteins from EPC's (Enantiomerically Pure Compounds, *i.e.*, the amino acids), certainly *also* in order to perform sophisticated tasks like chiral recognition, etc., but in zeroth order, to avoid the problem of producing up to 2^N isomers from N chiral, racemic building blocks.

Enantioselective (asymmetric) catalysis operates almost always with metal centers being the reactive sites, determining the stereochemical course of the reaction. Although considerable progress has been made in the design of catalysts of this type

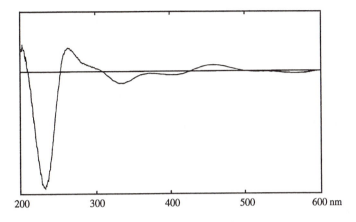

Figure 1. Survey CD-Spectrum of a small crystal of the original sample of $[Co(NO_2)_2(en)_2]Br$ prepared by Edith Humphrey in 1900

(856 mg dissolved in 5.0 ml H_2O; c = 4.9 • 10^{-4} mol L^{-1})

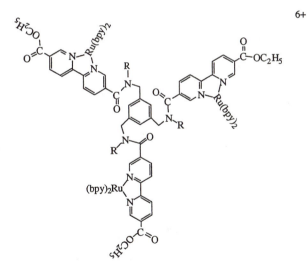

Figure 2. Structure of a trinuclear ruthenium complex (Adapted from ref.11)

(*12*), there is a strong need for chirally stable coordination units which survive many catalytic cycles.

M(L^L)$_2$X$_2$ Chiral Building Blocks. In 1913 Werner (*13*) reported that certain optically active metal complexes undergo substitution reactions without racemization. He described, *e.g.*,. the following reactions :

$$\left[\begin{array}{c} Cl \\ {}^{Cl} \end{array} Co\, en_2 \right] Cl + K_2CO_3 = 2KCl + \left[\begin{array}{c} O \\ OC \quad Co\, en_2 \\ O \end{array} \right] Cl$$

laevorotatory   dextrorotatory

$$\left[\begin{array}{c} Cl \\ SCN \end{array} Co\, en_2 \right] X + NaNO_2 = NaCl + \left[\begin{array}{c} O_2N \\ SCN \end{array} Co\, en_2 \right] X$$

laevorotatory   dextrorotatory

$$\left[Co\, en_2 \begin{array}{c} \cdot\, O_2\, \cdot \\ \cdot NH_2 \cdot \end{array} Co\, en_2 \right] X_4 \longrightarrow \left[en_2\, Co \begin{array}{c} \cdot\, OH\, \cdot \\ \cdot NH_2 \cdot \end{array} Co\, en_2 \right] X_4$$

laevorotatory   dextrorotatory

$$\left[en_2\, Co \begin{array}{c} \cdot\, O_2\, \cdot \\ \cdot NH_2 \cdot \end{array} Co\, en_2 \right] X_4 \longrightarrow \left[en_2\, Co \begin{array}{c} \cdot NO_2\, \cdot \\ \cdot NH_2 \cdot \end{array} Co\, en_2 \right] X_4$$

laevorotatory   dextrorotatory

He tacitly assumed that the configuration is retained in these cases. Several researchers have since studied substitution reactions in optically active OC-6 coordination species (*14,15*), but the subject is far from being exhaustively investigated. We have found that, *e.g.*, [Ru(o-phen)$_2$(py)$_2$]$^{2+}$ and [Ru(bpy)$_2$(py)$_2$]$^{2+}$ are useful *chiral building blocks* for the diastereoselective synthesis of dinuclear complexes with various types of bridging ligands (*16*). Figure 3 shows formulas and NMR-spectra of two dimeric compounds synthesized stereospecifically, using chiral building blocks of this type. Although this strategy yields the desired molecules, the enantiomerically pure building blocks still have to be obtained through resolution of a racemic mixture, and, in addition, some racemization can occur under severe reaction conditions. Both limitations are essentially avoided by using another approach.

Chiragen[X], a New Ligand Family, Predetermining Chirality at the M-Center. The two bidentate ligands in a complex M(L^L)$_2$ in OC-6 determine the Δ or Λ-helical chirality of the metal. If the two L^L ligands are connected by a conformationally rigid, chiral bridge, one of the two configurations around the metal will be predetermined. This strategy is schematically depicted in Figure 4.

The introduction of chirality centers in the ligands leads to a strong discrimination of the Λ and Δ configuration, respectively, if the two halves of the ligand are connected with a chiral bridge.

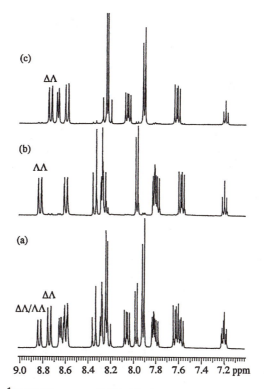

Figure 3. ^{1}H-NMR spectra (300 MHz in acetonitrile-d$_3$) (a) the mixture of ΔΔ/ΛΛ- and ΔΛ-[Ru(phen)$_2$-bpym-Ru(phen)$_2$]$^{4+}$; (b) ΛΛ-Ru(phen)$_2$-bpym-Ru(phen)$_2$]$^{4+}$ and (c) ΔΛ-[Ru(phen)$_2$-bpym-Ru(phen)$_2$]$^{4+}$.

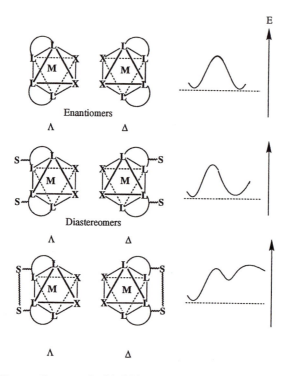

Figure 4. Energy diagrams for bis(bidentate) ligands coordinated *cis* in OC-6. The introduction of chirality centers in the ligands leads to a strong discrimination of the Λ and Δ configuration, respectively, if the two halves of the ligand are connected with a chiral bridge.

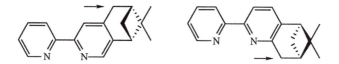

Figure 5. 4,5-Pinene-2,2'-bipyridine (I) and 5,6-Pinene-2,2'-bipyridine (II)

The realization of this concept has been achieved through the synthesis of pyridine molecules with substituents derived from pinene (*17*), the so called pinene-pyridines (Figure 5).

These molecules react easily at the indicated positions in a completely stereospecific way, in which the connection of two bipyridine moieties is easily achieved using dihalides as bridging molecules. Up to now, aliphatic bridges like $-(CH_2)_n-$ have been introduced, and a large variety of bridging groups can be applied.

Chiragen[X], where X=n, is a ligand with three asymmetric carbon centers in each pinene moiety of the molecule. Being synthesized from enantiomerically pure chiral pool precursors, like myrthenal or pinene itself, the ligand is obtained as an enantiomerically pure compound with known absolute configuration. Pinene is obtainable in both enantiomeric forms from natural sources. Therefore both enantiomers of the chiragen molecules are accessible. Model considerations show that a given configuration of the chiragen molecule can coordinate with an OC-6 metal only in one of the two helical forms Δ or Λ. The configuration of the metal is therefore completely predetermined through the configuration of the ligand (Figure 6). Even if precursors with low enantiomeric purity are used for the synthesis of the ligand, the complex is obtained in much higher configurational purity, since a ligand with "mixed" configuration cannot bind to the metal. Therefore a *chiral amplification* is achieved. If all the steps of the synthesis yield statistical ratios of the various isomers, the enantiomeric excess of the final complex M(chiragen[X]) is $(ee)_{complex} = (ee)_P/0.5(1 + (ee)_P^2)$, if the enantiomeric excess of the precursor material is $(ee)_P$. So far, chiragen[6] and chiragen[5] have been used for the synthesis of complexes of the type Ru(chiragen[X])(Diimine)$^{2+}$ (*18*).

NMR-analysis shows clearly the diastereoisomeric purity of the complex. The X-ray structure determination and the CD-spectra prove the expected absolute configuration of the complexes. These ligands will now be optimized for their use in producing metal complexes with predetermined absolute configurations.

New Forms of Chiral Metal Complexes. The tris(bidentate) OC-6 complexes are the classical representatives of three-bladed helical chirality in coordination chemistry. But even two bidentate ligands in OC-6 define a helix, provided the two remaining ligand positions are *cis*. As recently shown, two skew bidentate ligands can also occur in SP-4 coordination. It was often assumed (*19*) that chirality can occur in SP-4 coordination only if either chiral ligands are coordinated to the central metal or if specially designed ligands are used, where the symmetry of the one ligand is broken by the other ligand, as, *e.g.*, in the historically well known resolution of 1,2-diamino-2-methylpropane(meso-1,2-diamino-1,2-diphenylethane)-platinum(II) chloride reported by Mills and Quibell in 1935 (*20*).

The Two-Bladed Helix in SP-4. Biscyclometallated complexes with C^N–Ligands of d^8 metal ions are known only in the cis configuration (*21-23*) unless the ligands are specially designed (*24*) so that the trans configuration is forced upon the ligands. This is the case even when the two ligands make a coplanar arrangement impossible, as , *e.g.*, in Pt(thq)$_2$ or Pt(Hdiphpy)$_2$ (Figure 7).

Crystal structure analysis and detailed NMR-spectroscopic investigations reveal clearly the chiral nature of these complexes (*25*). Separation of these uncharged enantiomers proved hitherto to be extremely difficult. A method for the specific synthesis of one enantiomer was therefore sought.

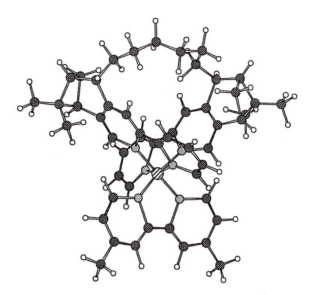

Figure 6. Model of the Δ-helical complex [Ru((+)-chiragen[6])(4,4'-dimethyl-2,2'-bipyridine)]²⁺

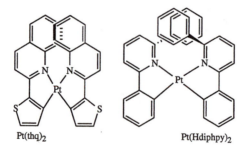

Pt(thq)₂   Pt(Hdiphpy)₂

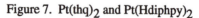

Figure 7. Pt(thq)₂ and Pt(Hdiphpy)₂

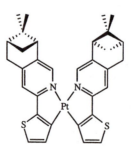

Figure 8. Pt(th-4,5ppy)₂

The Versatility of the Pinene-Pyridines. The pinene bipyridines described above are only a special case of the larger class of the pinene pyridines. A number of new pinene pyridines with the thiophenyl or phenyl group as substituent instead of the second pyridine ring, suitable for forming cyclometallated complexes, have now been synthesized. Easy cyclometallation is obtained, *e.g.*, with the ligands th-4,5-pinenepyridine or th-5,6-pinenepyridine, where th=2-(thiophen-2-yl) (Figure 8).

Pt(th-4,5ppy)$_2$ is a crystalline compound, which shows, according to X-ray analysis, Λ chirality. Again the absolute configuration is obtained directly from the structure analysis because the absolute configuration of the pinene moiety is known. NMR- and CD-analysis shows that the Λ-configuration is preserved in solution. Oxidative addition reactions of cylometallated Pt(II) complexes (Suckling, A.; von Zelewsky, A. *Inorg. Chem.*, submitted) with organic halides RX yield chiral Pt(IV), OC-6 complexes. Generally, both forms (the Λ and the Δ) of the octahedral complex Pt(th-4,5ppy)$_2$(R)(X) are produced (only one stereoisomer, the C,C'-*cis*; N,N'-*cis*, C(chelate), X-*trans* isomer is formed), but interestingly, they have very different solubilities, making separation an easy task. By this pathway, a new synthetic route to enantiomerically pure bis(bidentate) OC-6 complexes from chiral SP-4 complexes has been established.

Acknowledgments:
The authors thank the Swiss National Foundation for financial support of this work. They also wish to thank Dr. Liz Kohl for help in the preparation of this manuscript.

Literature Cited

(1) Kelvin, Lord. In *Baltimore Lectures* ; Clay C. J. & Sons, Ed.; Cambridge University Press Warehouse, 1904.

(2) Kelvin, Lord. In *Baltimore Lectures* ; Clay C. J. & Sons, Ed.; Cambridge University Press Warehouse, 1904, Appendix H, p 619.

(3) Werner, A.; Vilmos, A. *Z. Anorg. Chem.* **1899**, *21*, 145.

(4) Werner, A. *Ber.* **1914**, *47*, 3087; for a discussion and annotated English translation see Kauffman, G. B. Classics in Coordination Chemistry, Part 1: The Selected Papers of Alfred Werner; Dover: New York, 1968; pp 175-184.

(5) Bernal, I.; Kauffman, G. B. *J. Chem. Educ.* **1987**, *64*, 604.

(6) Kauffman, G. B.; Bernal, I. *J. Chem. Educ.* **1989**, *66*, 293.

(7) Yoneda, H. *Book of Abstracts 205th ACS National Meeting*, Denver, CO, March 20th-April 2nd, 1993, HIST. 35.

(8) Pyle, A. M.; Barton, J. K. *Prog. Inorg. Chem.* **1990**, *38*, 413.

(9) *Supramolecular Chemistry* ; Balzani , V.; De Cola, L., Eds.; Kluwer: Dordrecht, The Netherlands, 1992.

(10) *Supramolecular Photochemistry* ; Balzani, V.; Scandola, F., Eds.; Ellis Horwood: Warwick, England,1991.

(11) De Cola, L; Belser, P.; Ebmeyer, F.; Barigelletti, F.; Vögtle, F.; von Zelewsky, A.;Balzani, V. *Inorg. Chem.* **1990**, *29*, 495.

(12) Noyori, R. *Science* **1990**, *248*, 1194.

(13) Werner, A. *Neuere Anschauungen auf dem Gebiete der Anorganischen Chemie* ; F. Vieweg & Sohn : Braunschweig, 1913; p 368.

(14) Archer, R.D. *Coord. Chem. Rev.* **1969**, *4*, 243.

(15) Bailar, Jr., J.C. *Coord. Chem. Rev.* **1990**, *100*, 1.

(16) Hua, X.; von Zelewsky, A. *Inorg. Chem.* **1991**, *30*, 3796.

(17) Hayoz, P.; von Zelewsky, A. *Tetrahedron Lett.* **1992**, *33*, 5165.

(18) Hayoz, P.; von Zelewsky, A.; Stoeckli-Evans, H. *J. Am. Chem. Soc.* **1993**, *115*, 5111

(19) *Inorganic Chemistry* ; Purcell, K. F.; Kotz, J. C., Eds.; W. B. Saunders Company: Philadelphia, London, Toronto, 1977; p 627.
(20) Mills, W. H.; Quibell, T. H. H. *J. Chem. Soc.* **1935**, 839; for a discussion and annotated reprint see Kauffman, G. B. Classics in Coordination Chemistry, Part 3: Twentieth-Century Papers (1904-1935); Dover: New York, 1978; pp 196-224.
(21) Chassot, L.; von Zelewsky, A. *Helv. Chim. Acta* **1983**, *66*, 2443.
(22) Chassot, L.; Müller E.; von Zelewsky, A. *Inorg. Chem.* **1984**, *23*, 4249.
(23) Cornioley-Deuschel, C.; Ward, T.; von Zelewsky, A. *Helv. Chim. Acta* **1988**, *71*, 130.
(24) Cornioley-Deuschel, C. Dissertation **1988** No 938, University of Fribourg: Fribourg, Switzerland.
(25) Cornioley-Deuschel, C.; Stoeckli-Evans, H.; von Zelewsky, A. *J. Chem. Soc., Chem. Commun* . **1990**, 121.

RECEIVED February 8, 1994

Chapter 25

Equilibrium Shift Mechanism for the Pfeiffer Effect

Stanley Kirschner, Thaddeus Gish, and Ulysses Freeman, Jr.

Department of Chemistry, Wayne State University, Detroit, MI 48202

The equilibria that have been proposed to exist (and to change) during the occurrence of the Pfeiffer Effect are described, along with proposals for the nature of these equilibria. In particular, the effects on the equilibria of changing concentrations and concentrations ratios of the environment substance to the racemic complex have been studied, in an effort to identify the equilibria that actually exist during the appearance of the Pfeiffer Effect.

The Pfeiffer Effect (1) is the change in optical rotation of a racemic mixture of an optically labile complex when it is placed into a solution containing one enantiomer of an optically active compound (known as the "environment substance"). For example, if an aqueous solution of *levo*-malic acid is added to a solution of a racemic mixture of an optically *labile* complex, such as D,L-[Ni(phen)$_3$]Cl$_2$ (phen = *ortho*-phenanthroline), a marked change in optical rotation of the system is observed (the "Pfeiffer Effect"), and this rotation continues to undergo change over a period of about 120 hours. It should also be noted here that if the same malic acid solution is added to a *racemic* mixture of an optically *stable* complex, e.g., [Co(en)$_3$]Cl$_3$ (en = ethylenediamine), no such change in optical rotation is observed.

The Equilibrium Displacement Mechanism

Dwyer (2) and Kirschner (3) have attributed this change in rotation to a shift in the equilibrium between the *dextro*- and *levo*-enantiomers of the complex, a shift that occurs because of the presence of an optically active "environment" (the *levo*-malic acid) around the complex. Such an equilibrium shift is not possible in the case of optically stable complexes because no equilibrium exists between such enantiomers, as evidenced by the tendency of enantiomers of such complexes to resist racemization in solution for very long periods of time. Other proposals have also been put forth for the mechanism of this Effect, which are described in excellent reviews by Gillard and Williams (4) and Schipper (5), but space does not permit discussing these here.

0097–6156/94/0565–0303$08.00/0

Further, it has been proposed (6) that a "micro-mechanism" involving hydrogen-bonding between the protons on the oxygen atoms of the environment substance and the π-electron clouds of the *ortho*-phenanthroline ligands is responsible for this equilibrium shift because *levo*-malic acid (the S-enantiomer) fits preferentially (over R-malic acid) into the *delta* propeller formed by the ligands of the $(-)_D$-complex. This preferential hydrogen-bonding between one enantiomer of the optically active environment substance and the π-electrons of the "propeller blades" formed by the ligands of one enantiomer of the complex inhibits the ability of that enantiomer to isomerize to the opposite enantiomer, thereby altering the equilibrium between the enantiomers, *i.e.*, causing the equilibrium shift to occur.

The Enantiomeric Equilibrium in the Presence of the Environment Substance

During the work on this equilibrium displacement mechanism, a question arose about whether the shift in the equilibrium (and in the equilibrium constant for this shift) can be altered by changing the concentrations and/or concentration ratios of the complex to the environment substance (7). The equation for this originally-proposed equilibrium shift is:

$$\Lambda(+)_D \text{-[Ni(phen)}_3]^{2+} = \Delta(-)_D \text{-[Ni(phen)}_3]^{2+} \tag{1}$$

and the equilibrium constant for this reaction is represented by the equation:

$$K = [\ \Delta(-)_D \text{-[Ni(phen)}_3]^{2+}]/[\ \Lambda(+)_D \text{-[Ni(phen)}_3]^{2+}] \tag{2}$$

A series of Pfeiffer-active systems was prepared, using *racemic* -[Ni(phen)$_3$]Cl$_2$ (8) and *levo*-malic acid, which differ in concentrations of the complex and the environment substance as well as in the ratios of the complex to the environment substance. For each system the equilibrium shift was observed experimentally, and the equilibrium constant was calculated (9), based on the optical rotations observed for the system and corrected for the presence of the optically active environment substance.

The Nature of the Pfeiffer Effect Equilibrium

It was noted (Table I) that the equilibrium constants calculated according to equation 2 (for the enantiomeric shift described by equation 1) were not, in fact, constant. Therefore, it is proposed that, while the equilibrium described by equation 1 can hold for solutions of racemic mixtures of optically labile complexes themselves, it does not accurately describe Pfeiffer Effect equilibria for such systems that also contain optically active environment substances.

Consequently, equilibrium constant calculations were made using equations to describe the equilibria in Pfeiffer-active systems, which utilized the concentrations of the environment substance as well as those of the complex enantiomers and of the hydrogen-bonded complex enantiomers (those having environment substance enantiomers hydrogen-bonded to them). Among the equations studied for these equilibria are:

$$\Lambda(+)_D \text{-[Ni(phen)}_3]^{2+} + \Delta(-)_D \text{-[Ni(phen)}_3]^{2+} +$$

$$2\ \underline{S}(-)_D \text{-malic acid} = 2\ \Delta(-)_D \text{-[Ni(phen)}_3]^{2+} \cdot \underline{S}(-)_D \text{-malic acid} \tag{3}$$

for which:

$$K = \frac{[\Delta(\text{-})_D\text{-}[Ni(phen)_3]^{2+}\cdot\underline{S}(\text{-})_D\text{-malic acid}]^2}{[\Lambda(\text{+})_D\text{-}[Ni(phen)_3]^{2+}][\Delta(\text{-})_D\text{-}[Ni(phen)_3]^{2+}][\underline{S}(\text{-})_D\text{-malic acid}]^2} \quad (4)$$

and:

$$\Lambda(\text{+})_D\text{-}[Ni(phen)_3]^{2+} \;+\; \Delta(\text{-})_D\text{-}[Ni(phen)_3]^{2+} \;+$$

$$2\,\underline{S}(\text{-})_D\text{-malic acid} \;=\; \Lambda(\text{+})_D\text{-}[Ni(phen)_3]^{2+}\cdot\underline{S}(\text{-})_D\text{-malic acid} \;+ \quad (5)$$

$$\Delta(\text{-})_D\text{-}[Ni(phen)_3]^{2+}\cdot\underline{S}(\text{-})_D\text{-malic acid}$$

for which:

$$K = \frac{[\Lambda(\text{+})_D\text{-}[Ni(phen)_3]^{2+}\cdot\underline{S}(\text{-})_D\text{-malic acid}][\Delta(\text{-})_D\text{-}[Ni(phen)_3]^{2+}\cdot\underline{S}(\text{-})_D\text{-malic acid}]}{[\Lambda(\text{+})_D\text{-}[Ni(phen)_3]^{2+}][\Delta(\text{-})_D\text{-}[Ni(phen)_3]^{2+}][\underline{S}(\text{-})_D\text{-malic acid}]^2}$$

$$(6)$$

Experimental

The *racemic*-[Ni(phen)₃]Cl₂ was prepared and resolved according to the method of Kauffman and Takahashi (8). The Pfeiffer systems were observed for optical activity using a Perkin-Elmer Model 241 Photoelectric Polarimeter. The concentrations and concentrations ratios of complex and environment substance used are indicated in the tables given below. The methods used for the calculations of the equilibrium constants are described elsewhere (9).

Results and Discussion

Table I shows the effects on the equilibrium constant calculated from equations 1 and 2. It should be noted that, not only does the "equilibrium constant" not remain constant as the concentrations of both the environment substance and the *racemic* complex increase (while keeping their ratio constant), but this "constant" also increases when the ratio of the concentrations of the environment substance to the complex increases. This implies that it may be possible for more than one molecule of the environment substance to undergo hydrogen-bonding simultaneously to one molecule of the complex, which is a matter that is currently undergoing careful scrutiny in this laboratory.

For the conditions under which the Pfeiffer Effect is observed, equation 3 produces a set of equilibrium constants that are constant within a relative standard deviation of 9.5%, as calculated according to equation 4. Table II shows the concentrations of the complexes and environment substances, as well as the equilibrium constants obtained using equation 4. Further, it was observed that the relative standard deviation of the equilibrium constant calculated using equations 5 and 6 was 14.1%, which is a significantly larger deviations than that calculated using equations 3 and 4.

Table I

The Effects of Changes in Concentrations and Concentration Ratios of Complex to Environment Substance on the Equilibria of a Pfeiffer Effect System[a]

Enantiomer Concentrations Complex : Envir.	Observed Pfeiffer Rotation (^{0})	Equilibrium Constant
0.020 M : 0.040 M	-0.210	1.018
0.020 M : 0.080 M	-0.381	1.040
0.040 M : 0.020 M	-0.205	1.009
0.040 M : 0.040 M	-0.434	1.019
0.040 M : 0.080 M	-0.760	1.033
0.040 M : 0.160 M	-1.579	1.069

[a] For the equilibrium: $\Lambda(+)_D$ -[Ni(phen)$_3$]$^{2+}$ = $\Delta(-)_D$ -[Ni(phen)$_3$]$^{2+}$, for which the equilibrium constant:

$K = [\Delta(-)_D$ -[Ni(phen)$_3$]$^{2+}]/[\Lambda(-)_D$ -[Ni(phen)$_3$]$^{2+}]$;

complex: Δ,Λ-(-)$_D$ -[Ni(phen)$_3$]Cl$_2$; environment substance: $\underline{S}$(-)$_D$ -malic acid; wavelength: 589 nm; temperature: 21^0 C.

Table II

Concentrations, Concentration Ratios of Complex to Environment Substance, and Equilibrium Constants Obtained from Equation (4)[a]

Enantiomer Concentrations Complex : Envir.	Equilibrium Constant
0.020 M : 0.040 M	0.1779
0.020 M : 0.080 M	0.1486
0.040 M : 0.020 M	0.1694
0.040 M : 0.040 M	0.1917
0.040 M : 0.080 M	0.1491
0.040 M : 0.160 M	0.1659
0.050 M : 0.050 M (10)	0.1785

[a] For the equilibrium given in equation 3; complex: Δ,Λ-(-)$_D$ -[Ni(phen)$_3$]Cl$_2$; environment substance: $\underline{S}$(-)$_D$ -malic acid; wavelength: 589 nm; temperature: 21^0 C.

Conclusions

Table I shows that, as the concentrations of the complex are held constant and the concentrations of the environment substance are increased, the magnitude of the Pfeiffer Effect increase, *i.e.*, there is conversion of more of the $\Lambda(+)$-enantiomer to the $\Delta(-)$-enantiomer, which is hydrogen-bonded by the environment substance. Also, as the concentration of the environment substance is held constant and the concentrations of the complex are increased, the magnitude of the Pfeiffer Effect decreases (i.e., there is a smaller conversion of the $\Lambda(+)$- enantiomer of to the $\Delta(-)$-enantiomer).

These observations support the proposed hydrogen-bonding mechanism (6) for the equilibrium displacement, since increasing the concentration of environment molecules (while keeping the concentration of complex ions constant) would be expected to result in a larger number of the *S*-environment molecules becoming hydrogen-bonded to the ligands comprising the *delta* propeller configuration of the complex (the preferred configuration for attachment of the *S*-environment substance). This would further stabilize this configuration and inhibit its return to the opposite enantiomer.

Further, it is proposed that, whereas equation 1 represents the enantiomeric equilibrium for fast-racemizing complexes in the absence of an environment substance, equation 3 most accurately describes the equilibrium in the presence of such a substance.

Dedication

The authors are most pleased to dedicate this paper to Professor George B. Kauffman on the occasion of his receiving the American Chemical Society George C. Pimental Award in Chemical Education, sponsored by Union Carbide Corporation.

Literature Cited

1. Pfeiffer, P.; Quehl, K. *Ber.* **1931**, *64*, 2667; **1932**, *65*, 560.

2. Dwyer, F. P.; Gyarfas, E. C.; O'Dwyer, M. F., *Nature* **1951**, *167*, 1036.

3. Kirschner, S.; Serdiuk, P. In *Stereochemistry of Optically Active Transition Metal Compounds*; Saito, K.; Douglas, B., Eds.; American Chemical Society: Washington, DC, 1980; p 239.

4. Gillard, R. D.; Williams, P. A. *Intl. Revs. in Phys. Chem.* **1986**, *5*, 301.

5. Schipper, P. E. *J. Am. Chem. Soc.* **1978**, *100*, 1079.

6. Kirschner, S.; Ahmad, N.; Munir, C.; Pollock, R. *Pure & Appl. Chem.* **1979**, *51*, 913.

7. Kirschner, S.; Freeman, Jr. U. Abstracts of Papers, 203rd National Meeting, American Chemical Society, San Francisco, CA, April 5-10, 1992, INOR 664.

8. Kauffman, G. B.; Takahashi, L. T. *Inorganic Syntheses* **1966**, *8*, 227.

9. Freeman, Jr. U.; Gish, T.; Kirschner, S. *J. Indian Chem. Soc.* **1992**, *69*, 510.

10. Ahmad, N. *The Pfeiffer Effect in Transition Metal Complexes*; Ph.D. Dissertation, Wayne State University: Detroit, Michigan, 1969.

RECEIVED February 14, 1994

Chapter 26

Mechanism of Optical Resolution of Octahedral Metal Complexes

Hayami Yoneda[1] and Katsuhiko Miyoshi[2]

[1]Department of Applied Science, Faculty of Science, Okayama University of Science, 1–1 Ridai-cho, Okayama 700, Japan
[2]Department of Chemistry, Faculty of Science, Hiroshima University, 1–3–1 Kagamiyama, Higashi-Hiroshima 724, Japan

The study of the chiral discrimination between antimony d-tartrate and a series of $[Co(N)_6]^{3+}$ complexes reveals that chiral discrimination is effected by this anion through hydrogen-bonding association along the C_2 or C_3 axis of the complex. Furthermore, it is clear that the antimonyl d-tartrate ion favors the Λ-complex in the C_3 association, while it favors the Δ-complex in the C_2 association. A concrete association model is proposed to account for the mode of chiral discrimination between the complex and the antimonyl d-tartrate ion. The fundamental idea underlying this association model has been applied to the interpretation of the chiral discrimination between complex cations and complex anions. The C_3^+ complex cation always favors the homochiral combination (Δ-Δ or Λ-Λ) with the complex anion, while the C_2^+ complex cation always favors the heterochiral combination (Δ-Λ or Λ-Δ) with the complex anion. This tendency is explained by assuming that straight N-H···O hydrogen bonds are formed in such a favorable pair. This hydrogen bonding association model has made possible the highly efficient optical resolution of electrically neutral tris(aminoacidato)cobalt(III).

For many years we have been interested in clarifying the mechanism of optical resolution of octahedral metal complexes. Since $[Co(en)_3]^{3+}$ (en = ethylenediamine) is a prototype of chiral metal complexes and d-tartrate and antimonyl d-tartrate ions are two main resolving agents for metal complexes, we concentrated on the system involving this complex and these and these two resolving agents. As a result, a unique face-to-face ion-pair structure was found as a common factor in three diastereomeric salts containing Λ-$[M(en)_3]^{3+}$ and d-tart^{2-} (1,2). We believe that the origin of chiral discrimination lies in this unique ion-pair structure. Based on this C_3

0097–6156/94/0565–0308$08.00/0

association model, successful chromatographic resolution by using d-tart^{2-} as eluent was achieved for several cobalt(III) complexes having a triangular face composed of three NH$_2$ groups.

As to the mode of chiral discrimination by [Sb$_2$(d-tart)$_2$]$^{2-}$, attention was paid to the shape of the channel formed between the en chelate rings. The channel between chelate rings in the Λ-configuration has a shape of L, and the channel between chelate rings in the Δ-configuration has a shape of J. It was assumed that the [Sb$_2$(d-tart)$_2$]$^{2-}$ ion has a skeleton which makes a good fit to an L-shaped channel of the Λ-complex, but not to the J-shaped channel of the Δ-complex. With this L-J model, we can explain the degree of optical resolution (separation factor) for a series of [Co(N)$_6$]$^{3+}$ complexes having different numbers of L-shaped channels (*3*). Our early investigations on the mode of chiral discrimination by these two resolving agents have been reviewed in two articles (*4*). While the C$_3$ association model for the d-tart^{2-} ion is based on the concrete ion-pair structure obtained by X-ray crystal analyses, the L-J model is a mere conceptual model of the "key-and-lock" relation applied to chiral discrimination of octahedral complexes. It does not tell anything about the concrete stereochemical features of the ion-pair composed of the complex and the chiral resolving agent. Concerning this point, we noticed that the CD spectra of cobalt(III) complexes are quite sensitive to the addition of counter ions. Analyses of CD changes caused by the resolving agent anion provided a clue to the solution of this problem. We proposed a new association model which accounts for how this chiral selector ion recognizes the chirality of [Co(en)$_3$]$^{3+}$ and related complexes in solution.

In this article, we review the studies concerning this association model and the succeeding studies concerning chiral interactions between complex cations and complex anions.

Chiral Discrimination by [Sb$_2$(d-tart)$_2$]$^{2-}$ in Solution

In order to deduce the direction of access of the resolving agent toward the complex , the effect of the resolving agent upon the CD spectra has been studied for a series of [Co(N)$_6$]$^{3+}$ of trigonal symmetry. These complexes have N-H bonds projecting outward along their C$_3$ and C$_2$ axes. They are classified into C$_3^+$, C$_2^+$ and (C$_3^+$ + C$_2^+$) complexes according to the kind of axis along which they accept the oncoming anion (Figure 1). As to the effect of [Sb$_2$(d-tart)$_2$]$^{2-}$ upon the CD peak in the d-d transition region of the spectrum of the complex, addition of this ion enhances the intensity of the A$_2$ component for all the C$_3^+$ and (C$_3^+$ + C$_2^+$) complexes examined, while it causes enhancement of the Ea component for the C$_2^+$ complex, [Co(sep)]$^{3+}$ (sep = 1,3,6,8,10,13,16,19-octaazabicyclo[6,6,6]eicosane) (*5*). Enhancement of the A$_2$ and Ea components can be taken to indicate axial and equatorial perturbations, respectively, which are caused by formation of hydrogen bonding between axial or equatorial NH hydrogens of the complex and oxygen atoms of the $[Sb_2(tart)_2]^{2-}$ ion. Here,DCD spectra are defined as the CD spectra with the anion minus the CD spectra without the anion. The pattern of the DCD spectra is quite similar to the pattern of the corresponding DCD with $tart^{2-}$ ions (6). Thus it is reasonable to assume that $[Sb_2(d\text{-}tart)_2]^{2-}$ also anchors the complex with N-H···O hydrogen bonds along the C_3 or C_2 axis.

The next problem is to determine which oxygen atoms of the $[Sb_2(d\text{-}tart)_2]^{2-}$ ion are utilized for hydrogen bonding. Examining the structure of this anion in detail, we found that one carboxylate oxygen atom (directly linked to the Sb atom) of one tartrate group and one alcoholic oxygen atom of the other tartrate group attached to the same Sb atom have their lone-pair electrons spatially disposed to interact with two N-H bonds of the complex, whether the association takes place along the C_3 or the C_2 axis. The $[Sb_2(d\text{-}tart)_2]^{2-}$ ion has four such pairs of oxygen atoms.

Now that the oxygen atoms of the anion for hydrogen bonding were determined, the mode of chiral discrimination by this anion can be visualized (7). The mode of chiral discrimination in the C_3 association will be described, taking $[Co(en)_3]^{3+}$ as an example (Figure 2). Two oxygen atoms of the anion are linked to two axial N-H bonds on a triangular face of the complex and form double hydrogen bonding. The remaining part of the anion is oriented to a position opposite to the third N-H bond on the triangular face so as to avoid steric repulsion.. In such an orientation, the carboxylate group attached to the distal Sb atom comes upon an opening between two chelate rings of the Λ-complex., while it comes close to a chelate ring of the Δ-complex.

In this way, it is clear that the d-anion forms a favorable ion-pair with the Λ-complex. Since there are two triangular faces in $[Co(en)_3]^{3+}$, each of which has three axial N-H bonds, and only two N-H bonds are required for ion-pair formation with the anion, six favorable associations are posssible for Λ-$[Co(en)_3]^{3+}$.

Different numbers of favorable associations are possible for other $[Co(N)_6]^{3+}$ complexes. The numbers of favorable associations are (6,0), (4,0), (3,0), (4,2), and (0,0) for the (Λ,Δ) pairs of $[Co(en)_3]^{3+}$, α-$[Co(trien)(en)]^{3+}$ (trien = triethylenetetramine), β-$[Co(trien)(en)]^{3+}$, u-fac-$[Codien)_2]^{3+}$ (dien = diethylenetriamine), and mer-$[Co(dien)_2]^{3+}$. These numbers coincide exactly with the numbers of L-shaped channels. The separation factors obtained experimentally are 1.45, 1.37, 1.28, 1.20, and 0 in this order (3).

Although the efficiency of enantiomer separation is low, the Δ-isomer is eluted earlier with $[Sb_2(d\text{-}tart)_2]^{2-}$ in all cis-$[CoX_2(en)_2]^+$ complexes. In this case , the anion is assumed to approach the complex along the C_2 axis and to form hydrogen bonding with the two N-H bonds located *trans* to the negative ligands X. In this C_2

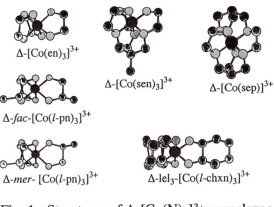

Fig. 1.   Structures of Δ-[Co(N)₆]³⁺ complexes
(All hydrogen atoms are omitted.)

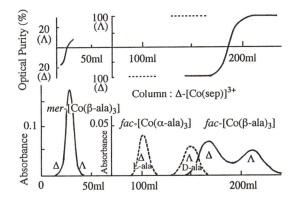

Fig. 2.   Elution curve of [Co(ama)₃] complexes,
column : saturated with Δ-[Co(sep)]³⁺

association the distal carboxylate group of the associated anion experiences steric repulsion by the chelate ring in the Λ-complex, but no such steric repulsion is experienced by the anion for the Δ-complex. Therefore it is not surprising that the $[Sb_2(d\text{-tart})_2]^{2-}$ ion forms a favorable ion-pair with the Δ-complex (Such C_3 association is , of course, possible also for $[Co(en)_3]^{3+}$. However, the contribution of the C_3 association is considered to surpass the contribution of the C_2 association. This interpretation was supported by the sign of the DCD spectra.).

An exception is the case of $[Co(sep)]^{3+}$, in which chiral discrimination takes place exclusively through the C_2 association, yet the Λ-complex is eluted earlier. Here the repulsion by the alkyl cap in the Δ-complex is stronger than that by the en chelate ring in the Λ-complex so that the Λ-complex is favored by the $[Sb_2(d\text{-tart})_2]^{2-}$ ion and is eluted earlier. The cases of cis-$[Co(N_3)_2(en)_2]^+$ and cis -$[Co(acac)(en)_2]^{2+}$ (acac = acetylacetonate) are also exceptions , in which the Λ-isomer is eluted earlier. It is evident that the "C_3 association" (the association corresponding to the C_3 association for $[Co(en)_3]^{3+}$) surpasses the C_2 association. Weak repulsion by the low effective negative charge of the N_3^- ion and the acac-O atom is not as effective in preventing the "C_3 association". The "C_3 association" is assumed to occur for cis-α-$[CoX_2(trien)]^+$, in which the C_2 association is completely prevented. It is therefore not surprising that the Λ-isomer is eluted earlier for cis-α-$[CoX_2(trien)]^+$.

Chiral Discrimination by Complex Ions (8)

Quite a few examples of optical resolution via formation of the less soluble diastereomeric salt have been reported in which the species to be resolved and the chiral selector are both complex ions. Thus elucidation of the mode of chiral discrimination between complex cations and anions is an important factor leading to the discovery of new efficient resolving agents. As one of several approaches to this subject, we carried out very simple experiments.

SP-Sephadex C-25 saturated with a Δ-complex cation was packed in a glass column. A racemic complex anion was eluted through the column with 30% aqueous ethanol. The eluate was fractionally collected, and the AB and CD spectra were recorded. The elution curves were obtained in this way when necessary, but the degree of resolution was generally so low that only the elution orders were determined in most cases.

The complexes used as resolving agents were those having NH or NH_2 groups which accept the anion along either the C_3 or the C_2 axis. The complexes to be resolved were those having COO^- groups which accept the complex cation along either the C_3 or the C_2 axis. They are classified as C_3^-,C_2^-, and $(C_3^- + C_2^-)$, according to the mutual disposition of carboxylate groups they use in ion-pair formation.

The result is shown in Table I, where the chirality of the second-eluted enantiomer is indicated. Since the second-eluted enantiomer is the one which forms a favorable pair with the chiral selector cation in the stationary phase, the result can be summarized as follows: The C_3^+ complex cation always favors the homochiral combination (Δ-Δ or Λ-Λ) with the complex anion, while the C_2^+ complex cation always favors the heterochiral combination (Δ-Λ or Λ-Δ), irrespective of the chirality of the complex anion . This seems to be a general tendency. In contrast, the (C_3^+ + C_2^+) complex represented by $[Co(en)_3]^{3+}$ favors the homochiral combination with some complex anions and the heterochiral combination with the other complex anions, and the degree of optical resolution achieved is very low.

These results can be explained by hydrogen-bonded association models. The C_3^+ complex directs its C_3 axis to the anionic complex to form double or triple hydrogen bonds between NH_2 and COO^-. Inspection of the molecular model reveals that nearly straight N-H$\cdots$O hydrogen bonds are formed in the homochiral combination of ions, while N-H$\cdots$O hydrogen bonds are fairly bent in the heterochiral combination. On the contrary, the C_2^+ complex directs its C_2 axis to the oxygen atoms of the anionic complex to form double hydrogen bonds with its NH hydrogens. Here nearly straight hydrogen bonds are formed in the heterochiral combination. It is natural to assume that the straight hydrogen bonds are more stable than the bent hydrogen bonds. In the case of the (C_3^+ + C_2^+) complex, the C_3 and the C_2 associations take place simultaneously , which offsets the effect of chiral discrimination of each other. Since the contribution of the C_3 association is not equal to the contribution of the C_2 association, a small fraction of the contribution remains either for the C_3 or the C_2 association. This explains the low efficiency of chiral discrimination by the (C_3^+ + C_2^+) complex.

Optical Resolution of Tris(aminoacidato)cobalt(III)Complexes

Although tris(aminoacidato)cobalt(III) complexes, $[Co(ama)_3]$, are electrically neutral, they can be regarded either as complex cations or as complex anions because they can anchor complex anions or complex cations with N-H$\cdots$O type hydrogen bonds along the C_3 or the C_2 axis. Therefore they may be expected to be resolvable either by chiral complex anions or by complex cations. We here describe the optical resolution of some $[Co(ama)_3]$ using Δ-$[Co(sep)]^{3+}$ and Δ-$[Co(chxn)_3]^{3+}$ (chxn = cyclohexane-diamine) as chiral selectors. The procedure is almost the same as described in the preceeding section.

Table I. Enantiomers Forming Favorable Pairs with Δ Cation Complexes

	C_2^-		C_3^-	$(C_2^- + C_3^-)$	
	/\		|	/\	
	1	2	3	4	5
C_2^+					
[Co(sep)]$^{3+}$	Λ	Λ	Λ	Λ	Λ
[Co(acac)(en)$_2$]$^{2+}$	Λ	Λ	Λ	Λ	Λ
[Co(ox)(en)$_2$]$^+$	Λ	Λ	Λ	Λ	Λ
cis-[Co(NO$_2$)$_2$(en)$_2$]$^+$	Λ	Λ	Λ	Λ	Λ
C_3^+					
lel$_3$-[Co(chxn)$_3$]$^{3+}$	Δ	Δ	Δ	Δ	Δ
$C_2^+ + C_3^+$					
[Co(en)$_3$]$^{3+}$	Δ[a]	Δ[a]	Δ	Λ	Λ
[Co(gly)(en)$_2$]$^{2+}$	Δ[a]	Λ[a]	Λ	Λ	Λ
[Co(sen)]$^{3+}$	Λ	Λ	Λ	Λ	Λ

a   very low resolution

1   [Co(ox)$_2$(en)]$^-$

2   C_2-[Co(ox)(gly)$_2$]$^-$

3   fac-[Co(β-ala)$_3$]

4   C_1-cis(N)-[Co(ox)(gly)$_2$]$^-$

5   [Co(ox)$_2$(gly)]$^{2-}$

A racemic complex is loaded at the top of the column and eluted with 30% aqueous ethanol. On the column containing Δ-[Co(sep)]$^{3+}$ very good resolution is achieved for *fac*-[Co(β-ala)3] and *fac*-[Co(α-ala)3] (ala = alaninate), while only poor resolution is attained for *mer*-[Co(β-ala)3] (Figure 2)(*9*). On the column containing Δ-[Co(chxn)3]$^{3+}$ *fac*-[Co(α-ala)3] and *fac*-[Co(β-ala)3] are resolved to some extent but the elution order is reversed. In these cases [Co(ama)3] behaves as a complex anion. This reconfirms that chiral discrimination between complex cations and complex anions is attained by hydrogen bonding along the C3 or the C2 axis and that the C3$^+$ complex favors the homochiral ion-pair, while the C2$^+$ complex favors the heterochiral ion-pair.

[Co(chxn)3]$^{3+}$ examined so far has three N-H bonds nearly parallel to the C3 axis and should be correctly designated as lel3-[Co(chxn)3]$^{3+}$. There is another isomer, ob3-[Co(chxn)3]$^{3+}$, which has three N-H bonds oblique markedly from the C3 axis. Using these two [Co(chxn)3]$^{3+}$ complexes as a chiral selector, optical resolution was attempted for *fac*-[Co(α-ala)3], *fac*-[Co(β-ala)3], [Co(ox)2(gly)]$^{2-}$ (ox = oxalate; gly = glycinate), C_1-*cis*(N)-[Co(ox)(gly)2]$^-$, and C_1-*cis*(N)-[Co(ox)(β-ala)2]$^-$. The elution curves obtained for *fac*-[Co(β-ala)3] are shown in Figure 3 (*10*). As seen in this figure, *fac*-[Co(β-ala)3] is not as well resolved by Δ-lel3-[Co(chxn)3]$^{3+}$, but yet it is seen that the Δ-isomer is eluted undoubtedly later, which means that the Δ-lel3 complex cation favors the Δ-isomer, as expected. In contrast to this, the same complex is completely resolved by the Δ-ob3 complex (the separation factor is extremely high, *ca.*.7), and the Λ-isomer is eluted later, which means that the Δ-ob3 complex favors the Λ-isomer. Therefore the heterochiral combination (Δ-Λ) is favored in this C3 association.

Why is the elution order of the resolved enantiomers reversed to that expected for the C3 association ? Figure 4 shows the schematic structures of the complexes involved in the present optical resolution. Three N-H bonds in a triangular face of Δ-ob3-[Co(chxn)3]$^{3+}$ are markedly inclined from the C3 axis so that they assume left-handed chirality. As a result, they form straight hydrogen bonds with the lone pairs of the three oxygen atoms of Λ-*fac*-[Co(ama)3] (For carboxylate oxygen , sp^2 hybridization is assumed). On the contrary, the lone pairs of the three oxygen atoms of Δ-*fac*-[Co(ama)3] have right-handed chirality so that they cannot form hydrogen bonds favorable for the stable association with left-handed Δ-ob3-[Co(chxn)3]$^{3+}$.

The remaining four complexes–*fac*-[Co(α-ala)3] and the three complex anions– were also resolved in the same way as was *fac*-[Co(β-ala)3]. The chirality of the second-eluted enantiomer was Δ on the column of the lel3-complex, and it was Λ on the column of the ob3-complex. Thus a similar hydrogen bonding is also probably operating in the optical resolution of these complexes.

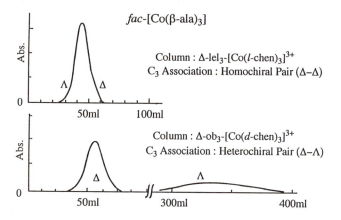

Fig. 3. Elution curve of *fac*-[Co(β-ala)₃],
column : saturated with Λ-lel₃-[Co(chxn)₃] ³⁺ and with
Δ-ob₃-[Co(chxn)₃]³⁺

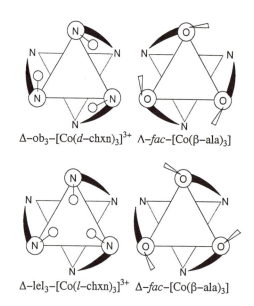

Fig. 4. Structures of Δ-ob₃- and Δ-lel₃-[Co(chxn)₃]³⁺ and Λ-*fac*-
and Δ-*fac*-[Co(β-ala)₃]

Acknowledgment

We would like to express our heartfelt thanks to Professor George B. Kauffman for the invitation to speak at the Coordination Chemistry Centennial Symposium.

Literature Cited

1. (a) Kushi,Y. ; Kuramoto, M. ; Yoneda, H. *Chem. Lett.* **1976**, 135; (b) Kushi, Y. ; Kuramoto, M. ; Yoneda, H. *Chem. Lett.* **1976**, 3396; (c) Kushi, Y. ; Tada, T. ; Yoneda, H. *Chem. Lett.* **1977**, 379.

2. The common existence of the face-to-face close contact was verified by succeeding crystal structure analyses of complexes:
(a) Λ-[Co(sen)]Cl(d-tart)·6H$_2$O: Okazaki, H.; Sakaguchi, U.; Yoneda, H.
Inorg, Chem. **1983**, *22*, 1539; (b)Δ,Λ-[Ni(en)$_3$]$_2$(d,l-tart)·H$_2$O:
Mizuta,T.; Yoneda, H.; Kushi, Y. *Inorg, Chim. Acta* **1987**, *132*, 11;
(c)Δ,Λ-[Co(en)$_3$]$_2$(d,l-tart)· 10H$_2$O, Λ-[Co(en)$_3$]$_2$(d-tart)$_3$· 19H$_2$O, and
Λ-[Co(en)$_3$]$_2$(d-tart)$_3$· 11.5H$_2$O: Mizuta, T.; Tada, T.; Kushi, Y.;
Yoneda, H. *Inorg. Chem.* **1988**, *27*, 3836;
(d)Λ-lel$_3$-[Co(chxn)$_3$]Cl(d-tart)·2H$_2$O and Δ-lel$_3$-[Co(chxn)$_3$]Cl
(d-tart)·2H$_2$O : Mizuta, T.; Toshitani, K.; Miyoshi, K.; Yoneda, H.
Inorg. Chem. **1990**, *29*, 3020.

3. Nakazawa, H.; Yoneda, H. *J. Chromatogr.* **1978**, *160*, 39.

4. (a)Yoneda, H. *J. Liq. Chromatogr.* **1979**, 2, 1157;
(b) Yoneda, H. *J. Chromatogr.* **1984**, *313*, 59.

5. Sakaguchi, U.; Tsuge, A.; Yoneda, H. *Inorg. Chem.* **1983**, *22*, 3745.

6. Sakaguchi, U.; Tsuge, A.; Yoneda, H. *Inorg. Chem.* **1983**, *22*, 1630.

7. Miyoshi, K.; Izumoto, S.; Yoneda, H. *Bull. Chem. Soc. Jpn.* **1986**, *59*, 3475.

8. Miyoshi, K.; Sakamoto, Y.; Ohguni, A.; Yoneda, H. *Bull. Chem. Soc. Jpn.*
1985,
58, 2239.

9. Miyoshi, K.; Nakai, K.; Yoneda, H. Presented at the Meeting of the Chugoku-
Shikoku
Branch of the Chemical Society of Japan , held in Tokushima, **1986**.

10. Miyoshi, K.; Yoshinaga, M.; Yoneda, H. Presented at the Annual Meeting of the
Chemical Society of Japan , held in Tokyo, **1987**.

RECEIVED December 27, 1993

COMPOUNDS OF VARIOUS ELEMENTS

Chapter 27

Nontraditional Ligands and Their Impact on Coordination Chemistry

Robert W. Parry

Department of Chemistry, University of Utah, Salt Lake City, UT 84112

Werner's introduction of new geometric patterns for a number of inorganic compounds established the need for a working bonding model. The electron-pair bonding model of Lewis (and Huggins) provided a means for rationalizing most coordination compounds. On the other hand, coordination compounds of the transition metal carbonyls introduced new bonding problems. These have now been resolved through the concept of π-acidity and back-bonding for ligands such as CO and F_3P. The idea of π-acidity for borane adducts of F_3P, CO, and PF_2H does not work well. A model based on a polarized sigma bond is developed here and is used to explain the following facts: (1) The shortest known P-B bond for a coordination compound containing 4-coordinate boron and phosphorus atoms is found in F_3PBH_3. That distance is even among the shortest for compounds containing 3-coordinate boron and phosphorus, yet this P-B bond dissociates very easily. (2) To date, all attempts to make CO or F_3P adducts of BCl_3 or BF_3 have been unsuccessful. (3) $AlCl_3$ will form a relatively stable F_3P adduct. (4) HF_2P forms a more stable adduct with BH_3 than does either F_3P or H_3P. Compounds containing three-center coordinate bonds of the form —B—H—→M are described. These compounds are compared to organic compounds containing "agostic" hydrogens.

0097–6156/94/0565–0320$08.00/0

Werner and the Birth of Coordination Chemistry

Structural chemistry started with the tetrahedral carbon atom, the planar benzene ring, and the planar double bond of the type found in C_2H_4. Many inorganic materials did not fit too well into these patterns. Of particular concern were well known compounds such as $Co(NH_3)_6Cl_2$, $Co(NH_3)_6Cl_3$, and related species whose formulas, reactions, and other properties did not resemble organic compounds at all (*1*). Werner solved the structural problem when he recognized that many substances can display geometric forms other than those common in carbon chemistry. Octahedral and square planar patterns were recognized as well as the traditional tetrahedral pattern of carbon. Other geometries for coordination numbers of 5, 7, 8, and 9 were to be recognized later. Modern coordination theory was born.

While Werner's structural proposals seemed reasonable to many chemists, his models seemed to do violence to the concepts of "valence" or bonding which were in use at the time. He had to try to rationalize the existence of these new geometries in terms of some "bonding theories" (*1b*). He responded in a very imaginative and noncommittal way. He simply *named* new bonding forms as "*primary valence*" and "*secondary valence.*" He admitted in his early papers that he could not characterize these new linkages very well, but they were needed to hold coordination compounds together. The concept of primary and secondary valence became closer and closer in Werner's mind as time went on and he finally concluded that there is no essential difference between the two (*1c*). The stage was set for the development of modern bonding concepts.

Electronic Models and Bonding: Sigma and Pi Bonds

It remained for G.N. Lewis (*2*), M.L. Huggins (*3*), and their contemporaries to interpret the primary and secondary valences of Werner in terms of the newly emerging electron patterns which were being used to explain "valence." Primary valences were normal covalent bonds (one electron from each bonded atom) and secondary valences were coordinate covalent bonds (two electrons for bond formation supplied by the ligand). The terms Lewis base and Lewis acid entered the literature. The Werner and Lewis-Huggins descriptions of structure and bonding were excellent for the metal-ammines, halo complexes of metals, and related species. Classical Werner coordination compounds fit the Werner-Lewis description, and there were many of such compounds.

On the other hand, some compounds, formally similar to classical Werner complexes, were prepared which caused conceptual problems. For example, metal carbonyls containing CO as a ligand such as $Ni(CO)_4$ and $Fe(CO)_5$ were prepared first by Mond and Langer in 1890 and 1891 (*4*).

Metal carbonyls were studied extensively by Hieber (5), and co-workers in Germany in the early part of the twentieth century. Compounds such as $[Rh(CO)_4]_2$ were first reported by Hieber. Blanchard (6) and co-workers in the U.S. also worked extensively on metal carbonyls. While CO does have a free electron pair on both the carbon and oxygen, the molecule differs sharply from conventional Lewis bases such as NH_3 in that the CO molecule does *NOT* combine readily with a proton to give a species such as HCO^+. NH_4^+ is, of course, very common. In only a few cases such as the combination of CO and O_2 to give CO_2, or CO and S to give COS does the electron pair on carbon engage in obvious Lewis-base type of chemical bonding. The existence of stable metal carbonyls required an extension of the Lewis model in order to conceptually differentiate coordination compounds of CO and NH_3 and to avoid the build-up of electron density on the central metal atom in species such as $Ni(CO)_4$ and $Fe(CO)_5$.

The idea of π-bonding or back-bonding, first implied as early as 1926 (7), was developed by many workers to avoid these problems. The modern model suggests that the Lewis electron pair on the ligand forms a sigma bond with the Lewis acid and that the d-electrons on the metal atoms (nominal Lewis acid) form linkages involving the empty antibonding orbitals on the CO molecules. For F_3P empty d-orbitals on the phosphorus may be used. This "back-bonding model" is still the most widely accepted explanation for metal carbonyls. Thus CO and F_3P in metal-CO or metal-F_3P complexes such as $Ni(CO)_4$ or $Ni(PF_3)_4$ are designated as π-*acids* since it is visualized that they *accept* electrons given to them by the metal to form bonds of π-symmetry. The π-bonding model works well for transition metal carbonyls in which d-electrons are available. It is widely used today.

On the other hand, the π-bonding model does have problems. For example, the compound borane carbonyl, $OCBH_3$, was prepared by Burg and Schlesinger (8) even though the boron atom contains no d-electrons to "back-bond" to the coordinated CO. It would appear that in this case the conventional π-bonding carbonyl explanation is not applicable. Similarly, the compound F_3PBH_3 was prepared by Parry and Bissot (9). Since the conventional d-electron donation from boron to ligand in each molecule was not possible, speculative "hyperconjugation" was proposed (10). One of the contributing resonance forms suggested is (10):

$$H^+$$

$$H-\overline{B}=C=O$$
$$\vert$$
$$H$$

It is significant that no acidic character due to H^+ has ever been observed (*11*) with H_3BCO. In contrast, H_3BCO reacts with KOH to give $K_2[H_3BCO_2]$. The negative OH^- attaches to the carbon of the H_3BCO to give:

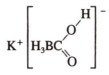

The acid proton of the OH group is then neutralized by a second molecule of KOH to give $K_2[H_3BCO_2]$. The overall equation is:

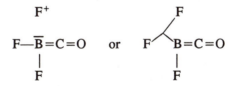

The process is comparable to the reaction of CO_2 and KOH. The salt $K_2[H_3BCO_2]$ is called potassium boranocarbonate (*12*).

It is also significant that all attempts to make F_3BCO or F_3BPF_3 have been <u>un</u>successful. This behavior has been rationalized using hyperconjugation models and is explained by the difficulty encountered in generating F^+ or multicentered bonds for the hyperconjugated form of the still unknown F_3BCO:

$$
\begin{array}{ccc}
F^+ & & F \\
| & & \diagup \\
F\!-\!\overline{B}\!=\!C\!=\!O & \text{or} & F^{\diagdown}\,\,\overline{B}\!=\!C\!=\!O \\
| & & | \\
F & & F
\end{array}
$$

Similarly, Cl_3BPF_3 and Cl_3BCO are unknown. Difficulty in generating Cl^+ has been used to rationalize this fact. Such an argument suggests that group III halides (*i.e.*, BF_3, $AlCl_3$, *etc.*) should not combine with F_3P at all since X^+ hyperconjugation or multicentered bonding would be needed. It is then significant that the halogen compound Cl_3AlPF_3 has been characterized at temperatures below -20°C (*13*). *At higher temperatures halogen interchange to give AlF_3 and Cl_3P occurs rather than dissociation back to $AlCl_3$ and F_3P* (*13*).

Some structural anomalies of the H_3B adducts of F_3P and R_3P (R = H or alkyl) are also of interest. Microwave studies by Kuczkowski and Lide (*14*) on F_3PBH_3 showed that the B-P bond is the *shortest* yet recorded for complexes containing 4-coordinate boron and phosphorus (1.836 ± .012

Å) while the enthalpy for the dissociation reaction at 25°C was very *low*. The equation is:

$$2F_3PBH_{3(g)} \longrightarrow 2F_3P_{(g)} + B_2H_{6(g)} \qquad \Delta H = 11 \text{ kcal}$$

For the process: $F_3PBH_{3(g)} \rightleftharpoons F_3P_{(g)} + BH_{3(g)}$ a ΔH value of 24 kcal was reported. These numbers are consistent with the equation shown below for the formation of B_2H_6 from $2BH_3$:

$$2BH_{3(g)} \longrightarrow B_2H_{6(g)} \qquad \Delta H = -38 \text{ kcal}$$

The estimate for the comparable P-B distance in $(CH_3)_3PBH_3$ is 1.901 Å. The enthalpy for the dissociation process is estimated as at least 22 kcal (*14b*):

$$2(CH_3)_3PBH_3 \longrightarrow 2(CH_3)_3P + B_2H_6 \qquad \Delta H = 22 \text{ kcal}$$

One sees the anomaly of the short bond in F₃PBH₃ being significantly weaker than the long bond in (CH₃)₃PBH₃ if dissociation energies are used as a reasonable criterion of bond strength.

The entire group of facts can be rationalized by a model in which both the CO and the F_3P bind to the BH_3 by relatively weak σ bonds. Because in both cases the coordinating electron pair on the ligand is pulled back toward the donor atom (*i.e.*, C in CO or P in F_3P) by the electronegative oxygen or fluorine atoms, this means that for an acid-base coordination reaction to occur, close approach of the electron donor atom to the electron acceptor atom is needed (short bond). The Lewis *acid species* which accepts the electron pair must be distorted to a new geometry to permit bonding. With BH_3 and BF_3 one needs distortion from a planar to a tetrahedral geometry. Because of the easy polarizability of the B-H linkages, the tetrahedral geometry required for F_3PBH_3 is acquired with relatively little expenditure of energy (*15,16*). However, the closer the acid and base must approach to get bonding interaction, the higher the energy required for distortion of the acid component. The higher the distortion energy required, the weaker the overall B-P bond will be. The F_3PBH_3 bond is *short*, but the dissociation energy (bond strength) is low because very close approach is needed to generate a bond, and distortion energy of the acid weakens the bond. The compound has low stability.

Because $AlCl_3$ requires *low* distortion energy to give the 4-coordinate structure (*i.e.*, Al_2Cl_6 shows Al with a coordination number of four), F_3P can approach close enough to generate a P-Al bond. Because BF_3 has a relatively high distortion energy, it will not combine with F_3P or CO. BF_3 will, however, combine with $(CH_3)_3P$ because the electron cloud on the phosphorus extends well out from the phosphorus nucleus (*13*). The foregoing arguments also explain the experimentally observed fact that HF_2P will displace F_3P from F_3PBH_3 quantitatively (*17*) — an observation which suggests that the $HF_2P\text{-}BH_3$ bond is stronger than its $F_3P\text{-}BH_3$ counterpart. Microwave data by Pasinski and Kuczkowski (*18*) show that the B-P bond distance in HF_2PBH_3 (1.832 ± .009 Å) is essentially identical to that in F_3PBH_3 (1.836 ± .012 Å). On the other hand, the presence of only 2 fluorine atoms to pull back the free-electron cloud on phosphorus in HF_2P rather than the three fluorine atoms in F_3P suggests that the bonding cloud should extend somewhat farther from the phosphorus nucleus in HF_2P. Under these circumstances a better overlap and stronger bond would be anticipated for the P-B bond in HF_2PBH_3 than for that in F_3PBH_3 (distances are equal). This model would also suggest the known fact that H_3PBH_3 is less stable than either HF_2PBH_3 or F_3PBH_3 because the cloud on the phosphorus of H_3P has not been pulled back and is too diffuse to interact effectively with the BH_3 group. Making the free cloud on phosphorus more constrained increases the bonding strength. The distortion energy of the Lewis acid acts to reduce the overall bond strength. At HF_2P a maximum overall bond strength is experimentally observed. These arguments are summarized in Tables I and II.

Coordination Compounds Containing Three Center B—H—→M Bridge Bonds

More recently another extension of Lewis-type coordination theory has been developed. In 1959 Parry and Edwards (*20*) suggested that the 3-center B—H—→B bond as found in boron hydrides could be viewed as an extension of the Lewis coordinate covalent bond. Then linkages such as B—H—→M, which exist in borohyrides, *etc.*, would represent further extensions of coordination theory. The authors wrote: "in such interaction the formation of a bridge bond replaces the conventional donor-acceptor bond of classic coordination theory." In opposition to this postulate one can argue very reasonably that linkages such as those found in $(\emptyset_3P)_2CuBH_4$, *etc.*, which contain B—H—→Metal linkages are really held together primarily by the polarized ionic attraction of Cu^+ for BH_4^-, and the concept of the B—H—→Cu bond as a "coordinate covalent" or "modified Lewis coordinate" bond is debatable. To answer such a criticism one needs an uncharged ligand capable of forming a B—H—→M linkage and a metal atom or ion to bind to the ligand. Such a ligand and metals to bind with it have been found, and a variety of such coordination compounds were prepared and characterized by Steve Snow in 1985 (*21*). Equations for the formation of typical compounds are:

Table I. Postulates

1. Bonds to F_3P require a close approach of the Lewis acid because the free-electron pair of phosphorus is pulled back by fluorine atoms attached to phosphorus.

2. As a Lewis acid moves in toward a free-electron pair of F_3P the Lewis acid geometry must be changed (*i.e.*, BH_3 or BF_3 must go from planar to modified tetrahedral geometry).

3. The closer the acid and base must approach to get bonding interaction, the higher the distortion energy which must be applied to the acid.

4. The distortion energy required to convert planar BH_3 to tetrahedral LBH_3 has been estimated from force constants to be about *10* kcal, while that for BF_3 is about *30* kcal (*15,19*). See reference (*16*) for other estimates; all show that the distortion energy of BF_3 is much larger than that of BH_3.

5. If one starts with B_2H_6, the energy of interest is *not* the distortion energy, but the displacement energy of one BH_3 group from another to give 2 BH_3 groups. The best estimates for this are about 17 to 19 kcal/BH_3.

Table II. Conclusions

1. BH_3, with a small distortion energy, can approach close enough to F_3P and HF_2P to form a bond which is *short* and fairly weak.

2. BF_3, with a large distortion energy, can *NOT* approach close enough to F_3P to bind.

3. $AlCl_3$, with a relatively low distortion (or displacement) energy, can approach F_3P closely enough to form a bond. Instability is a result of halogen exchange, not cleavage of the Al-P bond.

4. It is suggested that the unusual stability of HF_2PBH_3 compared to F_3PBH_3 is due to the fact that the two fluorines on HF_2P do not pull the electron pair back as far as the three fluorines on F_3P. As a result stronger overlap occurs at the bonding distance (1.83 Å). When H_3P is the Lewis base instead of F_3P, the electron cloud is so diffuse and the distance is so large that the bond is weak.

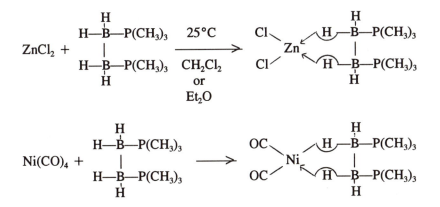

A copper chloride complex was also prepared of formula:

$$(\varnothing_3P)(Cl)Cu(B_2H_4 \cdot 2PCH_3)_3$$

The nature of the bonding is of interest. Since pi-bonding is not really a major item for zinc complexes, one would surmise that the B—H—→M linkage with $ZnCl_2$ is a bond of distorted sigma symmetry. More direct evidence of the sigma nature of the B—H—→M bond was provided by the infrared spectrum of $(OC)_2Ni[H_4B_2 \cdot 2P(CH_3)_3]$. Conventional wisdom (*22a-22b*) indicates that a ligand such as F_3P which is a poor electron donor but a good π-acceptor when it is bound to the metal should accept electrons *from the metal*. The resulting decrease in electron density on the metal should reduce the ability of the metal to donate electrons to the antibonding CO orbital of the remaining coordinated CO (*i.e.*, electrons flow from metal to PF_3 not to CO). Thus the antibonding CO orbital is less populated, and the CO stretching frequency should approach that of free CO. The structure would be represented better as M—C≡O rather than as M=C=O. The infrared data for $Ni(CO)_4$ and $Ni(CO_2)(PF_3)_2$ illustrate the point. For $Ni(CO)_4$ the asymmetric CO stretching frequency (*22-23b*) (CCl_4 solvent) is 2044 cm⁻¹ (for gaseous $Ni(CO)_4$ the absorption is at 2057 cm⁻¹). The corresponding asymmetric CO stretching frequency (*22-23b*) (hydrocarbon solvent) for $Ni(CO)_2(PF_3)_2$ is 2052 cm⁻¹. Because PF_3 is a better π-acid than CO, the PF_3 receives the antibonding CO electrons and the frequency rises. Conversely, if the ligand bonded to the metal is a good sigma-donor but a less effective π-acceptor, the metal, receiving σ electrons from the ligand, should have more π electron density to contribute to the antibonding orbitals of CO, and the bonding should approach the double bond of CO. The infrared frequencies should then show that the CO bond is closer to a double bond pattern than to a triple bond. Under these circumstances the bonding would be represented better as $M=C=O$ rather than as $M-C\equiv O$. When two molecules of the good sigma-donor $(CH_3)_3P$ replace two CO molecules of $Ni(CO)_4$, the asymmetric CO frequency of the remaining CO molecules should fall. Data confirm this expectation. The asymmetric stretching frequency (23) falls from 2044 cm^{-1} for $Ni(CO)_4$ to 1940 cm^{-1} for $Ni(CO)_2[P(CH_3)_3]_2$. Strong sigma-bonding ligands shift the CO stretching frequency down toward the $C=O$ region.

For $Ni(CO)_2[H_4B_2\cdot 2P(CH_3)_3]$ the CO asymmetric stretching frequency is found at 1909 cm^{-1}, indicating clearly that a distorted sigma interaction is being observed.

More recently, Shimoi (24) and his co-workers have prepared and structurally characterized by X-ray diffraction a number of transition metal carbonyls such as $Cr(CO)_4L$, $Mo(CO)_4L$, and $W(CO)_4L$ where L is the didentate ligand $H_4B_2\cdot 2P(CH_3)_3$. They have also characterized $Cr(CO)_5L$ and $W(CO)_5L'$ where L' is *monodentate* $H_4B_2\cdot 2P(CH_3)_3$. They also prepared a number of other monodentate ligand carbonyls such as $Cr(CO)_5(H_3B\cdot Base)$ and $W(CO)_5(H_3B\cdot Base)$ where bases attached to BH_3 were $P(CH_3)_3$, $N(CH_3)_3$, and $P(C_6H_5)_3$. In every case the bonding is of the form:

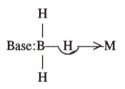

Structures of $Cr(CO)_4[H_4B_2\cdot 2P(CH_3)_3]$ and $Cr(CO)_5[H_4B_2\cdot 2P(CH_3)_3]$ taken from reference (24b) are shown in Figures 1 and 2.

Finally, the relationship between these Boron—H—>Metal interactions and the so-called *agostic* (25) Carbon—H—>Metal bonds is of interest since these *agostic* interactions are currently considered to be of interest in some processes such as hydrogen activation by metals. In general, agostic hydrogen interactions giving C—H—>M linkages seem to be significantly weaker than the B—H—>M bonds noted above. To the best of the author's knowledge, there are no cases in which stable ligand bonding through a C—H—>M agostic bond serves as the sole linkage of a ligand in a stable coordination compound. In essentially all agostic hydrogen linkages the ligand is held by a normal covalent bond or a coordinate covalent bond between ligand and metal, and the agostic hydrogen bond is a secondary interaction which alters the strength of the C-H linkage. One can rationalize the greater strength of the B—H—>M bond as compared to the C—H—>M bond in terms of the lower nuclear charge on boron as compared to carbon and the greater polarizability of the B-H bond, which results from the smaller nuclear charge on boron.

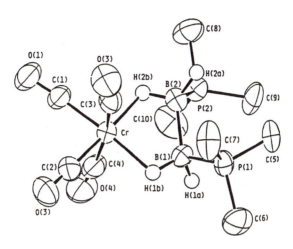

Figure 1. ORTEP diagram of $Cr(CO)_4[H_4B_2 \cdot 2P(CH_3)_3]$. Note that the borane ligand is didentate. (Reproduced from reference 24b. Copyright 1992 American Chemical Society.)

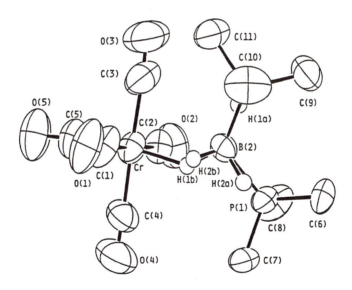

Figure 2. ORTEP diagram of $Cr(CO)_5[H_4B_2 \cdot 2P(CH_3)_3]$. Note that the borane ligand is monodentate. (Reproduced from reference 24b. Copyright 1992 American Chemical Society.)

The key to the formation of an [element—\A⟋—⟶M] linkage in which A is an element such as hydrogen or fluorine would appear to be related to the polarizability of the electron cloud in the [element—⟶A] bond. An [element—⟶A] bond of low polarizability will not form a multicenter bond with a metal. Only a bond of high polarizability can form a link to the metal. Thus in a species such as LBH_3 one would expect that the most stable metal complex should be one in which L is a very good electron donor and the H-B bond would be easily polarized or distorted. The idea is yet to be tested systematically. Other concepts of coordination theory can be applied.

Literature Cited

1. Bailar, Jr., J.C. In *Chemistry of the Coordination Compounds*; Bailar, Jr., J.C.; Busch, D.H., Eds.; American Chemical Society Monograph 131; Reinhold Publishing Corp.: New York, NY, 1956; (a) pp 100-108; (b) pp 108-118; (c) 110.

2. Lewis, G.N. *J. Am. Chem. Soc.* **1916**, *38*, 778.

3. Huggins, M.L. *Science* **1922**, *55*, 459.

4. (a) Mond, L.; Langer, C.; Quincke, F. *J. Chem. Soc.* **1890**, *57*, 749.

    (b) Mond, L.; Langer, C.; Quincke, F. *J. Soc. Chem. Ind.* **1895**, *14*, 945.

    (c) Mond, L.; Langer, C. *J. Chem. Soc.* **1891**, *59*, 1090 (for a more complete listing of carbonyl references see reference 5(a)).

5. (a) Mattern, J.A.; Gill, S.J. In *Chemistry of the Coordination Compounds*; Bailar, Jr., J.C.; Busch, D.H., Eds.; American Chemical Society Monograph 131; Reinhold Publishing Corp.: New York, NY, 1956; pp 509-546 (see references 1-154 for more historical background).

    (b) Hieber, W.; Lagally, H. *Z. Anorg. Chem.* **1943**, *251*, 96.

6. (a) Blanchard, A.A. *Chem. Rev.* **1937**, *21*, 3.

    (b) Blanchard, A.A.; Gilmont, P. *J. Am. Chem. Soc.* **1940**, *62*, 1192.

7. Keller, R.N.; Parry, R.W. In *Chemistry of the Coordination Compounds*; Bailar, Jr., J.C.; Busch, D.H., Eds.; American Chemical Society Monograph 131; Reinhold Publishing Corp.: New York, NY, 1956; p 192 (see footnote (*) for an early historical note).

8. Burg, A.B.; Schlesinger, H. *J. Am. Chem. Soc.* **1933**, *55*, 4009.

9. Parry, R.W.; Bissot, T.C. *J. Am. Chem. Soc.* **1956**, *78*, 1524.

10. For a summary see: Wade, K. In *Electron Deficient Compounds*; Waddington, T.C., Ed.; Studies in Modern Chemistry 8; Appleton-Century Crofts, Education Division, Meredith Corp.: New York, NY, 1971; p 81.

11.   One of the referees pointed out, quite properly, that the representation of H_3BCO with an ionic hydrogen is extreme. A more reasonable representation would involve a multicentered bond of the form:

The same arguments about acidity are still applicable because the acidity of bridge hydrogens in boron hydrides is well established.

12.   (a) Malone, L.J.; Parry, R.W. *Inorg. Chem.* **1967**, *6*, 817.
       (b) Carter, J.C.; Parry, R.W. *J. Am. Chem. Soc.* **1965**, *87*, 2354.
       (c) Parry, R.W.; Nordman, C.E.; Carter, T.C.; TerHaar, G.L. In *Boron-Nitrogen Chemistry*; Gould, R.F., Ed.; Advances in Chemistry Series 42; American Chemical Society: Washington, D.C., 1964, pp 302-311.

13.   Alton, E.R.; Montemayor, R.G.; Parry, R.W. *Inorg. Chem.* **1974**, *13*, 2267.

14.   (a) Kuczkowski, R.L.; Lide, D.R. *J. Chem. Phys.* **1967**, *46*, 357.
       (b) Bryan, P.S.; Kuczkowski, R.L. *Inorg. Chem.* **1972**, *11*, 553.

15.   Alton, E. R. *The Reaction of Phosphorus Trifluoride with Aluminum Chloride and Related Studies*; Ph.D. Dissertation, University of Michigan: Ann Arbor, MI, 1960; (a) pp 37-48; (b) pp 72-75.

16.   Cassoux, P.; Kuczkowski, R.L.; Serafini, A. *Inorg. Chem.* **1977**, *16*, 3005.

17.   Rudolph, R.W.; Parry, R.W. *J. Am. Chem. Soc.* **1967**, *89*, 1621.

18.   Pasinski, J.P.; Kuczkowski, R.L. *J. Chem. Phys.* **1971**, *54*, 1903.

19.   Parry, R.W. In *Coordination Compounds Containing Fluorophosphine Ligands*; Kirschner, S., Ed.; Coordination Chemistry: Papers Presented in Honor of Professor John C. Bailar, Jr.; Plenum Press: New York, NY, 1969; pp 207-216.

20.   Parry, R.W.; Edwards, L.J. *J. Am. Chem. Soc.* **1959**, *81*, 3554.

21.   Snow, S.A. *Studies of Boranes and Metalla Boranes: I. Metal Complexes of Bis(trimethylphosphine)-Diborane(4). II. Synthesis and Characterization of 2-(Dichloroboryl)pentaborane(9)*; Ph.D. Dissertation, University of Utah: Salt Lake City, UT, 1985; see Abstract.

22.   (a) Cotton, F.A.; Wilkinson, G. *Advanced Inorganic Chemistry: A Comprehensive Text*; 5th Ed.; John Wiley & Sons: New York, NY, 1988; pp 57-71.
       (b) Jones, L.H. *Inorganic Vibrational Spectroscopy*; Marcel Dekker Inc.: New York, NY, 1971; Vol. 1; pp 159-161.

23.   Braterman, P.S. *Metal Carbonyl Spectra*; Organometallic Chemistry
      12; Academic Press: London, England, 1975; (a) p 190 (for
      Ni(CO)₄); (b) p 206 (for Ni(CO)₂L₂).
24.   (a) Shimoi, M.; Katoh, K.; Tabita, H.; Ogino, H. *Inorg. Chem.*
      1990, *29*, 814.
      (b) Kato, K.; Shimoi, M.; Ogino, H. *Inorg. Chem.* **1992**, *31*, 670.
25.   Brookhart, M.; Green, M.L.H.; Wong, L.L. In *Carbon-Hydrogen-
      Transition Metal Bonds*; Lippard, S.J., Ed.; Progress in Inorganic
      Chemistry; John Wiley & Sons: New York, NY, 1991; Vol. 36;
      pp 1-124.

RECEIVED February 2, 1994

Chapter 28

Electron-Deficient Boranes as Novel Electron-Donor Ligands

Norman N. Greenwood

School of Chemistry, The University of Leeds, Leeds LS2 9JT, England

The history of the concept of "boranes as ligands" will be traced from its origins in the mid-1960s to the present day. Polyhedral boranes and their anions are often classified as electron deficient species but, in fact, many can act as excellent polyhapto ligands to metal centers, thereby forming coordination complexes that are often more stable than the parent borane species. Furthermore, many of the binary boranes and their anions can themselves usefully be regarded as coordination complexes of a borane ligand and a borane acceptor, *i.e.*, a borane-borane adduct. The exciting synthetic, structural, and bonding implications of these ideas will be outlined by referring to key compounds and reactions in the literature.

Boron is the element immediately preceding carbon in the periodic table and so has one less electron than orbitals available for bonding. As a consequence, many of its molecular compounds are "electron deficient" in the sense that there are insufficient electrons to form two-center two-electron bonds between each contiguous pair of atoms. The classic examples of this situation are the binary boron hydrides and the carbaboranes, which relieve their so-called electron deficiency by forming polyhedral cluster molecules in which pairs of electrons simultaneously bond more than two atoms by means of three-center or polycenter bonds (*1*). In diagrams depicting the structure of these compounds it is important to note that straight lines between the atoms do not necessarily indicate pairs of electrons but merely delineate the geometrical shape of the cluster. Another way of relieving the electron deficiency is for the borane species to act as an electron-pair acceptor by interaction with a Lewis base, L, *e.g.*, LBH_3 where L = CO, Me_2O, Me_2S, Me_3N, H^-, *etc.*

About 30 years ago, in the mid-1960s, we began to realize that, far from being deficient in electrons, many boranes and their anions could act as very effective

0097–6156/94/0565–0333$08.00/0

polyhapto ligands. That is, they could form donor-acceptor complexes or coordination compounds in which the borane cluster was itself acting as the electron donor or ligand. This paper explores the implications of this astonishing concept and shows how it extended enormously the range of boron hydride cluster compounds that can be made. By 1974 the first review on the topic appeared (2), and by 1978 the 19th ICCC in Prague was able to hold a microsymposium on "Boranes as Ligands" (3). Numerous other reviews of various aspects of the subject have appeared during the ensuing 15 years, of which the following are representative (1, 4-12). A parallel literature exists on carbaborane species as ligands (e.g., 13-19). Indeed, an important harbinger of the evolving idea of binary boranes as ligands was M. Frederick Hawthorne's seminal recognition in 1965 that the *nido*-dicarbaborane anion, $C_2B_9H_{11}{}^{2-}$, could act as a pentahapto analogue of the cyclopentadienide ligand $C_5H_5{}^-$ in organometallic compounds (20). However, there were no examples (until the following year) of binary boranes themselves acting in this way in the absence of carbon. Still less was it recognized that the B-H bond itself could donate its electron pair to form a (bent) B-H-M three-center two-electron bond.

It is now clear that, whether by forming B-H-M bonds or by forming direct boron-metal bonds, binary borane species can act as excellent ligands, and all hapticities from η^1 to η^6 (and occasionally beyond) are known. Typical examples will be described in the following sections, but the treatment is intended to be illustrative rather than exhaustive. In particular, emphasis will be placed on the chemical significance of the various examples selected.

The Tetrahydroborate Anion as a Ligand

The simplest binary borane species is the tetrahydroborate anion, $BH_4{}^-$, which is isoelectronic with CH_4 and $NH_4{}^+$. Many salts of this anion, such as $LiBH_4$ and $NaBH_4$, are essentially ionic and have been used for more than 50 years as versatile reducing agents (1, 21). However, $BH_4{}^-$ can also react by ligand displacement to form covalently bonded complexes in which it acts as a monohapto, dihapto, or trihapto ligand:

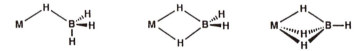

Examples are $[Cu(\eta^1\text{-}BH_4)(PMePh_2)_3]$, $[Cu(\eta^2\text{-}BH_4)(PPh_3)_2]$, and $[Zr(\eta^3\text{-}BH_4)_4]$. The hapticity is influenced both by the steric requirements of the coligands (e.g., $PMePh_2$ is less demanding than PPh_3) and by the size of the central metal atom (e.g., Zr, which can accommodate simultaneous ligation by 12 H atoms). The dihapto and trihapto modes are much more common than the monohapto mode, and there are also examples of more complex bonding patterns, such as the polymeric complexes $[M(BH_4)_4]$ of the very large actinide elements Th, Pa, and U, in which the 14-coordinate central metal atom is surrounded by two $\eta^3\text{-}BH_4{}^-$ and four bridging bis(didentate) μ-η^2,η^2-BH_4^- ligands. A particularly important example of the dihapto mode is the octahedral Al complex, $[Al(\eta^2$-$BH_4)_3]$, which was the first covalent borohydride to be characterized (1940) and also the first compound in which the now widespread phenomenon of fluxionality was observed, all 12 H atoms being equivalent on the nmr timescale (*22*).

Now comes another important new idea: if binary borane species can act both as electron-pair donors and electron-pair acceptors, can we consider the boron hydrides themselves to be coordination complexes of borane ligands and borane acceptors, *i.e.*, as borane-borane complexes? For example, B_2H_6 could formally be regarded either as a coordination complex of η^2-BH_4^- with the notional cation BH_2^+ or as a dimer formed by the mutual coordination of two monodentate BH_3 units. Replacement of the η^2-BH_4^- or η^1-BH_3 ligands by suitable (stronger) Lewis bases would then result in the well known diborane reactions of unsymmetrical (heterolytic) and symmetrical (homolytic) cleavage, respectively (*23, 24*). This is illustrated in Figure 1.

B-H-M Interactions Involving Higher Boranes

The ideas developed in the preceding section can readily be extended to the higher boranes. The "butterfly" structure of B_4H_{10} can be considered as a coordination complex formed either by the mutual donor-acceptor interaction of the monodentate ligand-acceptor moieties BH_3 and B_3H_7 or as the chelation of BH_2^+ by the known anion $B_3H_8^-$ acting as a dihapto ligand. Accordingly, B_4H_{10} can be cleaved by stronger ligands either homolytically (*e.g.*, with L = NMe_3) to give LBH_3 plus LB_3H_7 or heterolytically (*e.g.*, with L = NH_3) to give $[L_2BH_2]^+[B_3H_8]^-$ (*24*). Likewise, ligand replacement reactions in which, for example, the halide ions of classical coordination complexes are displaced by the $B_3H_8^-$ anion can lead to a variety of metallated derivatives of tetraborane such as the tetrahedrally coordinated copper complex, $[Cu(\eta^2$-$B_3H_8)(PPh_3)_2]$, in which the wing-tip $\{BH_2^+\}$ group has been subrogated by the isolobal $\{Cu(PPh_3)_2^+\}$ group (*25*). The octahedrally coordinated manganese complex, $[Mn(\eta^2$-$B_3H_8)(CO)_4]$, is an even more instructive example of this structure type since, when it is heated to 180°C or irradiated with ultraviolet light, it loses one of the four CO ligands and a further B-H group coordinates to give the trihapto complex, *fac*-$[Mn(\eta^3$-$B_3H_8)(CO)_3]$ (*26*). Treatment of this product with an excess of CO under moderate presssure results in the reformation of the original dihapto species. These reactions are all clearly best represented as straightforward ligand replacement reactions.

A more complex example of trihapto coordination of a borane to a metal centre via B-H-M bonds is afforded by *fac*-$[Mn(\eta^3$-$B_8H_{13})(CO)_3]$, in which a triangular face of three B-H groups coordinates to the octahedral manganese atom (*27*). Chelation and bridging via pairs of edge-related B-H groups are also known. For example, the *closo*-dianion, $B_{10}H_{10}^{2-}$, acts as a bis(didentate) bridging ligand in the binuclear Cu(I) complex, $[\{Cu(PPh_3)_2\}_2(\mu$-η^2,η^2-$B_{10}H_{10})]$ (*28*). This type of bonding had first been noted in the closely related polymeric structure of $[Cu_2B_{10}H_{10}]$ (*29*), which features octahapto tetra-bridging coordination of *closo*-$B_{10}H_{10}^{2-}$, involving ligation from four diagonally related pairs of axial-equatorial B-H bonds: (1)(2), (1)(4), (6)(10), and (8)(10).

So far, all the examples discussed of boranes acting as ligands have involved B-H-B or B-H-M bonding. The following section broadens the discussion by considering compounds in which there is direct ligation from boron to metal, thereby leading to a variety of coordination compounds in which the borane moiety displays hapticities from η^1 upwards.

Direct η^1 and η^2 Bonding from Borane Ligands to Metals

Notional replacement of a terminal H atom in a borane by a metal center results in the

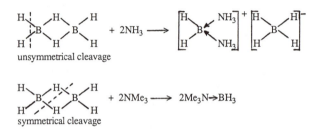

Figure 1. The classical diborane reactions of unsymmetrical and symmetrical cleavage, seen as ligand displacement reactions involving NH_3 or the sterically more demanding NMe_3

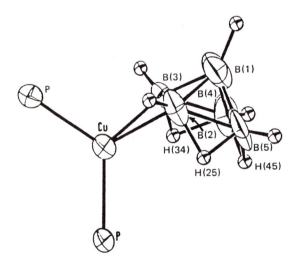

Figure 2. The molecular structure of $[Cu(\eta^2-B_5H_8)(PPh_3)_2]$

formation of an η^1-bonded metalloborane. Perhaps the first structurally characterized example of this bonding mode was in the octahedral Ir(III) complex, [Ir(η^1-2-B_5H_8)Br_2(CO)(PMe$_3$)$_2$], which arose, curiously, from the oxidative addition of either 1- or 2-BrB$_5H_8$ to [*trans*- Ir(CO)Cl(PMe$_3$)$_2$] (*30*). Other early representatives were [M(η^1-2-B$_5$H$_8$)(CO)$_5$] (M = Mn or Re), made by direct reaction of 1- or 2-XB$_5$H$_8$ with NaM(CO)$_5$ (*31*) and the Fe(II) complex, [Fe(η^1-2-B$_5$H$_8$)(η^5-C$_5$H$_5$)(CO)$_2$], which was made in almost quantitative yield by treating the corresponding organometallic iron iodide with KB$_5$H$_8$ (*32*). Examples of η^1-B-M bonding are now also known for several other boranes (*11*).

The B$_5$H$_8^-$ ion more frequently gives η^2 complexes when used in ligand-replacement reactions. Thus deprotonation of B$_5$H$_9$, by means of KH, for example, leads to B$_5$H$_8^-$, in which two basal boron atoms are joined by a direct 2-center 2-electron B-B bond. This can donate its electron density to a proton, thereby regenerating B$_5$H$_9$, or it can donate to an isolobal metal center via a ligand displacement reaction involving, typically, a metal-halogen bond. One of the first examples of this bonding mode was [Cu(η^2-B$_5$H$_8$)(PPh$_3$)$_2$] (*33*), and the structure of the complex, as determined by X-ray diffraction analysis (*34*), is shown in Figure 2. Similar complexes of Ni, Pd, Pt; Ag, Au; Cd, Hg;Si, Ge, Sn, Pb, and even B are now also known (*8*).

A particularly important example of this type of behavior by a neutral borane molecule is the reaction of *nido*-B$_6$H$_{10}$ with Zeise's salt (*35*):

$$K[Pt(\eta^2\text{-}C_2H_4)Cl_3] + 2B_6H_{10} \xrightarrow{\hspace{1cm}} \textit{trans-}[Pt(\eta^2\text{-}B_6H_{10})_2Cl_2] + C_2H_4 + KCl$$

The molecular structure of the yellow crystalline product is shown in Figure 3 (*36*). A significant feature of the B$_6$H$_{10}$ ligands in this square-planar Pt complex is that, within the triangular 3-center B-Pt-B bond, the B-B distance of 182 pm is considerably longer than that of the basal B-B bond in the uncoordinated ligand (162 pm) and is now typical of the B-B distances in triangulated polyhedral borane clusters. Related complexes of Rh and Ir can be prepared by similar reactions (*35*). Likewise, the reaction of B$_6$H$_{10}$ with [Fe$_2$(CO)$_9$] at room temperature results in the smooth elimination of [Fe(CO)$_5$] and the formation of [Fe(η^2-B$_6$H$_{10}$)(CO)$_4$] as a volatile yellow solid (*35, 37*).

An imaginative application of the η^2-ligation of B$_6$H$_{10}$ to prepare previously unknown boron hydrides was devised by Riley Schaeffer. For example, it was known that the *nido*-borane B$_9$H$_{13}$ has only transient existence unless stabilized by a ligand such as SMe$_2$; could B$_6$H$_{10}$ play this rôle? In the event it was shown that B$_6$H$_{10}$ reacted essentially quantitatively with *i*-B$_9$H$_{15}$ to form the new *conjuncto*-borane B$_{15}$H$_{23}$ with the loss of one mole of H$_2$ (*38, 39*), and subsequent X-ray analysis showed (*40*) that the structure was indeed composed of a B$_6$H$_{10}$ unit bonded by η^2 donation to a B$_9$H$_{13}$ unit as shown in Figure 4. Similarly, the novel *conjuncto*-borane B$_{14}$H$_{22}$ was made by addition of B$_6$H$_{10}$ to B$_8$H$_{12}$ (*39*).

Trihapto-B_3M Bonding

The best characterized trihapto-bonded metallaborane is the white, air-stable Ir complex $[Ir(\eta^3\text{-}B_3H_7)(CO)H(PPh_3)_2]$, which was prepared by the stoichiometric reaction of TlB_3H_8 with $trans\text{-}[Ir(CO)Cl(PPh_3)_2]$ (41). There is an effective transfer of a H atom from the borane to the Ir atom, and the structure (Figure 5) (42) is best visualized as that of B_4H_{10} with one of its two hinge {BH} groups replaced by a {$Ir(CO)H(PPh_3)_2$} group. The first examples of this structure type were with Ni, Pd, and Pt (43), but severe disorder problems beset the structural analysis at that time. A detailed discussion of bonding implications is in ref. 42, and further examples are in ref. 11.

Trihapto ligation was also observed when $B_5H_8^-$ was reacted with trans-$[Ir(CO)Cl(PPh_3)_2]$: instead of the expected displacement of Cl by $\eta^2\text{-}B_5H_8^-$ as in the preceding section, further reaction ensued to give $[Ir(\eta^3\text{-}B_5H_8)(CO)(PPh_3)_2]$, which, as shown in Figure 6, is a structural analogue of $nido\text{-}B_6H_{10}$ with one basal {BH_tH_μ} group replaced by the Ir(III) center (44).

Tetrahapto B_4M Bonding

One of Alfred Stock's classic ways of making boranes in the 1920s was by the thermolysis or cothermolysis of smaller boranes. We therefore thought that cothermolysis of boranes with volatile coordination complexes might provide a viable route to metallaboranes in which the borane moiety could be regarded as a polyhapto ligand. The first example of this technique was the cothermolysis of $nido\text{-}B_5H_9$ with $Fe(CO)_5$ in a hot/ cold reactor at $220° /20°C$ (45). The yield of the orange liquid product, $[Fe(\eta^4\text{-}B_4H_8)(CO)_3]$ (Figure 7), depends critically on conditions, and the reaction probably proceeds via the intermediate formation of {$Fe(B_5H_9)(CO)_3$}, followed by disproportionation to the product and the unstable $[Fe(B_6H_9)(CO)_3]$. The compound clearly has the same structure as $nido\text{-}B_5H_9$ but with the apical {BH} unit replaced by the isolobal {$(Fe(CO)_3$} unit. Compounds (prepared by other routes) in which further {BH} units have been replaced are also known, right through to the metal-metal cluster compound $[Fe_5C(CO)_{15}]$, i.e., $[\{Fe(CO)_3\}_5C]$, cf. $[(BH_t)_5(H_\mu)_4]$, (i.e., B_5H_9). It is also worth noting that $[Fe(\eta^4\text{-}B_4H_8)(CO)_3]$ is precisely isoelectronic with the well known cyclobutadiene complex $[Fe(\eta^4\text{-}C_4H_4)(CO)_3]$, both compounds being examples of the stabilization of otherwise fugitive ligands by coordination. The isoelectronic compound $[Co(\eta^4\text{-}B_4H_8)(\eta^5\text{-}C_5H_5)]$ (46) should also be mentioned because its preparation actually predated that of the iron analogue.

Clusters of Clusters

In all the compounds mentioned so far the metal acceptor has been isolobal with either {H}, or {BH_t}, or {BH_tH_μ}. But, we argued, other cluster vertex bonding geometries might be possible, for example, that provided by square-planar Pt(II), which would permit the construction of previously unknown cluster geometries. The first example of this (47) was $[Pt_2(\mu\text{-}\eta^3\text{-}B_6H_9)_2(PMe_2Ph)_2]$, in which each of the two pentagonal

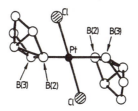

Figure 3. The molecular structure of *trans*-[Pt(η^2-B_6H_{10})$_2$Cl$_2$]

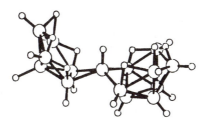

Figure 4. The molecular structure of $B_{15}H_{23}$ showing (on the left) a B_6H_{10} unit η^2-bonded to a boron atom of the B_9H_{13} unit (on the right) via a 3-center BBB bond

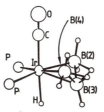

Figure 5.   The molecular structure of [Ir(η^3-B_3H_7)(CO)H(PPh$_3$)$_2$] with the phenyl groups omitted for clarity

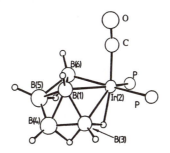

Figure 6.   The molecular structure of [Ir(η^3-B_5H_8)(CO)(PPh$_3$)$_2$] showing the result of cluster expansion to give a structural analogue of *nido*-B_6H_{10}

pyramidal *nido*-$B_6H_9{}^-$ units is attached in a trihapto bridging manner to the linear L-Pt-Pt-L system. The structure and bonding in the centrosymmetric complex are shown in Figure 8, the *anti* configuration of the two B_6 clusters being in contrast to their known *syn* disposition in the "isoelectronic" *conjuncto*-borane $B_{14}H_{20}$.

Even more complex macropolyhedral clusters can be obtained by appropriate thermolysis of $[Pt(\eta^3\text{-}B_8H_{12})L_2]$ (L = PMe_2Ph), *e.g.*, the red 17-vertex $[LPt(\eta^6\text{-}B_{16}H_{18}L)]$ (*48, 49*) and the green triplatina 17-vertex $[Pt_3(\mu_3\text{-}\eta^2,\eta^3,\eta^3\text{-}B_{14}H_{16})L_4]$ (*49, 50*). The original references should be consulted for details of the crystal structures and bonding. Likewise, discussion of still larger (19- and 20-vertex) macropolyhedral metallaboranes based on *syn-* and *anti*-$B_{18}H_{22}{}^{2-}$ as borane ligands, together with full references are given in several reviews (*e.g., 11*). Related to these, though simpler, are the series of complexes based on *nido*-$B_{10}H_{12}{}^{2-}$ as the ligand, and these will now be discussed.

The earliest (1965) work in this area (*51*) was done in an attempt to parallel, with Al and Ga, the well known cluster expansion reactions of decaborane when treated with borane adducts such as LBH_3. $B_{10}H_{14}$ was found to react readily with $[AlH_3(NMe_3)]$ in ether to give the novel, highly reactive anion $[AlB_{10}H_{14}]^-$, which is now probably best regarded as an analogue of $B_{11}H_{14}{}^-$ or $[H_2Al(\eta^4\text{-}B_{10}H_{12})]^-$. The reaction was subsequently extended to give a variety of derivatives of Mg; Zn, Cd, Hg; Al, Ga, In, Tl; Si, Ge, Sn(II), and Sn(IV) (*5, 8, 11*). For example, reaction of $B_{10}H_{14}$ with metal alkyls gave important classes of clusters in which the metal centers can be regarded as being chelated or bridged by the tetradentate η^2,η^2-$B_{10}H_{12}{}^{2-}$ ligand. Examples of new structural types obtained in this way, together with a representation of the bonding in terms of the ligand/ coordination chemistry model, are in Figures 9(a)-(e). More than a hundred such complexes have now been prepared, mainly by ligand replacement reactions, including those of transition metals such as Cr, Mo, W; Fe; Co, Rh, Ir; Ni, Pd, Pt, and many have been fully characterized by X-ray structural analysis, *etc.*

Conclusion

In this brief account I have been able to give only a general overview of this fascinating and important new approach to boron hydride cluster chemistry. As a result of the application of coordination chemistry principles dozens of new structural types have been synthesized in which polyhedral boranes or their anions can be considered to act as ligands which donate electrons to metal centers thereby forming the novel metallaborane clusters (*1-12*). Some forty metals have so far been found to act as accepors in this way. Even new binary boranes can be synthesized by a combination of notional borane ligands and borane acceptors to give borane-borane complexes. The ideas have also proved to be particularly helpful in emphasizing the close interconnections between several previously separated branches of chemistry, notably boron hydride cluster chemistry, metallaborane and carbaborane chemistry, organometallic chemistry, and metal-metal cluster chemistry--all are now seen to be

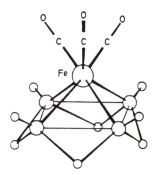

Figure 7. The proposed structure of $[Fe(\eta^4\text{-}B_4H_8)(CO)_3]$

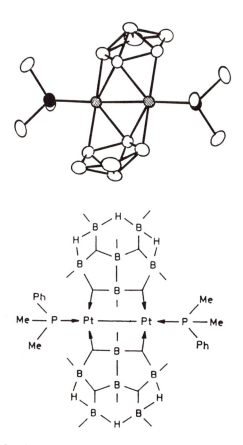

Figure 8. The molecular structure and localized bonding representation of $[Pt_2(\mu\text{-}\eta^3\text{-}B_6H_9)_2(PMe_2Ph)_2]$ showing the approximately square-planar distribution of bonding orbitals around the Pt vertices

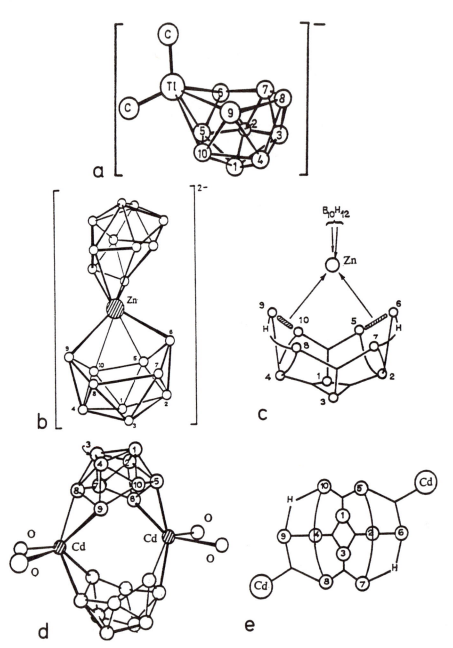

Figure 9. (a) Structure of the anion $[Tl(\eta^4\text{-}B_{10}H_{12})Me_2]^-$ (52, 53) (b) Structure of the *commo* dianion $[Zn(\eta^4\text{-}B_{10}H_{12})_2]^{2-}$ (54) (c) A three-dimensional representation of the bonding in $[Zn(B_{10}H_{12})_2]^{2-}$ in terms of the bis(dihapto) chelating ligand $(\eta^2,\eta^2\text{-}B_{10}H_{12})^{2-}$ (d) Structure of the bridged dimeric complex $[\{Cd(\mu\text{-}\eta^2,\eta^2\text{-}B_{10}H_{12})(OEt_2)_2\}_2]$ showing positions of the non-H atoms (e) Topology of part of the Cd dimer (d) emphasizing the bridging bis(dihapto) nature of the borane ligand

parts of a coherent whole. As a final example, which incorporates all these aspects within a single molecule, we can contemplate Figure 10, which illustrates the molecular structure of $[Ru_3\{\eta^6\text{-}B_{10}H_8(OEt)_2\}(\eta^6\text{-}C_6Me_6)_2(\mu_2\text{-}H)_3(\mu_3\text{-}H)]$ (*56*). This bright red, air-stable compound was formed in 32% yield by heating under reflux an ethanolic solution of the four-vertex metallaborane cluster, $[Ru(B_3H_8)(C_6Me_6)Cl]$, and *closo -* $B_{10}H_{10}^{2-}$. The Ru_3 triangle is unique in being the first (and only) such cluster to have no attached CO ligands, the pendant groups being a hexahapto B_{10} ligand and two hexahapto hexamethylbenzene ligands together with three edge-bridging H atoms and one semi-capping (triply bridging) H atom. It is therefore a *commo*-metallaborane which incorporates within a single compound a metal-metal cluster, a polyhapto borane ligand, and two organometallic moieties. Many further exciting developments are expected in the coming years.

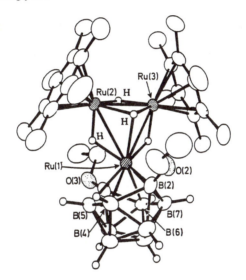

Figure 10. Structure of the compound $[Ru_3(\eta^6\text{-}B_{10}H_8(OEt)_2\}(\eta^6\text{-}C_6Me_6)(\mu_2\text{-}H)_3(\mu_3\text{-}H)]$, see text (*56*)

Literature Cited

1. Greenwood, N. N.; Earnshaw, A. *Chemistry of the Elements;* Pergamon Press: Oxford, UK, 1984; pp 171-220 and references cited therein.
2. Greenwood, N. N.; Ward, I. M. *Chem. Soc. Revs.* **1974**, *3*, 231-271.
3. Greenwood, N. N. (Convener), "Microsymposium on 'Boranes as Ligands'", *Proc. XIX Internat. Conf. Coord. Chem.* Prague, Czechoslovakia, 1978; Vol.1; pp 71-85.
4. Wegner, P. A. In *Boron Hydride Chemistry;* Muetterties, E. L., Ed.; Academic Press: New York, NY, 1975; pp 431-480.

5. Greenwood, N. N. *Pure Appl. Chem.* **1977**, *49*, 791-801.

6. Housecroft, C. E.; Fehlner, T. P. *Adv. Organomet. Chem.* **1982**, *21*, 57-112.

7. *Metal Interactions with Boron Clusters;* Grimes, R. N., Ed.; Plenum Press: New York, NY, 1982; pp *xiv* , 327.

8. Greenwood, N. N.; Kennedy, J. D. Chap 2 in ref. 7, pp 43-118.

9. Gaines, D. F.; Hildebrandt, S. J. Chap 3 in ref. 7, pp 119-143.

10. Greenwood, N. N. In *Inorganic Chemistry Toward the 21st Century;* Chisholm, M. H., Ed.; ACS Symposium Series 211, American Chemical Society: Washington, DC, 1983; pp 333-347.

11. Kennedy, J. D. *Progr. Inorg. Chem.* **1984**, *32* , 519-679; **1986**, *34*, 211-434.

12. Fehlner, T. P. In *Inorganometallic Chemistry;* Fehlner, T. P., Ed.; Plenum Press: New York, NY, 1992; pp 13-71.

13. Dunks, G. B.; Hawthorne, M. F. In *Boron Hydride Chemistry*; Muetterties, E. L., Ed.; Academic Press: New York, NY, 1975; pp 383-430.

14. Callahan, K. P.; Hawthorne, M. F. *Adv. Organomet. Chem.* **1976**, *14*, 145-186.

16. Todd, L. J. Chap 4 in ref. 7, pp 145-171.

17. Bresodola, S. Chap 5 in ref. 7, pp 173-237.

18. Grimes, R. N. Chap 7 in ref. 7, pp 269-319.

19. Grimes, R. N. In *Comprehensive Organometallic Chemistry,* Wilkinson, G.; Stone, F. G. A.; Abel, E. W., Eds.; Pergamon Press: Oxford, UK, 1982; pp 459-542.

20. Hawthorne, M. F.; Young, D. C.; Wegner, P. A. *J . Amer. Chem. Soc.* **1965**, *87,* 1818-1819.

21. Greenwood, N. N. In *Comprehensive Inorganic Chemistry,* Bailar, J. C.; Emeléus, H. J.; Nyholm, R. S.; Trotman-Dickenson, A. F., Eds.; Pergamon Press: Oxford, UK, 1973; pp 665-991 and references cited therein.

22. Ogg, R. A.; Ray, J. D. *Disc. Faraday Soc.* **1955**, *19*, 239-246.

23. Shore, S. G.; Parry, R. W. *J. Am. Chem. Soc.* **1958**, *80*, 8-10.

24. Parry, R. W.; Edwards, L. J. *J. Am. Chem. Soc.* **1959**, *81*, 3354-3360.

25. Lippard, S. J.; Melmed, K. M. *Inorg. Chem.* **1969**, *8*, 2755-2762.

26. Hildebrandt, S. J.; Gaines, D. F.; Calabrese, J. C. *Inorg. Chem.* **1978**, *17*, 790-794.

27. Calabrese, J. C.; Fischer, M. B.; Gaines, D. F.; Lott, J. W. *J. Am. Chem. Soc.* **1974**, *96*, 6318-6323.

28. Gill, J. T.; Lippard, S. J. *Inorg. Chem.* **1975**, *14*, 751-761.

29. Dobrott, R. D.; Lipscomb, W. N. *J. Chem. Phys.* **1962**, *37*, 1779-1784.

30. Churchill, M. R.; Hackbarth, J. J.; Davison, A.; Traficante, D. D.; Wreford, S. S. *J. Am. Chem. Soc.* **1974**, *96*, 4041-4042.

31 Gaines, D. F.; Iorns, T. V. *Inorg. Chem.* **1968**, *7,* 1041-1043.

32. Greenwood, N. N.; Kennedy, J. D.; Savory, C. G.; Staves, J.; Trigwell. K. R. *J. Chem. Soc., Dalton Trans.* **1978**, 237-244.

33. Brice, V. T.; Shore, S. G. *J. Chem. Soc., Chem. Commun.* **1970**, 1312-1313; *J. Chem. Soc., Dalton Trans.* **1975**, 334-336.

34. Greenwood, N. N.; Howard, J. A.; McDonald, W. S. *J. Chem. Soc., Dalton Trans.* **1976**, 37-39.

35. Davison, A.; Traficante, D. D.; Wreford, S. S. *J. Am. Chem. Soc.* **1974**, *96*, 2802-2805.

36. Brennan, J. P.; Schaeffer, R.; Davison, A.; Wreford, S. S. *J. Chem. Soc., Chem. Commun.* **1973**, 354.

37. Davison, A.; Traficante, D. D.; Wreford, S. S. *J. Chem. Soc., Chem. Commun.* **1972**, 1155-1156.

38. Rathke, J.; Schaeffer, R. *J. Am. Chem. Soc.* **1973**, *95*, 3402.

39. Rathke, J.; Schaeffer, R. *Inorg. Chem.* **1974**, *13*, 3008-3011.

40. Huffman, J. C. *Indiana Univ. Mol. Struct. Center Rept.* No. 81902, Sept. 1981.

41. Greenwood, N. N.; Kennedy, J. D.; Reed, D. *J. Chem. Soc., Dalton Trans.* **1980**, 196-200.

42. Bould, J.; Greenwood, N. N.; Kennedy, J. D.; McDonald, W. S. *J. Chem. Soc., Dalton Trans.* **1985**, 1843-1847.

43. Guggenberger, L. J.; Kane, A. R.; Muetterties, E. L. *J. Am. Chem. Soc.* **1972**, *94*, 5665-5673.

44. Greenwood, N. N.; Kennedy, J. D.; McDonald, W. S.; Reed, D.; Staves, J. *J. Chem. Soc., Dalton Trans.* **1979**, 117-123.

45. Greenwood, N. N.; Savory, C. G.; Grimes, R. N.; Sneddon, L. G.; Davison. A.; Wreford, S. S. *J. Chem. Soc., Chem. Commun.* **1974**, 718.

46. Miller, V. R.; Grimes, R. N. *J. Am. Chem. Soc.*, **1973**, *95*, 5078-5080; Miller, V. R.; Weiss, R.; Grimes, R. N. *J. Am. Chem. Soc.* **1977**, *99*, 5646-5651.

47. Greenwood, N. N.; Hails, M. J.; Kennedy, J. D.; McDonald, W. S. *J. Chem. Soc., Chem. Commun.* **1980**, 37-38; *J. Chem. Soc., Dalton Trans.* **1985**, 953-972.

48. Beckett, M. A.; Crook, J. E.; Greenwood, N. N.; Kennedy, J. D.; McDonald, W. S. *J. Chem. Soc., Chem. Commun.* **1982**, 552-553.

49. Beckett, M. A.; Crook, J. E.; Greenwood, N. N.; Kennedy, J. D. *J. Chem. Soc., Dalton Trans.* **1986**, 1879-1893.

50. Beckett, M. A.; Crook, J. E.; Greenwood, N. N.; Kennedy, J. D. *J. Chem. Soc., Chem. Commun.* **1983**, 1228-1230.

51. Greenwood, N. N.; McGinnety, J. A. *Chem. Commun.* **1965**, 331-332.

52. Greenwood, N. N.; Thomas, B. S.; Waite, D. W. *J. Chem. Soc., Dalton Trans.* **1975**, 299-304.

53. Greenwood, N. N.; Howard, J. A. *J. Chem. Soc., Dalton Trans.* **1976**, 177-180.

54. Greenwood, N. N.; McGinnety, J. A.; Owen, J. D. *J. Chem. Soc. (A)* **1971**, 809-813.

55. Greenwood, N. N.; McGinnety, J. A.; Owen, J. D. *J. Chem. Soc., Dalton Trans.* **1972**, 989-992.

56. Bown, M.; Fontaine, X. L. R.; Greenwood, N. N.; MacKinnon, P.; Kennedy, J. D.; Thornton-Pett, M. *J. Chem. Soc., Chem. Commun.* **1987**, 442-443; *J. Chem. Soc., Dalton Trans.* **1987**, 2781-2787.

RECEIVED December 6, 1993

Chapter 29

Lanthanide Aminopolycarboxylates

Gregory R. Choppin and Pamela J. Wong

Department of Chemistry, Florida State University,
Tallahassee, FL 32306–3006

Complexation by aminopolycarboxylate ligands has been a major area of research in lanthanide chemistry for almost five decades. From the 1950's, when the use of EDTA and HEDTA in ion exchange separations first provided multigram amounts of lanthanides of high purity to GdDTPA, which is currently utilized as an MRI agent, aminopolycarboxylates and f-elements have had a close association. Data from thermodynamics, kinetics, NMR, and luminescence are discussed to reflect the present understanding of the role of the carboxylate and nitrogen donors, of the number and size of chelate rings, and of hydration in these complexes. Results of recent studies of complexes of bis(amide) derivatives of DTPA are also discussed to illustrate further the significant factors in lanthanide-aminopolycarboxylate complexation.

The lanthanide family of elements has played an important role which can be expected to continue in the development of coordination chemistry. In the early history of these elements, their close chemical similarity in the stable tripositive oxidation state made the task of achieving high purity for individual elements very difficult. Although the entire lanthanide series had been discovered by 1907 (with the exception of Pm) and mixtures of lanthanides had been found in more than a hundred minerals, it was not until efficient separation methods were developed that detailed and diverse studies of their coordination chemistry could be undertaken.

The coordination chemistry of the lanthanide elements has interesting aspects because of the relatively high charge density of the cations, the strongly electrostatic nature of their bonding, and the variety of coordination numbers attained in different complexes. The regular, relatively small decrease in ionic radii across the family from La(III) through Lu(III) (the lanthanide contraction) results in relatively small differences in chemical properties between the elements in the same oxidation state. Although similar radial contractions are observed for other rows of metals in the periodic table (*e.g.*, 3d transition elements), the lanthanide contraction is much smaller. The effective ionic radius (*1*) of $_{57}La^{3+}(4f^0)$ is 1.216 Å (CN = 9) and that of $_{71}Lu^{3+}(4f^{14})$, 0.977 Å (CN = 8), respectively. While the ionic radii of Ca^{2+} (r = 1.00 Å, CN = 6) and Na^+ (1.02 Å, CN = 6) fall in the range of tripositive lanthanide radii and also exhibit strongly ionic bonding, the lanthanide cations have higher charge

0097–6156/94/0565–0346$08.00/0

densities and, correspondingly, form stronger complexes. Their paramagnetic and optical properties enable lanthanides to serve as effective probes for metal ion binding effects of biological systems.

Electrostatic bonding allows for a wide variety of coordination numbers (CN = 6-12) for lanthanide complexes. Steric, electrostatic, and solvation effects are the dominating criteria in determining the geometry of lanthanide complexes. The primary sphere hydration is not constant across the series; hydration numbers are 9.0-9.3 for the larger, lighter lanthanides and 7.5-8.0 for the heavier, smaller lanthanides (2).

Lanthanides behave as hard Lewis acids and bind most strongly to hard bases such as oxygen and fluorine, thus explaining their high affinity for water. Successful complexing agents frequently involve multiple lanthanide-oxygen bonds. Aminopolycarboxylate ligands (Figure 1) contain two or more carboxylate groups and at least one bonding secondary or tertiary amine. Although nitrogen is a softer base than oxygen, lanthanide-nitrogen bonds are important factors in the stability of these complexes. Because of their high coordination numbers, lanthanides commonly have water molecules remaining in their primary coordination sphere even when the ligand is polyfunctional.

Before the evolution of an efficient, economical means of separation, lanthanides were primarily utilized in the form of Mischmetall, an alloy of thorium and lighter rare earths, for the production of incandescent gauzes and mantles for gas lamps (3,4). The advent of aminopolycarboxylates in lanthanide separation methods provided the impetus for a true lanthanide industry, and lanthanides are currently being utilized in such diverse fields as biochemistry (*e.g.*, NMR shift reagents), nuclear medicine (*e.g.*, MRI contrast agents), optical sensing (*e.g.*, television tubes), industrial catalysis, organic synthesis, and high temperature superconductivity.

Early Studies

Initial efforts in lanthanide separations employed successive fractional crystallization (5). This process was both exhaustive and inefficient; as many as 10,000 recrystallizations could be required to produce milligram amounts. As part of the Manhattan Project during World War II, researchers at the Ames Laboratory and the Oak Ridge National Laboratory discovered the efficacy of ion exchange resins in lanthanide separation. Several lanthanides appeared as uranium fission products, and a method of separating these elements was needed for the proper measurement of their yields in the fission process. A research group headed by Spedding at the Ames Laboratory discovered that cation exchange resin, with a 0.1% citrate solution as the eluant, separated spectroscopically pure yttrium (6a) and neodymium from cerium(6b) and praseodymium in gram quantities. At the same time, Harris *et al.* (7) at Oak Ridge separated tracer quantities of lanthanum, cerium, praseodymium, and neodymium at tracer levels from cation exchange resin by elution with citrate solution. Significant contributions by Ketelle and Boyd improved the understanding of the parameters controlling lanthanide separation with these ion exchange resins (8). While ion exchange had been studied previously in the separation of lanthanides, elution with noncomplexing solutions achieved no greater separation than fractional crystallization. When a complexing agent was used as the eluant solution, the relative stabilities of the lanthanide complexes greatly enhanced the separations.

Just prior to these discoveries, a series of patents had been issued in Zürich (1937-1940) concerning a class of compounds which contained at least two carboxylate groups bonded to an amine nitrogen. These compounds exhibited extraordinary complexing ability with alkaline earth metals in aqueous solution (9). In 1942 Brintzinger reported complexation between ethylendiaminetetraacetic acid (EDTA) and lanthanum, neodymium, and thorium. Studies by Pfeiffer (10,11) and

Brintzinger (12,13) indicated strong metal-ligand bonding interactions through the amine nitrogen.

Schwarzenbach began investigating the chelating abilities of NTA and IDA (9). In 1953 he and Wheelwright observed greater stability and improved lanthanide separation using cation exchange resin with aminocarboxylate elution compared to citrate elution (14). The researchers initially investigated the strength of hydrazino-N,N-diacetic acid and EDTA as complexing agents and found the latter much superior. In 1951 Fitch and Russell had reported the use of NTA and IDA elution from cation exchange resin for macroseparation of lanthanides on the gram scale(15). Upon discovery of EDTA as a particularly effective lanthanide complexing agent, many researchers sought to understand and optimize the ion exchange separations using this ligand (16-19). A variety of other related aminopolycarboxylate ligands such as NTA (15), HEDTA (20,21), and DTPA (22-24) were also explored for their potential in lanthanide separations.

In contrast to the polycitrate complexes which formed, EDTA was shown to form a 1:1 complex with the lanthanide cations during elution (25). Unfortunately, the low solubility of the complexed species (e.g., HLnEDTA) during ion exchange posed a problem(26). Wheelwright, Spedding, and Schwarzenbach (14,20,27) solved this problem by using an ion exchange resin in the Cu^{2+} form rather than the H^+ form . The competitive copper aminocarboxylate complex was more soluble than H_4EDTA. The enhanced solubilities of NTA (15), HEDTA (20,21) and DTPA (22-24) made these attractive ligands for further study.

To improve the separations with aminocarboxylates, researchers turned their efforts toward understanding the nature and characteristics of aminopolycarboxylate-lanthanide complexation. The ensuing wealth of information resolved fundamental questions about the structure of the complexes, ligand effects, bonding characteristics, and hydration.

Chemical behavior of Lanthanide Aminopolycarboxylates

Wheelwright, Spedding, and Schwarzenbach conducted an intensive investigation on the formation of the lanthanide complexes (14). In 1953 they observed that LnEDTA complexes exhibited increasing stability with decreasing ionic radii, and they noted that the curve for the stability constants of these 1:1 species as a function of lanthanide atomic number was discontinuous, with a break appearing at Gd(III) (Figure 2). From potentiometric and polargraphic evidence, they postulated hexacoordination for the Ln(EDTA) complex through the two nitrogen atoms and four carboxylate oxygen atoms and, consequently, a coordination number of 6 for the lanthanide cation. To explain the break in the log ß vs. Z curve, they postulated that the smaller ionic radii of the heavier lanthanides and the steric hindrance of four carboxylate groups induced a change from hexadentate to pentadentate coordination. The assumption of a maximum coordination number of six for the lanthanides would continue through the1960s, reflecting the mistaken assumption that lanthanide behavior resembled the more extensively studied 3d transition elements.

Moeller was the first to examine these complexes with IR spectroscopy in 1950 (25). In 1955 he used results from X-ray diffraction studies to propose pentadentate complexation for the 1:1 species of all the lanthanide cations with EDTA via coordination through the two amines and three of the four carboxylate groups (28). The presence of an uncomplexed carboxylate group suggested the incorporation of water in the primary coordination sphere, thus achieving the assumed total coordination number of 6. The assumption of uncomplexed carboxylate was supported by infrared data showing absorptions of the C=O stretch around 1600 cm^{-1} (complexed -CO_2^-) and 1700 cm^{-1} (uncomplexed -CO_2^-). The relative intensities of these bands agreed with the hypothesis of three bonded and one free carboxylate groups.

I R_1–N$\begin{smallmatrix}R_2\\\\R_3\end{smallmatrix}$

(a) $R_1 = H$, $R_{2-3} = CH_2CO_2H$
Iminodiacetic acid, IDA
(b) $R_1 = CH_3$, $R_2 = R_3 = CH_2CO_2H$
Methylaminodiacetic acid, MIDA
(c) $R_{1-3} = CH_2CO_2H$
Nitrilotriacetic acid, NTA

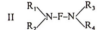

II $\begin{smallmatrix}R_1\\\\R_2\end{smallmatrix}$N–F–N$\begin{smallmatrix}R_3\\\\R_4\end{smallmatrix}$

(a) $R_1 = CH_3$, $R_{2-4} = CH_2CO_2H$, $F = C_2H_4$
N-Methylethylenediaminetriacetic acid, MEDTA
(b) $R_1 = C_2H_5OH$, $R_{2-4} = CH_2CO_2H$, $F = C_2H_4$
N-Hydroxyethylethylenediaminetriacetic acid, HEDTA
(c) $R_{1-4} = CH_2CO_2H$, $F = C_2H_4$
Ethylenediaminetetraacetic acid, EDTA
(d) $R_{1-4} = CH_2CO_2H$, $F = C_6H_{10}$
Trans-1,2,diaminocyclohexanetetraacetic acid, DCTA
(e) $R_{1-4} = CH_2CO_2H$, $F = C_3H_6$
1,3-Trimethylenediaminetetraacetic acid, TMDTA
(f) $R_{1-4} = CH_2CO_2H$, $F = C_4H_8$
1,4-Tetramethylenediaminetetraacetic acid, TMEDTA
(g) $R_{1,3} = CH_2CO_2H$, $R_{2,4} = C_2H_4CO_2H$, $F = C_2H_4$
Ethylenediamine-*N,N*'dipropionic$_2$-*N,N*'diacetic acid, EDPDA

III $\begin{smallmatrix}R_1\\\\R_2\end{smallmatrix}$N–F–N–F–N$\begin{smallmatrix}R_3\\\\R_4\end{smallmatrix}$
                     |
                     R_5

(a) $R_{1-5} = CH_2CO_2H$, $F = C_2H_4$
Diethylenetriaminepentaacetic acid, DTPA
(b) $R_{1,5,3} = CH_2CO_2H$, $R_{2,4} = CH_2CONHCH_3$
Diethylenetriaminepentaacetic acid *bis* (methylamide), DTPA-BMA

Figure 1.   Aminopolycarboxylate ligands.

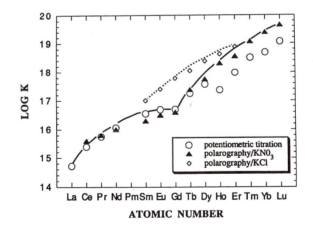

Figure 2. Log K, the formation constant, of LnEDTA complexation vs Atomic Number. A smooth curve has been drawn through the more accurate data; potentiometric values for the lighter lanthanides and polaragraphic/KNO3 values for the heavier lanthanides. The dotted line represents polaragraphic/KCl data. Adapted from ref. 19.

In 1959 Betts and Dahlinger conducted thermodynamic experiments and concluded that lanthanide-EDTA complexation decreased from hexadentate to pentadentate coordination from La to Gd and from pentadentate to tetradentate coordination for cations heavier than Gd (29). The changes in free energies, enthalpies, and entropies of LnEDTA complexation were measured for the entire lanthanide series. From their results Betts and Dahlinger concluded that the extraordinary stability of the LnEDTA complexes was attributable almost entirely to the large and positive change in entropy during complex formation. ΔS values could be divided into two groups, corresponding to the two distinct curves (La to Gd and Tb to Lu, respectively) in the plots of log ß vs. Z. Chaberek and Harder reached similar conclusions concerning entropy changes and coordination for HEDTA as well as EDTA complexation (23). However, from the similarity in plots of log ß vs. Z for NTA and EDTA (two distinct curves for both ligands), Schwarzenbach and Gut considered it unlikely that the break in the curves could be attributed to changes in coordination because it was unlikely that NTA would change from tetra- to tridentate coordination (30).

The break at Gd in the plots of log ß vs. Z plot engendered a great deal of interest in the 1950s and 1960s, and explanations for the two distinct curves abounded. In 1958 Stavely and Randall proposed that successive occupation of the 4f orbital and crystal field effects induced by the ligand contributed to the differences in EDTA complex stability across the lanthanide series (31). A thermodynamic study of the heats of formation of various metal-EDTA complexes led these investigators to consider it unlikely that when a critical size of cation was reached, there was an abrupt change in coordination. They pointed out that the break in continuity of the log ß plot appeared at Gd ($4f^7$) where no crystal field splitting would be present. Also, although Nd^{3+} ($4f^2$) has a smaller ionic radii than Ce^{3+} ($4f^1$), the heat of formation for Nd^{3+} was greater than that of Ce^{3+} by 120 J/mole. Furthermore, holmium and yttrium have similar ionic radii, yet the stability constant of holmium exceeds that of yttrium. At that time, the geometries of these complexes had not been investigated so no direct information on the coordination geometry was available.

Diamond, Street, and Seaborg proposed f orbital hybridization, postulating that although 4f orbitals of the correct geometries existed for covalent bonding, the energies of these orbitals were high. Hybridization of the 4f orbitals would lower the energy and, consequently, enable the 4f orbitals to participate in covalent bonding (32). In 1962, however, the hypothesis of f orbital crystal field effects was disproven by the work of Mackey, Powell, and Spedding (33). Calorimetric studies of LnEDTA and of LnAc complexation revealed similar trends in the molar complexation enthalpies as well as the entropies for both ligands. If crystal field effects were significant, EDTA and acetate complexation should have shown quite different thermodynamic effects from each other.

Covalency is somewhat a matter of definition, but there exists evidence of some degree of orbital overlap in lanthanide complexes. Lewis, Jackson, Lemons, and Taube examined the ^{17}O NMR shifts of aqueous solutions of lanthanide perchlorates (34). These researchers attributed the magnitude and direction of the chemical shift to the formation of covalent bonds between the 2s or the hybridized 2s and 2p orbitals of oxygen in water and the 6s orbitals of the hydrated lanthanide. No evidence of geometry-dependent f orbital overlap was observed. The degree of covalency in lanthanide aminocarboxylate complexation is still uncertain, but it is obvious that electrostatic interaction is a satisfactory model to explain the major features in lanthanide aminopolycarboxylate complexation.

The assumption that the coordination of the lanthanides resembles the hexacoordinate d transition metals was shaken by Thompson and Loraas who demonstrated that six-coordinate transition metals did not form mixed chelates with HEDTA and IDA ligands (35). Such M(HEDTA)(IDA) species could require octacoordination sites around the metal. By contrast, the formation of

[Ln(HEDTA)(IDA)]$^{2-}$ and [Ln(EDTA)(IDA)]$^{3-}$ was confirmed. From these results, these investigators concluded that the coordination number for the lanthanides was eight, assuming full hexacoordination of EDTA. However, the evidence for the hexadentate nature of EDTA was not conclusive at this time.

In 1965 Hoard, Lind, and Lee (*36*) determined the crystal structure of HLaEDTA to be monoclinic with P2$_1$/a symmetry. These crystal structures revealed coordination of EDTA in one hemisphere of the cation and four coordinated waters in the other hemisphere, thus indicating a total coordination number of ten. Four carboxylate oxygens are coordinated to La^{3+} in a trapezoidal planar array (r_{La-O} = 2.537) while the bond distance for the fifth carboxylate is 2.609Å. This fifth carboxylate group is also bound to a proton. The two nitrogen-metal bonds are 0.77 Å out of plane (r_{La-N} = 2.865 Å). The potassium salt of LaEDTA crystallized in the space group Fdd2 and exhibited increased chelation strength, reflected in the shortening of the lanthanide-oxygen and -nitrogen bonds (r_{La-O} = 2.507 Å; r_{La-N} = 2.755 Å). Consistent with an electrostatic model of bonding, this shortening was attributed to the loss of the acidic hydrogen which increased the overall charge of the complex, pulling the ligand closer about the metal. The KLaEDTA had 3 water molecules attached to the La for a total coordination number of 9. Hoard *et al* proposed that replacement of La^{3+} with smaller lanthanides in the series would induce a change in the coordination number of the complex, accompanied by a loss of water. They further postulated that this change would be accompanied by a sharp increase in the standard entropy and a decrease in the standard enthalpy values.

The question of the denticity of LnEDTA in aqueous LnEDTA complexes was clarified by Baisden, Choppin, and Garrett in 1977 (*37*) using NMR methods introduced by Day and Reilley (*38,39*). The spectra of LnEDTA revealed splitting patterns which indicated equal bonding to all four carboxylate groups as well as to both nitrogen atoms, therefore confirming hexacoordination of EDTA to lanthanide metals (Figure 3). However, no conclusions about the total coordination number of lanthanides in LnEDTA complexes in aqueous solution could be drawn from these NMR studies.

In 1964 Edelin de la Praudière and Staveley concluded from calorimetric data that ligand field effects could not account for the trends with cation atomic number in the thermodynamic parameters. They proposed a change in the state of hydration of the lanthanide cations in the complexation reaction (*40*). During spectroscopic investigations of LnEDTA in 1969, Geier and Karlen noticed a strong temperature dependence on the absorbance of EDTA complexes of Sm, Eu, and Gd (*41*). This temperature dependence of these three lanthanides (corresponding to the minimum in the Edelin de la Praudière and Stavely plot) was attributed to a change in hydration number of the complex; Geier and Karlen assigned a constant coordination number for the aquated cation across the series. Grenthe and Ots investigated the temperature dependence of the thermodynamic functions of LnEDTA complexation across the series in 1972 (*42,43*). From the temperature dependence of the heat capacities, they concluded that differences in the geometry of the coordination sphere of the complexes (*i.e.*, the number of water molecules coordinated to the complexed cation) were responsible for the difference in stabilities between the light and heavy lanthanide complexes.

In these studies the enthalpies and entropies of lanthanide complex formation had been measured and the enthalpies and entropies of hydration estimated from theoretical and semiempirical equations. Bertha and Choppin (*44*) examined the hydration thermodynamics of the lanthanide ions directly and obtained molar entropies of hydration of 338±4 J/mol-K and 401±4 J/mol-Kfor La-Pr and for Dy-Lu, respectively, and a range of 351-392 J/mol-K for Nd-Tb. These observations supported the existence of two differently sized hydration spheres corresponding to the La-Pr and Dy-Lu groups, with Nd-Tb comprising the transition between them. Spedding *et al.* (*45*) later confirmed the ΔS_{hyd} data of Bertha and Choppin.

The correlation between entropy and hydration number is illustrated by the ratio of ΔS_{hyd} between La and Lu. Bertha and Choppin estimated this ratio to be 1.2. From activity coefficient measurements, Glueckauf (46) reported outer sphere hydration numbers of 7.5 and 8.7 for La and Lu, respectively, and also a La/Lu ratio of 1.2. Choppin and Graffeo (47) calculated hydration numbers from conductance data and reported a La/Lu hydration ratio of 1.1, while Padova measured molar volumes to obtain a La/Lu hydration ratio of 1.2. The consistency in this ratio calculated by four different methods supports the correlation between the total hydration numbers, the S°_{hyd} values, and the atomic number.

While X-ray crystallography had determined the coordination of LnEDTA in the solid state, luminescence, proton relaxation NMR, and hydration experiments have provided information on the extent of ligand coordination in solution. The proton NMR spectral studies of Choppin (37) and Reilley (38,39) of lanthanide aminocarboxylate complexes reflected the relative lifetimes of the metal-oxygen and metal-nitrogen bonds. The ^{1}H spectra of diamagnetic LnEDTA complexes revealed an AB quartet splitting corresponding to the methylenic protons of the acetate arms and a singlet corresponding to the ethylene protons of the amine chain (Figure 3). Day and Reilley attributed this characteristic splitting profile to the existence of long-lived metal-nitrogen bonds (>1 ms) and relatively short-lived metal-oxygen bonds (<1 ms). Baisden, Choppin, and Garrett demonstrated that the magnitude of the chemical shift was dependent on the effective charge density of the metal which reflected the shielding effects of the cation on the protons, further supporting the electrostatic bonding nature of the lanthanides.

Through methods developed by Kropp and Windsor (48), Horrocks and Sudnick (49) correlated the inverse lifetime of lanthanide luminescence decay and the average number of water molecules coordinated to the cation. Recently, Brittain, Barthelemy, and Choppin examined Eu and Tb aminopolycarboxylate complexes and, using Horrocks' model, determined the average number of water molecules in the primary coordination sphere of the metal in the complex (49). These results did not correspond to expectations based on the hypotheses of crystal field effects postulated by Staveley and Randall and the changes in hydration of the complexes postulated by Edelin de la Praudière, Geier, and Grenthe. For EuEDTA, Brittain and Choppin determined that 2.6 ± 0.5 water molecules were coordinated to Eu (50). Given that EDTA is hexacoordinate and Eu has a total coordination between 8 and 9, a simple explanation is an equilibrium between $(EuEDTA \cdot 2H_2O)^-$ and $(EuEDTA \cdot 3H_2O)^-$. Luminescence lifetime experiments with DTPA, which has 8 coordination sites, revealed 1.1 average water molecules coordinated to the metal, indicating a total coordination number of 9 for Eu in $(EuDTPA \cdot H_2O)^2$.

Proton relaxation NMR using Gd(III) support these conclusions. Spin-lattice relaxation (T_1) values are inversely related to the average number of inner sphere hydration waters of the lanthanide cation (50-52). Chang reported hydration values of 3 and 1 for Gd complexes of EDTA and DTPA, respectively. Both NMR relaxation studies and luminescence lifetime data indicate a total coordination number between 8 and 10 for the lanthanides in aminopolycarboxylate complexes.

Thermodynamics of Complexation

Thermodynamic studies of lanthanide-aminopolycarboxylate complexation have been quite valuable in the understanding of lanthanide coordination chemistry. The same experimental conditions of Grenthe (42,43) were utilized by Choppin, Goedken, and Gritmon(53) in 1977 to investigate the thermodynamic parameters of a variety of aminopolycarboxylate ligands across the lanthanide series. Variations in the enthalpy and entropy curves vs. Z were attributed to different degrees and patterns of dehydration as well as the increasing polydentate nature of the complexation. The linear relationship between log ß and ΣpK_a of aminopolycarboxylate ligands which form 5-membered chelate rings emphasizes the ionic nature of the metal-ligand bonds (*53*). The Ln-N bond is a strong contributor to the high stability of the aminopolycarboxylate complexes. Figure 4 illustrates the effect of the analogous Ln-S, Ln-O, and Ln-N interaction in the complexes with ethylenedithiodiacetate (EDSDA), ethylenedioxydiacetate (EDODA), and ethylendiaminediacetate (EDDA) (*54*). The more positive enthalpy change in conjunction with the commensurate smaller value for the free energy indicate much weaker, if any, interaction between the metal and the sulfur. The negative ΔH for EDDA exphasizes the strength of the Ln-N bonding. NMR studies also support these conclusions(*55*).

The stability constant order EDTA > TMDTA > TMEDTA reflects the increased stability of the five membered ring. TMDTA and TMEDTA form 6 and 7 membered N-Ln-N chelate rings compared to the analogous 5 membered EDTA chelate rings. Similarly, EDPDA forms two 5-membered and two 6-membered O-Ln-N chelate rings. Relative to EDTA, the decrease in complex stability due to the presence of the two 6-membered rings is not as great as the decrease in stability of LnTMDTA due to the expansion of the single N-Ln-N ring. This is in agreement with the NMR data, which indicates that the Ln-N bonds are longer-lived and therefore, more significant in the stabilization of the complexes(*37-39*).

The complementary trends in the values of ΔH and ΔS vs. Z (*40-43,56*) suggested that these changes in enthalpy and entropy were dominated by the dehydration effects which accompanied ligand complexation(*57*). This has been termed a "compensation effect" in which the positive enthalpy and entropy contributions from dehydration of the lanthanide are much larger than the corresponding negative contributions from ligand-metal interactions; however, the enthalpy and entropy hydrational effects compensate, and ΔG reflects mainly metal-ligand interactions(*58*) .

Figure 5 illustrates the correlation between ΔS and the number of carboxylate groups of the ligand. The slope of this line, 60 $JK^{-1}mol^{-1}$, is identical with the ΔS_{Ac} of formation of $SmAc^{2+}$, thus supporting a $\Delta S \approx n(\Delta S_{Ac})$ relationship, where n = the number of carboxylate groups. This relationship suggests that the enthalpy change may also reflect the lanthanide-carboxylate bonding (*i.e.*, $\Delta H \approx n(\Delta H_{Ac})$). When $n(\Delta H_{Ac})$ is subtracted from the measured ΔH, the residual value, $\partial \Delta H_N$, presumably reflects the Ln-N bonding, which should be proportional to the basicity of the N donor sites and, consequently, described by the pK_a values of the N donor atoms. Figure 6 illustrates the correlation between $\partial \Delta H_N$ and $\Sigma pK_a(N)$. The deviations of MEDTA, TMEDTA, and EDPDA from this fit reflect the destabilizing effects of their 6-membered rings, consequently resulting in weakened Ln-N bond interactions. The good correlation of the N-Ln-N systems with the 5-membered rings support this analysis of the thermodynamic parameters.

Thus thermodynamic investigations indicate that 5-membered rings provide the greatest stability in lanthanide aminopolycarboxylate complexation. The variation of log ß with ΣpK_a of the ligands supports strongly ionic bonding between the metal and donor atoms, and the stronger metal-nitrogen interactions play larger roles in the relative stabilities of these complexes, as reflected in the more negative enthalpies and the ^1H-NMR patterns.

Kinetics of Complexation

Because of the overwhelming predominance of electrostatic interactions, the kinetics of lanthanide complexation are relatively rapid and conform to the Eigen-Tamm mechanism(*59*):

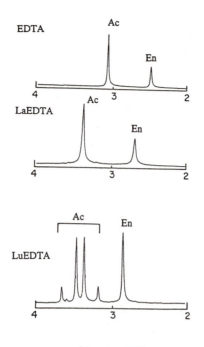

←δ (ppm) vs. DSS

Figure 3. Proton resonance spectra (90 MHz) of EDTA, LaEDTA, and LuEDTA; En=ethylene group, Ac=acetate group. A resolved quartet pattern is observed for the Ac protons in LaEDTA at higher fields. Adapted from ref. 36.

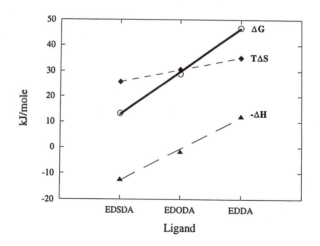

Figure 4. Thermodynamic parameters of Sm(III) complexation with selected aminopolycarboxylate ligands. Adapted from ref. 54.

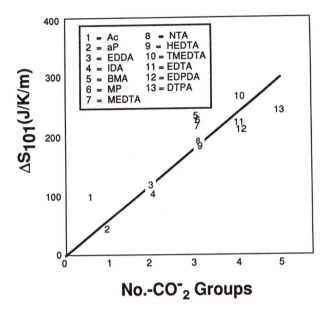

Figure 5. Correlation between ΔS_{101} and the number of carboxylate groups on various aminopolycarboxylates for Sm(III) complexation. Adapted from ref. 66.

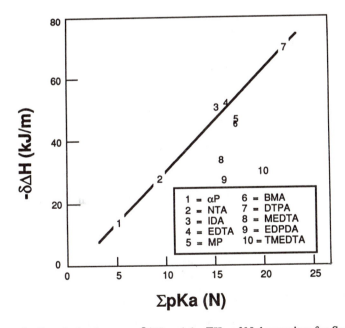

Figure 6. Correlation between $\delta \Delta H$ and the ΣK_a of N donor sites for Sm(III) complexation to various aminopolycarboxylates. Adapted from ref. 66.

$$\underset{(1)}{\text{Ln}_{(aq)} + \text{L}_{(aq)}} \underset{k_1}{\rightleftharpoons} \underset{(2)}{\text{Ln}_{(aq)}\text{L}_{(aq)}} \underset{k_2}{\rightleftharpoons} \underset{(3)}{\text{Ln}(\text{H}_2\text{O})_n\text{L}} \underset{k_3}{\rightleftharpoons} \text{LnL}_{(aq)}$$

where $k_1 > k_2 > k_3$. Step 1 represents the diffusion of the ligand into the vicinity of the metal in solution which is followed by the formation of the outer sphere species (Step 2). Depending on the basicity of the ligand, it may proceed to form the inner sphere species, as shown in Step 3. This last reaction, involving the dissociation of water from the lanthanide, is usually the rate-determining step.

The exchange of a radioisotopic lanthanide for a stable lanthanide metal chelated to an aminopolycarboxylate ligand has been found to proceed via two mechanistic pathways(60). Below pH 6, an acid-catalyzed mechanism predominates in which protonation catalyzes decomposition of the lanthanide-aminopolycarboxylate complex. The uncomplexed ligand chelates with free lanthanide cations in solution. The mechanism below illustrates the acid-catalyzed pathway, where A = aminopolycarboxylate ligand (‡ denotes the activated complex; Ln* is the radioactive tracer).

$$\text{LnA}^- + \text{H}^+ \underset{\text{fast}}{\rightleftharpoons} \text{HLnA} \underset{\text{slow}}{\rightleftharpoons} (\text{LnHA})^‡$$

$$(\text{LnHA})^‡ + (n-1)\text{H}^+ \underset{\text{fast}}{\rightleftharpoons} \text{HnA} + \text{Ln}$$

$$\text{Ln}^* + \text{HnA} \underset{\text{fast}}{\rightleftharpoons} \text{Ln}^*\text{HA} + (n-1)\text{H}^+$$

$$\text{Ln}^*\text{HA} \underset{\text{slow}}{\rightleftharpoons} (\text{HLn}^*\text{A})^‡ \underset{\text{fast}}{\rightleftharpoons} \text{Ln}^*\text{A} + \text{H}^+$$

Above pH 6 an acid-independent pathway exists in which complex decomposition is accomplished via the formation of a binuclear complex. The mechanism below illustrates the acid independent pathway .

$$\text{LnA} + \text{Ln}^* \underset{\text{fast}}{\rightleftharpoons} \text{LnALn}^* \underset{\text{slow}}{\rightleftharpoons} (\text{LnALn}^*)^‡ \underset{\text{fast}}{\rightleftharpoons} \text{Ln} + \text{ALn}^*$$

The rate constants of the acid catalyzed pathway reflect the stability of the chelate rings as they follow the sequence (2):

$$\text{TMEDTA} > \text{TMDTA} > \text{EDPDA} > \text{EDTA} > \text{DCTA}$$

The strength of the N-Ln-N ring relative to the N-Ln-O ring is reflected in the order of rate constants of TMDTA and EDPDA. Steric effects and the rigidity of the resultant chelate ring decrease the rate of decomposition, which explains the kinetic stability of DCTA.

Polyazapolycarboxylic Acids

A new type of aminocarboxylate ligands is the polyazapolycarboxylic acids. These cyclic ligands form a cavity via a ring of ethylene-spaced nitrogen donors with carboxylate "arms" which chelate to the metal above and below the plane of the cavity (Figure 7). Complex stabilities of these polyazapolycarboxylates relative to EDTA follow the sequence:

$$\text{DOTA} > \text{EDTA} > \text{TETA} > \text{NOTA}$$

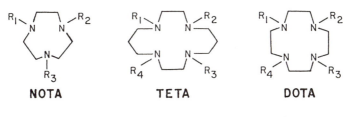

NOTA	TETA	DOTA

NOTA 1, 4, 7-triazacyclononane-N, N', N"-triacetic acid
TETA 1, 4, 8, 11-tetraazacyclotetradecane-N, N', N", N"'-tetraacetic acid
DOTA 1, 4, 6, 10-tetraazacyclododecane-N, N', N", N"'-tetraacetic acid
       R= CH₂COOH

Figure 7. Polyazapolycarboxylate ligands.

The encapsulating cavity serves to protect the metal from dehydration and destabilization of the complex. Despite their difference in cavity sizes, there is little difference between the complex stabilities of NOTA and DOTA for the same lanthanide cation. This suggests very little correlation between cationic diameter and that of the cavity. Thermodynamic, NMR, and fluorescence studies seem to indicate mixed penta- and hexadentate coordination with the lighter rare earths and hexacoordination beyond Sm for NOTA and DOTA. Chelation strength is determined by the pKa values of the donor ligands, the encapsulating nature of the multiple chelate rings, and the size and rigidity of the cavity. Not surprisingly, these complexes exhibit slow dissociation kinetics which make these complexes attractive as medical imaging agents.

Applications

The paramagnetic and optical properties as well as the hard Lewis acid character and high coordination numbers of lanthanides make their complexes attractive for utilization in such diverse applications as NMR shift reagents, medical imaging contrast agents, and EPR probes. Hinckley first reported the use of lanthanide complexes as shift reagents to resolve complicated ^{1}H spectra 20 years ago (*61*). The electrostatic nature of lanthanides induces an NMR pseudocontact shift (via through space interactions) (*62*), which has resulted in the use of $Ln(EDTA)(H_2O)_3$, $Ln(NTA)(H_2O)_5$, $Ln(NOTA)(H_2O)_3$, and $Ln(DOTA)(H_2O)$ to obtain structural information in solution. By contrast, covalent character is manifested in the contact shift induced by Gd(III).

The clinical use of MRI (magnetic resonance imaging) has become increasingly popular. This noninvasive imaging technique utilizes a paramagnetic probe to detect the proton resonances of water in the body and produce a three-dimensional *in vivo* tissue characterization. Longitudinal and transverse relaxation times, T_1 and T_2, govern signal intensity, and T_1 is inversely proportional to the signal intensity(*52*). Because the paramagnetic behavior of Gd(III) decreases the proton relaxation time by a factor of approximately 10^6, Gd complexes enhance the signal and, consequently, the detection of the image. $GdDTPA(H_2O)$ was the first clinical MRI imaging agent approved in the United States. Its high paramagnetism ($4f^7$) and its high complex stability makes Gd(III) an excellent choice(*63*). GdDPTA in MRI has found wide use in the detection of cancer, Alzheimer's disease, cardiovascular diseases, and epilepsy, among other diseases.

The half-filled 4f orbital and lack of orbital angular momentum enable Gd(III) to also serve as an effective EPR probe of metal binding sites in biological systems such as Ca(II), which has no UV-vis spectrum or EPR signal and a very broad NMR signal (^{43}Ca, natural abundance $\approx 0.1\%$) (64). The relatively long electron relaxation time of Gd(III), 10^{-9}-10^{-10} s, relative to other Ln(III) ions allows for observable EPR signals of GdEDTA complexes at room temperature(65). Because of the aqueous nature of most biological systems, however, the Gd EPR spectra is complicated by line broadening due to spin-spin relaxation.

Conclusion

Because of the lanthanide contraction, the family of trivalent lanthanide cations exhibit quite similar behavior. The use of aminopolycarboxylates in separation provided an economical, efficient means of isolating individual lanthanides. Almost five decades of research on lanthanide aminopolycarboxylatecomplexes have revealed a wealth of information about the coordination chemistry of the lanthanides. Electrostatics dominates the bonding nature of these complexes, and there is little evidence of any directional f orbital overlap in bonding. While the coordination of the ligand was found to be constant across the lanthanide series, the hydration number changes according to the size of the cation. Through thermodynamic, spectroscopic, and structural studies, the primary coordination is approximately nonadentate for the lighter lanthanides and octadentate for the heavier lanthanides. The Ln-N bond contributes strongly to the high stability of the lanthanide-aminopolycarboxylate complexes, as does the formation of 5-membered chelate rings. The thermodynamics of the complexation reaction may be described by a "compensation effect" in which positive enthalpy and entropy contributions from dehydration of the lanthanide often exceed in magnitude the ligand-metal interactions, which is reflected in the change in free energy. The kinetics of the system conforms to the Eigen-Tamm mechanism. The slower dissociation kinetics of the lanthanide-aminopolycarboxylates coupled with its paramagnetic and optical properties make these complexes attractive as NMR shift agents, MRI contrast probes, and EPR probes.

Acknowledgment

The preparation of this paper was assisted by a Grant from the USDOE-OBES Division of Chemical Sciences.

Literature Cited

(1)   Shannon, R.D. *Acta Cryst.* **1976**, *A32*, 751.
(2)   Rizkalla, E.N.;Choppin, G.R. *J. Coord. Chem.* **1991**, *23*, 33-41.
(3)   Trifonov, D.N. *The Rare-Earth Elements;* second ed.; Pergamon Press: New York, 1963.
(4)   Topp, N.E. *Chemistry of the Rare-Earth Elements*; Elsevier: Amsterdam, 1965; Vol. 4.
(5)   Svec, H.J. In *Two Hundred Year Impact of Rare Earths on Science*; Gschneidner, Jr, K.A., Ed.; North Holland: New York, 1988; Vol. 11; pp 14.
(6)   (a)Spedding, F.H.;Voight, A.F.;Gladrow, E.M.;Sleight, N.R.;Powell, J.E.;Wright, J.M.;Butler, T.A.;Figard, P. *J. Am. Chem. Soc.* **1947**, *69*, 2786-2792. (b) *ibid.* **1947**, *2777*.
(7)   Harris, D.H.;Tomkins, E.R. *J. Am. Chem. Soc.* **1947**, *69*, 2792-2799.
(8)   Ketelle, B.H.;Boyd, G.E. *J. Am. Chem. Soc.* **1947**, *69*, 2800-2812.
(9)   Schwarzenbach, G.;Kampitsch, E.;Steiner, R. *Helv Chim. Acta.* **1945**, *28*, 828-840.

(10) Pfeiffer, P.;Simons, H. *Ber.* **1943,** *75,* 1.

(11) Pfeiffer, P.;Offermann, W. *Ber.* **1942,** *76,* 847.

(12) Brintzinger, H.;Thiele, H.;Muller, U. *Z. Anorg. Chem.*. **1943,** *251,* 285-294.

(13) Brintzinger, H.;Hesse, G. *Z. Anorg. Chem.* **1942,** *249,* 299.

(14) Wheelwright, E.J.;Spedding, F.H.;Schwarzenbach, G. *J. Am. Chem. Soc.* **1953,** *75,* 4196-4200.

(15) Fitch, F.T.;Russell, D.H. *Can. J. Chem.* **1951,** *29,* 363-371.

(16) Vickery,R.C. *J.Chem.Soc.* **1952,** 4357-4363.

(17) Holleck, L.;Hartinger, L. *Angew.Chem.* **1956,** *68,* 411-412.

(18) Loriers, J. *Compt. Rend.* **1955,** *240,* 1537-1540.

(19) Mayer, S.W.;Freiling, E.C. *J. Am. Chem. Soc.* **1953,** *75,* 5647-5649.

(20) Spedding, F.H.;Powell, J.E.;Wheelwright ,E.J. *J. Am. Chem. Soc.* **1954,** *76,* 612-613.

(21) Moeller, T.;Horwitz, E.P. *J. Inorg. Nucl.Chem.* **1959,** *12,* 49-59.

(22) Chaberek, S.;Frost, A.E.;Doran, M.A.;Bicknell, N.J. *J. Inorg. Nucl.Chem.* **1959,** *11,* 181-196.

(23) Harder, R.;Chaberek, S. *J. Inorg. Chem.* **1959,** *11,* 197-209.

(24) Moeller, T.;Thompson, L.C. *J. Inorg. Nucl. Chem.* **1962,** *24,* 499-510.

(25) Moeller, T.;Brandtley, J.C. *J. Am. Chem. Soc.* **1950,** *72,* 5447.

(26) Moeller, T. *The Chemistry of the Lanthanides*; Reinhold: New York, 1963.

(27) Spedding, F.;Powell, J.E.;Wheelwright, E.J. *J. Am. Chem. Soc.* **1954,** *76,* 2557-2560.

(28) Moeller, T.;Moss, F.J.;Marshall, R.H. *J. Am. Chem. Soc.* **1955,** *77,* 3182-3186.

(29) Betts, R.H.;Dahlinger, O.F. *Can. J. Chem.* **1959,** *37,* *91-100*

(30) Schwarzenbach, G.;Gut, R. *Helv. Chim. Acta.* **1956,** *39,* 1589.

(31) Staveley, LA.;Randall ,T. *Disc. Far. Soc.* **1958,** *26,* 157-163.

(32) Diamond, R.M.;Street, K.;Seaborg, G.T. *J. Am. Chem. Soc.* **1954,** *76,* 1461-1469.

(33) Mackey, J.L.;Powell, J.E.;Spedding, F.H. *J. Am. Chem. Soc.* **1962,** *84,* 2047-2050.

(34) Lewis, W.B.;Jackson, J.A.;Lemons, J.F.;Taube, H. *J. Chem. Phys.*. **1962,** *36,* 694-701.

(35) Thompson, LC.;Loraas, J.A. *Inorg. Chem.* **1963,** *2,* 89.

(36) Hoard, J.L.;Lee, B.;Lind, M.D. *J. Am. Chem. Soc.* **1965,** *87,* 1612-1613.

(37) Baisden, P.A.;Choppin, G.R.;Garrett, B.B. *Inorg. Chem.* **1977,** *16,* 1367-1372.

(38) Day, R.J.;Reilley, C.N. *Anal. Chem.* **1964,** *36,* 1073-1076.

(39) Day, R.J.;Reilley, C.N. *Anal. Chem.* **1965,** *37,* 1326-1338.

(40) Edelin de la Praudière, P.L.;Staveley, L.K. *J. Inorg. Nucl. Chem.* **1964,** *26,* 1713-1719.

(41) Geier, G.;Karlen, U.;Zelewsky, A.V. *Helv. Chim. Acta.* **1969,** *52,* 1967-1975.

(42) Grenthe, I.;Ots, H. *Acta Chem. Scand.* **1972,** *26,* 1217-1228.

(43) Grenthe, I.;Ots, H. *Acta Chem. Scand.* **1972,** *26,* 1229-1242.

(44) Bertha, S.L.;Choppin, G.R. *Inorg. Chem.* **1969,** *8,* 613-617.

(45) Spedding, F.H.;Rard, J.A.;Habenschuss, A. *J. Phys. Chem.* **1977,** *81,* 1069-1074.

(46) Glueckauf, E. *Trans Faraday Soc.* **1955,** 1235-1244.

(47) Choppin, G.R.; Graffeo, A. *Inorg. Chem.* **1965,** *4,* 1254.

(48) Kropp, J.L.;Windsor, M.W. *J. Chem. Phys.*. **1966,** *45,* 761.

(49) Horrocks, W.D.;Albin, A. In *Progress in Inorganic Chemistry*; L. SJ, Ed.; John Wiley & Sons: New York, 1984; Vol. 31; pp 1-104.

(50) Brittain, H.G.;Choppin, G.R.;Barthelemy, P.P. *J. Coord. Chem.* **1992,** *26,* 143-153.

(51)   Chang, C.A.;Brittain, H.G.;Tweedle, M.F. *Inorg. Chem.* **1990,** *29,* 4468.
(52)   Tweedle, M.F. In *Lanthanide Probes in Life, Chemical and Earth Sciences*;
        Choppin, G.R.; Bunzli J-C,G, Ed.; Elsevier: Amsterdam, 1989; pp 127-180.
(53)   Choppin, G.R.;Goedken, M.P.;Gritmon, T.F. *J. Inorg. Nucl. Chem.* **1977,** *39,*
        2025-2030.
(54)   Grenthe I.;Gardhammar, G. *Acta Chem. Scand.* **28,** 125 (1975).
(55)   Choppin, G.R.;Kullberg, L. *Inorg. Chem.* **1980,** *19,* 1686-1688.
(56)   Spedding, F.H.;Pikal, M.D.;Ayers, B.O. *J. Phys. Chem.* **1966,** *70,* 2440-2449.
(57)   Choppin, G.R. *Pure Appl. Chem.* **1971,** *27,* 23.
(58)   Ives, D.J.G.;Marsden, P.D. *J. Chem. Soc.* **1965,** 649.
(59)   Eigen, M.;Tamm, K. *Z. Electrochem.* **1962,** *66,* 93-107.
(60)   D'Olieslager, W.;Oeyen, A. *J. Inorg. Nucl. Chem.* **1978,** *40,* 1565-1570.
(61)   Hinckley, C.C. *J. Am. Chem. Soc.* **1969,** *91,* 5160.
(62)   Sherry, A.D.;Geraldes, C.F.G. In *Lanthanide Probes in Life, Chemical and
        Earth Sciences Theory and Practice*; Choppin, G.R.; Bunzli, J-C.G., Ed.;
        Elsevier: Amsterdam, 1989; pp 93-179.
(63)   Aime, S.;Botta, M.;Dastru ,W.;Fasano, M.;Panero, M. *Inorg. Chem.* **1993,** *32,*
        2068-2071.
(64)   Stephens, E.M. In *Lanthanide Probes in Life, Chemical and Earth Sciences*;
        Choppin, G.R.; Bunzli, J-C,G., Ed.; Elsevier: Amsterdam, 1989; pp 181-209.
(65)   More, K.M.;Eaton, G.R.;Eaton, S.S. *Inorg. Chem.* **1986,** *25,* 2638-2646.
(66)   Choppin, G.R. *J. Alloys Comp.* **1993,** *192,* 256-261.

RECEIVED December 6, 1993

Chapter 30

Hydrolysis Behavior of the Transuranium Elements

Bernd Jung[1], Norman M. Edelstein[1], and Glenn T. Seaborg[2]

[1]Chemical Sciences Division, [2]Nuclear Sciences Division, Lawrence Berkeley Laboratory, University of California, Berkeley, CA 94720

The chemistry of the actinides is still a fruitful area of research more than half a century after the discovery of plutonium by Seaborg *et al.* However, the fields of interest have shifted from the pioneering work establishing the basic inorganic and extraction chemistry to areas that are more focused on environmental aspects and long-term storage of transuranium materials. The hydrolysis behavior of actinide cations is basic to all environmental studies. This paper reviews recent work in this area.

It has been more than fifty years since the discovery of the transuranium elements. The initial activities in this field established the fundamental solution and solid-state chemistry of the first two of these elements and their compounds under the auspices of the Manhattan Project. New separation methods including solvent extraction techniques and uranium isotope separation played a leading role in these programs. Tracer techniques were widely used to determine solubilities (or solubility limits) of transuranium compounds as well as to obtain information about the coordination chemistry in aqueous solution. A little later, special solvent extraction and ion-exchange techniques were developed to isolate pure transplutonium elements on the milligram and smaller scale. The second edition of *The Chemistry of the Actinide Elements*, published in 1986 (*1*), covers most of these topics. A detailed overview of the history of transuranium chemistry is given in *Transuranium Elements: A Half Century* (*2*).

In recent years much of the research emphasis has shifted from the classical inorganic chemistry of the transuranium elements to topics which are more related to environmental aspects. These research projects are driven mainly by the need to evaluate initiatives for the safe disposal of nuclear waste in geologic repositories. One major concern is the possible leakage of nuclear waste containers and thus the contamination of the disposal site and surrounding environment and groundwater with radioactive material. Hence the question arises as to the identification of migrating species and the solubility-controlling phases under ambient conditions at a disposal site. In principle, this question can be answered only by a thorough description of the solution and heterogeneous equilibria under geochemical conditions, coupled with an accurate thermodynamic database for geochemical modeling and transport calculations.

0097–6156/94/0565–0361$08.00/0

Despite efforts during the past five decades, definitive thermodynamic data are not available. The purpose of this review is to summarize recent work on the hydrolysis of transuranium ions for the oxidation states found in aqueous solution, An^{3+}, An^{4+}, AnO_2^+, and AnO_2^{2+}.

An(III): Hydrolysis and Solubility Product

The actinide ions, An^{3+}, undergo hydrolysis in aqueous solution in a similar fashion to the trivalent lanthanides:

$$An^{3+} + n\,H_2O \rightleftharpoons An(OH)_n^{(3-n)+} + nH^+ \qquad n = 0 \text{ to } 3 \qquad (1)$$

The extent of hydrolysis, and thus the actual amount of hydrolyzed species, $An(OH)_n^{(3-n)+}$, is a function of pH as well as temperature. At higher pH, "$An(OH)_3$" precipitates and is the solubility-controlling solid phase in the system $An(III)/H_2O/OH^-$. However, the measured solubility quotient varies as a function of time and depends on the crystallinity of the sample. Many hydrolysis studies for trivalent actinides have focused on Am^{3+} because Am(III) is rather stable toward oxidation under various experimental conditions and has an isotope — ^{243}Am — with a relatively long half-life ($t_{1/2} \sim 7380$ y). The measured hydrolysis constants and solubility products found for Am^{3+} in the recent literature are summarized in Table I.

As several studies have emphasized, CO_2 is a contaminant as it is absorbed from the atmosphere in experiments run under ambient conditions. Because of comparable stability constants for An(III) carbonate complexes and hydrolysis reactions, carbonate complexation is a competing side reaction (3-4). This might explain some discrepancies among the hydrolysis constants found in the older literature, as given in (5).

Stadler et al. (6) recently reinvestigated the hydrolysis of Am^{3+} under CO_2-free conditions. Experiments were performed with different electrolytes (NaClO$_4$ and NaCl) and at various ionic strengths (0.1 to 3 M) in the pH range 7 to 13. The solution speciation was performed by absorption spectroscopy at higher concentrations, laser-induced photoacoustic spectroscopy (LIPAS) for lower concentrations, and the total Am concentration was determined by liquid-scintillation counting. Stadler et al. also examined the influence of radiation ($0.1 < Ci\,l^{-1} < 5$) on the solubility product. Radiation damage of the crystal lattice—regardless if it occurs in the crystal lattice itself or via the reaction of radiolysis products on the surface, e.g., OH, H_2O_2, and ClO—tends to favor a less crystalline material and therefore results in a higher solubility:

$\log K_{sp} = -25.7$ (0.1M NaClO$_4$; 0.1 $Ci\,l^{-1}$);

$\log K_{sp} = -25.0$ (0.1M NaClO$_4$; 1.2-5 $Ci\,l^{-1}$) for the following reaction

$$An(OH)_3 + 3H^+ \rightleftharpoons An^{3+} + 3H_2O \qquad (2)$$

Similar results have been mentioned in previous reports (7) for Am(OH)$_3$. In addition, ^{241}Am(III) hydroxide undergoes autoradiolytical oxidation to AmO_2^+ in concentrated saline solution and higher activity ($\sim 10^{-3}$ M; ~ 1-2 $Ci\,l^{-1}$). The oxidation process is completed within one week. This corroborates earlier work by the same group (8-10) and is of particular interest in evaluating the solubility-controlling solid phase in a possible contamination of groundwaters or brines with highly active nuclear waste at a repository site.

The only known solubility product ($\log K_{sp} = -25.4$) for crystalline Am(OH)$_3$ is reported by Silva (3), who used ^{243}Am to reduce radiation effects. However, more drastic conditions have to be employed to obtain crystalline samples. The initially amorphous material could be converted into microcrystalline $Am(OH)_3$ by evaporating a 5 M NaOH solution to near dryness over a period of 3 to 4 days. The solubility product of crystalline $Am(OH)_3$ is about a factor of 30 smaller compared to the value reported earlier for $Am(OH)_3(am)$ [am = amorphous] by the same group (11). Even though Rai et al. (4) obtained X-ray powder patterns for some of their "$Am(OH)_3$" samples (log K_{sp} = -24.5; I = 0), they assumed that most of the precipitates remained amorphous, and consequently their solubility product is for $Am(OH)_3$ (am). This statement can be extended to all the other values listed in Table I. Remarkably, the time needed to reach steady-state concentrations is apparently quite different for amorphous and crystalline $Am(OH)_3$. Rai et al. (4) observed the same total concentration from over- and undersaturation experiments after 7 days, while Silva (3) reported a minimum of 28 days to reach steady-state conditions.

The formation of polynuclear species, shortly before $Am(OH)_3$ precipitates, has been excluded by several authors. Stadler et al. (6) confirmed the absence of polymers by filtration (pore size 13Å) and carried out direct speciation by absorption spectra of the supernatant solution. Moreover, on the basis of computer refinements Silva (3) demonstrated that species similar to those proposed for Nd (12), $Nd_2(OH)_2^{4+}$, and $Nd_3(OH)_5^{4+}$ are just minor hydrolysis species in the case of Nd^{3+} and Am^{3+}, respectively, if they occur at all. The solubility and speciation experiments could be explained by the assumption of a stepwise hydrolysis: $Am(OH)^{2+}$, $Am(OH)_2^+$, and $Am(OH)_3$, respectively. Previous reports (13,14), assuming the formation of the $Am(OH)_4^-$ hydroxy complex in strongly alkaline solution, could not be confirmed. Rai et al. (4) and Stadler et al. (6) among others (15) rejected the formation of $Am(OH)_4^-$ up to pH 13 based on their solubility studies. The total Am concentration became independent of pH at pH > 12. This was corroborated by migration experiments (16).

Even though the reported literature values for the first and second hydrolysis constants for Am^{3+} agree within half an order of magnitude, the deviation for log β^*_3 (-24 to -27) covers 3 units on a logarithmic scale. An average value of log β^*_1 = -7.5(5) and log β^*_2 = -15.5(5) for the formation constant of $Am(OH)^{2+}$ and $Am(OH)_2^+$ in 0.1 M $NaClO_4$ solution can be accepted. The large discrepancies among the values for log β^*_3 might be an intrinsic problem, since the total Am concentration (10^{-10} to 10^{-11} M) in the pH range where $Am(OH)_3$ contributes most— pH > 10.5—is near or beyond the detection limit, and thus the accuracy of log β^*_3 is directly related to the reliability of the Am sampling and the accuracy of log K_{sp}. Stadler et al. pointed out that even the contamination of the reaction vessel with an Am aerosol from the glove box might lead to erroneous results. However, Rai et al. (4) set an upper limit of log β^*_3 = -27 because of the uncertainty in log β^*_3. Rai et al. (17) later revised this value and set an even lower upper limit for log β^*_3 ~ -28.6 based on the results obtained for the solubility of $AmOHCO_3$. The value for log β^*_3 reported by Silva (3) is derived from the data in the pH range 7 to 9.5 where $Am(OH)_3$ contributes only as a minor species and thus might be erroneous.

The information about the hydrolysis of trivalent transuranium elements other than Am is limited because of the redox instability of the trivalent oxidation state (Pu and Np) or the restricted availability of the isotope (elements beyond Cm). The available data refer mostly to the older literature, given in Ref. 25.

Felmy et al. (18) investigated the solubility of $Pu(OH)_3$ under reducing conditions in deionized water and brine solution. They derived a much lower solubility product (log K_{sp} = -26.2) (see Table I) than the value (log K_{sp} = -19.6) reported in the literature (19). However, the solubility in brines [I ~ 6 and I ~ 10] was found to be larger than that in deionized (I = 0) waters. The solubility of $Pu(OH)_3$ in brines was accurately predicted with the Pitzer ion-interaction model using only the parameters for binary interactions between Pu^{3+} and Cl^-.

Recently, Wimmer et al. (20) used time-resolved laser fluorescence spectroscopy (TRLFS) to reinvestigate the hydrolysis reaction of Cm^{3+} in 0.1 M

Table I. Solubility Products and Hydrolysis Constants

$$An^{3+} + n\,H_2O \rightleftharpoons An(OH)_n^{(3-n)+} + n\,H^+$$

Ion	$\log \beta^{*}_1$	$\log \beta^{*}_2$	$\log \beta^{*}_3$	$\log K_{sp}{}^a$ $An(OH)_3$	Method[b]	Medium I [M]	T [°C]	Ref.
Pu	-	-	-	-26.2.(8)	sol.	0	23	18
Am			-28.6					17
	-6.9(2)	-	-23.8(9)	-	mig.	0.1; NaClO₄	25	16
	-7.5(3)	-15.4(4)	-26.9(5)	-25.7(3)	sol.	0.1; NaClO₄ 0.1 CiL⁻¹	25	6
	-7.5(2)	-15.4(3)	-26.9(2)	-25.0(3)	sol.	0.1; NaClO₄ 1.2-5 CiL⁻¹	25	6
	-7.8(4)	-15.4(5)	-26.5(5)	-25.1(5)	sol.	0.1; NaCl; 2-5 CiL⁻¹	25	6
	-8.1(3)	-15.8(4)	-27.0(5)	-25.0(1)	sol.	0.6; NaCl; 2-5 CiL⁻¹	25	6
	-6.6	-14.6	-23.6	-	calc.	S.I.T.; 0	25	63
	-6.8	-	-	-	ext.	0.5; NaClO₄	25	64
	-6.3(8)	-13.6(6)	-22.9(5)	-27.2(5)	sol.	0.1; NaClO₄	25	15
	<-8.2	-17.1(5)	<-27	-24.5(3)	sol	corr. to 0	23	4
	-7.5(2)	-	-	-	ext.	0.7; NaCl	21	65
	-	-16.0(7)	-24.3(3)	-25.4(4)[c]	sol.	0.1/NaClO₄	25	3
	-	-14.7	-	-	ext.	0.2/NaClO₄	25	66
	-7.0	-	-	-	sol.	1; NaClO₄	25	67
	-7.5(3)	-	-	-	ext.	1; NaClO₄	25	68
Cm	-7.1(2)	-15.5(3)			TRLF	0.1; NaClO₄	25	20
	-7.7(3)	-		-	titr.	0.1; KCl	25	22
Cf	-6.2(2)	-	-18.6(3)	-	mig.	0.1; NaClO₄	25	23

$^a An(OH)_3(s) \rightleftharpoons An^{3+} + 3OH^-$

[b] sol., solubility; mig., migration; calc., calculated; ext., solvent extraction; TRLF, time-resolved laser fluorescence; titr., titration

[c] crystalline

$NaClO_4$ at 25° C. The experiments were carried out at pH 6 to 10 for Cm^{3+} concentrations ranging from 1.21 x 10^{-7} to 3 x 10^{-7} M. The fluorescence emission line from the excited state $^6(P,D)_{7/2}$ to the $^8S_{7/2}$ groundstate at approximately 594 nm (pH = 5.95) was used to determine the different concentrations of the hydrolyzed Cm^{3+} species in solution (*20, 21*, and references therein). The peak maximum shifted from 593.8 nm for Cm^{3+} at pH 5.95 to 598.8 at pH 9.84 for $Cm(OH)_2^+$. The mean lifetimes for all three Cm species in solution were almost identical, and thus Wimmer *et al.* concluded that the quenching efficiency caused by OH^- was similar to that of water. The maxima of the emission bands and the average lifetimes for $Cm(OH)_n^{(3-n)+}$ species are given in Table II. Contamination with CO_2 and thus the formation of carbonate complexes was excluded, based on the lifetime discrimination. Additionally, the expected lifetime of any Cm carbonate complex should be larger than 85 ms. The spectroscopic results indicate a mixture of different hydrolyzed species, primarily $Cm(OH)^{2+}$ and $Cm(OH)_2^+$, which change as a function of pH in the pH range 5.95 to 9.84; the relative amounts of each Cm^{3+} species were numerically deconvoluted from the spectra at a given pH. Total Cm concentration was determined by liquid scintillation counting. The hydrolysis constants, $\log \beta^*_1$ and $\log \beta^*_2$, for the first and second hydrolysis reactions, respectively were found to be:

$$\log \beta^*_1 = -7.1(2); \qquad \log \beta^*_2 = -15.5(3).$$

A comparison of the first hydrolysis constant reported by Wimmer *et al.* with those reported in the older literature reveals a large discrepancy. However, assuming a ratio $[Cm(OH)^{2+}]/[Cm^{3+}] = 1$ and using $\log \beta^*_1 = -7.1(2)$, one would expect a considerable amount of hydrolyzed species above a pH of about 6. Moreover, these hydrolysis constants, $\log \beta^*_1$ and $\log \beta^*_2$, agree well with those recommended for Am^{3+} [-7.5(5) and -15.5(5)] and the value for $\log \beta^*_1$ (-7.7) derived by Bucher *et al.* (*22*) from potentiometric titrations.

Rösch *et al.* (*23*) reported the first and third hydrolysis constants for Cf^{3+} derived from electromigration experiments in 0.1 M $NaClO_4$. The values for $\log \beta^*_1$ and $\log \beta^*_3$ were calculated (I = 0.1) from the individual ion mobilities, $\mu°$ $[Cf(OH)^{2+}] = + 3.4(4) \times 10^{-4}$ and $\mu°[Cf^{3+}] = +5.1(2) \times 10^{-4}$ $cm^2 s^{-1} V^{-1}$, to be -6.2 and -18.6, respectively. No evidence was found for the existence of an anionic hydroxy complex, $Cf(OH)_4^-$, up to a pH of 12.8.

An(IV): Hydrolysis, Solubility Product and "Polymer" Formation

The hydrolysis of tetravalent plutonium in aqueous solution is dominated by the formation of colloids. Even in 1 M $HClO_4$ polymer formation occurs, and reliable hydrolysis constants are found only for the first step with n = 1 (*5*):

$$An^{n+} + nH_2O \; \rightleftharpoons \; An(OH)_n^{(4-n)+} + nH^+ \qquad n = 0 \text{ to } 4 \qquad (3)$$

Nitsche *et al.* (*24*) recently confirmed the value of $\log \beta^*_{11}$ [-1.6(2)] reported in the literature (*5, 25*) by absorption spectroscopy. The experiments were conducted as a function of ionic strength (0.032 to 0.5 M $H/NaClO_4$), pH (0.42 to 1.83), and time (5 min to 24 h). No peak shifts were observed with increasing pH, although the intensity decreased considerably.

Table II.   Lifetime and Fluorescence Maxima of $Cm(OH)_n^{(3-n)+}$
(From Reference 20)

Ion	emission maximum [nm]	average lifetime [ms]
Cm^{3+}	593.8 $(7.7)^a$	65 ± 2
$Cm(OH)^{2+}$	598.8 (11)	72 ± 2
$Cm(OH)_2^+$	603.5 (11)	80 ± 10

[a]Number in parentheses is the full width at half maximum.

Colloid formation can be prevented in concentrated nitric acid (> 3 M) via nitrate complexation, and the formation of $An(NO_3)_6^{2-}$ has been proposed; compare, for example, $An(NO_3)_6^{2-}$ in $(NH_4)_2An(NO_3)_6$ (An = Pu and Np) (26). The conditions which lead to Pu polymer are summarized elsewhere (see References 1 and 27). Moreover, there is evidence that Np(IV) also forms polymeric colloids under certain conditions. The formation of colloids during the dissolution of $NpO_2 \cdot xH_2O$ in near neutral saline solution (0.1 and 5 M NaCl) during a period of ten months was reported (Gehmecker, H.; Bucher, J.; Edelstein, N. *Radiochim. Acta*, to be submitted). The yellow Np polymer could be separated by filtration (92 nm Centricon) but remained in solution after centrifugation (~7000 rpm). The colloid was characterized by absorption spectroscopy in solution, and the FTIR spectra of the dried solid (between KBr pellets) showed similarities to that of Pu polymer.

Triay *et al.* (28) investigated the particle size of colloidal Pu polymers using autocorrelation photon spectroscopy (APS). The size of the colloid particles present in freshly prepared Pu polymer solutions strongly depends on the method of preparation. Results are summarized in Table III. The smallest Pu colloids were obtained by diluting well-characterized Pu stock solution with water. The average particle size ranged from 20 to 60 Å, whereas peptization of precipitated $PuO_2 \cdot xH_2O$ in 0.02 M HNO_3 yielded a mixture of Pu colloids. The diameters of the particles were found to be 140 and 3700 Å, respectively. The largest particles, with an average size of 320 Å, were found for the slow self-oxidation of Pu(III) solutions. In all cases the particle size changed very little during a period of 223 days.

Thiyagarajan *et al.* (29) used small angle neutron scattering (SANS) in aqueous media to investigate the morphology of aged Pu(IV) polymers. The scattering data could best be fitted with an elipsoidal model for the colloidal particles. The average particle is about 47 Å in diameter and longer than 1900 Å. These results differ from electron microscope investigations of dried Pu polymer obtained from the sol-gel process (30): the samples were either amorphous or primarily crystalline domains. Additionally, X-ray diffraction studies by Thiyagarajan *et al.* on Pu(IV) polymer in aqueous solution indicated a fluorite-type structure (a = 5.398 Å), with d-values resembling those reported for dried Pu(IV) polymer and crystalline PuO_2 (a = 5.3960). The minimum dimension derived from the X-ray line broadening—due to scattering on particles smaller than 10^3 Å—is about 60 Å and agrees well with the SANS results. However, particle sizes of samples polymerized in the presence of UO_2^{2+}—or UO_2^{2+} added after the initial formation of Pu polymer—are considerably smaller than those obtained without UO_2^{2+}. This confirms earlier reports (27) that UO_2^{2+} inhibits the polymerization process.

Even though many experimental and theoretical values for the solubility product of $PuO_2 \cdot xH_2O$ and PuO_2 have been reported in the recent literature, the discrepancies are huge and cover almost 12 orders of magnitude (see Table IV).

Table III. Pu Colloids: Particle Size Determined by Various Methods

Method	Preparation	Molarity[a] [M]	pH	Age[b] [days]	Diameter[c] [Å]	Length [Å]
APS[d]	Dropwise addition of Pu(IV) stock solution[e] to distilled water	0.01	1.3	223	37	
	Dropwise addition of Pu(IV) stock solution to distilled water; heated at 75-80° C for 4.5 hours	0.01	1.3	20, 223	64, 57	
	Pu(IV) stock solution taken to near dryness and 0.001 M HCl added to form colloid	0.015	1.6	216	30	
	Dropwise addition of Pu(IV) stock solution to distilled water with aging at room temperature	0.003	2	606	26	
	Dropwise addition of Pu(IV) stock solution to distilled water at 70° C	0.003	1.65	45	21	
	Pu(IV) in 0.5 M HNO$_3$ diluted to 0.08 M acid followed by near neutralization using 1 M NaOH	0.008	1.7	14	15	
	Pu(IV) stock solution added to 0.7 M NaH(O$_3$) with stirring; precipitate washed with water and peptized in 0.02 M HNO$_3$ at 80° C for 2.5 h, then centrifuged and suspended in 0.01 M HNO$_3$.	0.05	2	26	140 + 3700	
	Slow auto-oxidation[f] of Pu(IV) at pH 4	0.006		227	320	
SANS[g]	Aged Pu(IV) polymer in H$_2$O				47 ± 2	> 1900
	Pu(IV) polymerized in the presence of UO$_2^{2+}$				25 ± 6	84 ± 14
	Pu(IV) polymerized and UO$_2^{2+}$ added later				22 ± 3	120 ±10

[a] Moles of Pu, as monomer, per liter; [b] Age at the time of the APS determination; [c] Most probable diameter; analyzed using regularization; [d] APS - Autocorrelation photon spectroscopy (from reference 28); [e] Typical stock solution was about 1 M in Pu and 3-4 M in HCl; [f] Alpha induced and/or air-oxidized; [g] SANS - Small angle neutron scattering (from reference 29)

Table IV. Solubility Products of Pu(OH)$_4$ and PuO$_2$

log K_{sp}[a]	Compound[b]	Medium	Method[c]	Ref.
-57.85(5)	Pu(OH)$_4$(am)	I = 0/exp. in 1 M HClO$_4$	sol.	31
-60.2(2)	PuO$_2$(c)	I = 0/exp. in 1 M HClO$_4$	sol.	31
-56.9(4)	PuO$_2$·xH$_2$O	pH = 0.5 - 4.90	sol.	32
-50.2	Pu(OH)$_4$(am)		sol.	69
-64(1)	PuO$_2$(c)	I = 0	calc.	31
-63.3	PuO$_2$(c)	I = 0	calc.	33
-62.5	Pu(OH)$_4$	I = 0	calc.	12

[a]Pu(OH)$_4$ $\rightleftharpoons$ Pu^{4+} + 4OH$^-$ or PuO$_2$(cr) + 2H$_2$O $\rightleftharpoons$ Pu^{4+} + 4OH$^-$
[b]am, amorphous; c, crystalline
[c]sol., solubility; calc., calculated

Recently, Kim and Kanellakopulos (*31*) redetermined the solubility product of crystalline PuO$_2$(cr) (log K_{sp} = -60.2) and amorphous PuO$_2$·xH$_2$O (am) (log K_{sp} = -57.85) in 1 M HClO$_4$. Based on the solubility product for PuO$_2$·xH$_2$O(am), the free energy of formation was estimated as -1440.9 kJmol^{-1}.

The speciation of the solution in contact with PuO$_2$·xH$_2$O (am) reveals that under anaerobic conditions Pu^{4+} and Pu polymer are present as well as small amounts of PuO$_2$$^{2+}$, while PuO$_2$$^+$ and Pu^{3+} could not be detected. However, the solubility product was not determined in 1 M NaClO$_4$ at pH > 1 because Pu colloids (more than 50%) are formed predominantly under these conditions and coexist with various amounts of Pu^{4+}, PuO$_2$$^+$, and PuO$_2$$^{2+}$. Because of these difficulties only an upper limit (log K_{sp} = -56.6) has been proposed for the solubility product of PuO$_2$·xH$_2$O (am) at a pH of 1.09 (*31*).

Rai (*32*) reported a different species distribution under oxic conditions. The speciation in solutions equilibrated in air shows that Pu(IV) dominates below pH 1.5; however, PuO$_2$$^+$ and PuO$_2$$^{2+}$ become the dominant species in the pH range from 1 to 3, while the concentration of Pu(IV) drops below the detection limit. The thermodynamic equilibrium constants (22° C) for the following reactions were calculated on the basis of the redox potentials and calculated activities for Pu^{4+}, PuO$_2$$^+$, and PuO$_2$$^{2+}$ at a given pH:

$$PuO_2 \cdot xH_2O \text{ (am)} \rightleftharpoons Pu^{4+} + 4OH^- + (x-2)H_2O \qquad (4)$$
$$\log K_{sp} = -56.9(4)$$

$$PuO_2 \cdot xH_2O \text{ (am)} \rightleftharpoons PuO_2^+ + e^- + xH_2O \qquad (5)$$
$$\log K = -19.5(2)$$

$$PuO_2 \cdot xH_2O \text{ (am)} \rightleftharpoons PuO_2^{2+} + 2e^- + xH_2O \qquad (6)$$
$$\log K = -35.6(4)$$

$$Pu^{4+} + 2H_2O \rightleftharpoons PuO_2^+ + e^- + 4H^+ \tag{7}$$
$$\log K = -18.6(2)$$

$$Pu^{4+} + 2H_2O \rightleftharpoons PuO_2^{2+} + 2e^- + 4H^+ \tag{8}$$
$$\log K = -34.8(4)$$

$$PuO_2^+ \rightleftharpoons PuO_2^{2+} + e^- \tag{9}$$
$$\log K = -16.2(4)$$

As expected, the solubility product for crystalline $PuO_2(cr)$ ($\log K_{sp} = -60.2$) is considerably lower than the value for amorphous $PuO_2 \cdot xH_2O$. After three years, the predominant species were analyzed as Pu^{3+}, coexisting with Pu^{4+} and PuO_2^{2+}; no Pu polymer was detected. Apparently, the Pu^{4+} concentration is too low to form an appreciable amount of polymer. However, the experimental value for $PuO_2(cr)$ is still higher than that estimated by Baes and Mesmer ($\log K_{sp} = -62.5$) (*12*). The value calculated by Morss (*33*), using existing thermodynamic data for the reaction:

$$PuO_2 + 2H_2O \rightleftharpoons Pu^{4+} + 4OH^- \tag{10}$$

is even lower but in good agreement with the estimate ($\log K_{sp} = -63.8$) deduced for the dissolution reaction of $PuO_2(cr)$ by Kim and Kanellakopulos:

$$PuO_2 + 4H^+ \rightleftharpoons Pu^{4+} + 2H_2O \qquad \Delta G_R = 44.8 \text{ kJ mol}^{-1} \tag{11}$$

Even though it has been repeatedly reported (*31, 32*) that steady-state conditions are reached after about 58 days, the solubility of $PuO_2 \cdot xH_2O$ strongly depends on the preparative conditions, the crystallinity of the solid phase, and the half-life of the Pu isotope (*34*). The reproducibility of the solubility experiments is even more difficult when Pu polymer is formed, and it has been emphasized (*35*) that Pu polymer cannot be considered as a thermodynamically well-defined phase. Thus the solubility of $PuO_2 \cdot xH_2O$ may vary considerably, depending on whether or not polymer is formed over a period of time. Moreover, Kim and Kanellakopulos pointed out that even the theoretical solubility product for crystalline PuO_2 might be unattainable because of the formation of "hydroxide" surface layers. Hence the solubility of PuO_2 is controlled by the hydrate oxide surface rather than the bulk material. Indeed, the formation of surface "hydroxide" layers and the importance of the surface structure have been experimentally verified for several oxides by XPS spectroscopy and other techniques (*36, 37*).

Delegard (*38*) investigated the solubility of hydrous Pu(IV) oxide as a function of NaOH activity under ambient conditions. The predominant species in solution under oxic conditions is pentavalent Pu, and from spectroscopic and electrochemical investigations this species has been proposed as a hydroxy complex, $PuO_2(OH)_4^{3-}$. The solubility was found to increase in concentrated NaOH solution with the square of the NaOH activity, while the solubility at a particular NaOH concentration (1-10 *M*) decreased after an initial period of 21 days. Steady-state conditions were not reached within 700 days. However, the solubility was lower under reducing conditions ($NaNO_2$), and the Pu(IV) remained stable in solution. Similar results were obtained for the dissolution of $NpO_2 \cdot xH_2O$ in brines (6 *M* NaCl) under oxic conditions (Gehmecker, H.; Bucher, J.; Edelstein, N. *Radiochim. Acta*, to be submitted). The solubility of $NpO_2 \cdot xH_2O$ is determined by the slow oxidation rate of Np(IV) to NpO_2^+. Apparently, this process is pH dependent.

Approximately 80% of the $NpO_2 \cdot xH_2O$ had been oxidized to NpO_2OH at pH 10, whereas very little oxidation had occurred at pH 4. Additionally, there was evidence for the formation of Np polymer at low pH. In analogy with Delegard's investigation, steady-state conditions had not been attained after 285 days. Moreover, the solid phase remained amorphous in brine solution. Contrary results were obtained by Strickert et al. (39), who observed that initially amorphous $NpO_2 \cdot xH_2O$ transformed into crystalline material within 30 days. However, Rai et al. (40) showed that the solubility of $NpO_2 \cdot xH_2O$ under reducing conditions (Zn, Fe, $S_2O_4^{2-}$) in the pH range from 6 to 14 is near or below the detection limit ($10^{-8.3}$ M) for Np. No evidence of amphoteric behavior was found, and therefore the existence of $Np(OH)_5^-$ —predicted by thermodynamic calculations—remains doubtful. However, their solubility product for $NpO_2 \cdot xH_2O$ (log K_{sp} = -53.5) seems to be too high with respect to the values reported for $PuO_2 \cdot xH_2O$ (see Table IV).

An(V): Hydrolysis and Solubility Product

The behavior of Np and Pu in nearly neutral aqueous solution is of particular interest because of their long halflives and large abundance in nuclear waste. In CO_2-free solution with OH^- as the only ligand the solubility of NpO_2^+ is determined by the solubility product of "NpO_2OH" and the formation of the hydroxo complexes, NpO_2OH and $[NpO_2(OH)_2]^-$:

$$NpO_2^+ + n\, H_2O \rightleftharpoons NpO_2(OH)_n^{(1-n)+} + nH^+ \qquad (12)$$

A similar reaction holds for PuO_2^+ in basic solution. For PuO_2^+ in acidic solution the disproportionation of PuO_2^+ becomes important as are radiolytic reactions for AmO_2^+. These topics are discussed below. Although the hydrolysis of NpO_2^+ in aqueous solution has been the subject of many studies in the past ten years, the hydrolysis constants derived by various methods differ by several orders of magnitude, and it is difficult to determine the best value. The literature data are summarized in Table V. Recent long-term studies on the speciation of Pu in natural groundwaters show that PuO_2^+ is the predominant soluble species at tracer level concentrations under a wide variety of environmental conditions (41, 42, and references therein).

Recently, Neck et al. (43) reinvestigated the hydrolysis reaction of NpO_2^+ in aqueous solution (pH 7 to 14) at various ionic strengths under a pure argon atmosphere. The morphology of the NpO_2^+ hydroxide, "NpO_2OH", strongly depended on the concentration of the electrolyte. Samples prepared at higher ionic strength (3 M $NaClO_4$) were more crystalline than those precipitated from 0.1 M $NaClO_4$ and thus resulted in different solubility products for amorphous and aged "NpO_2OH". The solubility product for the amorphous sample (log K_{sp} = -8.76) agrees well with those reported recently by Itagaki et al. (44) and the older data in literature (see Table V). Surprisingly, the crystal structure and therefore the true composition of "NpO_2OH" are not known. Nitsche et al. (41) reported d-values for crystalline samples, but they were unable to assign the powder pattern to any known compound of similar composition. The log β^*_1 values for the first hydrolysis constant reported by the Munich group (43,45) [-11.3(2); I =0.1] as well as that by Itagaki et al. (44) [-10.7(4); I = 0] and Rösch et al.[-10.5(3); I= 0.1] (46) are consistently lower than the ones given by others (~ -9.1). The reason for the discrepancies needs to be further investigated, since, for example, the survey reported by Neck et al. seems to be thoroughly conducted and the log β^*_1 reported by Bucher et al. (22) was obtained from over 200 individual potentiometric titrations in the pH range of 3 to 10.

Table V. AnO_2^+ (An: Np, Pu and Am): Hydrolysis Constants and Solubility Products

$$AnO_2^+ + nH_2O \rightleftharpoons AnO_2(OH)_n^{(1-n)+} + nH^+$$

Ion	$\log K_{sp} (AnO_2OH)^a$	$\log \beta^*_1$	$\log \beta^*_2$	Method[b]	Medium	T [°C]	I [M]	Ref.
NpO_2^+	-8.7(3)	-10.7(4)	-22.4(3)	sol.	$NaClO_4$	25	SIT; 0	44
	-9.4(1) (aged)	-11.3(2)	-23.6(2)	sol.	$NaClO_4$	25	SIT; 0	43
	-8.76(5) (am)[c]			sol.		25	SIT; 0	43
	-10.7	-8.3	-19.4	sol.	$NaNO_3$	RT	0.01	71
		-8.0	-18.1	mig.	$KCl\text{-}H_3BO_4\text{-}NaOH$	RT	0.1	72
		-8.3	-18.8	mig.	$KCl\text{-}H_3BO_4\text{-}NaOH$	RT	0.005	72
		-10.5(3)	-22.0(4)	mig.	$NaClO_4$	25	0.1	46
		-9.8		ext.	$NaClO_4/NaHCO_3$	25	0.2	73
	-8.81	-11.47	-22.71	sol.	$NaClO_4$	25	1.0	45
		-9.12		sol.	$NaClO_4/NaHCO_3$	25	1.0	48
		-9.05(2)		pot. titr.	KCl	RT	0.1	22
		-9(1)		pulse rad.		25		70
PuO_2^+		-9.7(1)		LIPAS	$NaClO_4$	RT	0	53
		<-9.7		pot. titr.		25		54
AmO_2^+	-9.3(5)	-12.3(6)		rad. titr.	NaCl	25	3	6

$^a AnO_2OH$ (solid) $\rightleftharpoons AnO_2^+ + OH^-$

^{b} sol., solubility; mig., migration; ext., solvent extraction; pot. titr., potentiometric titration; pulse rad., pulse radiolysis; LIPAS, laser-induced photoacoustic absorption spectroscopy; rad. titr., radiometric titration

^{c} am, amorphous

Sullivan *et al.* (*47*) determined the reaction enthalpy, ΔH, for the hydrolysis of NpO_2^+ to NpO_2OH (equation 13) in 1.0 M NMe_4Cl solution using the calorimetric titration:

$$NpO_2^+ + H_2O \rightleftharpoons NpO_2OH + H^+. \tag{13}$$

ΔH was found to be + 35.8 kJmol^{-1}, and based on a log β^*_1 value of -9.12(6) (*48*), the entropy change, ΔS, was calculated as 16(5) JK^{-1}mol^{-1}. The log β^*_1 value of -11.57 reported by Lierse *et al.* (*45*) was rejected on the basis of a graphical correlation of ΔQ_{corr} (corrected enthalpy change) versus the number of moles, n, of NpO_2OH formed during the calorimetric titration. The fit with log β^*_1 = -9.11 was linear, whereas that with log β^*_1 = -11.57 showed a significant curvature. Assuming that ΔC_p is close to zero for the hydrolysis reaction (13), they estimated the first hydrolysis constant at 90° and 130° C, respectively. This rough calculation—the temperature dependence of K_w was not considered—gave a tenfold increase in hydrolysis at 90° C and a 40-fold increase at 130° C. However, Nitsche (*35*) has recently shown that Np_2O_5 is obtained at 90° C (pH 6) in CO_2-rich groundwaters, which suggests that Np_2O_5 might be the thermodynamically stable dehydration product of "NpO_2OH" at elevated temperatures:

$$2 \text{ "}NpO_2OH\text{"} \rightarrow Np_2O_5 + H_2O \tag{14}$$

Unfortunately, the solubility product of Np_2O_5 is not known.

Various groups independently confirmed the formation of $[NpO_2(OH)_2]^-$ at pH > ~11.5. However, the scatter of the data for the second hydrolysis constant, log β^*_2, is rather large (-16.9 to -22.5). In some cases where CO_2 has not been strictly excluded, carbonate complexation may have led to erroneous results for log β^*_1 as well as for log β^*_2. Neck and coworkers demonstrated that an exposure of the original CO_2-free solution to air for only 5 min. leads to considerable CO_2 absorption. The absorption peak for "NpO_2OH" at 980 nm (pH 9.9) decreases substantially and a second band at 993 nm appears. In the older literature (*49*) this absorption peak was erroneously assigned to a hydrolyzed NpO_2^+ species (*41, 43*).

The hydrolysis of PuO_2^+ has not been thoroughly investigated. This might be attributed to the fact that PuO_2^+ is thermodynamically unstable with respect to disproportionation below pH 2. Capdevila *et al.* (*50*) determined electrochemically the disproportionation constant, $K°[PuO_2^+]$, of PuO_2^+ to PuO_2^{2+} and Pu(III) at pH 1 in perchlorate media as a function of ionic strength (0.1 to 3 M):

$$2H^+ + 3PuO_2^+ \rightarrow 2PuO_2^{2+} + Pu^{3+} + 2HOH. \tag{15}$$

Pu(IV) could not be spectroscopically detected. The ionic strength dependency of log $K°[PuO_2^+]$ [2.6(5)] indicates that PuO_2^+ is most stable at low ionic strength, while the highly charged species (Pu^{3+} and PuO_2^{2+}) are favored at higher electrolyte concentration. Moreover, the stability of PuO_2^+ was estimated to decrease with increasing temperature. However, the mechanism for the disproportionation of PuO_2^+ is more complicated than described in equation 15, and contradictory results for the disproportionation products are reported. For example, Madic *et al.* (*51*) used Raman spectroscopy to investigate the disproportionation of PuO_2^+ [0.01 M $^{242}PuO_2^+$; I = 1] at pH 3.7 as a function of time. Raman spectroscopy has the advantage that it is much more sensitive with respect to PuO_2^+ than conventional absorbance spectroscopy ($\epsilon \sim 20$ M^{-1} cm^{-1}). Contrary to the results reported by Capdevilla *et al.*, Madic *et al.* observed the formation of Pu(IV) polymer and PuO_2^{2+} after an initial period of

6 days. During this initial period the PuO_2^+ concentration remained almost constant. The simple disproportionation reaction (equation 16)

$$4H^+ + 2PuO_2^+ \rightarrow PuO_2^{2+} + Pu^{4+} + 2HOH \qquad (16)$$

is accompanied by the rapid equilibrium:

$$PuO_2^+ + Pu^{4+} \rightleftharpoons PuO_2^{2+} + Pu^{3+} \qquad (17)$$

and a second slow reaction to yield Pu(IV):

$$2HOH + PuO_2^+ + Pu^{3+} \rightarrow 2Pu^{4+} + 4OH^- \qquad (18)$$

The intensity of the band for the symmetrical stretching vibration in PuO_2^+ at 748 cm^{-1} slowly decreased over a period of 20 days, and a second peak at 833 cm^{-1} grew, which was assigned to the symmetrical stretching vibration of PuO_2^{2+}. Additionally, evidence was found that the pH increased during the experiment because of the disproportionation of PuO_2^+. Simultaneously with the growth of the peak at 833 cm^{-1}, a second peak appeared at 817 cm^{-1}, which was attributed to $(PuO_2)_2(OH)_2^{2+}$. Steady-state conditions for PuO_2^{2+} were reported to be reached after 20 days. Although Newton et al. (52) reported similar results with respect to the disproportionation products of PuO_2^+ at pH 1.3 to 2.2 [I = 0.01 to 1.0 M $NaCl/LiClO_4$], steady-state conditions were reached very slowly and required more than 400 days. Moreover, $^{239}Pu(IV)$ colloid is slowly oxidized to PuO_2^+ and PuO_2^{2+}, while $^{239}PuO_2^{2+}$ is reduced by α-radiation to PuO_2^+:

$$2PuO_2^{2+} + H_2O \rightarrow 2PuO_2^+ + 2H^+ + 1/2O_2 \qquad (19)$$

Contrary to the results of Capdevilla et al., there was no experimental evidence for Pu(III). At present, the reasons for the different results remain unclear and need to be investigated further.

Bennett et al. (53) investigated the hydrolysis reaction of PuO_2^+ in 0.1 M $NaClO_4$ solution with laser-induced photoacoustic spectroscopy (LIPAS). The effect of hydrolysis on the absorption spectrum of PuO_2^+ (569 nm; 19 M^{-1} cm^{-1}) between 530 and 590 nm was studied in the pH range from 3 to 10.5. No hydrolysis was observed below pH 8, and the PuO_2^+ concentration was far below the detection limit above pH 10. The data were fitted to the hydrolysis reaction:

$$PuO_2^+ + n\,H_2O \rightleftharpoons PuO_2(OH)_n^{(1-n)+} + nH^+ \qquad (20)$$

The results can be explained solely with the formation of PuO_2OH, and a two-complex model does not improve the accuracy of the first hydrolysis constant. The hydrolysis constant for PuO_2^+ in 0.1 M $NaClO_4$ derived from the spectroscopic speciation confirms the lower limit (> -9.7) estimated earlier by Kraus (54). A comparison of this value and log β^*_1 for NpO_2^+, assuming a smaller effective positive charge for PuO_2^+ compared to NpO_2^+, leads to a slightly higher first hydrolysis constant for NpO_2OH and thus suggests a value close to about -9.2 for log β^*_1 (NpO_2OH).

Raman spectroscopic investigations by Guillaume et al. (55) give evidence for the self-association of pentavalent actinyl ions in acid solution (pH ~ 0; I = 0.1 to

3.5 M H/NaClO$_4$). The formation of NpO$_2^+$ polynuclear complexes is favored at concentrations of NpO$_2^+$ greater than 0.1 M. Dimers have been suggested in the concentration range 0.1 < [NpO$_2^+$] < 1 M based on the red shift (-29 cm^{-1}) of the symmetrical stretching frequency, ν_1, of the linear group AnO$_2^+$ compared to the band at lower NpO$_2^+$ concentration (< 0.1 M; 767 cm^{-1}). A further increase in NpO$_2^+$ concentration leads to fairly complicated spectra in the range 650 to 850 cm^{-1}, and hence a mixture of polynuclear complexes has been proposed. Moreover, upon addition of UO$_2^{2+}$ to aqueous perchlorate solutions of either NpO$_2^+$ or AmO$_2^+$, the formation of mixed UO$_2^{2+}$/NpO$_2^+$ and UO$_2^{2+}$/AmO$_2^+$ complexes, respectively were observed. The complexation of AnO$_2^+$ was monitored via the shift of ν_1 (Np: 767 → 741 cm^{-1}; Am: 732 → 719 cm^{-1}) of the actinyl ion. The symmetrical stretching frequency, ν_1, of the linear group UO$_2^{2+}$ was rather insensitive to the formation of mixed polynuclear species. The stability of the UO$_2^{2+}$/AnO$_2^+$ complex decreases from Np to Am.

Little is known about the formation of mononuclear hydroxide complexes of AmO$_2^+$. However, it has been demonstrated that AmO$_2^+$ can be radiochemically generated and stabilized in concentrated saline solution (6, 8, and references therein). The autoradiolytical oxidation (~1 Cil^{-1}) of ^{241}Am(III) hydroxide in 3 M NaCl solution at pH 8.3 was completed within one week. Stadler et al. (6) derived the solubility product of AmO$_2$OH and the first hydrolysis constant, log β*$_1$ [-12.3(6)], from radiometric titrations at pH 8.3 to 13 (I = 3 M NaCl). Their value for log K$_{sp}$ [-9.3(5), I = 3 M] is in reasonable agreement within those reported for "NpO$_2$OH" (see Table V), although the solid phase has not been further characterized. However, the first hydrolysis constant seems to be rather small, and it is not clear whether this might be attributed to the high activity—as mentioned by the authors—or perhaps related to the difficulties of pH measurements in high ionic strength solutions.

Tananaev (56) recently reported the synthesis and characterization of ternary alkali hydroxide complexes of AmO$_2^+$ in AOH solution (A = Li, Na, or K). Two different ternary hydroxy compounds were obtained, depending on the concentration of the base: yellow AAmO$_2$(OH)$_2$·xH$_2$O (A = Li to K) precipitated from 0.1-0.5 M AOH solution, while brown A$_2$AmO$_2$(OH)$_3$·xH$_2$O (A = Na and K) was isolated from solutions with AOH concentrations higher than 2 M. Both compounds were isostructural with the corresponding NpO$_2^+$ compounds.

An(VI): Hydrolysis

The data available on the hydrolysis of NpO$_2^{2+}$ and PuO$_2^{2+}$ in aqueous solution are scarce and are based mainly on surveys done in the late fifties and sixties, as compiled in references 5 and 12. Because of this lack of reliable hydrolysis data for NpO$_2^{2+}$ and PuO$_2^{2+}$, a complete picture cannot be given, and one must refer to the extensive literature on UO$_2^{2+}$ hydrolysis. Thus (Pu/Np)O$_2^{2+}$ undergoes hydrolysis in nearly neutral to basic solution:

$$m \, PuO_2^{2+} + n \, H_2O \; \rightleftharpoons \; (PuO_2)_m(OH)_n^{(2m-n)} + nH^+ \qquad (21),$$

and in analogy to UO$_2^{2+}$, the following complexes have been proposed:

$$[(Np,Pu)O_2]_m(OH)_n \text{ with } (m,n) = (1,1); \, (1,2); \, (2,2); \, (2,3); \, (3,5); \text{ and } (4,7)$$

(5, 12) (see Table VI). Speciation studies by various groups over a longer period of time (42, 57, and references therein) show that PuO$_2^{2+}$ is—besides PuO$_2^+$—one of the major species in solution under steady state conditions in oxic systems at tracer concentrations, regardless of the initial oxidation state.

Table VI. AnO_2^{2+} (An: Np and Pu): Hydrolysis Constants

$$m(AnO_2^{2+}) + nH_2O \rightleftharpoons (AnO_2)_m(OH)_n^{(2m-n)+} + nH^+$$

Ion	$\log \beta^*_{11}$	$\log \beta^*_{22}$	$\log \beta^*_{35}$	$\log \beta^*_{47}$	Method[a]	I [M]	T [°C]	Ref.
NpO_2^{2+}	-5.4(1)				pulse rad.	0	25	60
	-5.17(3)	-6.68(2)	-18.25(2)		pot. tit.	1; NaClO$_4$	25	74
PuO_2^{2+}	-5.2(2)	-8.0	-21.3[b]		LIPAS	0.1; NaClO$_4$	21	58
				-29.3	ram.	1; NaClO$_4$/NaCl	RT	51
	-6.3(1)				pulse rad.	0	25	60
	-5.97(5)	-8.51(5)	-22.16(3)		pot. tit.	1; NaClO$_4$	25	59

[a] pulse rad., pulse radiolysis; pot. tit., potentiometric titration; LIPAS, laser-induced photoacoustic spectroscopy; ram., Raman spectroscopy

[b] (3,5) complex used in fit instead of (4,7) species

Okajima *et al.* (*58*) used LIPAS as well as conventional absorption spectroscopy to study the hydrolysis of PuO_2^{2+} in perchlorate media (0.1 M; 21° C) in the pH range 1 to 7. An estimate of the first hydrolysis constant was made based on the shift of the absorption band at 622 nm as a function of pH. A value of log $\beta_1^* =$ -5.2(2) was proposed, which is slightly higher than those found in the older literature (-5.6, -5.7, and -5.9) (*12, 54, 59*) and that derived from pulse radiolysis and transient conductivity experiments [-6.3(1)] by Schmidt *et al.* (*60*). However, Schmidt *et al.* reported in the same study (*60*) a value of log $\beta_1^* = -5.4(1)$ for NpO_2^{2+}, which is almost identical to the value given by Okaijama *et al.* for Pu (see Table VI). Additionally, evidence for the existence of a polynuclear species was found at pH < 6. Choppin and co-workers (*61*) calculated the relative amount of monomeric and polymeric species $(PuO_2)_m(OH)_n^{(2m-n)}$ with (m,n) = (1,1); (1,2); (2,2); and (3,5) based on literature hydrolysis constants for UO_2^{2+}, NpO_2^{2+} and PuO_2^{2+}. According to their calculations, equal amounts of monomeric and polymeric hydrolyzed species are found at pH 7 when $[UO_2^{2+}]$ is about 10^{-7} M. Equal amounts of monomeric and polymeric species occur at concentrations lower than 10^{-5} M and 10^{-3} M for $[NpO_2^{2+}]$ and $[PuO_2^{2+}]$, respectively at pH 7. The large discrepancies between UO_2^{2+}, NpO_2^{2+}, and PuO_2^{2+}, respectively are not understood and may be due to the lack of reliable hydrolysis constants for NpO_2^{2+} and PuO_2^{2+}. However, Raman spectroscopic studies by Madic *et al.* (*51*) indicate that PuO_2^{2+} behaves differently from UO_2^{2+} in aqueous solution. The symmetrical stretching vibration, v_1, for 0.1 M PuO_2^{2+} in solution with an ionic strength of 1 (NaCl and ClO_4^-) at pH 1 to 7 consists of four bands which have been assigned to three different species:

PuO_2^{2+}: 833 cm^{-1}; < pH 3.7;

$(PuO_2)_2(OH)_2^{2+}$: 817 cm^{-1}; 3.8< pH < 5.5; log $\beta^*_{22} = -8.0$

$(PuO_2)_4(OH)_7^+$: 826 and 805 cm^{-1}; pH > 4.5; log $\beta^*_{47} = -29.3$.

The suggestion of the (4,7) polynuclear complex rather than a (3,5) species is based on a different Raman spectrum compared to UO_2^{2+}, assuming that the two bands at 805 and 826, respectively for PuO_2^{2+} are solely due to one species. In the case of UO_2^{2+} the (3,5) complex exhibits only one band for the symmetrical stretching vibration. Moreover, the different polarization factors $[I_T(obs\|)/I_T(obs\perp)]$ of the two bands (805: 0.47; 826: 0.06) indicate that either PuO_2^{2+} occupies two different sites with different symmetry or has a more complex structure.

Similar results were obtained for NpO_2^{2+} in the narrow pH range of 1.8 to 3.7. The symmetrical stretching frequency for NpO_2^{2+} remains unchanged below pH 3. A further increase in pH results in a broadening of the band at 834 nm and a ternary sodium NpO_2^{2+} hydroxide precipitated at a pH higher than 3.7.

Tananaev (*62*) reported the existence of hydroxy complexes of PuO_2^{2+} as well as AmO_2^{2+} in strongly basic alkali hydroxide solution. The formation of lemon yellow $PuO_2(OH)_4^{2-}$ and yellow $AmO_2(OH)_4^{2-}$ ([An(VI)] < $1 \cdot 10^{-3}$ M) has been observed spectrophotometrically at a base concentration higher than 1 M and 1.2 × 10^{-2} M, respectively.

Summary and Conclusion

Recent work on the hydrolysis constants of actinide ions and the solubility products of the actinide hydroxides has been reviewed. The data in the literature cover a wide range of ionic strengths and are usually obtained at ambient temperature or 25° C. A number of solubility studies have been reported, approached both from undersaturation and oversaturation, but in most cases the solid materials that result from these studies are not well-characterized. In solutions with high concentrations of salts, radiolysis effects play a major role in the determination of the oxidation states of the actinide ions under steady-state conditions. The times reported to reach steady-state conditions in some measurements reported in this survey vary from days to years, even with experiments run under similar conditions. New spectroscopic methods have now been utilized to determine directly the speciation in solution and to verify the species derived from potentiometric and solubility data.

In general, experimenters performing recent hydrolysis studies have recognized some of the problems in earlier work (such as the need to work under an inert atmosphere in order to avoid carbonate contamination), and these studies should be more reliable, especially when coupled with direct spectroscopic methods for species determination. The major problems associated with solubility studies remain the attainment of steady-state conditions, the identification of solid phases, and control of the redox conditions. Measurements of hydrolysis reactions and solubilities at temperatures other than 25° C or ambient are almost nonexistent.

Acknowledgments
We thank Dr. Lester Morss for his helpful comments. This work was supported by the Director, Office of Energy Research, Office of Basic Energy Sciences, Chemical Sciences Division of the U.S. Department of Energy under Contract No. DE-AC03-76SF00098.

Literature Cited

1. *The Chemistry of the Actinide Elements*; Katz, J. J.; Seaborg, G. T.; Morss, L. R., Eds.; 2nd Edition; Chapman and Hall: London, 1986; Vols. 1 and 2.
2. *Transuranium Elements, A Half Century*, Morss, L. R. and Fuger, J., Eds.; American Chemical Society: Washington, D.C., 1992.
3. Silva, R. J. LBL-15055; Lawrence Berkeley Laboratory: Berkeley, CA, 1983.
4. Rai, D.; Strickert, R. G.; Moore, D. A.; Ryan, J. L. *Radiochim. Acta* **1983**, *33*, 201.
5. Fuger, J.; Khodakovsky, I. L.; Sergeyeva, E. I.; Medvedev, V. A.; Navratil, J. D. *The Chemical Thermodynamics of Actinide Elements and Compounds* Part 12; IAEA: Vienna, 1992.
6. Stadler, S.; Kim, J. I. *Radiochim. Acta* **1988**, *44/45*, 39.
7. Ziv, D. M.; Sheotakova, I. A. *Radiokhim.* **1965**, *7*, 175.
8. Magirius, S.; Carnall, W. T.; Kim, J. I. *Radiochim. Acta* **1985**, *38*, 29.
9. Büppelmann, K.; Magirius, S.; Lierse, C.; Kim, J. I. *J. Less-Common Met.* **1986**, *122*, 329.
10. Kim, J. I.; Lierse, C.; Büppelmann, K.; Magirius, S. *Mater. Res. Soc. Symp. Proc.* **1987**, *84*, 603.
11. Edelstein, N.; Bucher, J.; Silva, R. J.; Nitsche, H. LBL-14325; Lawrence Berkeley Laboratory: Berkeley, CA, 1983.
12. Baes, Jr., C. F.; Mesmer, R. E. *The Hydrolysis of Cations*; John Wiley & Sons: New York, NY, 1976.
13. Allard, B. In *Actinides in Perspective* ; Edelstein, N., Ed.; Pergamon Press: New York, NY, 1982; p 553.
14. Phillips, S. L. LBL-14313; Lawrence Berkeley Laboratory: Berkeley, CA, 1985.
15. Kim, J. I.; Bernkopf, M.; Lierse, C.; Koppold, F. *ACS Symp. Ser. No. 246*, **1984**, 115.
16. Rösch, F.; Reimann, T.; Buklanov, G. V.; Milanev, M.; Khalkin, V. A.; Dreyer, R. *J. Radioanal. Nucl. Chem* . **1989**, *134*, 109.
17. Felmy, A. R.; Rai, D.; Fulton, R. W. *Radiochim. Acta* **1990**, *50*, 193.
18. Felmy, A. R.; Rai, D.; Schramke, J. A.; Ryan, J. L. *Radiochim. Acta* **1989**, *48*, 29.

19. Busey, H. M.; Cowan, H. D. LAMS-1105; Los Alamos Scientific Laboratory: Los Alamos, NM, 1950.
20. Wimmer, H.; Klenze, R.; Kim, J. I. *Radiochim. Acta* **1992**, *56*, 79.
21. Beitz, J. V. *Radiochim. Acta* **1991**, *52/53*, 35.
22. Bucher, J.; Edelstein, N. LBL-15150 annual report, 1982; Lawrence Berkeley Laboratory: Berkeley, CA, 1983.
23. Rösch, F.; Reimann, T.; Buklanov, G. V.; Milanev, M.; Khalkin, V. A.; Dreyer, R. *Radiochim. Acta* **1989**, *47*, 187.
24. Nitsche, H.; Silva, R. J. LBL-30195A; Lawrence Berkeley Laboratory: Berkeley, CA, 1990.
25. Fuger, J. *Radiochim. Acta* **1992**, *58/59*, 81-91.
26. Spirlet, M. R.; Rebizant, J.; Apostolidis, C.; Kanellakopulos, B.; Dornberger, E. *Acta Cryst.* **1992**, *C48*, 1161.
27. Toth, L. M.; Friedman, H. A.; Osborne, M. M. In *Plutonium Chemistry*; American Chemical Society Symposium Series 216; ACS: Washington, D.C., 1983; Chap. 15.
28. Triay, I. R.; Hobart, D. E.; Mitchell, A. J.; Newton, T. W.; Ott, M. A.; Palmer, P. D.; Rundberg, R. S.; Thompson, J. L. *Radiochim. Acta* **1991**, *52/53*, 127-31.
29. Thiyagarajan, P.; Diamond, H.; Soderholm, L.; Horowitz, E. P.; Toth, L. M.; Felker, L. K. *Inorg. Chem.* **1990**, *29*, 1902-7.
30. Lloyd, M. H.; Haire, R. G. *Radiochim. Acta* **1978**, *25*, 139-48.
31. Kim, J. I.; Kanellakopulos, B. *Radiochim. Acta* **1989**, *48*, 145-50.
32. Rai, D. *Radiochim. Acta* **1984**, *35*, 97.
33. Morss, L. R. In *The Chemistry of the Actinide Elements*, 2nd Ed.; Chapman and Hall: London, England, 1986; Vol. 2, p 1289.
34. Rai, D.; Ryan, J. L. *Radiochim. Acta* **1982**, *30*, 213-6.
35. Nitsche, H. *Mat. Res. Soc. Symp. Proc.* **1992**, *257*, 289.
36. Luckner, N. Ph.D. Thesis; Technische Universität: Munich, Germany, 1988.
37. Greiling, H. D.; Lieser, K. H. *Radiochim. Acta* **1984**, *35*, 79-89.
38. Delegard, C. H. *Radiochim. Acta* **1987**, *41*, 11.
39. Strickert, R. G.; Rai, D.; Fulton, R. W. "Effects of Aging on the Solubility and Crystallinity of Np(IV) Hydrous Oxide." In *Geochemical Behavior of Disposed Radioactive Waste*; American Chemical Society Symposium Series 246; ACS: Washington, DC, 1984; pp 135-145.
40. Rai, D.; Ryan, J. L. *Inorg. Chem.* **1985**, *24*, 247.
41. Nitsche, H.; Edelstein, N. *Radiochim. Acta* **1985**, *39*, 23.
42. Choppin, G. R. *J. Radioanal. Nucl. Chem.*, **1991**, *147*, 109.
43. Neck, V.; Kim, J. I.; Kanellakopulos, B. *Radiochim. Acta* **1992**, *56*, 25.
44. Itagaki, H.; Nakayama, S.; Tanaka, S.; Yamawaki, M. *Radiochim. Acta* **1992**, *58/59*, 61.
45. Lierse, C.; Kim, J. I.; Treiber, W. *Radiochim. Acta* **1985**, *38*, 27.
46. Rösch, F.; Milanov, M.; Hung, T. K.; Ludwig, R.; Buklunov, G. V.; Khalkin, V. A. *Radiochim. Acta* **1987**, *42*, 43.
47. Sullivan, J. C.; Choppin, G. R.; Rao, L. F. *Radiochim. Acta* **1991**, *54*, 17.
48. Maya, L. *Inorg. Chem.* **1983**, *22*, 2093.
49. Sevost'yanova, E. P.; Khalturin, G. V. *Sov. Radiochem.* **1976**, *6*, 738.
50. Capdevila, H.; Vitorge, P.; Giffaut, E. *Radiochim. Acta* **1992**, *58/59*, 45.
51. Madic, C.; Begun, G. M.; Hobart, D. E.; Halin, R. L. *Inorg. Chem.* **1984**, *23*, 1914.
52. Newton, T. W.; Hobart, D. E.; Palmer, P. D. *Radiochim. Acta* **1986**, *39*, 139.
53. Bennett, D. A.; Hoffman, D.; Nitsche, H.; Russo, R. E.; Torres, R. A.; Baisden, P. A.; Andres, J. E.; Palmer, C. E. A.; Silva, R. J. *Radiochim. Acta* **1992**, *56*, 15.

54. Kraus, K. A.; Dam, J. R. "Hydrolytic Behavior of Plutonium (V)." In *The Transuranic Elements, IV-14B*; Seaborg, G. T.; Katz, J. J.; Manning, W. M. Eds.; McGraw Hill: New York, 1949; pp 478-499.
55. Guillaume, B.; Begun, G. M.; Haber, R. L. *Inorg. Chem.* **1982**, *21*, 1159.
56. Tananaev, I. G. *Radiokhim.* **1990**, *32*, 4.
57. Nitsche, H; Lee, S. C.; Gatti, R. C. *J. Radioanal. Nucl. Chem* **1988**, *124*, 171.
58. Okajima, S.; Reed, D. T.; Beitz, J. V.; Sabu, C. A.; Bowers, D. L. *Radiochim. Acta* **1991**, *52/53*, 111.
59. Cassol, A.; Magon, L.; Portanova, R.; Tondello, E. *Radiochim. Acta* **1972**, *17*, 28.
60. Schmidt, K. M.; Gordon, S.; Thompson, R. C.; Sullivan, J. C.; Mulae, W. A. *Radiat. Phys. Chem.* **1983**, *21*, 321.
61. Choppin, G. R.; Mathur, J. N. *Radiochim. Acta* **1991**, *52/53*, 25.
62. Tananaev, I. G. *Radiokhim.* **1989**, *31*, 46.
63. Moulin, V.; Robouch, P.; Vitorge, P.; Allard, B. *Radiochim. Acta* **1988**, *44/45*, 33.
64. Rao, V. K.; Mahajan, G. R.; Natarajan, P. R. *Inorg. Chim. Acta* **1987**, *128*,131.
65. Caceci, M. S.; Choppin, G. R. *Radiochim. Acta* **1983**, *33*, 101.
66. Bidoglio, G. *Radiochem. Radioanal. Lettters* **1982**, *53*, 45.
67. Nair, G. M.; Chander, K.; Joshi, J. K. *Radiochim. Acta* **1982**, *30*, 37.
68. Lundquist, R. *Acta Chem. Scand.* **1982**, *A36*, 741.
69. Rai, D.; Serne, R. J.; Moore, D. A. *Soil Sci. Soc. Am. J.* **1980**, *44*, 490.
70. Schmidt, K. M.; Gordon, S.; Thompson, R. C.; Sullivan, J. C. *J. Inorg. Nucl. Chem.* **1980**, *42*, 611.
71. Nakayama, S.; Arimoto, H.; Yamada, N.; Moriyama, H.; Higashi, K. *Radiochim. Acta* **1988**, *44/45*, 179.
72. Nagasaki, S.; Tanaka, S.; Takahashi, Y. *J. Radioanal. Nucl. Chem.* **1988**, *174*, 383.
73. Bidoglio, G.; Tanet, G.; Chatt, A. *Radiochim. Acta* **1985**, *38*, 21.
74. Cassol, A.; Magon, L.; Tomat, G.; Portanova, R. *Inorg. Chem.* **1972**, *11*, 515.

RECEIVED February 17, 1994

APPLICATIONS

Chapter 31

Solvent Extraction of Metals *Is* Coordination Chemistry

Michael Laing

Department of Chemistry, University of Natal, Durban 4001, South Africa

The aim of solvent extraction of metals is to remove quantitatively and selectively metal cations present in very low concentration in an aqueous medium into an organic medium, thereby greatly increasing the concentration of the metal. This is achieved by forming a neutral species containing the metal of interest — a coordination compound.

The Principles Involved

Solvent extraction of metals embodies all aspects of coordination chemistry: rates, equilibria, stereochemistry, crystal field theory, covalent bonding, hard-soft acid-base theory, hydrogen bonding, steric hindrance, enthalpy and entropy. All of these basic principles can link together to produce pure metals on an industrial scale from dilute aqueous solutions — a remarkable achievement of elegant coordination chemistry. To achieve this result it is only necessary to form within the aqueous medium a neutral species containing the metal to be extracted.

Bonding and Structure

It is important to understand that there are NO differences in principle between the bonding and structure of a "solvent-extractable" complex and a typical familiar coordination compound. In fact, a large number of the solvent-extractable species are merely complexes well known in gravimetric and colorimetric analysis but suitably modified so as *not* to precipitate readily from aqueous medium but rather to dissolve preferentially in the organic phase by the simple expedient of adding long chain alkyl groups to the organic ligand.

0097–6156/94/0565–0382$08.00/0

Classes of Extractable Species

There are two broadly different classes of extractable metal compounds:
(a)   Simple molecules
(b)   Coordination compounds.

In class (a) fall compounds like $GeCl_4$ and RuO_4. The latter is extracted into CCl_4 as a vital step in the industrial process for the production of pure ruthenium metal.

However, it is the solvent-extractable coordination complexes which are of immediate interest. These can be further categorized as:
(1)   Neutral covalently bonded molecules, usually with chelating ligands
(2)   Ion pairs, also called ion-association complexes.

In class (1) fall the chelate compounds of monoprotic organic acids such as:
i   OXINE; 8-hydroxyquinoline; and derivatives substituted by an alkyl chain at C(7),
ii   D2EHPA; di(2-ethylhexyl)phosphoric acid,
iii   ACAC, acetylacetone; and beta-diketonate derivatives variously substituted at the two keto carbon atoms (Figures 1(a) to 1(e)).

In this class also fall the so-called "solvated" complexes, compounds of "oxygen-loving" metals from which coordinated water molecules are displaced by hydrophobic "solvating" ligands containing groups with powerfully electron-donating oxygen atoms. Typical ligands involved here are ethers, ketones like methyl isobutyl ketone (MIBK), tributyl phosphate (TBP), $O=P(OC_4H_9)_3$, and tri-octylphosphine oxide (TOPO), $O=P(C_8H_{17})_3$ (Figures 2(a) and 2(b)). A good example of an extractable complex of this type is that formed by uranium(VI), $[UO_2(NO_3)_2(TBP)_2]$(Figure 3). TBP is a very versatile solvent ligand because not only is its oxygen atom a powerful electron pair donor to hard-acid metal cation centers, but the lone pair is also a potent proton-acceptor, readily forming the positively charged $[HOP(O\text{-}But)_3]^+$ species, which in turn forms neutral solvent-extractable ion-pair species. Most of these have the metal as the heart of a complex anion whose charge is neutralized by the large cation, either a protonated amine or oxygen-donor molecule. Some examples are:
(i)   Fe(III) in $[FeCl_4]^-[HO(C_2H_5)_2]^+$
(ii)   Ir(IV) in $[IrCl_6]^{2-}[HN(C_8H_{18})_3]_2^+$
(iii)   Au(III) in $[AuCl_4]^-[HMIBK]^+$
(iv)   Ir(IV) in $[IrCl_6]^{2-}[HOP(O\text{-}But)_3]_2^+$

(Actually, the cationic species present can contain molecules of water with the ether or ketone, but these have been omitted from the formulae for simplicity.)
Note the differences in coordination geometry:
$[FeCl_4]^-$ is tetrahedral; $[IrCl_6]^{2-}$ is octahedral; $[AuCl_4]^-$ is square planar.

Earliest Uses : Ether Complexes

It is interesting to remember that it was complexes of this class, neutral ion pairs, which were the first compounds of transition metals to be solvent-extracted. One of the earliest recorded uses of solvent extraction is by Rothe, who, in 1892, exploited the quantitative extraction of Fe(III) from an HCl medium into diethyl ether as the $[FeCl_4]^-[HO(C_2H_5)_2]^+$ ion pair. The Fe^{3+} ion has a high-spin d^5 electron configuration, which results in a formal crystal field stabilization energy of zero for both octahedral and tetrahedral geometries. However, the entropy contribution favors the formation of the tetrahedral $[FeCl_4]^-$ species over the octahedral $[Fe(OH_2)_6]^{3+}$ cation in an HCl medium :

$$[Fe(OH_2)_6]^{3+} + 4Cl^- \rightarrow [FeCl_4]^- + 6H_2O.$$

Gallium behaves in the same way and can be extracted into ether as the d^{10} Ga(III) complex $[GaCl_4]^-[HO(C_2H_5)_2]^+$.

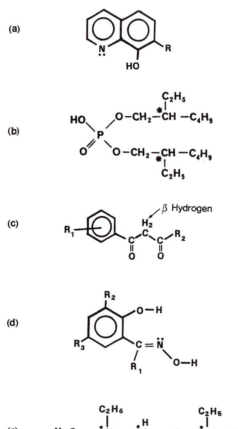

Figure 1. Monoprotic organic acids; chelating ligands
(a)   C(7)-substituted derivatives of 8-hydroxyquinoline, OXINE
(b)   Di(2-ethylhexyl)phosphoric acid, D2EHPA
(c)   Variously substituted β-diketonates related to acetylacetone, ACAC
(d)   Variously substituted β-hydroxyoximes; LIX 65N has R_1 = phenyl, R_2 = H, R_3 = C_9H_{19}
(e)   Aliphatic oximes; this particular example is LIX 63. The stars indicate chiral centers

Even earlier, in 1842, Péligot had observed that uranyl nitrate readily dissolves in diethyl ether with the formation of the neutral complex $[UO_2(NO_3)_2(ether)_2]$; 100 years later this phenomenon was the basis for the refinement of uranium for the Manhattan Project.

Gold and Ketones

In the presence of HCl gold is readily oxidized by chlorine gas to the square planar 16-electron $[AuCl_4]^-$ species. This anion forms ion pairs with protonated ketones and esters. It is, for example, easily extracted into methyl isobutyl ketone as the species $[AuCl_4]^-[HMIBK]^+$, which can be reduced directly to metallic gold.

Acetylacetone and Its Relatives

All beta-diketone compounds exist in the enol form and readily form stable anions with electron delocalization over the O—C—CH—C—O chain, which chelates the cations via the two oxygen atoms. The "oxygen-loving" metals whose cations are hard acids form stable chelate complexes of the form $[M(n)(ACAC)_n]$ (Figure 4). Beryllium is particularly well extracted as the neutral $[Be(ACAC)_2]$ complex because the combination of the tetrahedral $[BeO_4]$ coordination geometry and the small radius of the Be^{2+} ion results in a very stable and compact molecule.

Complexes with Derivatives of Oxine, 8-Hydroxyquinoline

Copper forms a very stable and insoluble compound with oxine: $[Cu(OXIN)_2]$, $K_{sp} = 10^{-29}$ (Figure 5), while aluminum forms the trisubstituted complex $[Al(OXIN)_3]$, $K_{sp} = 10^{-32}$, which is commonly used for the gravimetric determination of aluminum.

Rate k vs Equilibrium K

While the ultimate stability of a complex is a matter of thermodynamics and is describable in terms of an equilibrium constant K_f, the rate of formation of the complex is in the realm of kinetics, and these two phenomena can be totally unrelated. A good example of this in the field of solvent extraction is seen in the behavior of Kelex 100 toward Cu(II) and Fe(III).

Kelex 100 is a well-known commercial extractant, a derivative of oxine, substituted at C(7) with a branched dodecyl hydrocarbon chain (Figure 1(a)). This ligand easily loses its proton and quickly reacts with copper(II) to form the square planar $[CuL_2]$ complex with a large formation constant, K_f. It also forms the neutral compound $[FeL_3]$ with Fe(III), but the large bulk of the ligands prevents them from readily satisfying the steric requirements for the formation of the octahedral complex. This effect is illustrated by the competition between Cu^{2+} and Fe^{3+} for the ligand in an aqueous solution. Copper can be quantitatively separated from the iron and removed from the solution if the extraction is carried out within an hour. However, if the solution containing Cu^{2+} and Fe^{3+} is allowed to stand in contact with the Kelex 100 for about a day, the copper "disappears" and cannot be extracted. The extremely stable $[FeL_3]$ complex is now formed to the exclusion of the copper complex (Figure 6).

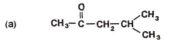

(a)

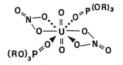

(b)

Figure 2. Molecules containing powerful electron-donating oxygen atoms
(a)   Methyl isobutyl ketone, MIBK
(b)   Tributylphosphate, TBP

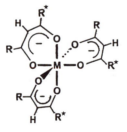

Figure 3. The neutral complex [UO$_2$(NO$_3$)$_2$(TBP)$_2$]
The uranium(VI) has achieved a maximum coordination number of 8 with the 6 oxygen atoms from the ligands lying in the equatorial plane. The two TBP ligands are *trans*, yielding a resultant dipole moment of zero for the molecule

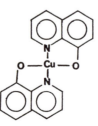

Figure 4. A typical neutral octahedral [M(III)(ACAC)$_3$] complex
The O—C—CH—C—O systems are planar with delocalized electrons, which impart aromatic properties to the unique H atoms. Groups R and R* can differ

Figure 5. The neutral square planar complex formed between Cu^{2+} and 8-hydroxyquinoline, [Cu(OXIN)$_2$]

The [CuL$_2$] complex is formed first because the two ligands can easily approach the aquated Cu^{2+} ion and displace the four coordinated water molecules to form the very compact square planar structure with its strong Cu—O and Cu—N bonds. The rate of formation of the [FeL$_3$] is extremely slow because of steric hindrance of the bulky ligands. The slow rate of formation of the [FeL$_3$] molecule is indicative of the steric struggle to fit the **third** ligand onto the relatively easily formed *trans*-[FeL$_2$(H$_2$O)$_2$]$^+$ ion. This step requires a major rearrangement of the chelating ligands already on the cation, simultaneously with the displacement of the two water molecules to form the neutral tris species (Figure 7).

The Fe(III) complex is extremely stable in a thermodynamic sense, as its K$_f$ is extremely large, which might be expected when one recalls that the Al(III) analogue [Al(OXIN)$_3$] is used for the gravimetric determination of aluminum. The slow rate of formation is an excellent example of steric hindrance controlling the rate of a reaction.

Complexes of Aromatic β-Hydroxy Oximes

Salicylaldoxime forms an extremely insoluble compound with Cu(II) and is therefore used for the gravimetric determination of copper. This planar [CuL$_2$] compound is extremely insoluble because there is no steric barrier to the close packing of the molecules, which thus can form a tightly packed crystal structure in which the Cu atoms can achieve a coordination number of 5 via a short intermolecular Cu···O separation. Nevertheless, β-hydroxyoximes have been successfully used to solvent-extract copper(II).

The Importance of Steric Hindrance

LIX 65N is a closely related derivative of salicylaldoxime commonly used for the extraction of copper (Figure 1(d)). As may be expected, it forms the complex [CuL$_2$], with the Cu—N and Cu—O bonds in a square plane. As before, the chelate rings are six-membered, and the structure is stabilized by a pair of strong O—H···O intramolecular hydrogen bonds. However, it is the presence of the two phenyl rings bonded to the C atoms of the C=N—OH moieties in the chelate rings that is critical to the efficacy of this ligand as a solvent for extraction. Once the [CuL$_2$] complex has formed, these two phenyl rings cannot lie coplanar with the rigid square planar bond system that contains the Cu atom because of collisions of their *ortho*-H atoms with those of the phenyl groups forming the chelate rings with the Cu atom. The two phenyl rings are thus forced by steric hindrance to rotate about the C—C bonds and to lie roughly perpendicular to the plane of the [CuN$_2$O$_2$] system. Their bulk prevents close packing of the complexes, thereby preventing association and the likelihood of crystallization of the complex, while simultaneously favoring the solubility of the complex in the organic hydrocarbon phase (Figure 8).

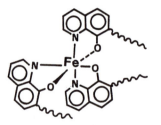

Time

Figure 6. Rates of appearance and disappearance of [Cu(OX)$_2$] and [Fe(OX)$_3$], where OX is KELEX 100

The square planar copper(II) complex forms quickly because it does not suffer from steric hindrance. The iron(III) complex inevitably forms at the expense of the copper compound because K_f of the iron complex is so much greater. The rate of formation is slow because of the steric problems involved in fitting the three ligands onto the Fe^{3+} ion

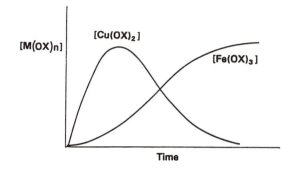

Figure 7. The octahedral [Fe(OX)$_3$] complex, where OX is KELEX 100

The planar aromatic 8-hydroxyquinoline rings give rigidity to the complex, and the long aliphatic hydrocarbon chains, shown as wavy lines, render the complex soluble in organic solvents

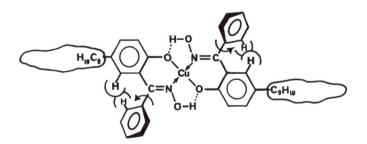

Figure 8. The complex of copper(II) with LIX 65N, a relative of salicylaldoxime

The [CuN$_2$O$_2$] system is square planar and rigid, and further stabilized by two O—H$\cdots$O hydrogen bonds. Note how the two phenyl rings are forced out of this plane by van der Waals repulsions

Chirality vs Racemates

The commercial product LIX 63 is an aliphatic hydroxyoxime widely used for the extraction of copper(II) (Figure 1(e)).

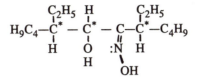

There are several important structural features about this compound that make it a very effective chelating agent for extraction. First, the aliphatic chain is flexible and subject to free rotation, which makes it easier for the chelation process to take place. Second, the chelate ring formed to the Cu metal center is five-membered and is thus more rigid and energetically favored. Third, the chain has *three* chiral centers. The commercially used product is therefore a mixture of 8 stereoisomers, and so there will be 64 coordination complexes of stoichiometry $[CuL_2]$, all of different stereochemistry.

It is therefore almost impossible for identical stereoisomers to associate, so crystallization will not occur. In addition, once the two chelate rings have formed, the C_7 hydrocarbon chain on the hydroxy ring carbon will NOT lie in the plane of the $[CuN_2O_2]$ system but will point up from it, thus preventing close association between molecules (Figure 9). The net effect is to give the molecule a surface of chaotic aliphatic hydrocarbon chains, and this simultaneously enhances the solubility of the complex in the organic diluent and accelerates the extraction.

Extraction of Uranium at Rössing, Randfontein, and ERGO

In one of the best known industrial solvent extraction processes, uranium is leached from its ore by dilute sulfuric acid to yield the uranium(VI) species $[UO_2(SO_4)_3]^{4-}$ in the solution. Addition of the tertiary trioctylamine, ALAMINE 336, to this acidic solution first forms the cationic species $[HN(C_8H_{17})_3]^+$, which in turn generates the neutral ion-pair $[UO_2(SO_4)_3]^{4-}[HN(C_8H_{17})_3]_4^+$ that is extracted into kerosene (Figure 10). The structure of the complex uranium-containing anion is typical of complexes formed by the uranyl species: a linear UO_2 group with six O atoms coordinated to the U atom around the equator. This geometry is found in several other extractable species of uranium(VI).

Hexachloro Species and the Separation of Rhodium from Iridium

The first step in the refining of the platinum group metals (PGMs) is the dissolution of the sulfide matte residue from the base metals refinery. This is done by reaction at high temperature with concentrated HCl and Cl_2 gas under pressure. Under these conditions the PGMs are oxidized and form stable chloro species. Iridium achieves the oxidation state Ir(IV), while rhodium attains only the 3+ state. The chloro species in solution are thus: $[AuCl_4]^-$, $[PdCl_4]^{2-}$, $[PtCl_6]^{2-}$, $[RhCl_6]^{3-}$, and $[IrCl_6]^{2-}$. Gold, palladium, and platinum are easily removed. The separation of rhodium from iridium can be achieved by solvent extraction with tributylphosphate from the acidic HCl medium because the behavior of the doubly charged $[IrCl_6]^{2-}$ species differs considerably from that of the triply charged $[RhCl_6]^{3-}$ ion. The tributylphosphate is protonated and forms the neutral species $[IrCl_6]^{2-}[HO=P(OC_4H_9)_3]_2^+$, which can be solvent-extracted, leaving behind in the aqueous layer the $[RhCl_6]^{3-}$ species, which does not form a neutral ion association species that is stable enough to be extracted. This process is critically dependent on both the concentration of Cl^- in solution and the acidity because the hexachloro rhodium species is susceptible to both aquation and hydrolysis, *e.g.*,

$[RhCl_6]^{3-} + H_2O \rightarrow [RhCl_5OH_2]^{2-} + Cl^-$
$[RhCl_5OH_2]^{2-} + H_2O \rightarrow [RhCl_4(OH_2)_2]^- + Cl^-$

Each of these steps has a well-defined forward rate k_f as well as a rate k_b for the reverse reaction, and thus an equilibrium constant K with the obvious implication that the distribution of rhodium species will alter with time. Also, the aqua species are acidic and can lose a proton, *e.g.*, $[RhCl_5OH_2]^{2-} + H_2O \rightarrow [RhCl_5OH]^{3-} + H_3O^+$

Unfortunately, all of the various doubly charged anionic rhodium species can form neutral ion pairs with the protonated TBP and so can be coextracted with the $[IrCl_6]^{2-}$. This will have disastrous consequences for the platinum metal refinery.

The use of the "solvating ligand" TBP in this way is unusual. Normally its task is to displace coordinated water molecules from the cationic center. An alternative extractant could be a tertiary base like ALAMINE 336, $N(C_8H_{17})_3$, commonly used for uranium extraction. This readily protonates to form the cationic tertiary ammonium species, which in turn will yield the neutral solvent-extractable ion pair $[IrCl_6]^{2-}[HN(C_8H_{17})_3]_2^+$.

What is Synergism ?

When a combination of two different extractants greatly exceeds the effect that would be expected from the simple sum of their separate efficacy as solvents, the phenomenon is termed synergism.

This effect is very dramatically illustrated in the extraction of uranium(VI) as $[UO_2]^{2+}$ from a nitric acid medium. The species initially present in the acid medium is $[UO_2(NO_3)_2(OH_2)_2]$, strongly hydrogen bonded to the surrounding hydration shell. Addition of tributylphosphate, by displacement of the two coordinated water molecules, yields the stable neutral bisolvated species $[UO_2(NO_3)_2(TBP)_2]$, which is extractable into the organic layer (Figure 3).

Alternatively, the chelating phosphoric acid derivative D2EHPA can be added to the aqueous layer to yield the neutral extractable species $[UO_2(D2EHPA)_2(OH_2)_2]$ by displacement of the two coordinated nitrate ions. If, however, D2EHPA and TBP are used in combination, the neutral species formed is $[UO_2(D2HEPA)_2(TBP)_2]$, which is 1000 times more extractable than either of the complexes containing only one of the phosphorus ligands (Figure 11). This is synergism.

The reasons are straightforward. The addition of D2EHPA alone gives a robust complex, thanks to the strongly bonded $[UO_2P]$ chelate rings. This complex is made hydrophobic by the two ethylhexyl aliphatic chains. However, the two coordinated water molecules can still participate in hydrogen bonding to the aqueous solution and hinder the extraction.

On the other hand, when the TBP alone is added, the coordinated water molecules are displaced, yielding a neutral hydrophobic species. However, the concentration of

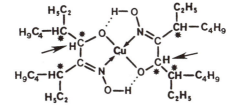

Figure 9. The complex formed between Cu^{2+} and the aliphatic hydroxyoxime LIX 63
The stars indicate chiral centers. The five-membered rings bonded to the Cu center are rigid, and the alkyl chains bonded to the chiral CH group, marked with an arrowhead, are forced out of the plane of the planar $[CuN_2O_2]$ system

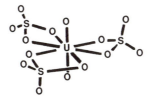

Figure 10. The $[UO_2(SO_4)_3]^{4-}$ anion
This robust species forms in dilute H_2SO_4 solution and is solvent extractable as a neutral ion pair with a protonated trialkyl amine

Figure 11. The neutral complex $[UO_2(D2EHPA)_2(TBP)_2]$
The surface of this complex is totally covered by hydrophobic alkyl chains, which impart the phenomenon of synergism to its extraction

negative charge on the O atoms of the nitrate groups still results in some hydrogen bonding to the solvent.

When the extractants D2EHPA and TBP are used in combination simultaneously: the maximum coordination number of 6 in the equatorial plane of the UO_2^{2+} ion is attained; all possible sources of hydrogen bonding are eliminated; the $[UO_2P]$ chelate moieties give rigidity and stability; the many alkyl chains on both TBP and D2EHPA greatly enhance the hydrophobicity (the complex is effectly a ball of kerosene), and the extractability into organic medium is enormously enhanced.

In summary, synergism will be observed when:

1. The metal atom in the complex achieves its maximum coordination number;
2. All coordinated water molecules are displaced by neutral solvating donor ligands, S, to render the complex hydrophobic; and
3. The chelating agent, HX, forms a singly charged bidentate anion, X^-, which neutralizes the positive charge on the cation, to form a neutral species with the metal, *i.e.*, for a cation M^{n+}, there will be n chelating groups X^- to give a species MX_nS_y.

The Separation of Co^{2+} from Ni^{2+} Explained by Crystal Field Theory

Seldom does one require an understanding of crystal field theory to understand an industrial process. However, the separation of Co^{2+} from Ni^{2+} by solvent extraction is beautifully explained by a simple application of crystal field theory.

At room temperature in an aqueous medium both metals exist as octahedral hydrated species : $[Co(OH_2)_6]^{2+}$ and $[Ni(OH_2)_6]^{2+}$. Addition of D2EHPA forms the neutral species $[M^{2+}(D2EHPA)_2(OH_2)_2]$ for both metals (Figure 12). The coordinated water molecules hydrogen-bond strongly with the aqueous solution making the complexes nonextractable. However, if the temperature of the solution is raised, the cobalt complex loses its coordinated water molecules, forming a neutral tetrahedral anhydrous molecule $[Co(D2EHPA)_2]$ (Figure 13). This species is hydrophobic and is extractable into organic (kerosene) medium. The nickel complex remains octahedral with its two coordinated water molecules, which hold the complex in the aqueous solution by the strong hydrogen bonding.

The reason is immediately obvious from a consideration of the electron configurations of the cations: Co^{2+} is d^7; Ni^{2+} is d^8, and the electron distributions in the octahedral species are:

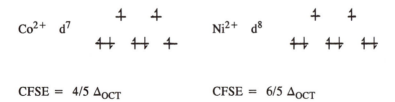

$$CFSE = 4/5\ \Delta_{OCT} \qquad\qquad CFSE = 6/5\ \Delta_{OCT}$$

The crystal field stabilization energy of the octahedral/tetragonal Ni^{2+} species is a maximum. However, if Co^{2+} adopts the tetrahedral geometry, its crystal field stabilization energy becomes more favorable :

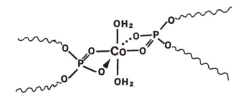

Figure 12. The octahedral hydrated species $[Co(D2EHPA)_2(H_2O)_2]$
The nickel(II) derivative has the same geometry. Both strongly hydrogen-bond to the water molecules of the aqueous solution

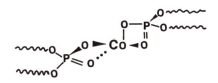

Figure 13. The tetrahedral anhydrous neutral molecule $[Co(D2EHPA)_2]$
Formation of this bright blue complex is favored at high temperatures (about 50°C), and it is easily extracted into kerosene

Co^{2+} d^7

CFSE $=$ 6/5 Δ_{TET}

The effect is enhanced by the entropy advantage of losing the two water molecules as the temperature rises:

$$[Co(D2EHPA)_2(OH_2)_2] \xrightarrow{\text{Heat}} [Co(D2EHPA)_2] + 2H_2O$$

The bright blue color of the organic extractant phase confirms that the extracted cobalt(II) species is indeed tetrahedral.

Conclusion

The various complex species involved in the industrial refining of metals by the technique of solvent extraction are fine examples of coordination chemistry in action. The metals come from all parts of the periodic table, and their atoms exhibit a wide range of coordination numbers and geometry. There are the tetrahedral $[FeCl_4]^-$ and $[GaCl_4]^-$ anions; the square planar $[AuCl_4]^-$ anion; the tetrahedral molecules $[Be(ACAC)_2]$ and $[Co(D2EHPA)_2]$; the square planar $[Cu(OXIN)_2]$ molecules; the octahedral anions $[IrCl_6]^{2-}$ and $[RhCl_6]^{3-}$; and the eight-coordinate molecule $[UO_2(D2EHPA)_2(TBP)_2]$. The seemingly mysterious processes involved in solvent extraction of metals can be readily explained by a knowledge of the basic principles of bonding and thermodynamics of coordination complexes.

Acknowledgment

I thank my friends at the extraction plants: for uranium — in Rössing, Namibia; ERGO, Springs; and Randfontein, Transvaal; and for the base metals and platinum metals at Rustenburg, Transvaal, who showed me the real world of large-scale solvent extraction of metals.

Bibliography

The topics presented above are mentioned in several places so for simplicity the citations are grouped under topics.

A. Organic Reagents
1.   Holzbecher, Z.; Divis, L.; Kral, M.; Sucha, L.; Vlacil, F. *Handbook of Organic Reagents in Inorganic Analysis*; Ellis Horwood: Chichester, **1976**.
2.   Cheng, K. L.; Ueno, K.; Imamura, T. *Handbook of Organic Analytical Reagents*; CRC Press: Boca Raton, Florida, **1982**.
3.   Fries, J.; Getrost, H. *Organic Reagents for Trace Analysis*; Merck: Darmstadt, 1977.

B. Solvent Extraction: Analytical Applications
4.   Vogel, A. I. *Quantitative Inorganic Analysis*; 3rd ed.; Longman: London, **1962**; pp 124-136; 890-907.
5.   Kolthoff, I. M.; Sandell, E. B.; Meehan, E. J.; Bruckenstein, S. *Quantitative Chemical Analysis*; 4th ed.; Macmillan: New York, **1969**; pp 335-375.
6.   Morrison, G. H.; Freiser, H. *Solvent Extraction in Analytical Chemistry*; John Wiley: New York, **1957**.
7.   Irving, H. M. N. H. In *Comprehensive Coordination Chemistry*; Wilkinson, G.; Gillard, R. D.; McCleverty, J. A.; Eds.; Pergamon: Oxford, **1987**; Vol. 1, pp 521-563.
8.   De, A. K.; Khopkar, S. M.; Chalmers, R. A. *Solvent Extraction of Metals*; Van Nostrand Reinhold: London, **1970**.

C. Solvent Extraction: Industrial Applications
9.   *Handbook of Solvent Extraction*, Lo, T. C.; Baird, M. H. I.; Hanson, C.; Eds.; J. Wiley: New York, **1983**; especially chapters 2·2; 24; 25·2; 25·7; 25·11.
10.   Habashi, F. *Principles of Extractive Metallurgy*; Gordon and Beach: New York, **1970**; Vol. 2, pp 331-430.
11.   Ritcey, G. M.; Ashbrook, A. W. *Solvent Extraction*; Elsevier: Amsterdam, **1984**; pp 88-171.
12.   Nicol, M. J.; Fleming, C. A.; Preston, J. S. in *Comprehensive Coordination Chemistry*; Pergamon: Oxford, **1987**; Vol. 6; pp 788-814.

RECEIVED March 15, 1994

Chapter 32

Coordination Chemistry in the Solvent Extraction of Metals
Developments from Russian Laboratories

Yu. A. Zolotov

**Kurnakov Institute of General and Inorganic Chemistry, Russian
Academy of Sciences, Moscow 117907, Russia**

Extractive separation of metals is usually based on complex formation with inorganic and
organic ligands. Therefore the use of the ideas, approaches, and methods of coordination
chemistry has always been a most fruitful approach to the extraction of the elements. History shows that many problems of selectivity
of separation or enhanced isolation have been
successfully solved by the rational application of coordination chemistry, e.g., the concept of hard and soft acids and bases. The efficiency of extraction depends on, inter alia,
the ratio of charge and coordination number of
metal ion. Study of this effect permitted the
development of ways to improve separation due
to changes in hydration of the species to be
extracted.

General requirements for extraction of inorganic species
from an aqueous solution into an immiscible organic phase
are known. These include electroneutrality, higher solubility in the organic solvent than in water, bulky molecules, and stability, i.e., a high enough stability constant.
    The main types of inorganic compounds used for ex-
tractive separation of metal ions include (1):

Neutral Compounds

1. Coordinatively nonsolvated neutral compounds, such as
   $HgCl_2$, AsI_3, or OsO_4.
2. Metal chelates.
3. Coordinatively solvated (mixed) neutral compounds,
   such as $AuCl_3L$, $UO_2(NO_3)_2L_2$, where L is a neutral

0097–6156/94/0565–0395$08.00/0

extractant within the inner coordination sphere of the complex.

Ion Associates

4. Coordinatively nonsolvated ion associates of the type Cat^+An^- where Cat^+ and An^- are a large and hydrophobic cation and anion, respectively, with predominantly electrostatic bonding between them.
5. Strong mineral acids.
6. Complex (metal-containing) strong acids, such as $HGaCl_4$, $HAuBr_4$, or $HNbF_6$; heteropoly acids are also included in this group.
7. Various compounds not included in groups 1-6 for one reason or another.

This list consists of two large groups: neutral compounds and ion associates.

Evidently, with rare exceptions, extracted substances are complex compounds. Complex formation in aqueous and nonaqueous solutions is indeed a decisive factor for the efficiency of solvent extraction. Correspondingly, one of the most fruitful developments in the theory of extraction has been the incorporation of the achievements of coordination chemistry. This is especially important for resolving selectivity problems -- the main problem in this field. Other problems are: a priori evaluation of the conditions required to extract a specific metal, purposeful synthesis of extractants, the use of modern techniques for evaluating the real state of metals in the phases, etc.

Complexation in extraction systems has been the subject of extensive research. The results obtained helped to establish the mechanisms of many extraction processes, to predict the type of compounds that must be formed in the system, and to evaluate the conditions for selective or group extraction. This review concentrates primarily on results obtained in our laboratories in Moscow.

Halide Complexes

One field of using coordination chemistry approaches is the solvent extraction of halide complexes of metals. The classical example is the extraction of Fe(III) from HCl solution with oxygen-containing solvents such as diethyl ether. It is of interest to consider historically how the chemistry of this extraction was studied (2).

The practical method was developed as early as a century ago (1892) (3). At the earliest stage of development there was no information on the nature of the compound extracted. Then, the proposal was made that $FeCl_3$ is extracted. Later, it was considered (4-6) that the extracted form is $HFeCl_4$. However, Nekrasov and Ovsyankina (7) made conclusion that several chloro complexes are extracted, mostly the higher ones ($FeCl_4^-$, $FeCl_5^{2-}$, $FeCl_6^{3-}$). Finally, it was found (8-10) that the only extracted form is $FeCl_4^-$ (as $HFeCl_4$). Another field of study was solvation and hydration of this species $HS^+FeCl_4^-$, then $\{H(H_2O)_xSy\}^+FeCl_4^-$, but this is not the topic of this paper. Now we know that extractable halides and related complexes are, in principle, coordinatively unsolvated species MX_m (M^{m+} is the metal ion; X^- is a halide or similar anion); coordinatively solvated compounds MX_mL_n (L is the extractant); or complex anions MX_{m+n}^{n-} incorporated as ion-association compounds. It has been shown (11,12) that it is possible to explain why and which complexes of these types are formed by using the approach of hard and soft acids and bases (HSAB).

For instance, the following fields were studied:
- comparison of the stability of complexes and their extraction;
- elucidation of the reasons and conditions for extraction of complexes of different types;
- study of coordination hydration.

Coordinatively saturated neutral complexes are weak acceptors (only to σ-antibonding molecular orbitals or the vacant 4d atomic orbital of germanium) such as the halides of Ge (GeX_4, where X is Cl, Br, or I), Hg, As, Sn, or Sb. Elements which form such complexes, whose extraction is very selective, are situated in one area of the periodic table -- at the end of the long periods (13).

For halide complexes of soft metals, it is possible to predict the change of extraction in a series: chlorides - bromides - iodides, because size, stability, and weakness of hydration of complexes are changed in a single direction. It is difficult to predict such behavior for halides of hard metal ions.

	MCl_3	MBr_3	MI_3
Soft ions			

Stability size
→

Weakness of hydration
→

Extraction
→

Hard ions

Stability
→

Size
←

Decrease of hydration
→

As to halide complexes of other types, a simple es-

timation of the possible coordination numbers of the metal atom and their involvement in extraction is often
useful. For instance, coordination aquation of inorganic
complexes has been studied (14,15). The results indicated the possibility of formation of intermediate complexes, hydrated in aqueous solutions, especially for
lanthanides and trivalent actinides.

Thus lanthanides are extracted from chloride solutions by long-chain amine salts as singly and doubly
charged tetra- and pentachlorocomplexes (MCl_4^-, MCl_5^{2-}),
respectively (16). Because the coordination numbers of
the lanthanides are high, it was possible to expect that
complexes will be hydrated in the inner coordination
sphere. If this is true, the degree of extraction can be
increased by adding to the system a second extractant,
capable of replacing water in the inner coordination
sphere. An example of such a compound is tributyl phosphate. This prediction was tested in the 1970s (17,18),
and it was found that a mixture of a cationic and a neutral extractant produced a considerable synergistic effect, and the distribution coefficients were increased.
This type of synergistic effect was employed to improve
the efficiency of lanthanide and actinide extraction
from chloride and nitrate solutions.

The same approach was used to increase extraction of
ion associates of complex metal anions with cationic
dyes as in the case of cadmium iodide complexes CdI_3
with such dyes, used in photometric analysis. $CdI_3(TBP)^-$
is extracted readily by benzene (19).

It was also possible to explain the known fact that
the indium chloride complex is purely extracted in comparison with the gallium and thallium chloride complexes.

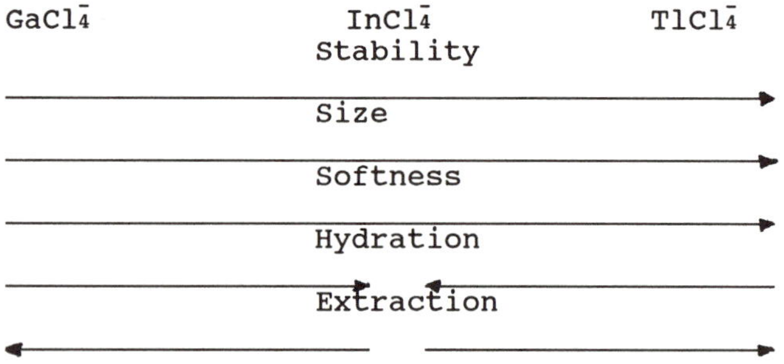

$GaCl_4^-$	$InCl_4^-$	$TlCl_4^-$
	Stability	

Size

Softness

Hydration

Extraction

This is not the case with the bromide complexes.

Chelates

A knowledge of coordination chemistry is especially important for the extraction of metal chelates. The composition, charge, and coordination saturation (or unsa-
turation) of a chelate can often be evaluated from four

parameters, viz., the charge on the central metal ion, the maximum coordination number of the metal ion, the number of ionogenic groups in the reagent molecule, and the coordination capacity of this molecule.

The extraction of chelates was known from the beginning of the century (20). At the beginning of the 1960s it was found that in addition to classical chelates, coordinatively unsaturated chelates are formed when the coordination number of the metal ion exceeds its double charge (for didentate ligands). Water molecules enter the inner coordination sphere in this case, and extraction can be increased if coordinatively active organic solvents are used. Instead of such solvents, a mixture of nonpolar organic solvents with special active additives can be employed, and in this case synergistic effects result (increasing the distribution coefficients when a mixture is used in comparison with distribution coefficients for individual extractants; more exactly -- nonadditive values of distribution coefficients).

The HSAB approach helped to find new effective extractants such as 2-octylaminopyridine for soft platinum metals (22), to select synergistic additives like the

soft triphenylphosphine (but not the hard triphenylphosphate) for silver chelates -- acetylacetonate, thenoyltrifluoroacetonate, or 8-hydroxyquinolinate. Parameters mentioned earlier were used to formulate conditions for the formation of charged chelates.

Cationic chelates are formed, for instance, when the coordination number is less than double the charge of a central atom, using didentate reagents. Tin(IV) bis(hydroxyquinolinate), gold(III) bis(dithizonate), or boron(III) bis(acetylacetonate) are examples of such metal chelates. Another reason for the formation of cationic chelates is steric hindrance as in the cases of alumin - um bis(2-methyl-8-hydroxiquinolinate), or cobalt(III) bis(dimethylglyoximate) (23,24).

A number of separation procedures for the extraction of metals by using cationic and anionic complexes have been developed. These complexes can be extracted by adding hydrophobic counter ions to the system.

Returning to extractants, we need to emphasize that the main problem is to look for selective extractants. The general principle that was used as the basis for designing selective extractants is maximum compatibility of the properties of the metal ion and the extractant molecule. This is the old approach which is valid for many different fields of application of coordination compounds. However, the approach is too general. In each

case many specific factors should be taken into consideration.

For example, for macrocyclic extractants (25), there
are several such factors, viz., the number and position
of the donor atoms, the cavity size, the presence of
flexible chains, the rigidity of the molecules as a whole, and the possibility for hydration of the central
atom. The effect of cavity size and rigidity of the molecule on extraction selectivity has been studied in
more detail. In the case of the rigid, small-cavity ni-
trogen-containing macrocycles (I-III) only one element
(silver) is extracted. If a flexible small-cavity ligand (IV and V) is used, three or four elements are extracted. With a flexible large-cavity ligand (VI) there
is no selectivity.

It is desirable to mention two interesting and unusual groups of extractants which have been suggested,
studied, and used in our laboratory.

The first group is organotin compounds, e.g., dino-
nyltin dinitrate. These are excellent, perhaps the best,
extractants for oxygen-containing anions such as PO_4^{3-} or
AsO_4^{3-} (26). Formally, the extraction process can be considered as an ion exchange, but complex formation actually takes place: the oxygen atom of, for instance, the
phosphate enters into the inner coordination sphere of
the tin atom.

The second group consists of spin-labeled chelate-
forming reagents (27). Extractants VII-VIII can be used
as examples.

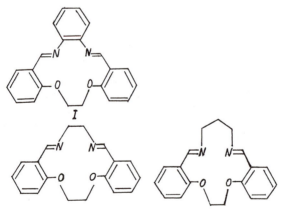

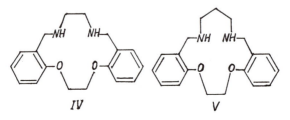

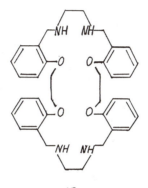

VI

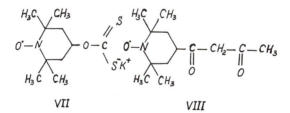

VII   VIII

Using these reagents made it possible to employ ESR for the determination of almost any metal ion, not only the paramagnetic ones.

As to methods for the investigation of the coordination chemistry of extracted compounds, Raman spectroscopy was shown to be an excellent technique (28,29). It is possible to study both phases of the extraction system, and it is not necessary to use high concentration of the extracted metal. Results for bismuth chloride complexes extracted by tributylphosphate are examples (28).

Conclusion

Even this short review of some solvent works made mostly in author's laboratories (in V.I.Vernadskii Institute of Geochemistry and Analytical Chemistry and M.V.Lomonosov Moscow State University in 1950-1980s) shows the importance of the coordination chemistry for extractive separation of metal ions. Of course, the consideration was not comprehensive; thus, the role of differences in formation constants on separation or entropy of chelate formation and other significant aspects of the problem were not considered.

Literature Cited

1. Zolotov,Yu.A. *Solvent Extraction of Inorganic Compounds;* Publishing House of Moscow University:Moscow, 1988 (in Russian).
2. Zolotov,Yu.A.; Iofa,B.Z.; Chuchalin,L.K. *Solvent Extraction of Metal Halide Complexes;* Nauka Publishing House: Moscow, 1973 (in Russian).
3. Rothe,J.W. *Chem. News* **1892,** *66,* 182.
4. Dodson,R.W.; Forney,G.J.; Swift,E.H. *J. Am. Chem. Soc.* **1936,** *58,* 2573.
5. Kato,S.; Ishii,R. *Sci. Papers Inst. Phys. Chem. Res. (Tokyo)* **1939,** *36,* 82.
6. Axelrod,J.; Swift,E.H. *J. Am. Chem. Soc.* **1940,** *62,* 33.
7. Nekrasov,B.V.; Ovsyankina,V.V. *Zh. Obshchei Khim.* **1941,** *11,* 573.
8. Mayers,R.J.; Metzler,D.E,; Swift,E.H. *J. Am. Chem. Soc.* **1950,** *72,* 3767.
9. Nachtrieb,N.H.; Conway,J.G. *J. Am. Chem. Soc.* **1948,** *70,* 3547.
10. Friedman,H.L. *J. Am. Chem. Soc.* **1952,** *74,* 5.
11. Petrukhin,O.M.; Spivakov,B.Ya.; Zolotov,Yu.A. *Dokl. AN SSSR* **1974,** *214(3),* 594.
12. Spivakov,B.Ya.; Petrukhin,O.M.; Zolotov,Yu.A. *Analit. Khim.* **1972,** *27(8),* 1584.
13. Zolotov,Yu.A. *Zh. Analit. Khim.* **1971,** *26(1),* 20.
14. Spivakov,B.Ya.; Stoyanov,E.S.; Zolotov,Yu.A. *Dokl. AN SSSR* **1975,** *220(2),* 392.

15. Spivakov,B.Ya.; Zolotov,Yu.A.; Myasoedov,B.F. et al. *Zh. Neorg. Khim.* **1972,** *17(12),* 3334.
16. Chmutova,M.K.; Myasoedov,B.F.; Kochetkova, N.E.; Spivakov,B.Ya.; Zolotov,Yu.A. *Radiokhimiya* **1974,** *16(5),* 702.
17. Myasoedov,B.F.; Spivakov,B.Ya.; Zolotov,Yu.A.; Kochetkova,N.E.; Shkina,V.M.; Chmutova,M.K. *Zh. Neorg. Khim* **1974,** *19(5),* 1379.
18. Chmutova,M.K.; Myasoedov,B.F.; Spivakov,B.Ya.; Kochetkova,N.E.; Zolotov,Yu.A. *J. Inorg. Nucl. Chem.* **1973,** *35,* 1317.
19. Kish,P.P.; Balog,I.S.; Spivakov,B.Ya.; Zolotov,Yu.A. *Zh. Analit. Khim.* **1976,** *31(6),* 1114.
20. Cazeneuve,P. *C. r. Acad. Sci.* **1900,** *131,* 346.
21. Baudisch,O; Furst,R. *Ber. Deut. Chem. Gesellsch.* **1917,** *50,* 324.
22. Borshch,N.A.; Petrukhin,O.M.; Zolotov,Yu.A.; Zhumadin,E.G.; Nefedov,V.I.; Sokolov, A.B.; Marov,I.N. *Koord. Khim.* **1981,** *7(8),* 1242.
23. Zolotov,Yu.A.; Petrukhin,O.M.; Kuz'min,N.M. In *Chemical Bases of Solvent Extraction Method of Element Separation;* Nauka Publishing House: Moscow, 1966; p 28.
24. Zolotov,Yu.A.; Vlasova,G.E. *Zh. Analit. Khim.* **1969,** *24(10),* 1542.
25. *Macrocyclic Compounds in Analytical Chemistry;* Zolotov,Yu.A.;Kuz'min,N.M.,Eds.; Nauka Publishing House: Moscow, 1993 (in Russian).
26. Spivakov,B.Ya.; Shkinev,V.M.; Zolotov,Yu.A. *Zh. Analit. Khim.* **1975,** *30(11),* 2182.
27. Nagy,V.Yu.; Petrukhin,O.M.; Zolotov,Yu.A. *Crit. Rev. in Anal. Chem.* **1987,** *17(3),* 265.
28. Stoyanov,E.S.; Spivakov,B.Ya.; Zolotov,Yu.A.; Gribov,L.A. *Koord. Khim.* **1975,** *1(1),* 59.
29. Stoyanov,E.S.; Spivakov,B.Ya.; Gribov.L.A.; Zolotov, Yu.A. *Koord. Khim.* **1975,** *1(2),* 228.

RECEIVED December 6, 1993

Chapter 33

Humic and Hydrous Oxide Ligands in Soil and Natural Water

Metal-Ion Complexation

Cooper H. Langford

Department of Chemistry, University of Calgary, 2500 University Drive NW, Calgary, Alberta T2N 1N4, Canada

Werner complexation and the coordination theory play an important role in understanding the chemical processes in natural systems in water and soil. However, the ligands are more complicated than those usually encountered in the laboratory because they are commonly polymeric, colloidal-scale, heterogeneous, and polyelectrolyte. The ideas required for understanding simple ligand equilibria and kinetics must be supplemented with a sensitivity to the effects of a polyfunctional distribution of sites that is subject to electrostatic effects arising from the polyelectrolyte character and conformational and aggregational equilibria, which may be coupled to the extent of metal ion coverage. This paper develops these points using soil humic substances and hydrous metal oxides as examples.

Natural Ligand Donor Sites

Although some of the complexing sites for metals that exist in natural waters and soil solutions are located on molecular ligands such as amino acids, simple oxo anions, and hydroxyacids, the main and most persistent ligand sites are on larger particles. In the few-nm domain we find organic ligands such as the water-soluble low-molecular-weight (*ca.* 10^3) humics known as fulvic acids and inorganic species with cation-binding capability such as $Al_{13}(OH)_{32}^{7+}$ and the somewhat larger $Fe_x(OH)_y$ polymers. Ascending the size scale we encounter the humic acids (larger cousins of the fulvic acids), polysaccharides, and small hydrous oxide colloids. Near the usual boundary for an experimental "definition" of dissolved matter (the 0.45-μm filter) lie typical $FeOOH$, MnO_2, and clay particles as well as cellular debris and metal sulfides. Organic compounds adsorbed on inorganic particles function as

0097–6156/94/0565–0404$08.00/0

members of the size class of the adsorbing substrate. Thus this coordination chemistry is focused on the domain between the molecular-level ligands and "suspended solids."

If the physical scale of the ligands introduces complexities which were not of concern to Werner, the nature of the donor sites on these intermediate scale ligands adds no complexities. Structures 1, 2, and 3 could stand as representatives of the vast majority of the complexing sites for metal ions, (M), with many O and a few N donor atoms. Sulfide sites and simple halide ions are the other donors of lesser importance. Consequently, we need to anticipate nothing in the bonding of the ligand to the metal that was not exhibited in the complexes known to Alfred Werner. This soil/water system chemistry is the chemistry of classical complexes, mainly between metal ions and hard base ligands.

Ligand Heterogeneity

A fulvic acid of number average molecular weight 1000 can yield as many as 30 different substituted mononuclear phenol-carboxylate structures in a mild oxidative degradation which accounts for 30% of the mass of the sample. Because only a small fraction of these structures would be needed to make a polymer of MW = 1000, it is clear that the fulvic acid is a complex mixture, yet the aliphatic components have not so far been considered. It may be asked if efforts should not concentrate on the resolution of the mixture, regardless of how difficult that task may be. However, the task in understanding natural system chemistry would then require modeling the interactions that occur in the mixture, which is, in my opinion, more difficult than studying them directly.

Buffle[1] has introduced an extremely helpful concept, that of a homologous group of ligands. He recognizes that there are polymeric, polyelectrolytic ligands which vary in details of molecular structure but have common features in the basic metal-ligand bonding. Such homologous groups will not be fully characterized at the molecular level, but it is useful to divide the study of coordination chemistry into the study of the various homologous groupings and to characterize the behavior of the contributing homologous groups. Examples of groups of homologous ligands are the fulvic acids, humic acids, natural polysaccharides, nonhomodisperse Fe(III) or Al(III) hydrous oxides, and clay particles.

In all of the homologous groups the three factors which complicate the treatment of metal ion equilibria and kinetics are: (1) the polyfunctional character and distribution of molecular type of donor site; (2) change in conformation on metal-ion binding, coiling, aggregation, and gel formation; and (3) polyelectrolyte character changes in electrostatic effects on metal-ion binding. Homologous ligand groups can be subdivided into three types, which differ mainly in the roles of these three factors: type I, dissolved polyfunctional complexant; type II, monofunctional polyelectrolyte (permeable

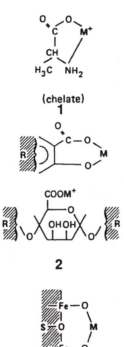

(chelate)
1

2

3

R & S are generalized groups

gel); and type III, monofunctional polyelectrolyte (surface reactant). Table I summarizes the differences among these types and gives examples of each.

Consider the specific cases of a fulvic acid, a simple polyfunctional complexant, and the surface of a hydrous ferric oxide particle, a typical monofunctional polyelectrolyte. In the first of these cases, the variety of ligand structures in the mixture will require recognition of a distribution of binding constants to describe the system. In addition, both conformational change in the polymer and variable electrostatic effects will make these component binding constants themselves appear to be functions of the degree of site coverage by the metal ion. The most useful approximation has been to assume that the distribution of component binding constants is a continuous function of site coverage.

In the hydrous oxide surface case, the treatment can be somewhat simpler. All of the complexing sites are variants of the surface hydroxyl type. Interest is usually directed to the binding of a trace metal, present in low concentration compared to the number of surface sites. This may be called the geochemical limit because it describes common natural concentration ratios. Only the most strongly complexing sites are likely to be important, and it is often a very good approximation to assume a single surface ligand binding constant for a metal and to correct only for polyelectrolyte effects.

Kinetically, the situation is not simple for either of the ligand types. Especially, the number of kinetically distinguishable sites seems to exceed the number distinguishable at equilibrium in the hydrous oxide surface case. This may well be a consequence of sites of similar character thermodynamically being kinetically differentiated by the length of the diffusion path into the particle interior, which must be traversed by a metal ion in the course of binding or dissociation.

Equilibrium and Kinetics of Binding to Hydrous Oxides

A typical study observes the uptake of trace metal ions on a dispersion of a hydrous oxide, yielding the percentage of metal adsorbed (% ads) as a function of pH. The pH range over which the variable % ads changes substantially is usually narrow, leading to the term "adsorption edge." Figure 1 shows data from Schindler(2), one of the pioneers of the interpretation, for uptake of Cu(II), Pb(II), and Cd(II) on hydrated rutile (TiO_2). The lines show a fit with the theory which assumes a single intrinsic binding constant, $^*K_{1int}$, for reaction (equation 1) and a single intrinsic constant, $^*\beta_{2int}$, for reaction (equation 2). Corrections have been applied for polyelectrolyte character based on the Gouy-Chapman-Stern theory.

$$Surf\text{-}OH + M^{z+} = Surf\text{-}OM^{(z-1)+} + H^+ \qquad (1)$$
$$2Surf\text{-}OH + M^{z+} = (Surf\text{-}O)_2M^{(z-2)+} + 2H^+ \qquad (2)$$

Figure 2 shows the persuasive correlation between the intrinsic constants used in the treatment of surface complexation and the stability constants of simple hydroxo complexes measured in homogeneous solution.

Table I Characteristics of the Main Homologous Complexing Groups

	I	II	III
	Dissolved polyfunctional complexant	Monofunctional polyelectrolyte (permeable gel)	Monofunctional polyelectrolyte (surface reaction)
Schematic structures			
Examples	fulvic compounds PROM AROM	little hydrated Fe, Al, Mn oxides clays	
		peptides proteins humic compounds polysaccharides cell walls	strongly hydrated metal hydroxides
		← metal oxides and clays covered by NOM →	
<u>a</u> polyfunctional character	strong	intermediate	intermediate-weak
<u>b</u> change in conformation	aggregation	gel formation spreading/coiling	coagulation
<u>c</u> polyelectrolyte character	important	important	important

The horizontal positioning of examples depends on their behavior, e.g., the intermediate position of proteins indicates that in this case properties I and II are both important. O.N.S. = coordinating atoms of complexing sites. (Ref. 1).

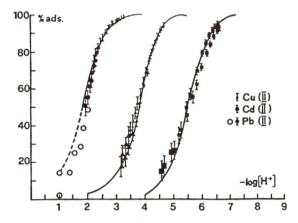

Figure 1. Adsorption of some divalent metal ions on TiO_2 (rutile). For each metal ion there is an interval of 1–2 pH units where the extent of adsorption rises from zero to almost 100%. The solid lines are calculated with the constants calculated for single-site model. (Reproduced with permission from *Chimia.* Copyright 1976 Chimia Abodienst.)

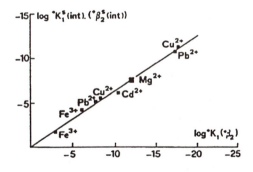

Figure 2. Correlation of stability constants $\log {}^*K^5_{1(int)}$ $({}^*\beta^5_{2(int)})$ of surface complexes at amorphous silica with stability constants $\log {}^*K_1$ $({}^*\beta_2)$ of hydroxo complexes. The solid line represents the equation $\log {}^*K^5_{1(int)}({}^*\beta^5_{2(int)}) = -0.09 + 0.62 \log {}^*K_1 ({}^*\beta_2)$. (Reproduced with permission from *Oester. Chem. Z.* Copyright 1985 Verlag Lorenz.)

A somewhat different experiment is illustrated in Figure 3(*3*). Here the adsorption isotherm for the uptake of Cu(II) ion on and into colloids of hydrous ferric oxide is reproduced. The Cu(II) ions were present in the acid solution prior to the neutralization steps leading to formation of the Fe(III) colloids. The isotherm is simple and can easily be fitted within experimental error to a single binding constant approximately consistent with those obtained in adsorption edge experiments despite the somewhat different uptake mechanism. However, a kinetic experiment in which the Cu(II) was stripped from the hydrous iron oxide with a Cu(II)-selective ligand required a minimum of four kinetically distinct Cu(II) types to account for the data. Figure 4(*4*) shows the trend in the concentrations of the four kinetic components as a function of the Fe:Cu ratio in the initial solution. The largest component (top curve in the figure) is the labile component comprised of free Cu(II) and Cu(II) bound to surface sites and accessible for reaction in less than 3 s. (The kinetic technique will be explained in the section on ligand exchange kinetics.)

Equilibrium of Binding to Humic Substances

The study of binding to humic substances has been greatly aided by the availability of ion-exchanged samples rendered into the protonated form with all other cations being rendered negligible. It is then possible to carry out stoichiometrically well-defined experiments on the titration of these ligands with metal ions. A classic example is illustrated by the three titrations shown in Figure 5(*5*). The titration of the Armdale soil fulvic acid with Cu(II) was monitored with three different experimental probes. The filled circles represent the monitoring of the Cu(II) bound by means of the measurement of "free" Cu(II) with a Cu^{2+}-ion-selective electrode (ISE). The open circles represent the monitoring of the quenching of the fulvic acid fluorescence at 465 nm. The triangles represent the monitoring of the increase of absorbance (and light scattering) at 465 nm as a result of Cu(II) complexation.

The quenching of fluorescence, Q (see left axis), is smooth and suggestive of a relatively simple binding function. The fraction of stoichiometrically characterized sites covered as indicated by ISE, χ_c (left axis), increases monotonically but clearly indicates more than one binding constant. The change of optical absorbance, A_{465} (right axis), has two inflection points, strongly suggesting some physical changes in the ligand during Cu(II) uptake. The last point will be documented by light scattering data (Figure 7). The fluorescence data suggest that the fluorescing sites are a subset of the stoichiometric total, probably associated with stronger binding constants. A separate experiment using gel filtration chromatography has shown that the fluorescence is associated with the lower-molecular-weight fractions separated from the mixture in the gel filtration process. The ISE experiments were treated quantitatively on the assumption of a continuous distribution of binding constant. The results in the form of a plot of the free energy of

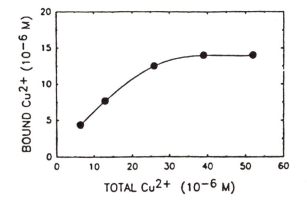

Figure 3. Example of Fe-hox binding of Cu(II) determined at pH 6.8. Fitting of these data to conventional isotherms shows good adherence to Langmuir theory. (Reproduced with permission from reference 3. Copyright 1989 National Water Research Institute.)

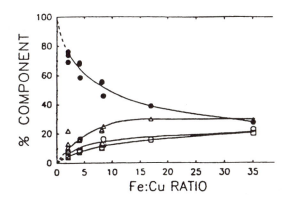

Figure 4. Results of applying kinetic speciation analysis to the equilibrium binding of Cu(II) by Fe-hox at pH 6.0. •, X component 1; 0, component 2; □, component 3; Δ, not recovered, component 4. A minimum of four kinetically distinguishable components are necessary to describe the system. (Reproduced from reference 4. Copyright 1993 American Chemical Society.)

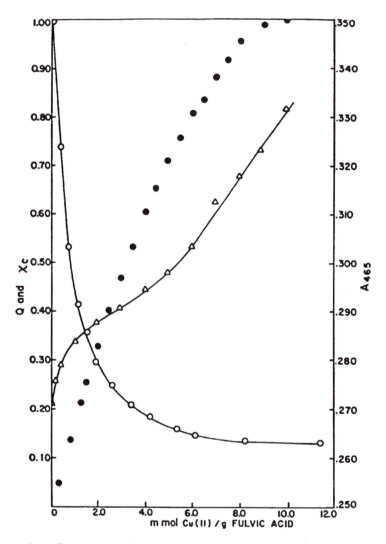

Figure 5. Comparison of three experimental probes of metal binding to the fulvic acid. x_c, the fraction of sites occupied by Cu^{2+} as calculated from ISE titration; Q, the relative quenching of FA fluorescence; A_{465}, the absorbance by fulvic acid at 465 nm as a function of Cu^{2+} added. See text for guide to curves. (Reproduced from reference 5. Copyright 1981 American Chemical Society.)

binding *vs.* fraction of sites covered is shown in Figure 6. In this figure the left axis divides the free energy of binding into two terms, an intrinsic chemical term, $\Delta G°$, and a term for the electrostatic effect expressed as a ratio of activity coefficients, Γ. The partition is conceptual only; it was not actually performed.

The role of conformation and aggregation is spectacularly indicated by the light scattering data shown in Figure 7. In this figure the rise of right-angle Rayleigh scattering as a function of the coverage of sites by Cu(II), estimated from ISE data, is shown. At all pH values the initial uptake of Cu(II) leads to little change in light scattering. At higher coverage uptake of Cu(II) leads to a sharp increase in aggregation with a slope steep enough to suggest a cooperative process. It was proposed that the initial portion reflects the uptake of Cu at didentate sites. As these become saturated, Cu is bound to monodentate sites. This allows the formation of bridged complexes of the type L-Cu-L. However, if L is a polymer which can interact with another L by hydrogen bonding and/or donor-acceptor interactions, the bridged complex may have a closed form as shown by the sketch in the figure. The interacting ligands can form a *pseudo-chelating* ligand, and this may account for the apparent cooperativity. The interaction is promoted by bridging, and it stabilizes the bridged species.

Kinetics of Ligand Exchange from Humic Substances

As mentioned in the context of hydrous oxide surfaces, the kinetic behavior of naturally occurring complexes is important. In addition to the fact that kinetics may show a different partition of complexing sites from equilibrium experiments, there is the issue that biological uptake of metal ions is a dynamic process and that kinetics may control the effects of metal ions on biological systems. Because there are few sensitive methods to detect complexation reactions in progress, we adopted ligand exchange between the metal complexed with a humic substance and a complex of that metal with a spectroscopically sensitive ligand (R, below)(6). This is not only experimentally beneficial, but it also relates to the biological situation where a metal ion is often taken up *via* a surface complex with a cell membrane.

Consider the complexes of Ni(II) with the acid-soluble soil humic fraction, fulvic acid (FA) as they react with a reagent R to form a spectrophotometrically detectable product. There is a set of competing reactions:

$$
\begin{aligned}
&\mathrm{Ni(aq)^{2+} + R \rightarrow NiR} \\
&\mathrm{FA_1\text{-}Ni + R \rightarrow NiR + FA_1} \\
&\mathrm{FA_2\text{-}Ni + R \rightarrow NiR + FA_2} \\
&\cdots \\
&\mathrm{FA_i\text{-}Ni + R \rightarrow NiR + FA_i.}
\end{aligned}
\tag{3}
$$

The subscripted FA_i represent the kinetically distinguishable groups of binding

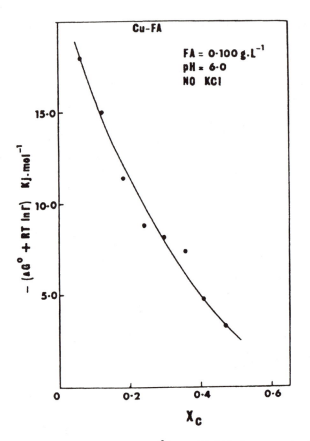

Figure 6. Results of titrations with Cu^{2+} at pH 6.00 shown as the free energy of binding derived from the K values. $\Delta G°$ is the intrinsic term. Γ represents the electrostatic term. (Reproduced from reference 5. Copyright 1981 American Chemical Society.)

sites on the FA mixture. The first reaction, with free aqua Ni(II) is the fastest. Provided that a large excess of R is used, all reactions will be experimentally pseudo-first order. More fundamentally, the reactions may often (but not always) be described by the following mechanism:

$$
\begin{array}{l}
\text{FA}_i\text{-Ni} \xrightarrow{\text{slow}} \text{FA}_i + \text{Ni}^{2+}(\text{aq}) \\
\text{Ni}^{2+}(\text{aq}) + \text{R} \xrightarrow{\text{fast}} \text{NiR}
\end{array} \tag{4}
$$

In this favorable case the measured first-order rate constant for each component will be the dissociation rate constant for the metal complex of the corresponding complexing site or site collection.

With all reactions reduced to pseudo-first order, the experimental rate law with excess R takes the form:

$$
d[\text{NiR}]/dt = \sum_i A_{o,i}[1 - \exp(-k_i t)] + X. \tag{5}
$$

In equation 5 $A_{o,i}$ represents the initial (time zero) concentration of the ith complex species expressed in units consistent with the measure of the product NiR, k_i is the rate constant for the ith species, and X is a time-independent term which contains the spectroscopic blank and the contribution of any species which reacts too rapidly for the time scale of the experiment. The blank can be measured independently. The task is to fit the experimental signal-time function to equation 5, identifying a set of rate constants, k_i, and initial species concentrations, $A_{o,i}$. Because the equation is nonlinear, this is not easy.

The process has three steps(7). First, it is necessary to identify objectively a suitable minimum number of kinetic components, i. Second, the best nonlinear fit to the data must be obtained. Finally, the analysis must be tested for its chemical validity. This must be done to assure that the calculations have accomplished more than simply a set of numerical parameters which form no more than a way to archive the data. (We cannot overemphasize the importance of this last step. It is extremely easy to generate numerical analysis artifacts of little chemical interest.) The first step is now accomplished using a numerical approximation to the Laplace transform of the signal-time function(8). The numerical differentiation used in the transform always requires data smoothing. The second step can be accomplished using nonlinear regression on the unsmoothed data. The chemical validation is usually accomplished by studying the kinetics over a range of concentration and pH conditions of samples. The indication of validity is the stability of the rate constants and the evolution of the concentrations as expected from mass action considerations.

Figure 8 shows the evolution of the four components identified in a study of the FA- Ni(II) system(9) as a function of FA:Ni ratio in the samples at pH = 4. C_1 is free aqua Ni(II), C_2 is the most labile complexes, and C_3 and C_4

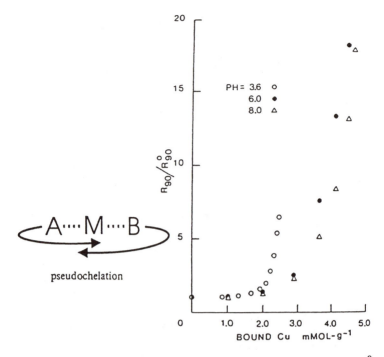

Figure 7. The behavior of relative right-angle light scattering R_{90}/R^0_{90} at various pH values as bound Cu^{2+} increases. Bound Cu^{2+} is calculated from ISE titration. (Reproduced from reference 5. Copyright 1981 American Chemical Society.)

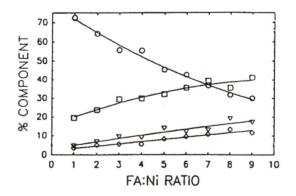

Figure 8. Changes in relative contribution of initial components (C_i) as a function of ligand-to-metal ratio following equilibration at pH 4.0. $\bigcirc$, C_i; $\square$, C_2; $\bigtriangledown$, C_3; and $\Diamond$, C_4. (Reproduced from reference 9. Copyright 1987 American Chemical Society.)

represent less labile sites. As expected if the least labile complexes are the strongest binding, their concentrations rise with increase of FA, the labile (weak) complexes increase but show signs of leveling off. Free Ni(II) declines. Rate constants for species 3 and 4 are approximately 2×10^{-2} s^{-1} and 3×10^{-3} s^{-1}, respectively, and are the only species whose lability may be too low for the dynamic bioavailability to be equal to that of free aqua Ni(II).

Conclusion

The novel features introduced into coordination chemistry by study of the colloidal ligands which are critical in soils and natural waters are three. First, we must deal with a distribution of similar but distinct binding sites. Second, polyelectrolyte effects are prominent. Third, the ligands undergo conformational and aggregational equilibria which are influenced by metal ion uptake.

Literature Cited

1. Buffle, J. *Complexation Reactions in Aquatic Systems: An Analytical Approach*; Ellis Horwood Limited: Chichester, England, 1988.
2. Schindler, P. E.; Stumm, W. In *Aquatic Surface Chemistry: Chemical Processes at the Particle-Water Interface*; Stumm, W., Ed.; Wiley: New York, NY, 1987; Chapter 4.
3. Gutzman, D. W.; Langford, C. H. *Can. J. Water Pol. Res. and Control* **1989**, *23*, 379-387.
4. Gutzman, D. W.; Langford, C. H. *Env. Sci. and Tech.* **1993**, *27*, 1388-1393.
5. Underdown, A.; Gamble, D. S.; Langford, C. H. *Anal. Chem.* **1981**, *53*, 2139-2140.
6. Langford, C. H.; Kay, R.; Quance, G. W.; Khan, T. R. *Anal. Letts.* **1977**, *10*, 1249.
7. Langford, C. H.; Gutzman, D. W. *Anal. Chim. Acta* **1992**, *256*, 183-201.
8. Olson, D. L.; Shuman, M. S. *Anal. Chem.* **1983**, *55*, 1103.
9. Lavigne, J. A.; Langford, C. H.; Mak, M. K. S. *Anal. Chem.* **1987**, *59*, 2616.

RECEIVED August 16, 1994

Chapter 34

Coordination Model of Metal-Ion Interactions with Water Hyacinth Plants

Dean F. Martin

Institute for Environmental Studies, Department of Chemistry,
University of South Florida, Tampa, FL 33620–5520

Waterhyacinth plants [*Eichhornia crassipes* (Mart.) Solms], which infest waterways in many parts of the world, were used as a model system for studying uptake of metal ions by floating aquatic plants. Observed rates of uptake of manganese(II)-54 and iron(III)-59 by water hyacinth compare favorably with the values expected based upon a coordination model: control by chelation with dihydrogen ethylenediamine- tetraacetate ion and control by metal(II) carbonate solubility. Agreement depends upon the assumption of the reduction of iron(III) species by the plant.

There is no evidence that Alfred Werner had a great concern for the environment, nor even for what might be called applied problems. This association with pure fundamental chemistry stands in contrast with Justus von Liebig, who served as a consultant for the German fertilizer industry and learned what elements were essential by a clever, practical method: he analyzed healthy plants for their constituents, especially nitrogen (N), phosphorus (P), and potassium (K) (1). These analyses led to the NPK numbers of fertilizer sacks, and represent a tribute to Liebig's practical nature. Liebig also developed baking powder, with the understanding that if soldiers had baking powder, clean water, and flour, they could bake their own bread and have no need for yeast, which generates carbon dioxide much more slowly (2).

On the other hand, the fact that Werner focused on significant inorganic chemical problems is hardly a criticism that would have validity in his time or ours. In fact, it was his principles that permitted studies of the effects of metal ions on plant growth. These

0097–6156/94/0565–0418$08.00/0

principles were applied to an understanding of problems of, and applications involving, waterhyacinth plants [*Eichhornia crassipes* (Mart.) Solms]. Werner's principles were thus combined with Liebig's approach.

Waterhyacinth Plants

Waterhyacinth plants were probably first introduced into the United States during a horticultural exposition in New Orleans in the late 1880s, and souvenir plants were given away (*3,4*). The plant floats on water, is called an emersed aquatic macrophyte, and it possesses a beautiful hyacinth flower that appears in the late spring or early summer. Owing to its attractive flower, the plant seemed a logical choice as an ornamental floating plant. It was thus deliberately and accidentally propagated in various lakes and waterways. It was introduced to the St. Johns River of Florida near Palatka, being grown deliberately, but then escaping as a consequence of the flooding of an ornamental pond.

A waterhyacinth plant can proliferate rapidly by virtue of its structure (Fig. 1), which has a stolon that leads to the formation of a daughter plant. Under optimum conditions of sunshine, temperature, and sufficient nutrients, waterhyacinth plants can proliferate at the rate of 1.8 daughter plants per parent plant per week (*5*). Plants are characterized in terms of their doubling time -- the time required for the population to double in number. The doubling time for waterhyacinth plants is something less than a week, in comparison with 20 minutes or less for many bacteria and 3.5 days for *Ptychodiscus brevis*, the Florida red tide organism (*6*).

Despite what may seem like a comparatively slow growing time, by the mid-1890s, waterhyacinth plants were a major problem on the St. Johns River because of the thick, nearly impenetrable mats that shaded the native plants, provided a nesting spot for mosquito larvae, and affected the navigability of the St. Johns River. This fact must have been noticed by many. It seems likely, however, that the Army Corps of Engineers, an agency with responsibility for maintaining navigability of waters, was stimulated into action by the interest of Henry Ford and Harvey Firestone. It is thought that these two leading industrialists wintered in Florida and used paddle boats on the St. Johns River as floating bases for fishing and hunting. The formation of floating mats must have seriously interfered with the progress of their paddle boats (*7*).

Management of Waterhyacinth Plants

Early efforts at managing waterhyacinth plants were limited to sodium arsenite, a herbicide not widely

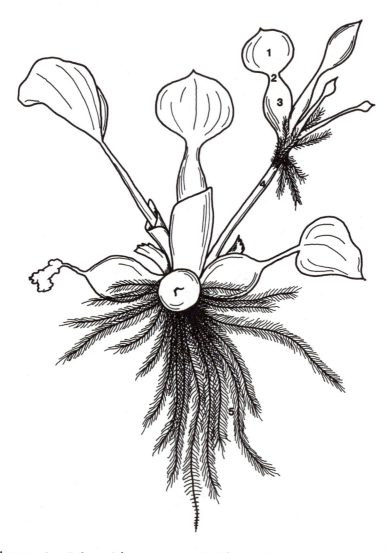

Figure 1. Schematic representation of a waterhyacinth,
showing daughter plant with 1, leaf; 2, isthmus, 3,
petiole; 4, stolon, and 5, roots of parent plant.

favored because Florida cattle were known to eat waterhyacinth plants adjacent to the banks. In Louisiana, sodium arsenite was manufactured on spray boats from raw stocks and was used for 35 years until 1937 when it was abandoned in the interests of safety (*7, 8*). The other alternative control chemical was copper sulfate, an all-purpose algicide that affected a plant's redox system (*9*). After World War II, 2,4-D (2,4-dichlorophenoxyacetic acid, supplied as a suitable salt) was the herbicide of choice for waterhyacinth plants (*9, 10*).

Mechanical removal by cranes was also a successful, albeit expensive, approach, most suited for control of waterhyacinths in flood control canals.

Subsequently, biocontrol proved to be successful in controlling waterhyacinth plants (*12*), particularly when undertaken in conjunction with chemical control, though not all types of waterhyacinth plants responded similarly(*cf. 13*). Appreciating the differences in the uptake of metal ions by the plant types was a key to understanding their characteristics, especially size and insect avoidance.

Waterhyacinth Plant Types

Waterhyacinth plants have a characteristic called "plasticity": they are able to attain notably different sizes in response to environmental characteristics, presumably differences in concentrations of nutrients or species of nutrients. This led to a designation of three groups or plant types,"small," "medium," and "super" (Table I). The largest plants are found in water with

Table I. Comparison of Waterhyacinth Plant Types[a]

	Small	Medium	Super
Height, cm	30-31	60-75	90-120
Dissolved oxygen at site,ppm	8.0	4.0	0.5
mg Fe/kg dry plant[a]			
leaves		557 ± 157[b]	837 ± 351[b]
stems		811 ± 154	904 ± 107
roots		711 ± 186	$12,039 \pm 1,567$
Dry wt/fresh wt	0.049 $\pm$ 0.008	0.059 $\pm$ 0.004	0.068 $\pm$ 0.004

[a] Refs *11* and *12*
[b] $\pm$ Standard error for 5 samples

low dissolved oxygen concentrations, where iron(II), a more soluble form, might be found. In contrast, medium and small waterhyacinth plants were found in water where ferric form would be the dominant species. Superhyacinth plants contained significantly more iron, an average of 2890 mg of iron per kg of dry plant (2890 ppm), compared with an average of 104 ppm for medium plants. Most of the differences were associated with differences in the iron content of the roots for the plant types (13).

The environmental characteristics that seemed most logical to explain the ecotypes and their differences were based upon trace metals and their coordination chemistry. There are at least two consequences of this hypothesis: insect avoidance and trace-metal uptake (for use in the "polishing" stages of sewage treatment).

Insect Avoidance

A cooperative program involving the U.S. Department of Agriculture and The Florida Department of Natural Resources (Bureau of Aquatic Plant Research and Control) was directed at finding natural enemies of waterhyacinth plants. Accordingly, a search was launched in Argentina, where waterhyacinth plants were thought to have originated, and after searching, a weevil (*Neochetinia eichhornia* Warner) was isolated, studied, brought to the United States, and subjected to additional study during quarantine (12). The insect stresses the plant by boring into the leaves and bulbous tissue and lays eggs; the affected waterhyacinth plants then are additionally stressed by fungi that have a pathway to the interior.

Unfortunately, by March, 1976 it was evident that the weevil avoided the obvious waterhyacinth plants, i.e., the 10 percent of the population that were superhyacinth plants. The reason was not plant adaptability (13). The basis seemed to be differences in the trace metal content of the ecotypes: most notably the superhyacinth plants had a greater concentration of iron in leaves and stem and roots. Though this might be expected on an absolute basis, the plants being notably larger (Table I), it was also true on a relative basis, i.e., when the concentration as expressed as mg metal per kg dry plant.

In addition, the superhyacinth plant leaves were thicker in texture, more leathery, and had 25 percent less chlorophyll per gram dry weight than the small plants (13). One can imagine that the superhyacinth plants would be unpalatable or would represent a more challenging situation to a borer insect. It is possible that the larger plants may generate defense chemicals, though this should be a genetically linked characteristic, and there was no evidence that the three waterhyacinth plant types are genetically distinct (14). On the basis of an assumption that insect avoidance was based upon differences in metal-ion concentrations, two studies were done.   First, we measured the metal concentrations in three plant types from a natural setting (Table I). Second, we measured the rate of uptake by small waterhyacinth plants in controlled conditions.

Trace-metal Content of Three Waterhyacinth Plant Types

Plants were collected from thriving infestations of each plant type (16), were dried, sectioned (into roots, leaves, stems), decomposed in nitric acid-perchloric acid mixtures using a modified microwave oven (15) to save time, and analyzed using atomic absorption spectrometry or an AutoAnalyzer II (for iron samples) (11,13,16). The elemental fraction of each plant segment was calculated, and the values were found to be a linear function of $-\log K_{sp}$ for the metal carbonate (11, 16). It should be noted that the values for leaves and stems were grouped about the descending linear portion and the roots were associated with the ascending branch. The data for small and   medium   waterhyacinth   plants   had   good   linear correlation coefficients for six metals and leaves and stems ($r = 0.97$, $P < 0.01$) and for six metals and for roots ($r = -0.98$, $P < 0.01$) (16). A similarly valid correlation was not observed for superhyacinth plants because this plant type had a significantly greater fraction of iron and cobalt in the leaves. The values for iron represented a significant deviation.   The metal ions examined were magnesium, calcium, zinc, manganese, copper, and iron. The metal carbonate values used were for the divalent species, which seems significant for iron in the light of previous discussion of the speciation of iron as a function of dissolved oxygen. No value for the K_{sp} for $CoCO_3$ was found, and subsequently, the data for cadmium did not fit (17). The model predicted 1% of the cadmium would be distributed in the leaves and stems; instead, 25% was the observed percentage,   and   there   was   good   evidence   that waterhyacinth plants could absorb and translocate cadmium ion (17).

These observations indicated the need for measuring the rates of uptake of selected metals by waterhyacinth plants and trying to obtain a better model for metal-ion uptake.

Metal-ion Uptake by Waterhyacinth Plants

Uptake Studies. Small waterhyacinth plants floating in a defined medium (Hoagland's solution) were treated with known amounts of $Mn(EDTA)^{2-}$ and manganese(II)-54, or $Fe(EDTA)^{-}$ and iron(III)-59, or phosphorus-32 as orthophosphoric acid and orthophosphate (18).   Rates of radioactivity were counted appropriately using portions of roots or portions of stems at various times. By

previous experiments, iron and manganese were known to
stimulate the growth of this plant, and waterhyacinth
plants were known to remove orthophosphate. Uptake of
all three elements followed first-order rate constants,
but the rates of translocations from the roots to the
leaves differed. Phosphorus as phosphorus-32 was taken
up by the roots and appeared in measurable levels in the
stems within 48 hours. Radio-iron was taken up rapidly
by the roots, but experimentally significant quantities
of the element appeared in the leaves only after about 21
days. Radio-manganese was taken up by the roots as
rapidly as iron, but the rate of appearance in the leaves
was ten times faster. The accumulation of phosphorus,
calculated on the basis of a 12-month growing season, was
consistent with estimates provided by others (18).

**Coordination Chemistry Model of Manganese and Iron
Uptake.** On the basis of available information concerning
metal uptake by waterhyacinth plants (17,18) certain
assumptions seem reasonable. First, the rate-determining
step is a function of the concentration of free metal
ion. This concentration is governed by an equilibrium
involving a more stable species, which is a sparingly
soluble complex salt (recalling the linear plots of
fraction vs log Ksp for metal carbonate species) or a
stable complex ion, e.g., the EDTA metal complex.
     Assuming that carbonate species determine the
concentration of trace metal ions, appropriate Ksp
expressions may be used and rearranged in terms of the
carbonate concentration:

$$(CO_3^{2-}) = K_{sp}/(Mn^{2+}) \tag{1}$$

$$(CO_3^{2-}) = K_{sp}'/(Fe^{2+}) \tag{2}$$

For a given pH in Hoagland's solution, the carbonate
ion concentration is constant, and right-hand portions of
the two equations are equal to each other. Thus,
rearranging yields

$$Log(Mn^{2+})/(Fe^{2+}) = \log K_{sp}/\log K_{sp}' \tag{3}$$

Given known values for Ksp for manganese(II) and iron(II)
carbonates [$10^{-9.41}$ and $10^{-10.46}$, respectively (20)], the
appropriate rate ratios from the last equation can be
calculated to be $10^{1.05}$.
     The agreement between calculated and observed rate
ratios ($10^{1.12}$) was quite good (19). Realistically, the
values of solubility product constants vary, and thus the
agreement depends upon which values are taken. The range
of predicted values thus was $10^{1.05}$ - $10^{1.20}$, and the
observed value is still within the predicted range.
     It was also recognized (19) that other ions might
control the free metal ion concentration. Sulfate might

be valid for some species, but manganese(II) sulfate is comparatively soluble. Phosphate ion is environmentally logical, especially in Florida, which is a source of phosphate deposits (*19*), but the predicted rate ratio would be $10^{0.5}$.

Finally, control by a chelating agent, EDTA, seems logical. A derivation indicated that the predicted rate ratios would be 10^{11}. The predicted rate ratio was improved using formation constants for protonated species for iron, but not manganese, and the chelation approach was not pursued further (*19*).

Conclusion

This has been a summary of the utilization of a field of chemistry founded by Alfred Werner and adapted to a problem that he could not possibly have anticipated. Excessive aquatic plant growth was not a problem in Switzerland in Werner's time, though now it is for some species. Waterhyacinth plants serve as a useful model for metal ion uptake because their characteristics lead to compartmentalization and because removal of metal ions from sewage by means of waterhyacinths has been recognized as a useful process (*21, 22*). It is an interesting irony that a few years ago, one of the insects specifically imported for waterhyacinth plant control was observed to focus on a sewage treatment plant where small waterhyacinth plants were growing as part of the water treatment process. A volatile substance , more prevalent in young tissue, serves as an attractant (*23*).

Acknowledgments

It is a pleasure to acknowledge financial support of the Florida Department of Natural Resources (now the Department of Environmental Protection) Bureau of Aquatic Plant Research and Control, the encouragement of Dr. A. P. Burkhalter, the collaboration of those listed in references cited, and the long-term encouragement and advice of Barbara B. Martin.

Literature Cited

1.   Liebig, J. F. von. *Organic Chemistry in Its Applications to Agriculture and Physiology* (edited from the manuscript of the author by L. Playfair); Bradbury and Evans: London, 1840.
2.   Jones, P.R. "Justus von Liebig, Eben Horsford, and the Baking   Powder Industry," Paper no. 18, Division of the History of   Chemistry, 205th ACS National Meeting, Denver, CO, March 28- April 2, 1993.
3.   Pierterse, A. H. *Abstrs. Trop. Agric.* **1978**, *4*, 9-42.
4.   Penfound, W. T.; Earle, T. T. *Ecol. Monogr.* **1948**, *18*, 448-472.

5. Ornes, W. H.; Sutton, D. L. *Hyacinth Contr. J.* **1975**, *13*, 56-58.
6. Doig, III, M. T.; Martin, D. F. *Mar. Biol.* **1974**, *24*, 223-228.
7. Tabita, A.; Woods, J. W. *Hyacinth Contr. J.* **1962**, *1*, 19-23.
8. Wunderlich, W.E. *Hyacinth Contr. J.* **1962**, *1*, 15-16.
9. Martin, D. F.; Martin, B. B. *J. Chem. Educ.* **1985**, 62, 1006.
10. *Herbicide Handbook*, 6th ed.; Weed Science Society of America: Champaign, IL, 1989.
11. Cooley, T. N.; Martin, D. F. *J. Environ Sci. Health* **1978**, A13, 469-479.
12. Perkins, B.D.; Lovarco, M. M.; Durden, W. C. *Fla. Ent.* **1976**, *59*, 352.
13. Cooley, T. N.; Martin, D. F.; Durden, Jr., W.C.; Perkins, B. D. *Wat. Res.* **1979**, *13*, 343-348.
14. Wain, R P.; Martin, D.F. *J. Environ. Sci. Health* **1980**, A15, 625-633.
15. Cooley, T. N., Martin, D. F.; Quincel, R. H. *J. Environ. Sci. Health* **1977**, A12, 15-19.
16. Cooley; T. N.; Martin, D. F. *J. Inorg. Nucl. Chem.* **1977**, *39*, 1893-1896.
17. Cooley, T. N.; Martin, D. F. *Chemosphere* **1979**, *2*, 75-79.
18. Cooley, T. N.; Gonzalez, M. H.; Martin, D. F. *Econ. Bot.* **1978**, *32*, 371-378.
19. Cooley, T. N.; Martin, D. F. *J. Inorg. Nucl. Chem.* **1980**, *42,*151-153.
20. Sillen, L. G.; Martell, A. E. *Stability Constants of Metal-Ion Complexes*; Special Publication No. 17; The Chemical Society: London, 1964.
21. Wolverton, W. *New Scient.* **1976**, *71*, 318-320.
22. Simmons, M. A. *Prog. Water Tech.* **1979**, *11*, 507-519.
23. Center, T. D.; Durden, W. C. "Studies on the Biological Control of Waterhyacinth with the Weevils *Neochetina eichhorniae* and *N. bruchi*," Misc.Paper A-84-4,APCRP, U.S. Army Waterway Experiment Station: Vicksburg, MS, 1984; pp. 85-98.

RECEIVED October 14, 1993

Chapter 35

Design of New Chelating Agents for Removal of Intracellular Toxic Metals

Mark M. Jones

Department of Chemistry and Center in Molecular Toxicology, Vanderbilt University, Nashville, TN 37235

The principles developed by Alfred Werner that determine the coordination number, net ionic charge, and stereochemistry of a coordination compound must be considered in the design of chelating agents which are to remove toxic metals from the mammalian body. There are also requirements which must be met to obtain chelating agents and metal complexes with the desired biological properties. The success of newly designed chelating agents *in vivo* is determined by how effectively these two separate but interdependent sets of requirements are reconciled. The way in which these factors can be manipulated to develop chelating agents for the removal of specific metals from designated organs is outlined.

Alfred Werner showed very clearly that the properties of metal ions could be drastically altered via carefully planned and executed changes in the coordination sphere of the metal ion (*1-3*). Although there is no indication that Werner ever had any biological applications of his principles in mind, his studies furnish part of the foundation for efforts to design new chelating agents for the treatment of metal intoxication. The design of chelating agents to remove toxic metals from *intracellular* sites starts from Werner's numerous experiments which demonstrated the enormous changes undergone by typical metal ions subsequent to their incorporation into a chelate complex. These ideas begin with his definition of the coordination number of a central metal ion and its relation to the other properties of the complex. Included also are the demonstration of how the ionic charge of a complex ion can be altered via changes in the nature of the coordinated groups. It is now known that the pharmacological properties of a metal complex are very dependent on its ionic charge and other molecular features. Werner's studies of the stereochemistry of various coordination geometries is another key feature underlying the overall process of the design of new chelating agents as well as the development of explanations for the mechanisms of action of known compounds. Werner's investigations furnished a firm conceptual foundation for the idea that the properties of metal ions could be manipulated far beyond the boundaries previously thought possible. He demonstrated that this could be accomplished by the use of hypotheses whose validity he proved in a long series of experimental studies (*1-3*). It must be noted that while Werner used chelating agents extensively in his studies, especially those on the types of isomerism which are present in metal complexes, the name "chelate" and the explicit formulation of the special properties of chelating agents originated in later studies by Morgan, Pfeiffer, Schwarzenbach, and many others.

Basic Properties of the Metal Ion and Metal Complex

Coordination Number. The design of chelating agents for a toxic metal ion must take into consideration the coordination number of that metal ion. Werner's studies (1) established characteristic coordination numbers for a large number of metal ions and showed several ways in which this could be accomplished. The demonstration of the advantages of using chelating agents to obtain complexes of enhanced stability and altered properties is seen in his studies establishing the chirality of inorganic complexes. The use of a chelating agent which will occupy more of the coordination positions of a metal ion will generally (but not always) give a complex of greater stability than is found for those complexes with chelating agents which occupy fewer positions. These metal chelate complexes will also have a reduced tendency to undergo exchange reactions once they are formed. However, it is frequently advantageous to use a preferred donor atom in a chelating agent of lower denticity. Thus *meso* 2,3-dimercaptosuccinic acid (DMSA) is preferable to EDTA in the treatment of lead intoxication (4,5). The later development of a detailed knowledge of coordination numbers and geometries from X-ray studies has led to the understanding that many metal ions have a variable coordination number.

Net Ionic Charge. One of the properties of a metal complex which is extremely important in governing its biological properties is its net ionic charge. The way in which this is determined was clearly set out for the first time by Werner and Miolati (6). This involves first, sorting out the ions or groups which are bound together in the complex ion from those independent counter-ions needed to obtain an electrically neutral solid. Then one simply sums up the contributions of the ionic charge on the complex ion from the central ion and all those charged groups bonded to it. Complex ions bearing a net positive charge may exhibit a curare-like activity, which is best avoided by an appropriate design of any chelating agent to be used *in vivo*. It is also necessary to keep in mind that the introduction of the chelating agent into any intracellular space requires that it pass through the cellular membrane. This passage, (Figure 1) can be accomplished either: (1) by passing through the lipid part of this membrane as an uncharged molecule or (2) via utilization of one of the anion or cation transport systems present in the membrane, which is possible for a chelating agent of appropriate design. The variation of effectiveness of a series of iron(III)-binding agents with structural changes which influence the solubility of the compound in the lipid part of the membrane has been clearly demonstrated (7,8). There is some support for the idea that large complex ions with a positive charge will pass out of a cell very slowly because of their inability to pass through either the lipid portion of the cellular membrane or the cation transport systems designed to move ions with a +1 or a +2 charge across this membrane(9).

Stereochemistry. The stereochemistry of the toxic metal ion is of considerable importance in the design of chelating agents which are to tie up all the coordination positions of a metal ion. This can easily be appreciated from the different requirements of square planar and tetrahedral complexes. Our current knowledge of the stereochemical requirements of various metal ions, again largely derived from X-ray structural studies, has added enormous detail to the picture originally sketched by the investigations of Werner and his students. The design of chelating agents to fit specific coordination geometries has been extensively studied and reviewed (10-18). Specific examples relevant to our purposes include the design and synthesis of

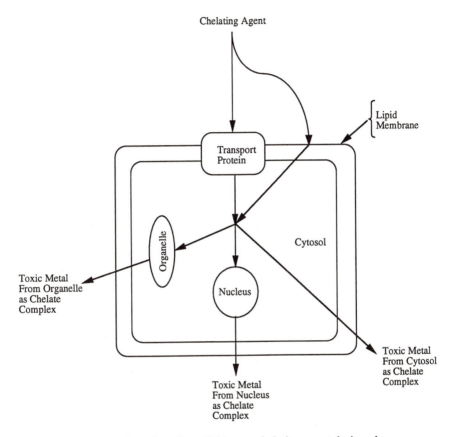

Figure 1. Pathways into the cell available to a chelating agent designed to remove a toxic metal from intracellular sites and pathways out for the metal chelate complex which is formed.

chelating agents which match the coordination geometry of Fe(III) (*19*), Pu(IV) (*20*), and the planar hexadentate structure of the UO_2^{2+}ion (*21*).

The use of chelating agents in medicine and the development of special chelating agents for this purpose is a very active field of research encompassing agents for the removal of toxic metals, in diagnostic applications, as imaging agents, antiinflammatory agents and as antiviral, antimicrobial, and antiparasitic agents (*22*). It must also be noted that many microorganisms synthesize very effective chelating agents to remove the iron that they need from their surroundings. These compounds, called siderphores, include enterobactin (*23*), a compound with a higher affinity for Fe(III) than all but a few recently synthesized chelating agents (*24*).

Biological Behavior and Metal Complex Properties

We may now examine the interaction between these properties and the design requirements for new chelating agents for a toxic metal ion as we effect changes in the following properties of the metal ion:

     (1) toxic properties
     (2) biological distribution
     (3) pathological effects
     (4) excretion patterns
     (5) stability
     (6) rate and extent of absorption.

Werner's theories must be combined with information on the pharmacological properties of known chelating agents and their metal complexes in order to design new chelating agents for the removal of toxic metal ions from specific intracellular sites. Such systems must be designed to satisfy several criteria: (a) transport of the chelating agent by an appropriate membrane transport system into the cells in which the toxic metal ion is concentrated, (b) formation of a stable complex with the intracellular toxic metal after removal of the toxic metal from its bonds to a biological binding site, and (c) formation of a metal complex whose properties will facilitate its excretion as well as decrease the toxicity of the toxic metal. These processes are outlined in Figure 1.

One expects, and finds, that the stability constants increase as the donor atoms of the ligand match up better with the preferred coordination number and stereochemistry of the metal ion. However, an effective chelating agent for toxic metal mobilization need not occupy all of the coordination positions of a metal ion or be designed for the coordination geometry of a single metal ion. With the exception of deferoxamine, none of the chelating agents used in the clinic as antagonists for toxic metals (Figure 2) satisfy these conditions rigorously for the metal ions for which they are used. Thus chelating agents capable of accelerating the excretion of Pb^{2+} include EDTA, DTPA, DMSA, DMPS, BAL, and DPA (Figure 2). It would appear from these examples that chelating agents can be effective (but not necessarily maximally effective) where a single molecule does not completely occupy all of the coordination positions of the metal ion or incorporate a molecular design specifically tailored for the stereochemistry of the metal ion.

Werner's studies led to a general realization that the typical properties of metal ions could be modified by binding them firmly to appropriate chelating agents, and it was a short step to extend these ideas to the biological properties of toxic metal ions. An early application of this in medicine was the use of the *d*-tartrate complex of antimony(III) in the treatment of parasitic diseases such as schistosomiasis (*25*). The parasite which causes this disease is readily poisoned by antimony(III) compounds, but for simple antimony compounds the margin of safety is too small and humans suffer from cardiotoxicity when such compounds are administered. The *d*-tartrate complex allows this to be somewhat more readily controlled and allows the more

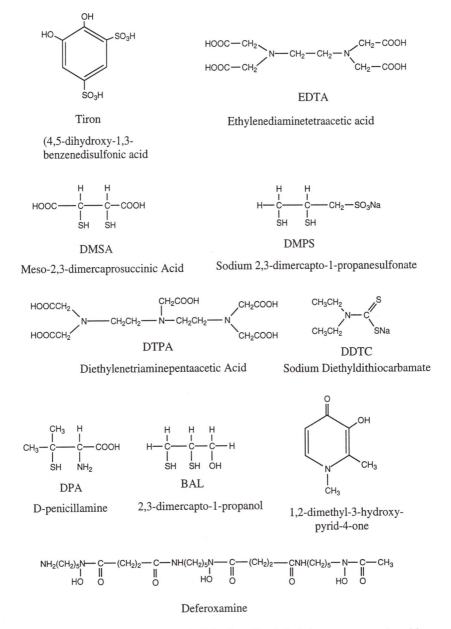

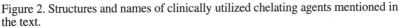

Figure 2. Structures and names of clinically utilized chelating agents mentioned in the text.

selective destruction of the parasite. The structure of this d-tartrate complex contains dimeric units in which Sb(III) ions are coordinated to a carboxyl oxygen and a hydroxyl oxygen from each of two tartrate ions (26). A more conscious application of these same ideas was the development of the more stable and effective -- and less toxic -- complex of antimony(III) with Tiron (sodium 4,5-dihydroxy-1,3-benzenedisulfonate), called stibophen, which was introduced in 1926 for the treatment of schistosomiasis (27). The structure of stibophen contains Sb(III) ions coordinated to both catechol-type oxygen atoms of two Tiron molecules (28). All four of such phenolic oxygens have lost their hydrogen atoms in the complex. The chelating agents used in both of these antimony compounds provide negatively charged oxygen donors in pairs.

It is useful to examine the six properties of metal ions listed above and to examine the manner in which they can be manipulated via changes in the types of atoms bonded to the central metal ion.

Toxic Properties. The toxic properties of metal ions such as Pb^{2+}, Hg^{2+}, Cd^{2+}, etc. are the result of the incorporation of some biological structure into the coordination sphere of the metal ion. The direct consequence of this is to alter the biological properties of these newly coordinated species, often by deactivation of any normal enzymatic activity. This can be clearly seen in the anemia characteristic of lead intoxication. Thus Pb^{2+} reacts with and reduces the activity of three of the enzymes involved in the synthesis of heme and reduces the ability of the stem cells in the bone marrow to form normal red blood cells (29).

Biological Distribution. The biological distribution of various nonessential metal ions is determined in considerable part by their ionic charge and coordination preferences, and we find that many ions follow the pathways of essential metal ions to which they have chemical similarities, e.g., Cd(II) follows the pathways used by Zn(II) (30), Pu(IV) uses the pathways normally use by Fe(III) (31), etc. The ability of these metal ions to fulfill the normal biological roles of the essential metals which they displace is limited and results in toxic effects. The coordination of a metal ion to a chelating agent which results in an uncharged, lipid soluble complex may result in the movement of the metal ion across lipid barriers which otherwise are impassable for it. Thus the reaction of sodium diethyldithiocarbamate with cadmium(II), copper(II), and lead(II) gives complexes which can move across the blood/brain barrier into the brain (32).

Pathological Effects. The pathological effects of a toxic metal can be drastically altered by changes in its coordination sphere. One of the principal effects of such changes is to alter the *in vivo* organ distribution of the toxic metal. A clear case is seen in the behavior of lead. Lead present in low molecular weight complexes in the serum is filtered at the glomerulus and passes into the renal tubule. If the lead is weakly complexed, it will be deposited at metal binding sites in the proximal tubule and interfere with essential processes which normally occur at those sites, such as the reabsorption of amino acids. If enough such lead is accumulated, characteristic lead bodies deposit within the cell nucleus, which can be detected on microscopic examination (33). If, on the other hand, the lead present in the serum is in a very stable complex, such as is formed with EDTA or DMSA, it is filtered at the glomerulus and passes through into the renal calyx without being deposited in the proximal tubule.

Excretion Patterns. The route of excretion of a toxic metal is largely determined by the nature of the species to which it is coordinated. Low-molecular-weight, water-soluble complexes usually undergo predominant excretion through the kidneys. As the molecular weight increases, especially above 500 for humans, there is a pronounced tendency for compounds to be sorted out on the basis of their polarity with the less polar compounds passing through the liver and undergoing biliary excretion following passage through hepatocytes. Compounds which are most effective in removing toxic metals from the liver are often amphipathic or relatively non-polar, while those which are most effective in removing toxic metals from the kidneys are more polar (but not highly charged) and more hydrophilic.

Stability of the Toxic Metal Complex. In order to obtain a metal complex of sufficient stability to be effective in removing a toxic metal from an organism it is necessary to incorporate the donor atoms into structures which will form four-, five-, or six-membered chelate rings when the metal bonds. The selection of the optimal donor atoms also has an enormous effect on the stability of the complexes which are formed, and this can be done most effectively via an examination of the stability constants for the complexes of the toxic metal ion which have already been characterized.

Rate and Extent of Absorption. It is very advantageous, but not always feasible, to have therapeutic chelating agents which can be given orally. To use this route of administration an appreciable fraction of the compound must be absorbed, and this places limitations on the donor groups which can be used as well as the total ionic charge. Thus chelating agents which bear a large negative charge, such as EDTA or DTPA, are only very slightly absorbed following oral administration so they are usually given by injection. Chelating agents such as deferoxamine (see Figure 2), which contain an amine group that is protonated at physiological pH values, are also absorbed to only a slight extent from the gastrointestinal tract (9). Further, in deferoxamine (generic drug name; chemical name: desferriferrioxamine-B), the hydroxamic acid groups are extensively altered metabolically in the gastrointestinal tract and the liver so this compound must also be administered parenterally. Deferoxamine wraps itself around an octahedral ferric ion so that each of the three hydroxamic groups chelates to the iron ion using both oxygen atom, and the protons are lost from these hydroxamic acid groups; the amine group at the end of the molecule is not directly bonded to the iron ion. Types of compounds which are more readily absorbed when given orally include amino acid derivatives, such as D-penicillamine, monoanions, such as DMPS, compounds which give dianions such as *meso*-2,3-dimercaptosuccinic acid, and finally, neutral polar compounds such as 1,2-dimethyl-3-hydroxypyrid-4-ones.

Design Procedures

There are numerous previous studies on the design of chelating agents to achieve specfic properties (10-21,34), including recent studies directed toward selective chelating agents directed more specifically toward new agents for iron(III) (19,24) and plutonium (20). These design studies were almost exclusively directed toward the development of chelating agents with greater effective stability constants for the metal ion of interest. On the other hand, because the design process has a great effect on the resulting pharmacological properties of the chelating agent, we find that design studies which emphasize the ability of the chelating agent to get across the lipid portion of the cellular membrane to gain access to the intracellular metal ion place considerable emphasis on the distribution coefficient of the chelating agent (35,36).

The steps in the design of a chelating agent to remove a specific toxic metal ion from an intracellular site can be seen in the procedure summarized below, which was used to design a chelating agent for the removal of lead from intracellular sites in certain organs which were accessible via certain monoanion transport systems, such as the kidneys and the brain.

(A) **Use of the Information on the Stereochemistry of the Metal Ion and Its Coordination Preferences in the Selection of the Donor Atoms.** Pearsons's concept of hard and soft acids and bases (*37,38*) can furnish a very useful guide to the selection of preferred donor atoms for a given metal ion. Lead(II) forms complexes in which it has a coordination number of four or more and forms bonds to sulfur and oxygen donor systems in preference to nitrogen. It also bonds to sulfur (as sulfide) in preference to oxygen. Oxygen donors, such as are present in EDTA, are effective in enhancing the excretion of lead from lead-intoxicated animals and humans, but the high negative charge on EDTA at physiological pH (where it exists as a mixture of H_2EDTA^{2-} and $HEDTA^{3-}$) and its size restrict this chelating agent to extracellular spaces. Unfortunately, the best oxygen donors for lead are those found in carboxylate groups after a proton has been removed. As we shall see below, the charge on the chelating agent must be limited to a value that can be accommodated by the anion transport systems. This makes multiple carboxylate donors less attractive than alternative donor systems involving sulfur, such as -SH groups, if one wishes to mobilize intracellular toxic metals. The -SH group has a higher pK_a value than the carboxylate group and is neutral (protonated) before it is complexed to the metal ion. Thus DMSA (Figure 2) exists as a -2 anion at physiological pH values but may also be able to use the succinate transport systems. Lead(II) also reacts to a much greater extent with thiol groups than it does with thioether groups. Lead(II) is also held more firmly by a group with a negative charge, but from considerations listed below, if the removal of a proton from the donor atom can be postponed until the actual coordination of the lead ion, this provides some advantageous flexibility in the design process. The net result of these considerations is the selection of sulfur in -SH as the donor atom, in part, because it has a higher pK_a value.

(B) **Selection of the Geometry of the Chelate Rings.** The generation of a chelate ring requires two donor atoms, and with sulfur as the donor atom this brings up two obvious choices: dithiocarbamates and vicinal dithiols. An examination of the comparable lead-mobilizing ability of $Na_2CaEDTA$ and two dithiocarbamates indicated that the dithiocarbamates of appropriate structure could surpass $Na_2CaEDTA$, at least under certain conditions (*39*). Unfortunately, this earlier data indicated that of the dithiocarbamates tested, the one which gave the most hydrophobic lead complex was the most effective. Further investigation revealed that, as expected, the generation of such hydrophobic lead complexes *in vivo* enhances the lead level of the brain (39). This suggested that a reexamination of the use of the thiol group was more promising and that dithiol groups might be more useful than other donor combinations. There are well documented examples of dithiol chelating agents which are effective in the mobilization of lead (40). A five- or six-membered chelate ring which incorporates the metal is generally preferable. Our previous studies with seven-membered chelate rings showed them to be ineffective. On the basis of ease of access we selected the five-membered chelate ring which was furnished by the chelation of Pb^{2+} with a *vicinal dithiol* group. An alternative choice of donor atoms is also possible, as can be seen in the recent study by Raymond and co-workers (41).
    The principal goal of parts A and B of the design process is to obtain a basic chelating agent donor structure core capable of removing the toxic metal from the binding sites which it occupies *in vivo*.

(C) **Incorporation of Structural Features to Facilitate Transport across Target Cellular Membranes.** In order to mcve the chelating agent across the cellular membrane so that it can react with intracellular deposits of lead we must design it to use a known mode of transport. We selected the monoanion transport system because this transport system is known to be present in the kidneys and several other organs which accumulate lead. The selection of the monoanion transport systems was based on earlier studies by Diamond and co-workers (*42*), who demonstrated that transport systems of this type present in the renal tubules were used by sodium 2,3-dimercaptopropane-1-sulfonate (DMPS). This limited the possible structures to those bearing a single negative charge. An examination of these requirements and their comparison with the recently reported monomethyl ester of meso-2,3-dimercaptosuccinic acid (*43*) suggested that compounds of this latter type might prove of considerable interest in that they contain both a single negative charge at physiological pH and a grouping of donor atoms which is known to be effective in the mobilization of lead. The structure of the lead complex formed is presumed to have both of the thiol groups of the chelating agent bonded to the Pb^{2+} ion to give complexes in which one or two such chelate groups are present. The charge on the complex containing a single chelate ring would be -2, and that on the complex containing two chelate rings would be -6. The 1:1 complex is consistent with the reported data on the reaction between Pb^{2+} and DMSA (*44*).

(**D**) **Incorporation of Structural Features to Adjust Toxicity.** It was decided to carry out the adjustment of the toxicity of the basic chelating structure after we had a prototype compound which functioned as desired for factors A, B, and C. The basic process in reducing the toxicity of a chelating structure usually involves alterations in the nonchelating portion of the structure to enhance the polarity without altering the effectiveness.

(**E**) **Reconciliation of the Various Structural Features and Requirements to Design Specific Candidate Compounds.** At this point it was apparent that appropriate vicinal dithiols were suitable candidates. Previous studies on the mono esters of meso-2,3-dimercaptosuccinic acid suggested that compounds of this sort would be good candidates, and the greater amount of experimental data on the mobilization of lead by both non-polar vicinal dithiols such as 2,3-dimercaptopropanol-1 (*40*) and more polar ones such as *meso*-2,3-dimercaptosucinic acid (*45*) (DMSA) suggested that this group was more promising than the dithiocarbamates.

(**F**) **Preparation and Testing of Candidate Structures.** Suitable candidate structures were then examined for relative efficacy in the mobilization of lead from lead-loaded mice. On the basis of previous studies in which we examined the relative ability of many compounds to remove cadmium from its aged deposits in the liver and the kidneys, six vicinal dithiols were selected, which were all monoesters of *meso*-2,3-dimercaptosuccinic acid. The compounds selected were the mono esters of *meso*- 2,3-dimercaptosuccinic acid with the following alcohols: *n*-propyl, *iso*-propyl, *n*-butyl, *iso*-butyl, *n*-amyl, and *iso*-amyl (Figure 3). The results of experiments comparing these compounds showed that the *n*-butyl, *iso*-butyl, *n*-amyl, and *iso*-amyl monoesters of *meso*-2,3-dimercaptosuccinic acid were the most effective compounds and that all were capable of reducing both kidney and brain lead levels in lead-intoxicated mice (*46*). The data collected on lead-intoxicated mice treated with the most effective of such compounds are compared with the results obtained with DMSA in Table I.

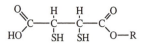

Figure 3. Monoesters of *meso*-2,3-Dimercaptosuccinic acid found to be most effective in the lead study. For M*n*-BDMS, R = -CH₂CH₂CH₂CH₃, M*i*-BDMS, R = -CH₂CH(CH₃)₂, M*i*-ADMS , R = -CH₂CH₂CH(CH₃)₂ ; for M*n*-ADMS; R = -CH₂CH₂CH₂CH₂CH₃. These compounds exist as monoanions at the physiological pH of 7.4.

Table I. Removal of Lead from the Kidneys of Lead-Treated Mice (Residual lead as % of untreated controls)

Compound	Kidney Pb (% Control)	Brain Pb (% Control)
DMSA	57	67
M*n*-BDMS	48	30
M*i*-BDMS	40	32
M*i*-ADMS	43	24
M*n*-ADMS	27	28

(The structures of these new compounds are shown in Figure 3.)

(G) **Reevaluation of Design Criteria.** The results indicated that these compounds were effective in removing lead from two organs to which access can be gained via monoanion transport systems: the kidneys and the brain (*46*). The brain is presumably accessible to these compounds because the tissue of the choroid plexus, which forms the cerebrospinal fluid, utilizes such transport systems in the generation of this fluid. The most effective compounds were found to be the mono-*n*-amyl and mono-*iso*-amyl esters of *meso*-2,3-dimercaptosuccinic acid, which were much more effective than an equal dosage of *meso*-2,3-dimercaptosuccinic acid (DMSA) in reducing lead levels of both the kidney and the brain. The next step in the development of more effective compounds of this sort will involve structural changes which will reduce the toxicity. While some hydrolysis of these esters by endogenous esterases was expected, the extent of such hydrolysis, even when the compounds were administered orally, was not sufficient to suppress the lead mobilization activity. The reason for the slow *in vivo* hydrolysis of these esters is not known.

At the present time there are obviously two different sets of structural requirements which are to be imposed on optimally effective chelating agents for toxic metals. The first of these arises from the chemical requirements of the toxic metal ion. The second arises from the nature of the biological systems which must be traversed by the chelating agent if it is to reach the site at which the toxic metal ion is present, react with the toxic metal and transform it into a new and readily excreted metal complex. The most effective way to combine these two sets of structural requirements in a general manner must be developed for each toxic metal ion and is dependent upon its preferred biological sites for deposition and its preferred donor atoms and their attendant requirements. Some further examples may be found in recent reviews (*47,48*).

Acknowledgments

I wish to acknowledge, with thanks, the support received for these studies from the National Institute of Environmental Health Sciences via grants ES-00268 and ES-02638.

Literature Cited

1. Werner, A. *Neuere Anschauungen auf dem Gebeite der anorganischen Chemie*, 3rd ed.; F. Vieweg & Sohn: Braunschweig, 1913.
2. Kauffman, G. B. *Alfred Werner: Founder of Coordination Chemistry;* Springer-Verlag: Berlin, 1966.
3. Bailar, Jr., J. C.; Busch, D.H. In *The Chemistry of the CoordinationCompounds;* Bailar, Jr., J.C., Ed.; American Chemical Society Monograph Series No. 131; Reinhold Publishing Corp: New York, NY, 1956; pp 1-99.
4. Willes, M. J.; Williams, D. R. *Inorg. Chim. Acta* **1985,** *106*, L21.
5. Cory-Slechta, D. A. *J. Pharmacol. Exp. Ther.* **1988,** *246*, 84.
6. Werner, A.; Miolati, A. *Z. physik. Chem.* **1893,** *12*, 35; **1894,** *14*, 506. For discussions and annotated English translations see Kauffman, G. B. *Classics in Coordination Chemistry, Part 1: The Selected Papers of Alfred Werner;* Dover Publications: New York, NY; pp 89-139.
7. Porter, J.B.; Gyparaki, M.; Huehns, E.R.; Hider, R.C. *Biochem. Soc. Trans.* **1986** *,14*, 1180.
8. Streater, M.; Taylor, P.D.; Hider, R.C.; Porter, J. *J. Med. Chem.* **1990,** *33*, 1749.
9. Kontoghiorghes, G. J.; Sheppard, L.; Chambers, S. *Arzneim.-Forsch./Drug Res.* **1987,** *37*, 1099.
10. Black, D. St. C.; Hartshorn, A. J. *Coord. Chem. Rev.* **1972-1972,** *9*, 219.
11. Clevette, D. J.; Orvig, C. *Polyhedron* **1990,** *9*, 151.
12. Hancock, R. D.; Martell, A. E. *Chem. Rev.* **1989,** *89*, 1875.
13. Hancock, R. D. *Acc. Chem. Res.* **1990,** *23*, 253.
14. Lehn, J.-M. *Acc. Chem. Res.* **1978,** *11*, 49.
15. Busch, D. H.; Stephenson, N. A. *Coord. Chem. Rev.* **1990,** *100*, 119.
16. Lindoy, L. F. *The Chemistry of Macrocyclic Ligand Complexes*; Cambridge Univsity Press: Cambridge, 1989.
17. Izatt, R. M.; Pawlak, K.; Bradshaw, J. S.; Breuning, R. L. *Chem. Rev.* **1991,** *91*, 1721.
18. Sargeson, A. M. *Pure & Appl. Chem.* **1986,** *11*, 1511.
19. Turowski, P.N.; Rodgers, S.J.; Scarrow, R.C.; Raymond, K.N. *Inorg. Chem.* **1988,** *27*, 474.
20. White, D.L.; Durbin, P.W. ; Jeung, M.; Raymond, K.N. *J. Med. Chem.* **1988,** *31*, 11.
21. Franczyk, T.S.; Czerwinski, K.R.; Raymond, K.N. *J. Am. Chem. Soc.* **1992,** *114*, 8138.
22. Bulman, R. A. *Struct. Bond.* **1987,** *67*, 92.
23. Raymond, K. N.; Müller, G.; Matzanke, B. F. *Top. Curr. Chem.* **1984,** *123*, 49
24. Moitekaitis, R. K.; Sun, Y.; Martell, A. E. *Inorg. Chim. Acta* **1992,** *198-200*, 421.
25. Plimmer, H.G.; Thomson, J.D. *Proc. Roy. Soc.* **1908,** *B 80*, 5.
26. Gress, M. A.; Jacobson, R. A. *Inorg. Chim. Acta* **1974,** *8*, 209.
27. Uhlenhuth, P.; Kuhn P.; Schmidt, H. *Arch. Schiffs-Tropen .Hyg.* **1925,** *29*, 623.
28. Sneader,W. In *Comprehensive Medicinal Chemistry*; Hansch, C.; Sammes, P. G.; Taylor, J. B., Eds.; Pergamon Press: Oxford, 1990; Vol. 1: p 52-53.
29. Moore, M.R.; Meredith, P.A.; Goldberg, A. in *Lead Toxicity*; Singhal, R.L.; Thomas, J.A., Eds..; Urban & Schwarzenberg: Baltimore, MD, 1980; pp 79-117.
30. Shaikh, Z.A.; Lucis, O.J. *Arch. Environ. Health* **1972,** *24*, 419.
31. Durbin, P.W. *Health Physics* **1975,** *29*, 495.

32. Koutensky, J.; Eybl, V.; Koutenska, M.; Sykora, J.; Mertl, F. *Eur. J. Pharmacol.* **1971**, *14*, 389.
33. Goyer, R.A.; Wilson, M.H. *Lab. Invest.* **1975**, *32*, 149.
34. Hancock, R.D.; Martell, A.E. *Comments Inorg. Chem.* **1988**, 6, 237.
35. Gyparaki, M.; Porter, J.B.; Hirani, S.; Streater, M.; Hider, R.C.; Huehns, E.R. *Acta Haemat.* **1987**, *78*, 217.
36. Hider, R.C.; Hall, A.C. *Perspect. Bioinorg. Chem.* **1991**, *1*, 209.
37. Pearson, R. G. *Survey Prog. Chem.* **1969**, *5*,1.
38. Wulfsberg, G. *Principles of Descriptive Inorganic Chemistry;* Brooks/Cole Publishing Co.: Monterey, CA, 1987; pp 266-305.
39. Gale, G.R.; Atkins, L.M.; Smith, A.B.; Jones, M.M. *Res. Commun. Chem. Pathol. Pharmacol.* **1986**, *52*, 29.
40. Chisolm, J.J. *J. Toxicol., Clin. Toxicol.* **1992**, *30*, 493.
41. Abu-Dari, K.; Hahn, F.E.; Raymond, K.N. *J. Amer. Chem. Soc.* **1990**, *112*, 1519.
42. Stewart, J. R.; Diamond, G. L. *Am. J. Physiol.* **1987**, *252* , F800.
43. Rivera, M.; Levine, D. J.; Aposhian, H. V.; Fernando, Q. *Chem. Res. Toxicol.* **1991**, *4*, 107.
44. Harris, W. R.; Chen, Y.; Stenback, J.; Shah, B. *J. Coord. Chem.* **1991**, *23*, 173.
45. Graziano, J. H.; LoIacono, N. J.; Moulton, T.; Mitchell, M. E.; Slakovich, V.; Zarate, C. *J. Pediatrics* **1992**, *120*, 133.
46. Walker, Jr., E. M. ; Stone, A.; Milligan, L. B.; Gale, G. R.; Atkins, L. M.; Smith, A. B.; Jones, M. M.; Singh, P. K.; Basinger, M. A. *Toxicology* **1992**, *76*, 79.
47. Jones, M. M. *J. Coord. Chem.* **1991**, *23*, 187.
48. Jones, M. M. *Comments Inorg. Chem.* **1992**, *13*, 91.

RECEIVED October 14, 1993

Chapter 36

Coordination Compounds of Metal Ions in Sol–Gel Glasses

Renata Reisfeld[1] and Christian K. Jørgensen[2]

[1]Department of Inorganic Chemistry, Hebrew University, 91904 Jerusalem, Israel
[2]Section of Chemistry, University of Geneva, 30 Quai Ansermet, CH 1211 Geneva 4, Switzerland

Sol-gel glasses (known since 1846) are a family of inorganic glasses or organic composites prepared at ambient temperature, usually from silicon alkoxides (sometimes mixed with other alkoxides). Hydrolysis and subsequent polycondensation of the precursor solution allows incorporation of specific cations or their pre-existing complexes. Detailed study of absorption and emission spectra, including lifetimes of their excited states, sometimes allows determination of the site symmetry and the coordination number, *e.g.*, for cobalt (II). The ruthenium (II) tris(2,2'-bipyridine) cation in glass shows a much higher yield of fluorescence than in solution, partly because of less triplet quenching. Lanthanides can show very high luminescence yields because no collisions occur between the species in an excited state and in its ground state.

A few minerals (mainly related to volcanic activity) are glasses, such as the dark brown obsidian readily broken up for cutting tools. 3000 to 3500 years ago, Phoenicians prepared glasses by melting white sand (SiO_2) with calcium, sodium, or potassium carbonate (derived from calcite, Egyptian natron lakes, and wood ashes, respectively). Ancient sources suggest that making transparent quartz crystals was the goal, although the hottest fires could not go much above 1100°C (slightly above the melting points of the coinage metals). Later, lead oxide decreased the temperature needed for glass making, but nearly all multicomponent glasses made today are still prepared above 500°C. Another early motivation was to make brightly colored glass beads (*1*) imitating gemstones, *e.g.*, found on Merovingian crowns 1500 years ago. For a long time such beads were objects of trade in all of Africa and the two Americas.

As recently reviewed by Hench and West (*2*), an entirely different technique developed since 1846, in which silicon alkoxides such as $Si(OCH_3)_4$ and $Si(OC_2H_5)_4$ [with the colloquial acronyms TMOS and TEOS] were hydrolyzed and subsequently underwent polycondensation under controlled conditions (losing alcohol vapor) to form quite clearly transparent, moderately viscous to almost vitreous materials. Such a glass has close relations to the limpid silica glass obtained today by melting quartz

0097–6156/94/0565–0439$08.00/0

above 1700°C and, to a certain extent, to glassy stoichiometric $NaPO_3$, obtained by dehydration of NaH_2PO_4. A difference is that colored chromium (III), cobalt (II), nickel (II), or copper (II) must be introduced as finely divided phosphates or oxides during the fusion of monosodium phosphate, whereas sol-gel glasses prepared by hydrolysis and subsequent polycondensation can incorporate organic dyestuffs and luminescent compounds in tiny (20 to 100 nm) cavities or crevices and be spared from caramelization and pyrolysis by very carefully avoiding heating the sol-gel sample above, e.g., 120°C or 200°C. The metaphosphate glasses (3) have absorption spectra showing the 3d-group ions to be roughly octahedral, with the exception of the Jahn-Teller unstable $3d^4$ manganese (III) and $3d^9$ copper (II). The behavior of octahedral $3d^3$ chromium (III) has been reviewed at length (4, 5) and compared with $3d^8$ nickel (II) in fluoride glasses (6).

By sufficient heating of a sol-gel glass, e.g., to 400 or 500°C, most absorption spectra of 3d- and 4f-group (lanthanide) ions become quite similar to conventional silicate glasses containing large ions Na(I), K(I), Ca(II), Ba(II), etc. as "network modifiers" (1, 7). At this point, the behavior of $3d^7$ cobalt (II) is atypical (8). Although pale raspberry-red (not fully dehydrated) samples have band positions closely similar to strawberry-red $[Co(OH_2)_6]^{2+}$, the intensities are 2 to 5 times higher. Progressive loss of water and alcohol (ROH) during heating provides a sky-blue color but only with some 3 times higher molar extinction coefficients than the raspberry-red solvated form. This is a much smaller difference than the ratio (600/4.6 = 130) between the highest band of $[CoCl_4]^{2-}$ in 12M hydrochloric acid and of $[Co(OH_2)_6]^{2+}$ in water. It cannot be argued that exactly one definite blue cobalt (II) species occurs in sol-gel glasses (9). However, the trend to three absorption bands close to 640, 590, and 525 nm (15600, 17000, and 19100 cm^{-1}) is also known for several (10, 11) L_2CoX_2 with two neutral ligands L (acetone, tryphenylphosphine, etc.) and two X = Cl, Br, or NCS. Actually, such spectra are quite similar to Co(II) in cesium borate glass (1). The three rather broad bands in the visible region cannot be described as a weak deviation from the cubic point-group T_d. As early as 1958 unexpected deviations from predicted T_d energy levels were reported (12) for cobalt(II) syncrystallized in several spinel-type mixed oxides and for $[CoCl_4]^{2-}$ and $[CoBr_4]^{2-}$ in various solvents (13, 14). In view of the Tanabe-Kamimura stability (5) of both the $^4A_2$ ground state of regular tetrahedral Co(II)X_4 and of the Pauli-related $3d^3$ Cr(III)X_6, the deviations in the former case are unexpected (and have largely been neglected in the literature despite their frequent and conspicuous appearance).

The "three-bands-in-the-red" syndrome cannot be more than half explained (15, 16) by nondiagonal elements of spin-orbit coupling between the highest of the two $^4T_1$ (six Kramers doublets) and the almost (14) superposed ^2E, $^2T_1$, and also $^2T_2$ (i.e., 2+3+3=8 Kramers doublets). In view of the identical symmetry type t_2 in T_d of all 3 orbitals of a given p shell and of the σ-antibonding orbitals [xy, xz, and yz] of the 3d shell, a weak amplitude of the inner shell 3p (known from photoelectron spectra (17, 18) to have ionization energies close to 68 to 72 eV) in the LCAO description of $(t_2)^3$ of Co(II)X_4 might easily enhance spin-orbit coupling ζ_{3p} is very close to 2 eV.

The observed spreading out of the excited Kramers doublets in the red would be assisted by the squared amplitude of the Hartree-Fock ground configuration $1s^2\ 2s^2\ 2p^6\ 3s^2\ 3p^6\ 3d^7$ today being recognized (19, 20) to be below 0.8 (if not below 0.7) in the Schrödinger solution. This effect would increase the importance of the interelectronic repulsion (*20*) operator mixing well-defined electron configurations to considerable amounts (in particular with $1s^2 2s^2 2p^6 3s^2 3p^4 3d^9$).

Parallel to the unavoidable configuration intermixing, the spin-allowed transitions of cobalt (II) in (fairly heated) sol-gel glasses suggest an energy dispersion of the three antibonding orbitals (xy), (xz), and (yz) that would be easy to accept in six-coordinate Mn(III) or copper(II) with a large intrinsic propensity to Jahn-Teller effect, or, for that matter, tetrahedral Cu(II). The observed 3-band spectra of Co(II) suggest a holohedrized symmetry (*7, 8, 21*) like D_{2h} but are also readily compatible with the point-groups C_{2v}; S_4 (or even as low as C_1) before holohedrization. This includes the maximum symmetry C_{2v} of L_2CoX_2. A possible scenario is the sol-gel glass providing short Si-O bonds, making the 3d shell of an adjacent cobalt(II) simultaneously σ- and π-antibonding, but also (probably nonlinear) Si-O-Si bonds not being linear ligators (*i.e.*, cylindrically symmetrical, at least to the extent that R_3SiO are) and hence the roughly coplanar Si_2OCo showing distinct energies of (xz), (yz), and (xy) orbitals. A plausible lower-symmetry scenario would be the intermediate between 4- and 5-coordinated $Co(II)O_N$ (in one extreme, all N distances are quite different).

It is generally true for glasses at cryogenic temperatures that bands due to partly filled 3d or 4f shells do not narrow almost to lines, whereas, *e.g.*, $3d^3$ chromium(III) on approximately octahedral sites in crystalline ruby $Al_{2-x}Cr_xO_3$ in emerald (beryl) $Be_3Al_{2-x}Cr_xSi_6O_{18}$, and in microcrystalline glass ceramics (*5*) doped with Cr(III) often show very sharp absorption and fluorescence lines below -150°C (*e.g.*, the lowest excited ^2E level). This physical distinction can be easily explained by the glass formed from a liquid mixture (at an absolute temperature T so high that it is not fundamentally different from a molten crystal) becoming rapidly more viscous by cooling. A crystal growing at a definite point of fusion immediately shows a narrow (*22*) distribution of internuclear distances Cr-O, Al-O, Mg-O, *etc.* with amplitudes (roughly proportional with the square-root of T) like "thermal vibrations". But in glasses, ions of metallic elements M are almost immobile (as far as M-M distances are concerned) at a high T slightly below the apparent melting point of the glass. By further cooling, the absorption (and, if detectable, luminescence) spectra will give the impression of T being almost as high as when the internuclear distances "froze" in. The far larger vibrational (and Stokes) width suggests, in a sense, a high T not decreasing by cooling the sample.

It is well known (*23*) that polytungstate $[W_{12}O_{40}]^{8-}$ and molybdate $[Mo_{12}O_{40}]^{8-}$ can wrap themselves around a central M such as Si(IV), P(V), or As(V) surrounded by a tetrahedral set of four oxygens at a short distance. Such heteropolytungstates (and -heteropolymolybdates) are also known, in which blue Co(II) can be oxidized to dark brown $Co(III)O_4$ and, with somewhat differing stoichiometry (*24*), octahedral purple $Cr(III)O_6$ and bluish green $Co(III)O_6$. Trivalent lanthanides can be incorporated as in $[LnW_{10}O_{35}]^{7-}$, but also (*25*) Ce(IV), Pr(IV), Tb(IV), and $5f^5$ Am(IV) are known in the quite heavy anions $[M(SiW_{11}O_{39})_2]^{12-}$. They may be fair models of polycondensed silicates.

In the last few years sol-gel glasses have been applied in many novel inventions (*2, 26*) such as tunable solid-state lasers (containing photochemical resistant organic colorants with very high luminescence yields (*27, 28*)); nonlinear optical effects due to colorants or other pigments with specific physical characteristics (*26, 29, 30*) [also to be discussed in another volume of *Structure and Bonding* edited by the writers R.R. and C.K.J.]; and very small (*31-34*) translucid semiconductor (ZnS, CdS, $CdSe_xS_{1-x}$, AgI, *etc.*), 1- to 10-nm size particles, colloquially named "quantum dots", for high resolution information storage (intended for various time-scales of duration). Organic colorants acting as antibases (Lewis acids) can be used as sensors (*35*) for analysis of minor constituents of air invading the sol-gel glass. In other instances (*26*) the behavior of strongly colored organic species is more comparable to large cationic complexes like the ruthenium(II)-2,2'-bipyridine species [$Ru(bipy)_3^{2+}$] showing much higher yields of luminescence (*36*) than in less viscous solvents allowing very frequent intermolecular collisions. For comparison with other ion-exchanging solids, the firm binding (*26*) of europium (III) species with differing fluorescence lines (*37*) to heated sol-gel glasses is similar to absorption in $Zr(HPO_4)_2$ or on argillaceous montmorillonite [and conceivable examples of complexes of multidentate oxo ligands], whereas organic ion-exchange resins are analogous to the large Malachite Green (*35, 38*) or six-coordinate Ru(II) and Ir(III) cations in faintly heated sol-gel glasses (*39*).

Many interesting areas of research remain open (*5, 8, 40, 41, 42*) for coordination chemists considering the immediately adjacent environment of the sol-gel glass as a multidentate ligand for colored and/or luminescent ions of transition elements and lanthanides.

Acknowledgments

We are grateful to the Swiss National Science Foundation for their grant (20.32127.91) making possible our collaboration and also that with the experimental staff in Jerusalem.

Literature Cited

1. Weyl, W. A. *Colored Glasses;* Dawson's of Pall Mall: London, **1959**.
2. Hench, L. L.; West, J. K. *Chem Rev.* **1990**, *90*, 33.
3. Brawer, S. A.; White, W. B. *J. Chem. Phys.* **1977**, *67*, 2043.
4. Reisfeld, R.; Kisilev, A. *Chem. Phys. Lett.* **1985**, *115*, 457.
5. Reisfeld, R.; Jørgensen, C. K. *Struct. Bonding* **1988**, *69*, 63.
6. Reisfeld, R.; Eyal, M.; Jørgensen, C. K.; Guenther, A.; Bendow, B. *Chimia* **1986**, *40*, 403.
7. Reisfeld, R.; Jørgensen, C. K. *Lasers and Excited States of Rare Earths*; Springer: Berlin and New York, **1977**.
8. Reisfeld, R.; Chernyak, V.; Eyal, M.; Jørgensen, C. K. In *Proceedings by the Second International School on Excited States of Transition Elements;* World Scientific: Singapore, 1992; pp 247-256.
9. Reisfeld, R.; Chernyak, V.; Eyal, M.; Jørgensen, C. K. *Chem. Phys. Lett.* **1989**, *164*, 307.
10. Fine, D. A. *J. Am. Chem. Soc.* **1962**, *84*, 1139.
11. Cotton, F. A.; Faut, O. D.; Goodgame, D. M. L.; Holm, R. H. *J. Am. Chem. Soc.* **1961**, *83*, 1780.

12. Schmitz-DuMont, O.; Brokopf, H.; Burkhardt, K. Z. *Anorg. Allg. Chem.* **1958**, *295*, 7.
13. Ballhausen, C. J.; Jørgensen, C. K. *Acta Chem. Scand.* **1955**, *9*, 397
14. Cotton, F. A.; Goodgame, D. M. L.; Goodgame, M. *J. Am. Chem. Soc.* **1961**, *83*, 4690.
15. Griffith, J. S. *Theory of Transition-metal Ions*; Cambridge University Press: Cambridge, 1961.
16. Horrocks, W. De W.; Burlone, D. A. *J. Am. Chem. Soc.* **1976**, *98*, 6512.
17. Jørgensen, C. K.; Berthou, H. *Mat. Fys. Medd. Dan. Vid. Selskab* (Copenhagen) **1972**, *38*, No.15.
18. Jørgensen, C. K. *Struct. Bonding* **1975**, *24*, 1.
19. Jørgensen, C. K. *Chimia* **1988**, *42*, 21.
20. Jørgensen, C. K. *Comments Inorg. Chem.* **1991**, *12*, 139.
21. Jørgensen, C. K. *Modern Aspects of Ligand Field Theory;* North Holland: Amsterdam, 1971.
22. Jørgensen, C. K. *Top. Current Chem.* **1989**, *150*, 1.
23. Pope, M. T. *Heteropoly and Isopoly Oxometalates;* Springer: Berlin and New York, 1983.
24. Schaeffer, C. E.; Jørgensen, C. K. *J. Inorg.Nucl.Chem.* **1958**, *8*, 143.
25. Jørgensen, C. K. In *Handbook on the Physics and Chemistry of Rare Earths;* Gschneidner, K. A.; Eyring, L., Eds.; North-Holland: Amsterdam, 1988; Vol. 11; pp 197-292.
26. Reisfeld, R.; Jørgensen, C. K. *Struct. Bonding* **1992**, *77*, 207.
27. Reisfeld, R.; Brusilovsky, D.; Eyal, M.; Miron, E.; Burstein, Z.; Ivri, J. *Chem. Phys. Lett.* **1989**, *160*, 43.
28. Reisfeld, R.; Seybold, G. *Chimia* **1990**, *44*, 295.
29. Vogel, E. M.; Weber, M. J.; Krol, D. M. *Phys. Chem. Glasses* **1991**, *32*, 231.
30. Klein, L. C., Ed. *Sol-gel Technology for Thin films, Fibers, Preforms, Electronics and Speciality Shapes.* Noyes Publishers: Park Ridge NJ, 1988.
31. Rosetti, R.; Hull, R.; Gibson, J. M.; Brus, L. E. *J. Chem. Phys.* **1985**, *82*, 552.
32. Henglein, A. *Chem. Rev.* **1989**, *89*, 1861.
33. Steigerwald, M. L.; Brus, L. E. *Acc. Chem. Res.* **1990**, *23*, 183.
34. Minti, H.; Eyal, M.; Reisfeld, R.; Berkovic, G. *Chem. Phys. Lett.* **1991**, *183*, 277.
35. Chernyak, V.; Reisfeld, R. *Sensors and Materials* **1993**, *4*, 195.
36. Reisfeld, R.; Brusilovsky, D.; Eyal, M.; Jørgensen, C. K. *Chimia* **1989**, *43*, 385.
37. Levy, D.; Reisfeld, R.; Avnir, D. *Chem. Phys. Lett.* **1984**, *109*, 593.
38. Reisfeld, R.; Chernyak, V.; Jørgensen, C. K. *Chimia* **1992**, *46*, 148.
39. Slama-Schwok, A.; Avnir, D.; Ottolenghi, M. *J. Am. Chem. Soc.* **1991**, *113*, 3984.
40. Avnir, D.; Braun, S.; Ottolenghi, M. In Supermolecular Architecture; Bein, T., Ed.; *A.C.S. Symposium Series* No. 499; Washington DC, 1992; pp 384-404.
41. Wu, S.; Ellerby, L.; Cohan, J. S.; Dunn, B.; El-Sayed, M. A.; Valentine, J. S.; Zink, J. I. *Chem. Mater.* **1993**, *5*, 115.
42. Rottman,C.; Ottolenghi, M.; Zusman, R.; Lev, O.; Smith, M.; Gong, G.; Kagan, M. L.; Avnir, D. *Material Letters* **1992**, *13*, 293.

RECEIVED December 27, 1993

Chapter 37

Coordination Compounds in New Materials and in Materials Processing

Herbert D. Kaesz

Department of Chemistry and Biochemistry, University of California, Los Angeles, CA 90024–1569

Coordination compounds are found in new materials such as one-dimensional conductors and are also utilized as volatile precursors for the deposition of metal oxides in thin superconducting films or deposition of metal carbides or nitrides for refractory protective coatings and other applications. Coordination compounds have also proven to be useful precursors for the deposition under relatively mild conditions of high purity thin metal films for microelectronic applications. Historically, we consider $Ni(CO)_4$ as the prototypical example discovered by Mond in 1898. Purely hydrocarbon complexes such as $(\eta^5\text{-}C_5H_5)Pt(CH_3)_3$ can be decomposed initially at around 120 °C. in an atmosphere of H_2 to give highly pure thin metal films of platinum metal. Deposition of electronically high-grade copper metal is achieved from precursors such as (hexafluoroacetylacetonate)Cu(cyclooctadiene) under vacuum conditions. The deposition of other metals is discussed including aluminum from the complex $(Me_3N)_2AlH_3$ and mixed metal films such as $PtGa_2$, PtGa, and CoGa from precursor coordination compounds.

Thin films with a large variety of uses such as conductors, semiconductors, and refractory materials have extensive importance in modern technology (1). The films can be deposited by using physical methods such as sputtering or evaporation. However, where conformal coverage is required for deposition into channels, chemical vapor deposition (CVD, including Vapor Phase Epitaxy, VPE) is the method of choice (2-4), in which volatile coordination compounds play a key role. CVD is carried out most simply by thermal decomposition of a metal complex that cleanly releases its ligands with negligible hetero-atom incorporation into the deposited thin metal film. Frequently, very clean deposition (less than 0.1 % impurities) is achieved in the presence of hydrogen gas. This article is largely an account of the work in our own laboratory with selected examples of other recent developments illustrating the role of coordination compounds in the deposition of thin films of metallic conductors.

Platinum Group Metals

Historically, tetracarbonylnickel, $Ni(CO)_4$, was the first organometallic complex to be used for metal deposition (5). The carbonyls usually decompose at relatively mild

0097–6156/94/0565–0444$08.00/0

temperatures; however, they are known to contaminate the metal films with traces of carbon and oxygen, owing to the dissociative chemisorption of CO on freshly formed (and highly reactive) metal films *(6)*. Hydrocarbon precursors, especially compounds with carbon-ring fragments that are more likely to be disengaged without leaving behind carbon fragments offer the purest films for metals on which adventitious surface carbon can be removed as CH_4 under a H_2 atmosphere. Where simple hydrocarbon derivatives are not available, chelating groups such as acetylacetonate and its fluorinated analogs have been used. These have appropriate volatility, but special safeguards must be undertaken to avoid hetero-atom incorporation into the films. Such safeguards are described in the sections which follow. The application of volatile transition metal precursors in MOCVD (or MOVPE as alternatively termed) has been recently reviewed *(7, 8)*, and a monograph is in preparation (Kodas,T.; Hampden-Smith, M. *Chemical Aspects of Chemical Vapor Deposition for Metallization*; VCH: Deerfield Beach, FL, in preparation).

Hydrocarbon complexes were found to be ideal precursors for the deposition of platinum group metals *(9-12)*. This is because they are volatile and also thermally decompose in the region of 100 to 300 °C. Adventitious traces of carbon arising from ligand decomposition are removed as methane (in the presence of hydrogen). This removal can be achieved only when the deposition temperatures are significantly below 450 °C. the temperature at which surface carbon will graphitize and cannot be removed by hydrogenolysis. The deposition of very clean platinum metal, for example, is achieved at 180 °C. from (η^5-methylcyclopentadienyl)(tris-methyl)platinum, accompanied by the formation of methane and methylcyclopentane (equation 1) *(9, 10)*:

$$(\eta^5\text{-}CH_3C_5H_4)Pt(CH_3)_3 + 4\,H_2 \longrightarrow Pt(s) + CH_3C_5H_9 + 3\,CH_4 \tag{1}$$

Kinetic studies show the existence of an induction period, which is found to be 80 minutes at 100 °C on Pyrex® glass. After induction the reaction becomes *autocatalytic* and proceeds without further external heating. An added trace of moisture in the gas stream will *shorten* the induction period, while the deposition of metal is *suppressed* by pretreatment of the glass with trimethylchlorosilane. This reagent converts surface hydroxyl groups to trimethylsiloxy groups. These results suggest that the principal initial reaction must be driven by an electrophilic interaction of the surface -OH groups (which are acidic) with the hydrocarbon metal complex. The intially attached platinum moieties, however, need to migrate together to form ensembles of some minimum size before a catalytically active deposit is achieved. There is much surface organometallic chemistry that needs to be explored to confirm such working hypotheses.

Clean deposits of Co, Rh, Ir, Ni, or Pd have also been achieved from similar hydrocarbon metal precursors *(11, 12)*. However, it would be more desirable to deposit a good conducting metal such as tungsten under these mild conditions. Tungsten is now deposited by reduction of the volatile hexafluoride, WF_6, in the presence of hydrogen. However, this is accompanied by production of the corrosive reaction product HF, whose presence cannot be tolerated by other components of delicate microelectronic circuits. The first attempts to deposit tungsten from the volatile hydrocarbon precursor (η^5-C_5H_5)$_2$$WH_2$ led to films with as much as 30% carbon incorporation *(13)*. This type of impurity greatly increases the resistivity of the metal film, causing it to be useless for integrated circuits. Tungsten is not catalytically active in the methanation of surface carbon. It was thus reasoned that a catalytic amount of platinum codeposited from a hydrocarbon precursor along with the tungsten could help to lower the carbon content. Indeed, this is what was found. A film of tungsten codeposited with a small amount of platinum was shown to contain less than 5% carbon *(13)*. The research groups are presently attempting to measure the rate of migration of carbon from tungsten onto the codeposited small platinum islands so that optimum carbon removal from the films can be achieved.

Copper

Deposition of high quality copper films with a resistivity of less than 2.0 $\mu\Omega$cm and deposition rates of 80-180 nm./min. is achieved at 250 °C. using the hexafluoroacetylacetonate complex Cu(hfac)$_2$ with H$_2$ at 760 torr as carrier gas and reducing agent (14):

$$Cu(F_3CC(O)CHC(O)CF_3)_2 + H_2 \text{------->} Cu + 2 F_3CC(O)CH_2C(O)CF_3 \quad (2)$$

The β-diketonate copper(I) phosphine complexes have also been found to be good precursors to the deposition of copper, which occurs through a disproportionation reaction (15):

$$2 (hfac)CuP(CH_3)_3 \text{------->} Cu + Cu(hfac)_2 + 2 P(CH_3)_3 \quad (3)$$

Since Cu(hfac)$_2$ is more volatile and more stable than the copper (I) precursor, it volatilizes with both of the organic groups intact, leaving behind pure copper. The advantage of using the Cu(I) complexes is that deposition occurs at temperatures as low as 150 °C., and contamination of the films with impurities derived from the ligands is avoided. The films show a resistivity of 1.7-2.2 $\mu\Omega$cm with no detectable impurities. Another advantage of (hfac)CuP(CH$_3$)$_3$ is that it is the only compound thus far known to deposit copper selectively on etched Teflon (Kodas, T; *et al. J. Electrochem. Soc.*, in press) (16); no copper deposits are found on the unetched Teflon. A simple three-step process of irradiation, chemical etching, and CVD of Cu is employed, replacing more involved processes for such selective deposition in various electronic devices.

Other Cu(I) precursors are the (β-diketonate)Cu(I) alkyne complexes (17). These are easy to synthesize (equation 4). The products are volatile and undergo the same disproportionation reaction depicted in equation 3 to produce highly pure copper films.

$$CuCl + excess\ alkyne + Na(hfac) \text{----->} (hfac)Cu(alkyne) + NaCl \quad (4)$$

Aluminum

This metal with a melting point of 646 °C. and a bulk resistivity of 2.74 $\mu\Omega$-cm is widely used in the fabrication of interconnects in electronic devices, lithographic patterns, highly reflective mirrors, and protective coatings of fibers and steel. Conventional precursors for CVD application are the pyrophoric aluminum alkyls such as Al(*i*-Bu)$_3$, which is the most frequently used (18). It is a colorless, pyrophoric liquid. Pyrolytic deposition for very large-scale integration (VLSI) has been carried out in horizontal hot-wall CVD reactors. Typical conditions are furnace temperatures of 200 - 300 °C. at up to 1 atmosphere pressure using H$_2$ as a carrier gas. A useful alternative to the pyrophoric trialkyl aluminum compounds are the alane amine adducts (R$_3$N)$_2$AlH$_3$. Deposition of films is catalyzed by traces of titanium, applied by exposure of surfaces to TiCl$_4$ (19). Films are then deposited at rates an order of magnitude faster than with Al(*i*-Bu)$_3$ and at lower lower temperatures (100-200°C.), which is useful for deposition on temperature-sensitive substrates.

Transition Metal Gallides

Almost all pure metals deposited on compound semiconductors such as GaAs or InP are unstable with repect to chemical reaction forming MM' (M' = a group 13 element) or MM" (M" = a group 15 element) phases (20). Conductors which are thermodynamically stable (and lattice matched) are thus needed to form stable metal contacts to the compound semiconductors. Towards this end, intermetallic films of CoGa and PtGa$_2$ have been deposited from the precursors (CO)$_4$CoGaCl$_2$.THF (21, 22) and platinumbis-

(dimethylglyoximato)bis(dimethylgallium), Pt(dmga·GaMe$_2$)$_2$ *(21)*, respectively. The nonstoichiometric intermetallic phase β-CoGa becomes lattice matched to GaAs *only* when it is gallium-rich (*i.e.*, 64 atom% Ga). Thus separate precursors CpCo(CO)$_2$ and GaEt$_3$ were employed in its deposition *(23)*. The desired atom% of Ga is achieved using a 300-fold excess in the mole fraction GaEt$_3$: CpCo(CO)$_2$ in the carrier gas stream and at deposition tempratures in the range of 260 - 300 °C.

I hope the above selected survey has given a glimpse of how coordination compounds are useful in new materials and in materials processing.

Dedicated to Professor Ekkehard Lindner at the occasion of his 60th birthday.

Literature Cited

1. Pierson, H. O. *Handbook of Chemical Vapor Deposition (CVD)*; Noyes Publications: Park Ridge, NJ, 1992.
2. Stringfellow, G. B. *Organometallic Vapor-Phase Epitaxy (OMVPE)*; Academic Press: New York , NY 1989.
3. Green, M. L.; Levy, R. A. *J. of Metals* **1985**, 63.
4. Yablonovich, E. *Science* **1988**, *246*, 347.
5. Mond, L. *J. Chem. Soc.* **1890**, *57*, 749 ;
    Carlton, H. E.; Oxley, J. H. *AIChE J.* **1967**,*13* , 86.
6. Goodman, D. W.; Kelley, R. D.; Madey, T. E.; Yates, Jr., J. T. *J. Catal.* **1980**, *63*, 226.
7. Girolami, G. S.; Gozum, J. E. *Mat. Res. Soc. Symp. Proc.* **1990**, *168* , 319.
8. Rubezhov, A. Z. *Nauka (Sci.) Moskva* **1986**, 95 [*CA 107:* 159372x].
9. Xue, Z.; Strouse, M. J.; Shuh, D. K.; Knobler, C. B.; Kaesz, H. D.; Hicks, R. F.; Williams, R. S. *J. Am. Chem. Soc.* **1989**, *111* , 8779.
10. Xue, Z.; Thridandam, H.; Kaesz, H. D.; Hicks, R. F. *Chem. Mater.* **1992**, *4* , 162.
11. Kaesz, H. D.; Williams, R. S.; Hicks, R. F.; Chen, Y.-J.; Xue, Z.; Xu, D.; Shuh, D. K.; Thridandam, H. *Mat. Res. Soc. Symp. Proc.* **1989**, *131*, 395.
12. Kaesz, H. D.; Williams, R. S.; Hicks, R. F.; Zink, J. I.; Chen,Y.-J.; Müller, J.-J.; Xue, Z.; Xu, D.; Shuh, D. K.; Kim, Y.-K. *New J. Chem.* **1990**, *14* , 527.
13. Niemer, B.; Zinn, A. A.; Stovall, W. K.; Gee, P. E.; Hicks, R. F.; Kaesz, H. D. *Appl. Phys. Lett.* **1992**, *61*, 1793.
14. Kaloyeros, A. E.; Feng, A.; Gargart, J.; Brooks, K. C.; Ghosh, S. K.; Saxena, A. N.; Luehrs, F. *J. Electronic Mater.* **1990**, *19* , 271.
15. Jain, A.; Farkas, J.; Kodas, T. T.; Shin, H.-K.; Chi, K. M. *Appl. Phys. Lett.* **1992**, *61*, 2662; Shin, H.-K.; Chi, K. M.; Farkas, J.; Hampden-Smith, M. J.; Kodas, T. T.; Duesler, E. N. *Inorg. Chem.* **1992**, *31* 424.
16. Dagani, R. *Chem. & Eng. News* **January 6, 1992**, *70* (1), 24.
17. Chi, K. M.; Shin, H.-K.; Hampden-Smith, M. J.; Kodas, T. T.; Duesler, E. N. *Inorg. Chem.* **1991**,*30*, 4293.
18. Sekiguchi, A.; Kobayashi, T.; Hosokawa, N.; Asamaki, T. *J. Vac. Sci. Technol. A* **1990**, 8 , 2976.
19. Gladfelter, W. L.; Boyd, D. C.; Jensen, K. F. *Chem. Mater.* **1989**, *1*, 339.
20. Liu, D. K.; Lai, A. L.; Chin, R. J. *Mat. Letters* **1991**,*10*, 318.
21. Chen, Y.-J.; Kaesz, H. D.; Kim, Y. K.; Müller, H.-J.; Williams, R. S.; Xue, Z. *Appl. Phys. Lett.* **1989**, *55*, 2760.
22. Maury, F.; Brandt, L.; Kaesz, H. D. *J. Organomet. Chem.* **1993**, *449*, 159.
23. Maury, F.; Talin, A. A.; Kaesz, H. D.; Williams, R. S. *Chem. Mater.* **1993**, *5* , 84.

RECEIVED February 2, 1994

Author Index

Affiliation Index

Subject Index

A